# The Canine as a Biomedical Research Model: Immunological, Hematological, and Oncological Aspects

Editors
Moshe Shifrine
Floyd D. Wilson

Laboratory for Energy-Related Health Research
University of California, Davis, California

1980

Prepared for the
Office of Health and Environmental Research
Assistant Secretary for Environment
U. S. Department of Energy

Published by
Technical Information Center/U. S. Department of Energy

NOTICE



**Library of Congress Cataloging in Publication Data**
Main entry under title:

The canine as a biomedical research model.

"DOE/TIC-10191."
Includes index.
1. Diseases—Animal models. 2. Dogs as laboratory animals. 3. Dogs—Diseases. 4. Medical research. I. Shifrine, Moshe. II. Wilson, Floyd D. III. United States. Dept. of Energy. Office of Health and Environmental Research.
RB125.C36 619'.7 80-24174
ISBN 0-87079-122-2

Available as DOE/TIC-10191 for $14.50 from

National Technical Information Service
U. S. Department of Commerce
Springfield, Virginia 22161

DOE Distribution Category UC-48

Printed in the United States of America
September 1980

To

**Leon Saul Rosenblatt,**

whose vision and many years of enthusiastic support
of the beagle as an important animal model for biomedical research
was a major stimulus for the creation of this book
and whose infectious enthusiasm, dedication, patience, and counsel
have been our mainstay in preparing this volume.

# Preface

The canine has been used as a biomedical research model in radiation studies by a number of laboratories supported primarily by the U. S. Department of Energy and its predecessors. These studies were unique in that they covered the life spans of the canines and permitted the collection of data from birth to death under controlled conditions. Since these were multiparametric studies, an extensive data base has been established, not the least of which are normative values covering all biologic systems, including immunohematology. The canine model has also been extensively used by other groups, such as transplantation biologists. The virtues of the canine as a model in these and many other endeavors are becoming increasingly more apparent with the passing of time.

Unfortunately the biomedical community has not been fully aware of the scope of information available from the canine. Thus there has been minimal interaction between researchers in radiation biology, transplantation biology, veterinary medicine, oncology, etc., and, as a result, the full potential of the canine model has not been achieved. For this reason, one of the primary goals of this volume was to compile the knowledge and experience of researchers using the canine model and to focus their expertise on furthering the use of the canine for studies in immunology, hematology, and oncology. It is our hope that this effort will serve as a catalyst for further interactions between canine research groups and also will enhance the appreciation of the unique aspects of the canine model by the biomedical research community in general and that this, in turn, will enhance our knowledge of both veterinary and human medicine.

We have attempted to present some of the contemporary, diverse uses of the canine in biomedical research, emphasizing immunologic endpoints, and also to present in detail some of the latest technology used in such studies.

This volume will be of value to researchers who use, or contemplate using, the canine for research by providing a comprehensive source of information on some of the important aspects of current canine research.

Moshe Shifrine

Floyd D. Wilson

# Acknowledgments

We wish to thank Jerry Azevedo for excellent technical editing of the chapters and Margaret Givens and Jean Smith, Technical Information Center, U. S. Department of Energy, for editing the manuscript for publication and for coordinating the processing of the book. We extend our thanks to Charles Baty and Nancy Hardaker for typing the final drafts and to Elfriede DeRock and Nancy Taylor for their help in coordinating communications with the authors and publisher. We also wish to thank Dr. M. Goldman, Director, Laboratory for Energy-Related Health Research, for his unflagging support.

Moshe Shifrine

Floyd D. Wilson

# Contents

# Introduction

# 1

**MOSHE SHIFRINE**
*and*
**F. D. WILSON**
*Laboratory for Energy-Related Health Research,*
*University of California, Davis*

The word "model," which conjures different images related to the discipline of the beholder, is used by researchers working with experimental animals to draw attention to the relevance of their work to human medicine. We choose to think of the canine model as serving a dual function, i.e., providing information on biomedical problems in humans and providing fundamental knowledge of benefit to the dogs themselves, thus providing a service to human and veterinary medicine.

The study of diseases or abnormal physiological phenomena using human subjects is quite limited for obvious practical and ethical reasons. Thus one must resort to the use of animals with the hope that the choice of a particular species is appropriate for the extrapolation of experimental results to humans. To aid in this endeavor, the Registry of Comparative Pathology of the Armed Forces Institute of Pathology recently published a handbook, *Animal Models of Human Disease,* which describes 150 models of animal disease relevant to the study of human diseases which are based mainly on comparative pathology. For total numbers of models, the dog takes second place only to the mouse and is ahead of the rat.

Although a particular disease model in animals may be homologous (i.e., the etiology and pathogenesis are identical to the human disease), more often animal models are isomorphic (i.e., the biological condition is similar in animal and man whereas the cause may be quite different for the two). If we assume the universality of the physiologic machinery for all creatures with some evolutionary modification to allow species to best fit into their respective niches, either of the above models is of value to man.

The more than two decades of energy-related research on beagle dogs in the various laboratories which was sponsored by the U. S. Department of Energy

(DOE) and its predecessors, the U. S. Atomic Energy Commission and the Energy Research and Development Administration, have contributed significantly to making the dog one of the most valuable models for biomedical research. The accumulated extensive data base on canine immunohematologic parameters, conjoined with both spontaneous and radiation-induced hematopoietic neoplasias and spontaneous mammary carcinomas, make the dog an ideal model for research on tumor immunology and hematopoietic neoplasia and for the study of the mechanisms underlining carcinogenesis. This is a crucial point since cancer is a major endpoint for both nuclear and nonnuclear (energy production by-products) toxicity programs of the DOE and other agencies.

The relatively large size and longevity of the dog are advantageous for sequential studies in individual animals related to etiology, pathogenesis, and therapy of malignancies. Indeed, such attributes are essential in the study of the pathogenesis of hematologic, immunologic, and oncologic disorders.

Unfortunately, relatively few scientists outside the DOE "Beagle Club" are aware of the unique value of the dog for biomedical research. Thus the efforts of canine researchers have not achieved full fruition. The purpose of this book, therefore, in part, is to apprise the biomedical research community of the value of the canine in biomedical research. We feel that this effort will enhance the use of the dog as a biomedical model and lead to greater use of dogs by researchers outside DOE.

As with any other publication, constraints of space dictated limitations of subject matters covered in this book. We are quite aware of some omissions, but we feel that any further increase in the scope of the book would require limiting the space allotted for each chapter and would thus detract from the total value of the publication.

Our choices of the various chapters were guided by our prejudices and penchant for the exciting and the relevant in biomedical research. The book includes chapters that review basic aspects and current knowledge of canine immunology, hematology, and oncology. This, hopefully, will provide a background for the researcher who is not using canine models to better understand the chapters that follow on transplantation and immunologic diseases of immunohematologic origin. We close the book with six canine tumor models that satisfy our above stated criteria.

As man reigns supreme, at least at this juncture in history, the dog becomes the model for man. Nevertheless, we should not lose sight of the fact that the dog is a unique creature that should be studied for its own sake. Information from dog studies is obviously of value to our understanding of the dog's physiology and medical care. Thus the use of the dog as a research animal also is providing useful information for veterinary clinicians involved in the treatment of dogs.

We take this opportunity to thank all the contributing authors for their chapters and cooperation in this endeavor and the U. S. Department of Energy for its support.

# The Canine as a Model for Studies on Stem and Progenitor Cells of Lymphohematopoiesis

# 2

F. D. WILSON
*and*
MOSHE SHIFRINE
*Laboratory for Energy-Related Health Research, University of California, Davis*

The extensive data base on canine immunohematological parameters coupled with the development of both spontaneous and radiation-induced forms of neoplasm in canines make that animal an ideal model for research on tumor immunology. No other large-animal species with a high incidence of malignancies (spontaneous, radiation-induced, and transplantable) has been the subject of such a large variety of techniques developed for the evaluation of both progenitor and effector cell components of the immune response. In addition, the relatively large size and longevity of the dog allow for sequential sampling and observations in individual animals, an advantage for studies related to etiology, pathogenesis, and therapy of hematopoietic and other malignancies.

In this chapter we will examine the canine as a model for immunohematological studies on lymphohematopoietic disorders as well as for investigation of hematopoietic and other forms of neoplasia. We will emphasize the various new cloning techniques that permit the quantitation of immunohematopoietic progenitor populations in vitro.

## CURRENT CONCEPTS OF LYMPHOHEMATOPOIESIS

### Model for Lymphohematopoiesis

Experimental study of mechanisms involved in hematopoietic disorders requires a basic understanding of the components of the physiological regulation of normal lymphohematopoiesis. A current model for hematopoiesis (Fig. 1) suggests at least a population of pluripotent stem cells (CFU-S or spleen colony

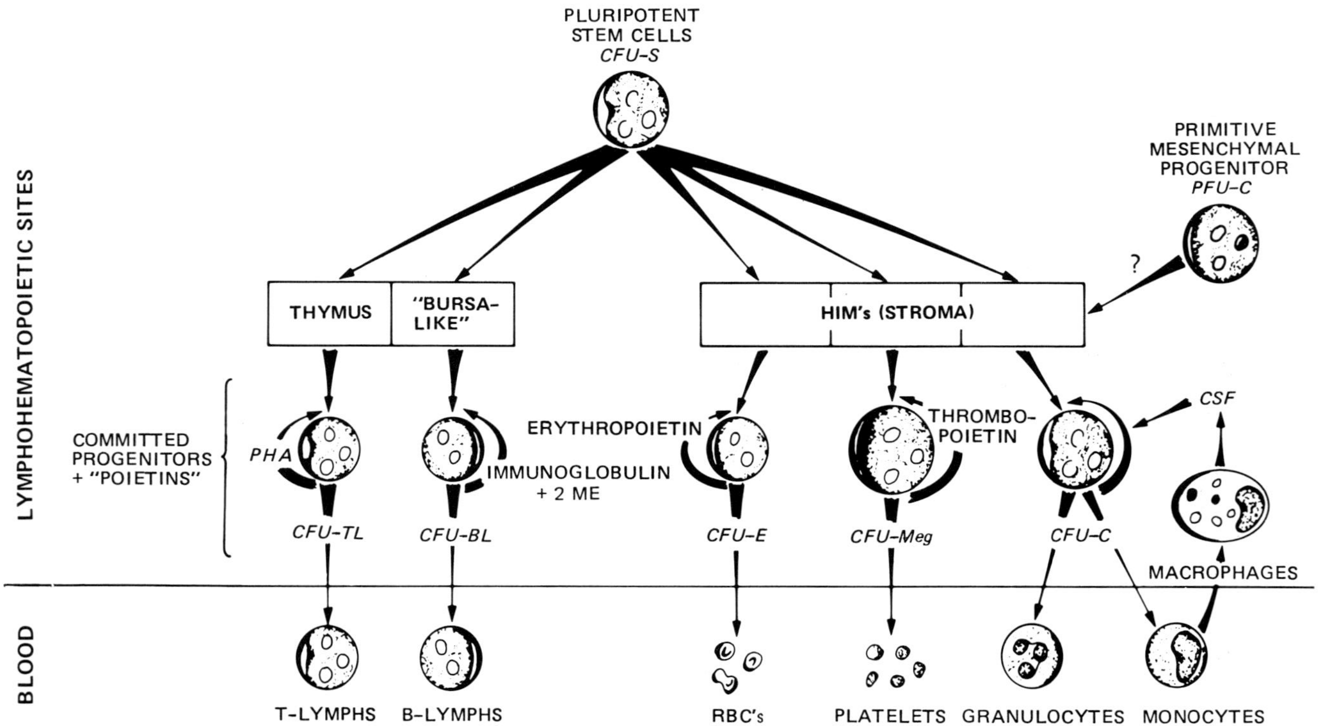


Fig. 1 Current concepts of lymphohematopoiesis. CFU-S, colony forming units, spleen, pluripotent stem cells; HIM, hematopoietic inductive microenvironments, "stromal" elements; PFU-C, plaque forming units in culture = "candidate" stromal cell progenitors; CFU-TL, colony forming units for T lymphocyte progenitors; CFU-BL, colony forming units for B lymphocytes; CFU-E, colony forming units for erythroid pathways; CFU-Meg, colony forming units for megakaryocytes; CFU-C, colony forming units in culture = progenitors for granulocytes and monocyte–macrophages; CSF, colony stimulating factor, "granulopoietin."

equivalent), stromal-associated hematopoietic and lymphopoietic inductive microenvironments, populations of "committed" stem or progenitor cells, and various humoral regulators that are thought to exert their influence at the level of the committed stem cells.

The lymphohematopoietic system in adult mammals is unique by virtue of its cell renewal properties. Since the mature cellular products of this system are constantly used or removed from the body, a finely regulated mechanism to balance production with cell loss is inherently necessary in the physiological maintenance of lymphohematopoiesis. In addition, mechanisms must exist for rapid expansion of lymphohematopoietic elements in response to the many recurrent pathological insults to the body.

Any system in which embryonic demands for cellular proliferation persist throughout the life of the animal requires a pool of stem cells, which, by definition, must exhibit the properties of self-replication or self-renewal to maintain their own kind. In addition, stem cells must provide a source of lymphohematopoietic progeny to replenish the cells that are constantly consumed; that process requires a pluripotent capacity. Those cells must maintain cellular populations which include a wide variety of morphologically and functionally diverse lymphohematopoietic elements. Thus populations of stem cells capable of both self-renewal and pluripotent cellular differentiation must be maintained throughout the life of an animal. Such cells are defined as pluripotent hematopoietic stem cells.

Although the exact morphology of pluripotent stem cells is not known, results of studies on highly enriched gradient fractions of stem cells have suggested a morphological "candidate" for pluripotent stem cells in mice,[1] rats, primates,[2] and dogs.[3] Exclusive of morphological features, pluripotent stem cells can be studied and quantitated with assays that incorporate the basic criteria for their dual functions for self-maintenance and pluripotentiality. The most widely used assay for pluripotent stem cells is the spleen colony technique of Till and McCulloch.[4] The investigators demonstrated that the intravenous injection of cells obtained from hematopoietic organs into lethally irradiated syngeneic mice resulted in the formation of hematopoietic colonies in the spleen after about 7 days. The relationship of the number of cells injected and the number of resultant spleen colonies specifies the quantity of pluripotent stem cells (colony forming units in spleen or CFU-S). In that assay erythroblastic, granulocytic, and megakaryocytic colonies are formed. Lymphocytic spleen colonies are not apparent, but transplantation of cells derived from the various myeloid colonies into lethally irradiated recipients results in the reconstitution of immunocytes as well as myeloid elements.

A major disadvantage of the spleen colony assay is its limitation to murine species only. Thus the development of cloning techniques for the growth of a variety of lymphohematopoietic colonies in semisolid culture systems for essentially any mammalian species, including canines and humans, was a major advancement in experimental hematology.[5] In those assays single cells are immobilized in semisolid media, such as agar or methylcellulose, and are induced to proliferate by various growth factors or mitogens. Repeated division of such target cells results ultimately in the production of lymphohematopoietic colonies, which can be quantitated. Numerous studies have indicated that parent cells of in vitro colonies (colony forming units in culture, or CFU-C) represent cells that are more differentiated than pluripotent stem cells and hence are not pluripotent but, rather, represent progenitors that are committed to specific pathways of lymphohematopoiesis.

Evidence that CFU-C are populations distinct from CFU-S comes from a variety of experiments, including their response to $^{3}$H–TdR suicide,[6] density gradient profiles,[7] and comparisons of the responses of CFU-C and CFU-S in animals subjected to marrow insults or with various hematological disorders.[8–10]

Since lymphohematopoietic colony forming progenitors are targets for various humoral regulators or poietins, the colony forming assays can also be used as a bioassay for humoral factors thought to regulate lymphohematopoiesis.

Other important factors hypothesized to be involved in the regulation of lymphohematopoietic stem and progenitor cells are the various stromal elements in lymphohematopoietic organs that are thought to exert microenvironmental inductive influences on the differentiation of stem cells (the hematopoietic inductive microenvironment, or HIM).[11–29] A classic demonstration of the influence of stromal elements on pluripotent stem cells was provided by Wolf and Trentin[29] in their studies on factors influencing spleen colony formation. In mice the bone marrow is a hematopoietic site that is predominantly granulopoietic, whereas the spleen, exclusive of lymphopoiesis, is primarily an erythropoietic myeloid organ. Wolf and Trenton implanted irradiated bone-marrow stroma into the spleens of mice. In subsequent spleen colony assays, spleen colonies developed at the interface of spleen stroma and the implanted marrow stroma. Histological examination of such colonies demonstrated that portions of the colony that developed within the spleen stroma were predominantly erythroid, whereas those regions of the same colony contained within the bone-marrow stromal implant were granulopoietic. The concept of stromal microenvironmental influence has also been supported by the demonstration that the genetically transmitted macrocytic anemia in mice of $S1/S1^{d}$ genotype results from an abnormality in hematopoietic stromal elements. Whole-spleen transplantation corrects anemia, whereas reconstitution with hematopoietic stem cells does not.[30]

Microenvironmental influences are also involved in the production of functional subpopulations of lymphocytes. The development of T lymphocyte is influenced by epithelial–reticular components of the thymus, and B-cell differentiation is influenced by the avian bursa or mammalian bursa equivalent.

Monocyte and macrophage products of CFU-C are important regulatory cells or components of the microenvironmental lymphohematopoiesis. Those cells are important accessory cells for a variety of immunological reactions. Macrophages are involved in the processing of antigens ("super antigen" production) and in the mediation of interactions between T and B lymphocytes.[28] Macrophages also may be involved in the regulation of myelopoiesis. A "nurse cell" function of macrophages for erythropoiesis has been inferred from the occurrence of "erythroblastic islands."[31] Monocytes and macrophages also produce colony stimulating factor (CSF)—a candidate poietin. Recent studies by Kurland et al.[32,33] have demonstrated that monocytes and macrophages produce prostaglandin E, which is an inhibitor of CFU-C, and Kurland[34] has proposed a unique regulator role for the monocyte and macrophage in the positive–negative feedback regulation of granulocyte precursors. The close interrelation of lymphopoiesis and myelopoiesis is underlined by the demonstration that lymphocytes also release CSF in response to mitogenic[35] or antigenic stimulation.[36] Increasing evidence also points to a major influence of the thymus on myeloid elements in the bone marrow.[37] A wide range of cellular interactions are involved in the regulation of lymphohematopoiesis.

The characterization of mechanisms involved in physiological, homeostatic, and pathological states of immunohematopoiesis would be facilitated by techniques that allow for the concurrent evaluation of the quantitative and functional status of stem and progenitor cells, of humoral and cellular regulators of these populations, including stromal components, and of the effector cell products of the stem-cell populations.

## Methods for Studies on Hematopoietic, Lymphopoietic, and Stromal Stem or Progenitor Cells in the Canine

No techniques equivalent to the spleen colony assay for studies on pluripotent stem cells exist for canines, but those cells can be studied indirectly with techniques involving transplantation. Also, focal proliferations of hematopoietic cells have been described in dogs recovering from radiation.[38] The homogeneous nature of such foci suggests a clonal origin and thus the possibility that such hematopoietic foci in dogs are equivalent to spleen colonies in mice.

Another approach to quantitation of more-primitive progenitors in dogs is the use of diffusion-chamber techniques. Colony forming progenitors from bone marrow of a variety of animal species will proliferate within diffusion chambers implanted in the peritoneal cavity of mice.[39] Recent results have indicated that the progenitors of colony forming units within diffusion chambers (colony forming units—diffusion chamber, or CFU-d) represent progenitors which are more primitive than CFU-C but which are more differentiated than the pluripotent stem cell.[40] Recent reports of "mixed colonies" in agar, from progenitors that are pluripotent but less primitive than CFU-S, suggest that in vitro assays of pluripotent progenitors may be possible for dogs in the near future.[41] Also, some quantitative data on stem cells can be obtained by testing the ability of hematopoietic elements to reconstitute lethally irradiated dogs (Chapter 9).

Another indirect approach for the observation of changes in pluripotent stem cells in dogs with hematopoietic disorders is to use simultaneous assays for committed stem cells. For example, simultaneous changes in CFU-C, CFU-E, and CFU-L would suggest that those alterations directly reflect changes at the pluripotent stem-cell level. Conversely, changes in only one of the committed progenitors would signify that the effect is distal to the level of pluripotent stem cells. That approach has been used by Dunn et al.[42] in studies comparing CFU-C and CFU-E in dogs with cyclic hematopoiesis.

In dogs injected with CFU-C mobilizing agents, such as dextran sulfate, the level of CFU-C in the blood closely parallels the ability of peripheral blood to reconstitute lymphohematopoietic elements in lethally irradiated recipients.[43]

Several techniques based on the colony forming potentials of committed stem cells in semisolid culture systems are available; the nature of the progenitor assayed in the culture system depends on the cell source coupled with various other factors, such as "poietins" or mitogens added to the cultural environment.[5] To date techniques for the assay of CFU-C,[24] T lymphocyte progenitors (CFU-L),[44] candidate "stromal" progenitors,[19] erythroid progenitors,[42] and alveolar macrophage progenitors[45] have been developed for the dog. Techniques developed for humans and other species possibly can be adapted for quantitation of canine pluripotent progenitors,[41] B lymphocyte progenitors,[46] eosinophil progenitors,[47] and megakaryocytic progenitors.[48] The morphological appearance of canine macrophage and of T lymphocyte colonies is shown in Figs. 2 and 3, respectively.

We have recently reported the formation of lymphocyte colonies from canine peripheral blood stimulated by Con-A or PHA[44] in a methylcellulose culture system that we developed for humans.[49] Those colonies, as with those forming in response to mitogens in mice and humans, were of T-cell type.[44] Lymphoid colony formation can be used as a sensitive biodosimeter for radiation injury.[50] Also, subpopulations that are immunologically specific can be examined (F. D. Wilson, unpublished results). Other studies have noted radioresistant subpopulations of T lymphocyte progenitors in the peripheral blood of beagles.[50] Further, such radioresistant subpopulations may replace the more-sensitive populations in beagles receiving continuous whole-body gamma radiation.[50] Those techniques may provide a powerful new method for examining a variety of immunological phenomena in the dog.

The increasing evidence of a functional role of stromal elements in the regulation of hematopoiesis and lymphopoiesis underlines the importance of development of assay systems for stromal elements, including potential stromal progenitors. Whereas much is known about hematopoietic stem and progenitor cells, relatively little is known about stromal components of hematopoietic tissues.

Although experiments in rodents have suggested a major functional role of hematopoietic stromal elements in the regulation of hematopoiesis,[12] essentially no knowledge of a similar function of stroma is currently available for other

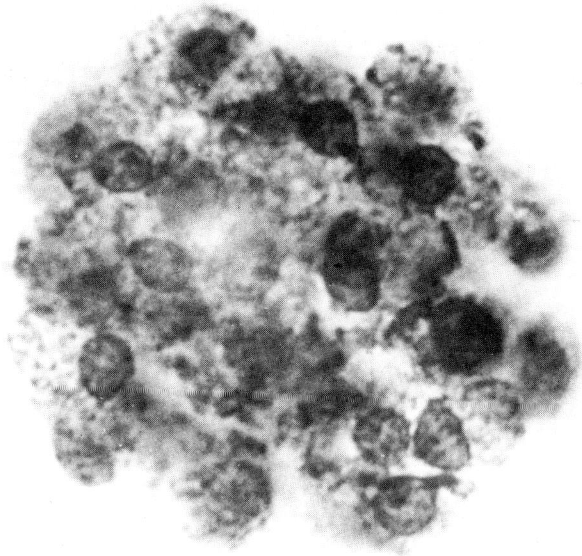

Fig. 2 Canine macrophage colony in methylcellulose culture of bone marrow.

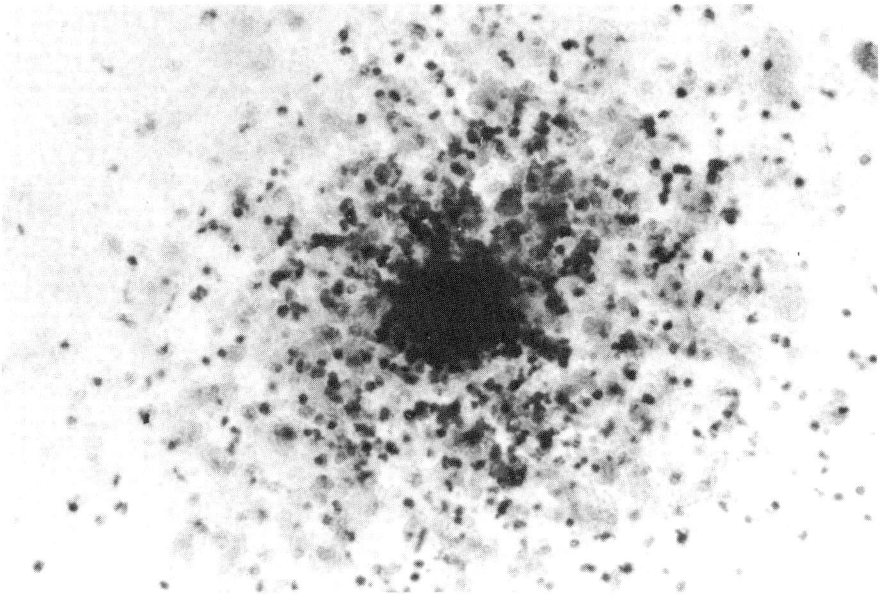

Fig. 3 T lymphocyte colony from culture of canine peripheral blood lymphocyte fractions. (Magnification, 400 ×.)

species, including humans. The absence of appropriate techniques for dissecting the role of stromal components from hematopoietic elements has hindered progress in specifying the functional role of stromal elements and consequently in understanding the pathogenesis of such disorders as myelofibrosis.

One clue in the search for a technique for separating stromal and hematopoietic components was the simultaneous development of surface-adherent colonies with fibroblastic and/or reticuloendothelial morphology (plaque forming units in culture, PFU-C) in addition to the formation of suspended hematopoietic colonies (CFU-C) in methylcellulose cultures in beagles (Figs. 4 and 5).[19] The responses of murine CFU-C and PFU-C to a variety of insults (radiation, radiomimetic drugs, and femoral medullary curettage) and the numbers of CFU-C and PFU-C in mice with congenital defects ($S1/S1^d$ and nu/nu genotypes) have been reported.[18,19,23] In similar studies[24] the responses of CFU-C and PFU-C in beagles receiving acute and chronic irradiation were compared. These fibroblastic colonies are similar to those reported by Friedenstein, Chailakhjan, and Lalykina[20] to form in liquid-phase culture systems of bone marrow of guinea pigs and mice. These investigators demonstrated that the fibroblastic elements, when harvested and transplanted into diffusion chambers and subsequently implanted into animal hosts, gave rise to stromal elements. Further, such elements reportedly can transfer inductive information for lymphohematopoiesis.[13]

Recently, we[22,26] have been able to confirm and extend those observations to humans through the production of mesenchymal (stromal) tumors in nude mice by the injection of fibroblasts isolated from a CML patient. Both the tumors and the in vitro fibroblasts were of normal human karyotype and, unlike the hematopoietic elements in that patient, were negative for the $Ph^1$ chromosome. Also, fibroblastic colonies from a male patient who had been successfully treated for aplastic anemia by bone-marrow transplantation from a histocompatible female donor and who was a demonstrated male–female hematopoietic chimera were of male origin and therefore of recipient karyotype.

In our studies of five patients with $Ph^1$-positive CML, the $Ph^1$ chromosome was absent in metaphases from the fibroblastic colonies but was present in metaphases from direct marrow preparations.[26] The marker chromosome is associated with all myeloid elements derived from hematopoietic stem cells.[51,52] The $Ph^1$ chromosome also marks CFU-C (considered to be progenitors for granulocyte and monocyte–macrophage pathways) and is present in macrophages in tissue cultures derived from $Ph^1$-positive CML patients.[53] The cytogenetic studies and nude-mouse transplantation studies both suggest that the adherent colonies are derived from progenitors that are not closely related to either hematopoietic stem cells or to other adherent populations, such as macrophages, despite morphological similarities. On the basis of our studies and those of Friedenstein, Chailakhjan, and Lalykina, we have postulated that the bone-marrow fibroblastic colonies are derived from

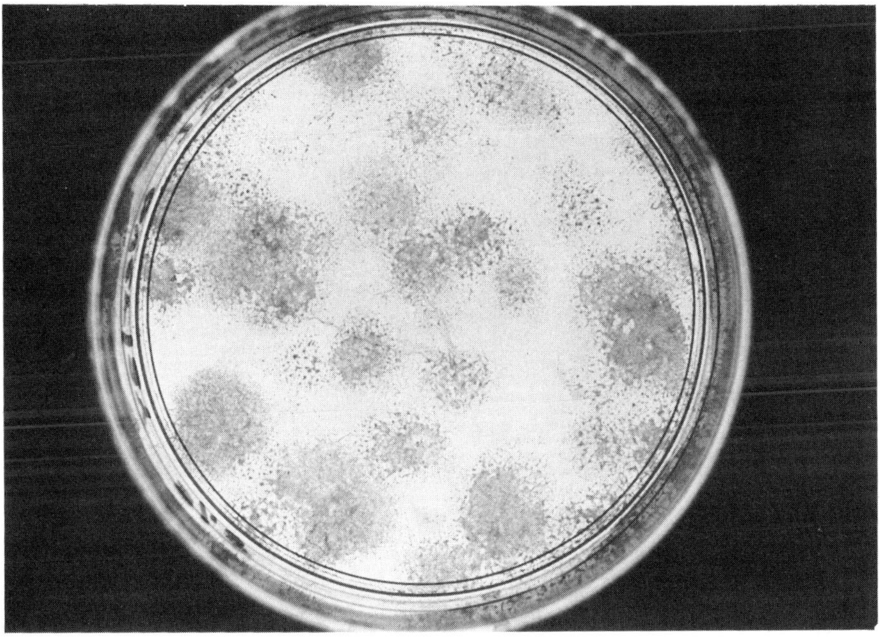

Fig. 4 Gross appearance of canine "stromal colonies" from bone-marrow cultures.

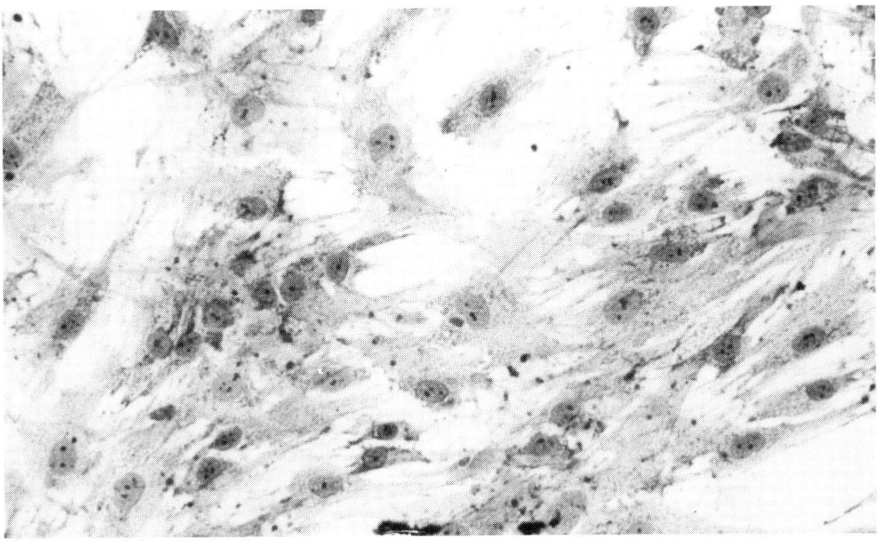

Fig. 5 Microscopic appearance of cells in bone-marrow stromal colonies.

marrow progenitors that are related more specifically to marrow stromal (fibroblast, reticuloendothelial, and/or osteogenic) progenitor populations.

Fibroblastic colonies also produce colony stimulating activity; in methylcellulose and agar cultures, they enhance the growth of CFU-C.[18,19] In this regard, irradiated fibroblasts are used routinely as feeder layers for CFU-C in some tissue-culture systems. Fibroblasts also enhance the growth of CFU-E.[54] Recent studies have indicated that the enhancing effects of the fibroblastic colonies appear to be local, totally different from and not synergistic with colony stimulating factor (F. D. Wilson, unpublished results). In addition, fibroblastic elements obtained from humans with acute myelogenous leukemia (AML) and myeloproliferative disorders (MPD) show abnormalities in their ability to stimulate CFU-C.[55] Thus the interaction of cells of stromal origin with hematopoietic progenitors may be important in the clinical expression of hematological malignancies.

The importance of surface-adherent populations in the regulation of hematopoiesis has been supported further by studies of Dexter and Moore,[16] who, like Friedenstein, Chailakhjan, and Lalykina, used liquid-phase bone-marrow culture systems. If cultures of adherent cells were maintained by supplements of fresh media, pluripotent and committed stem cells would persist in culture. Proliferation and differentiation of stem cells did not proceed without feeder layers of the attaching populations. Dexter and Moore[56] have also demonstrated functional defects in the adherent cells of mice of $S1/S1^d$ genotype. The attaching cells described by these investigators consisted of three morphological types: adherent phagocytic mononuclear cells, "epitheloid" cells, and fat cells. Petrakis, Davis, and Lucia[57] also have reported that human leukocytes, when placed into subcutaneously implanted diffusion chambers, will differentiate into histiocytes, fibroblasts, and fat cells. Cells with both macrophage and fibroblastic morphology will also support hematopoietic colony formation in cellulose acetate membranes implanted into the peritoneal cavity of mice.[58]

Although data from a variety of studies have suggested the importance of attaching cells in the regulation of hematopoiesis, the exact morphology and functional nature of such adherent cells remain obscure. A variety of names, including macrophages (polyblasts), fibroblasts, fibroblastic plaques, and RE-like or mesenchymal elements, have been applied to adherent cells. More-recent studies have suggested that attaching cells comprise a triad of morphologies, including mononuclear phagocytic cells, epitheloid cells, and fat cells.[16] Therefore the attaching cells in culture may contain a diverse population of cellular elements with differing morphological and functional characteristics.

To better characterize attaching cells, we[59] used transmission and scanning electron microscopy to compare attaching cells in bone-marrow cultures from humans, dogs, and mice. The attaching cells in canines and humans (Fig. 6A) consist of relatively homogeneous populations of cells with fibroblastic or epitheloid morphology. The cells are flattened or spread out with uniform

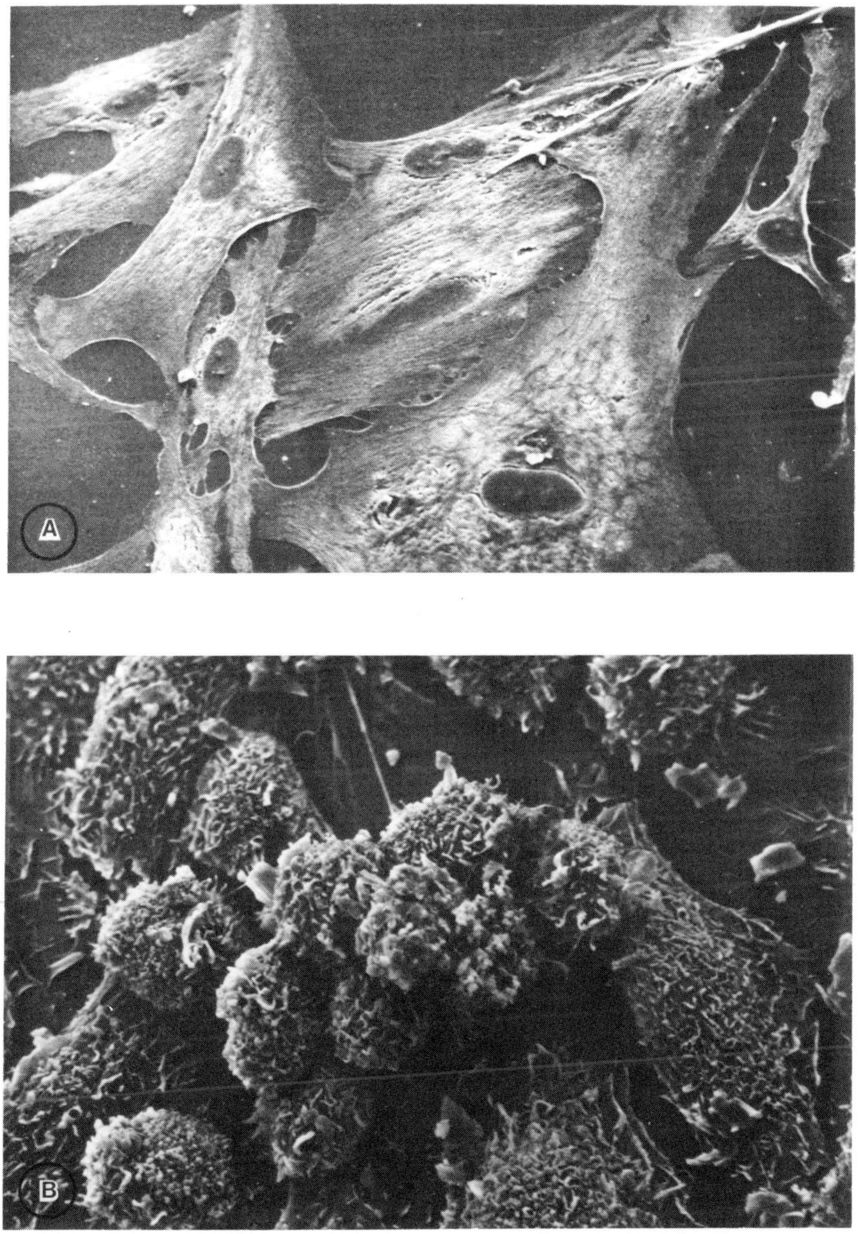


Fig. 6 Scanning electron micrograph of attaching cells from bone-marrow cultures of humans (A) (magnification, 355 x) and mice (B) (magnification, 1790 x).

surface morphology. The cells typically have abundant cytoplasm containing extensive networks of intracellular fibers and large ovoid nuclei. In contrast, the attaching populations in mice (Fig. 6B) are much more heterogeneous, containing many more rounded cells, resembling macrophages, intermixed with the flattened fibroblastic cell type. The macrophage-like cells are characterized by variation in surface morphology, with diffuse folding, ruffling, and formation of villous structures in the membranes.

Transmission electron microscopic (TEM) studies of the attaching cells from humans and dogs demonstrated that the flattened types of cells (Fig. 7) are characterized by their uniform or stellate shapes and by a large, typically elongated, primitive nucleus containing finely dispersed chromatin and nucleoli. The abundant cytoplasm exhibits extensive networks of microtubules and submembranous microfilaments and the formation of ribosome-associated

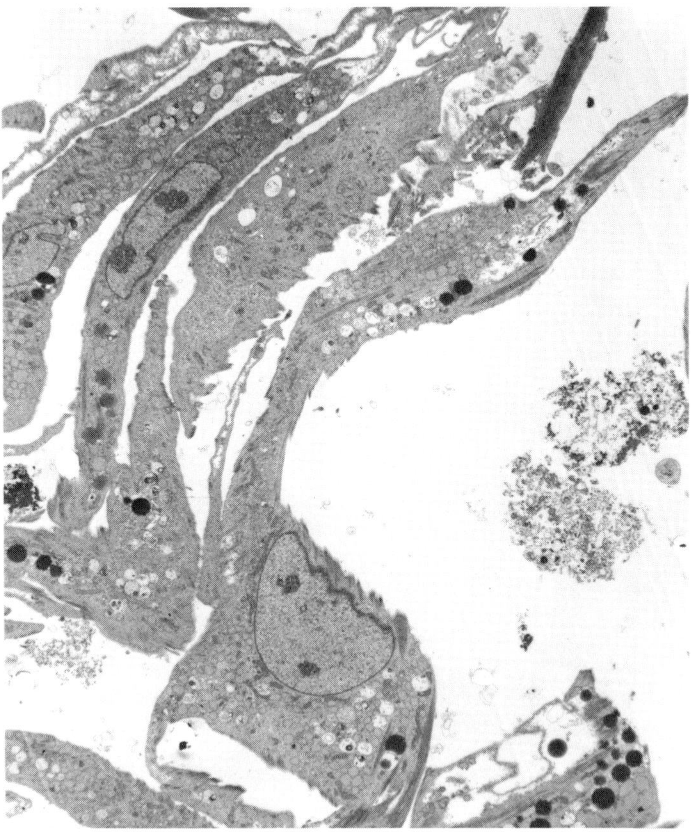

**Fig. 7 Transmission electron micrograph of canine bone-marrow "stromal" cells. (Magnification, 5000 x.)**

"secretory" bodies. Pinocytic activity was evident. At higher magnification (Fig. 8) the extensive formation of parallel bundles of microfilaments and electron-dense secretory bodies are readily apparent. Microfibers extend into extracellular areas. The TEM appearance of a murine bone-marrow macrophage, shown for comparison in Fig. 9, shows extensive phagosomes but no microfilaments.

The uniform populations of attaching cells in humans and canines, which we relate to the "epitheloid" cells of Dexter and Moore[16] and the "fibroblastic" cells of Friedenstein, Chailakhjan, and Lalykina[20] and Wilson et al.,[19] may provide a means to further segregate the function(s) of specific populations of attaching cells and delineate their role in hematopoiesis. Those cells, unlike

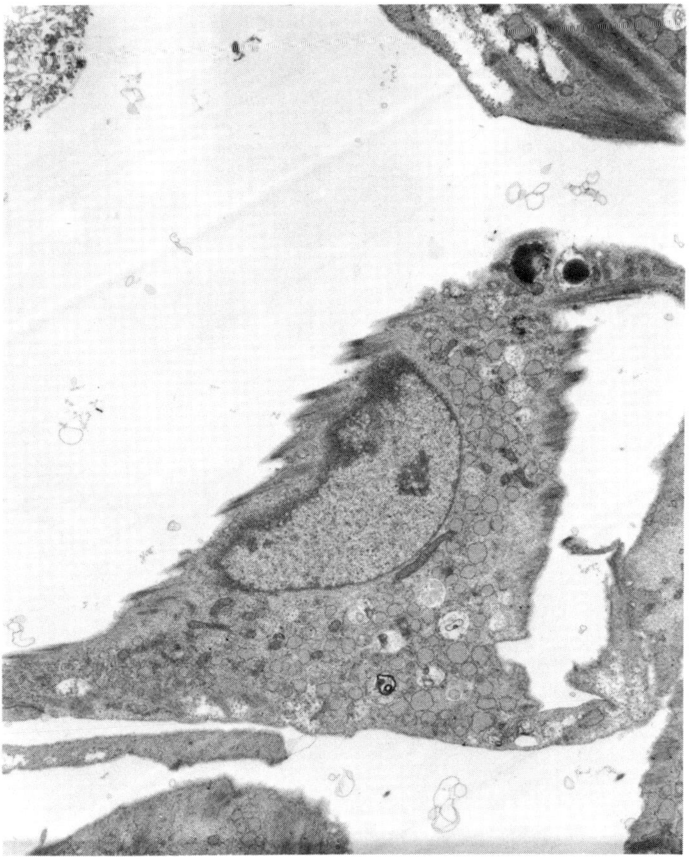

Fig. 8 Transmission electron micrograph of canine attaching cell. Note the extensive formation of parallel bundles of microfilaments and electron dense secretory bodies. Note also the extension of microfibers into extracellular areas. (Magnification, 10,000 x.)

macrophages, are negative for lipase activity and generally are not phagocytic.[60] Also, the homogeneous fibroblastic or epithelial type of attaching colonies in humans and dogs will augment granulopoiesis in vitro, and humans with myeloproliferative disorders may show defects in that function.[55]

Those studies collectively suggest that the adherent cells forming in culture are characterized by mesenchymal morphology (fibroblastic or RE-like), are radioresistant,[24] produce factors in vitro that stimulate or enhance the growth of CFU-C and CFU-S, give rise to stromal elements when injected into mice, and are candidate in vitro equivalents to the hematopoietic inductive microenvironment.

A similarly radioresistant population with epithelial morphology has been isolated in culture from the thymus. Those thymic cells appear capable of providing microenvironmental inductive influence for the conversion of null-cell

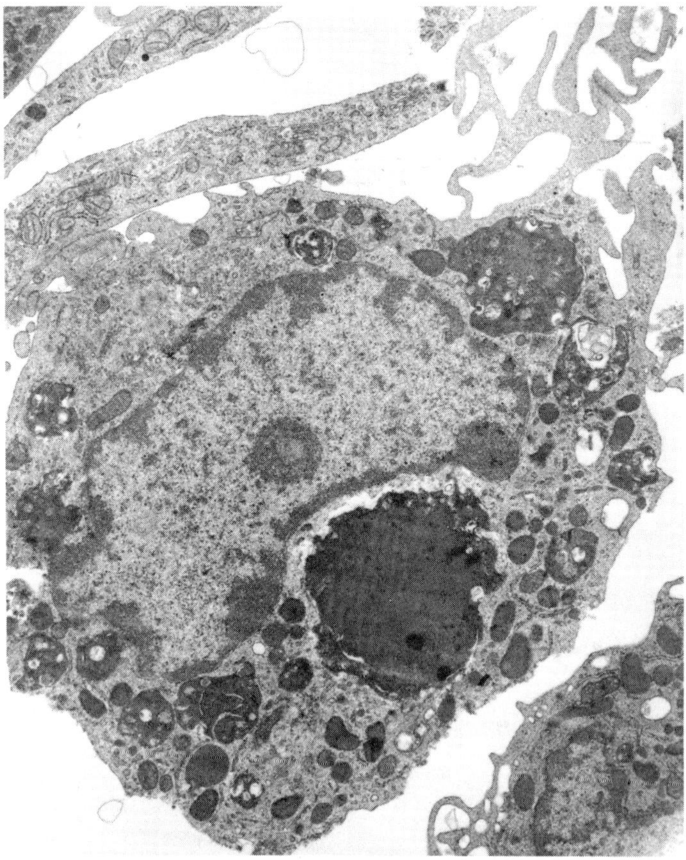

Fig. 9 Transmission electron micrograph of murine macrophage. (Magnification, 20,000 x.)

lymphocytes into functional T lymphocytes.[61,62] In this exciting new research area, the nature of cellular interactions between hematopoietic and lymphopoietic stem cells and their microenvironments can be studied in vitro, and potential changes in the interactions in hematological and immunological disorders as well as application of such knowledge for therapy can be investigated.

## New Directions in Colony Forming Techniques

**Formation of Lymphocyte Colonies in Mixed-Lymphocyte Cultures.** Although many sensitive systems have been developed for the assay of humoral immunity, only a few limited techniques can be applied to the study of cell mediated immunity (CMI). Most studies on CMI have used such techniques as stimulating lymphocyte blastogenesis by T-cell mitogens, such as PHA and Con-A, and measuring the incorporation of $^3$H–TdR into the DNA of dividing lymphocytes (Chapter 3). Although such methods are valuable for the identification of subpopulations of lymphocytes, the relevance of the lymphocyte blastogenic response to plant mitogens as a measurement of cellular immunity is questionable.

Another approach to the measurement of CMI involves stimulation of lymphocyte blastogenesis through genetic differences. In this case, differences in cell-surface markers (histocompatibility antigens) trigger the proliferation of T lymphocytes. Such mixed-lymphocyte (ML) stimulation is used as a major tool for identification of histocompatibility between donor–recipient pairs in transplantation immunology (Chapter 8).

Although the standard technique for the measurement of ML reactivity involves the incorporation of $^3$H–TdR, we have developed an in vitro colony forming assay for the measurement of mixed-lymphocyte reactions in humans and dogs.[63] The lymphocyte colony forming techniques have the advantage over $^3$H–TdR incorporation in allowing for the isolation of specific colonies of lymphocytes and potentially for characterization of their function. Peripheral blood lymphocytes or bone marrow from two genetically different individuals are first allowed to react in liquid-phase culture for an average of 5 days. Cells are then plated into semisolid cultures and incubated for an additional 7 days before the colonies are counted.

Direct comparisons of the colony forming and $^3$H–TdR incorporation techniques have indicated that the cloning method is far more sensitive in measuring allogenic reactions and produces less autologous response. The appearance of a lymphocyte colony from a canine ML is shown in Fig. 10. The ultrastructural appearance is shown in Fig. 11. Characterization studies of ML colonies with rosetting techniques have indicated that the majority of the cells obtained from such ML colonies are T-cells.

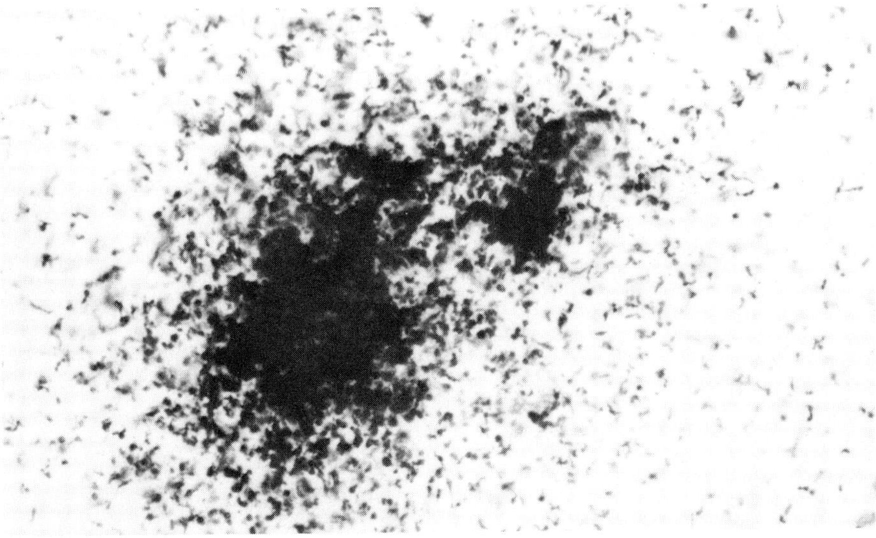

**Fig. 10** Canine "mixed-lymphocyte" colony in agar stained with new methylene blue. (Magnification, 40 x.)

We also observed in the ML colonies large numbers of surface-adherent macrophages and an occasional multinucleated giant cell (Fig. 12). Macrophages reportedly are essential components of the ML colony reaction, and the ML colony technique potentially can be applied in the investigation of macrophage components and macrophage–lymphocyte interactions in CMI and graft-rejection phenomena. We are currently determining if the ML colony forming technique can be used as a better measurement of CMI than standard techniques using plant lectins for mitogenic stimulation of lymphocytes.

**Whole-Blood Lymphoid Colonies.** Most techniques that use lymphocyte blastogenesis as a measurement of immunity routinely separate cellular fractions on gradients of such materials as Ficoll–Hypaque (FH). Those separations enrich lymphocytes and conversely deplete the other cellular elements. Numerous humoral and nonlymphoid cellular factors exert both positive and negative influences on lymphocyte blastogenesis. For example, lymphocytes release colony stimulating factor in response to antigenic or mitogenic stimulation. In another case, macrophages produce factors that either enhance lymphocyte blastogenesis, such as lymphocyte activating factor, or inhibit that reaction, such as prostaglandin E. Also, serum contains a wide variety of factors influencing lymphocyte blastogenesis, including immunoglobulins and colony stimulating factors. Therefore techniques for the measurement of lymphocyte stimulation in whole blood would involve the full complement of serum and circulating blood cells and thus would more closely approach the in vivo milieu.

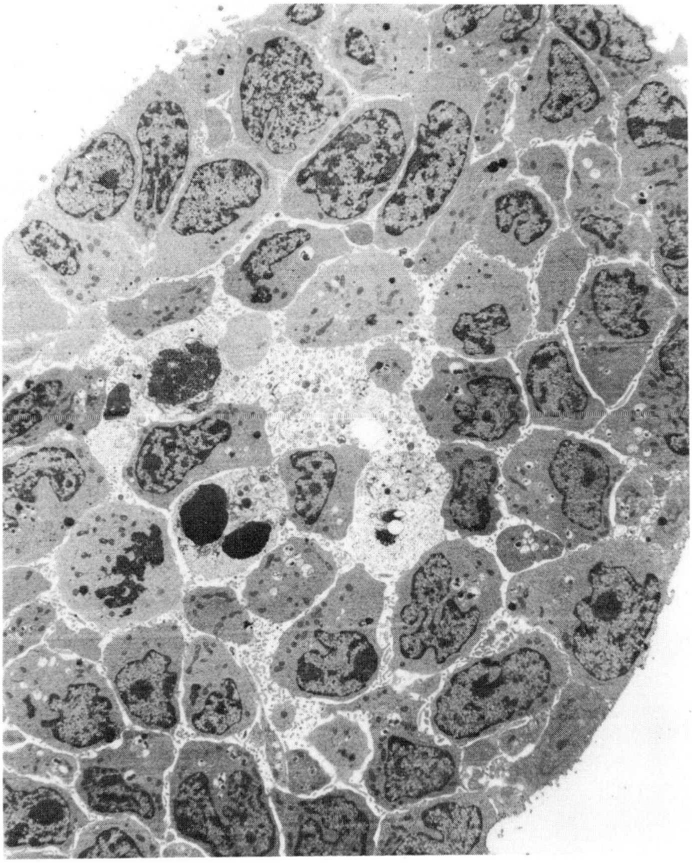

**Fig. 11** Ultrastructure of a single ML colony in agar. Note both lymphoblastic and lymphocytic elements as well as a mitotic figure. (Magnification, 5000 x.)

We have adapted the T lymphocyte colony technique to a whole-blood culture system (F. D. Wilson, unpublished results). Stimulation of canine peripheral whole blood by Con-A produces colonies in vitro. Unlike the FH technique, where a constant number of lymphocytes ($10^6$ lymphocytes) are plated with 50 $\mu$g of Con-A, the whole-blood cultures are set up with 3% whole blood. The optimum concentration of 125 $\mu$g/ml of Con-A required for whole-blood colony formation translates to about 2000 $\mu$g Con-A/$10^6$ lymphocytes, which is much higher than for the gradient technique. That increased requirement most likely results from the uptake of mitogen by nonlymphoid elements. We are currently comparing the morphological and functional nature of colonies forming in the whole-blood and gradient-enriched culture systems.

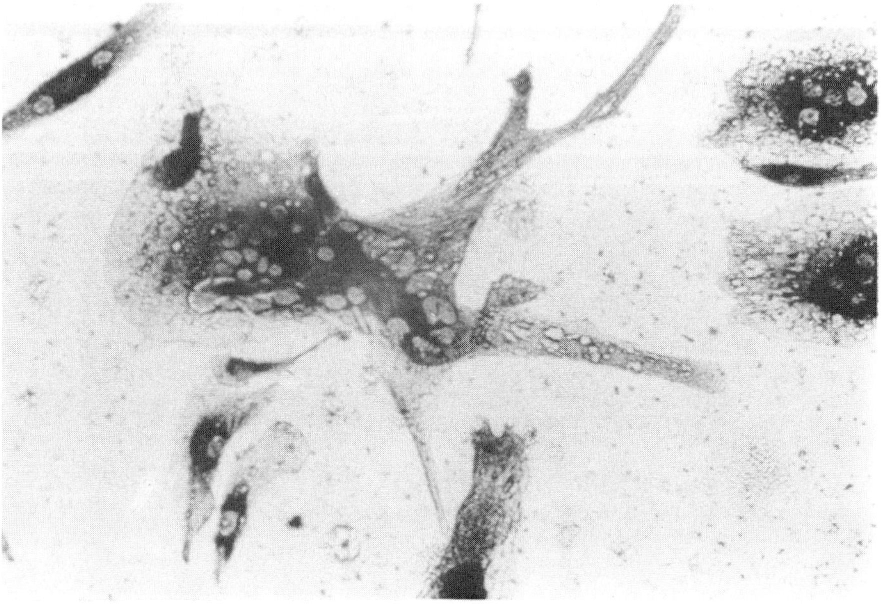

**Fig. 12 Adherent multinucleated giant cells in ML colonies. (Magnification, 100 x.)**

**Spontaneous Colonies.** Another unique observation on hematopoietic colonies in the dogs is the occasional occurrence of large numbers of colonies in agar cultures of FH lymphocyte-enriched fractions of peripheral blood not stimulated with mitogens. Such spontaneous formation was observed in the basic culture system used for the source of T lymphocyte colonies from the peripheral blood of dogs.[44] Thus cultures designed to produce no-add (NA) or background controls for mitogen-stimulated cultures unexpectedly occasionally contained up to several hundred colonies (Fig. 13). The microscopic appearance of the spontaneous colony, the progenitor for which we have termed CFU-NA, is shown in Fig. 14. Although the mechanism responsible for such colony formation is not known, preliminary studies suggest an increased tendency for "spontaneous" colony formation from lymphocytes from older dogs.

The anomaly of "spontaneous" colony formation underlines the importance of control cultures in assessing the true influence of mitogens or growth factors on colony formation.

**Methods for the In Situ Morphological Characterization of Lymphohematopoietic Colonies in Agar.** Although techniques for culturing hematopoietic cells in a semisolid medium have been available for many years, the examination of cells derived from such colonies has been difficult. The most common examination methods involve either plucking colonies or clusters out of the plate

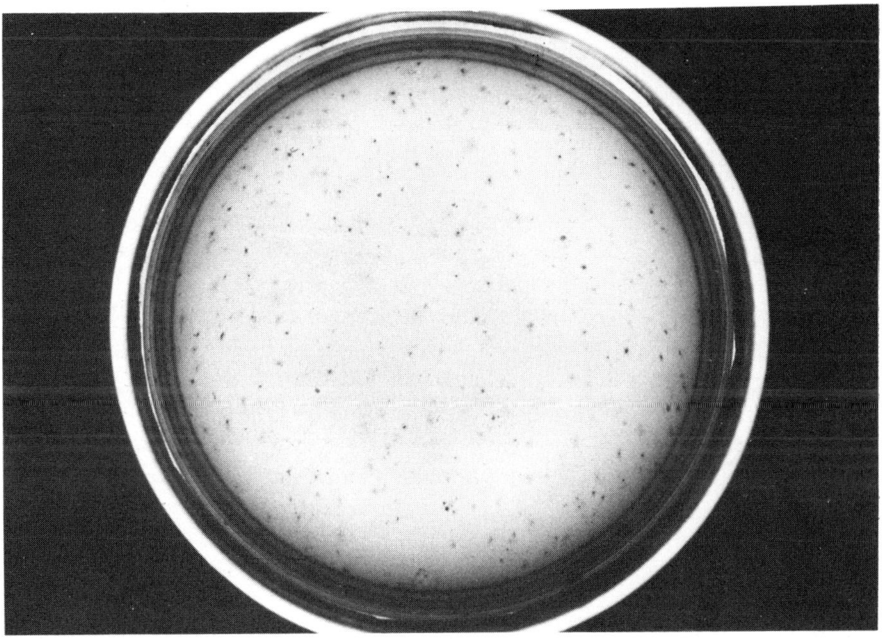

Fig. 13 Gross appearance of canine "spontaneous colonies" in agar culture stained with new methylene blue.

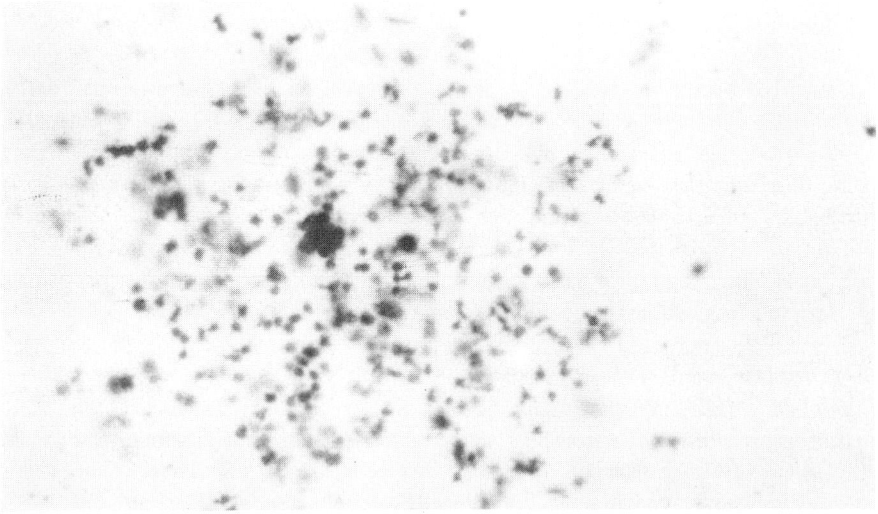

Fig. 14 Canine "spontaneous colony." (Magnification, 40 x.)

and depositing them on a slide or pooling colonies, washing them free of medium, and subsequently smearing the cell pellet on a slide. With both methods many of the extremely fragile cultured cells are destroyed in the process of taking them out of culture and processing them for deposition onto a slide. Additionally, neither method accurately portrays the in situ situation. The importance of identification of individual colonies is illustrated by our finding that multiple-colony types formed when canine bone marrow was used as a target of "T lymphocyte" mitogens.[44] We have developed a technique for in situ morphological analysis of lymphohematopoietic colonies in agarose. Our method is similar to that described previously by Zucker-Franklin and Grusky.[64] The agarose containing the colonies is fixed with methanol, which produces a solid pellicle of agar that can be removed from the culture plate. A variety of cytochemical stains can be applied to that permanent preparation of colonies. Colonies can also be embedded in paraffin and sectioned with a microtome (Fig. 15), which allows a wide variety of special staining procedures. The technique has wide application for the identification and quantitation of morphologically diverse colonies in situ. In addition, preliminary data suggest that an immunoperoxidase technique can be applied to the system. Therefore the identification of immunologically specific functional subpopulations of lymphocyte clones (e.g., B-cells and T-cells) may be possible in the near future.

If the cultures are fixed with 2% glutaraldehyde rather than methanol, individual clones can be analyzed by electron microscopy. The colonies are marked by using a dissecting microscope, and 3-mm squares of agarose containing individual colonies are cut out with a sharp blade. The sections are then embedded, sectioned, and processed with standard methods for transmission electron microscopy. Figure 11 shows an electron micrograph of a single colony prepared by this procedure.

**Bone-Marrow Cellularity.** One of the major obstacles to the interpretation of quantitative changes in the colony populations from bone marrow of large-animal species, including man, has been the lack of appropriate techniques for the determination of bone-marrow cellularity. Without such techniques, relative numbers of progenitors (number of progenitor cells in culture) cannot be equated with or converted to absolute values (number of progenitors per organ or bone). For example, the relative number of cultured bone-marrow progenitors could be markedly decreased in an animal, but, if bone-marrow cellularity is profoundly increased, the absolute number of progenitors in vivo could be normal or even elevated. Such a situation could be envisioned in leukemic animals, which, in general, are poor producers of in vitro colonies but which have greatly increased numbers of cellular elements in their bone marrow and other myeloid organs.

One useful approach to the qualitative evaluation of marrow cellularity in beagles is $^{52}Fe$ positron scanning with the Anger camera.[65] This noninvasive technique allows analysis of changes in the distribution of red marrow in animals

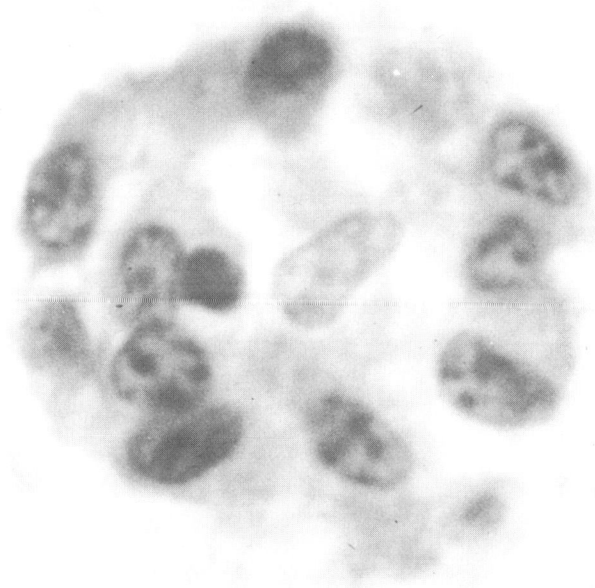

Fig. 15 Thin section of bone-marrow macrophage colony in agar. (Magnification, 900 x.)

with leukemia and other hematological abnormalities. In $^{52}Fe$ scans for a normal beagle and a beagle with radiation-induced chronic myelomonocytic leukemia (Figs. 16A and 16B), the leukemic dog has completely compensated for erythroid elements entirely in extramedullary organs. That scanning technique provides nondestructive qualitative analysis of hematopoietic activity but does not provide quantitative data on nucleated marrow cells for the entire body.

For the routine quantitation of bone-marrow cellularity in dogs, we[24] have developed a method based on the techniques described by Suit.[66] Ten microcuries of $^{59}Fe$ is injected into each dog 24 hr before the dog is irradiated. At the time the dog is sacrificed, bone marrow is flushed from one humerus, and the nucleated marrow cells per milliliter are determined with an electronic cell counter. A 1-ml sample of the marrow is also counted for $^{59}Fe$ activity with a gamma well counter. In addition, representative bones are removed in their

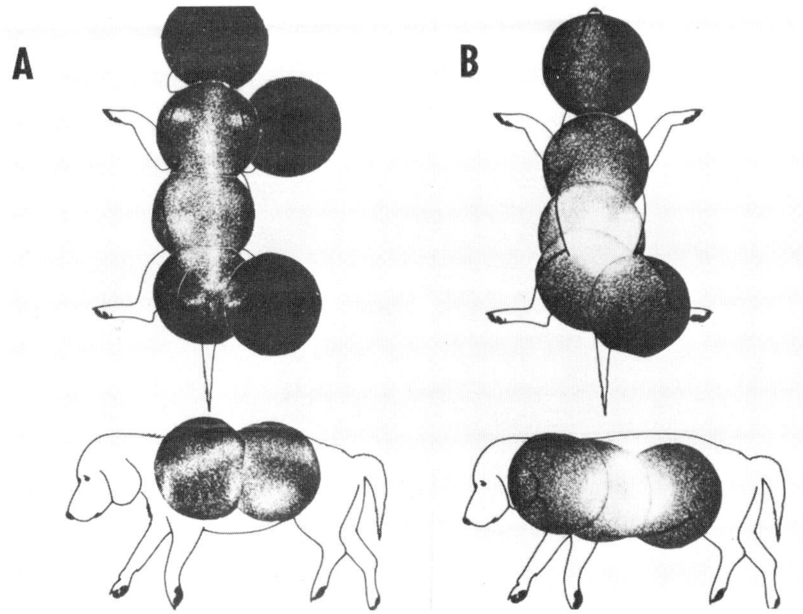


Fig. 16 Iron-52 positron scan of normal beagle (A) and a beagle with myelomonocytic leukemia (B).

entirety and are counted separately in the whole-body counter for $^{59}Fe$ activity. An identical standard in the gamma and the whole-body counters allows conversion of all data to whole-body equivalents. From that data, nucleated marrow cells per bone can be calculated from

$$\frac{\text{Nucleated cells}}{\text{bone}} = \frac{\text{cpm}}{\text{bone}} \text{(whole-body counter)} \times R \times \frac{\text{nucleated cells/ml marrow}}{\text{cpm/ml marrow}}$$

where R is the ratio of $^{59}Fe$ standard counts of gamma well counter to whole-body counter.

Values obtained with the $^{59}Fe$ labeling technique compare well with previously published values of Calvo et al.,[67] who used a more time-consuming morphometric approach.

The results of preliminary studies with the $^{59}Fe$ cellularity technique for dogs of various ages and for a variety of bones (Table 1) show that the greatest changes with age occur in long bones and, in particular, in the tibia and ulna, a reflection of the significant fatty involution in those bones. In red marrow sites, such as the ribs and vertebrae, cellularity either increases or remains constant with increasing age. The technique therefore appears to offer a rapid and reliable method for the determination of bone-marrow cellularity in the dog.

TABLE 1

AGE-RELATED CHANGES IN BONE-MARROW CELLULARITY IN NORMAL BEAGLES

| Age | Cells/g bone | | | | | |
|---|---|---|---|---|---|---|
| | Ribs* | Vertebrae* | Humeri* | Femurs* | Ulnae* | Tibiae* |
| 76 days | $0.68 \times 10^8$ (8) | $2.85 \times 10^8$ (3) | $3.92 \times 10^8$ (8) | $3.96 \times 10^8$ (8) | $2.53 \times 10^8$ (8) | $2.59 \times 10^8$ (8) |
| 299 days | $1.82 \times 10^7$ (6) | $2.83 \times 10^8$ (3) | $2.65 \times 10^8$ (6) | $2.40 \times 10^8$ (6) | $0.17 \times 10^8$ (6) | $0.56 \times 10^8$ (6) |
| 3 years | $1.14 \times 10^7$ (1) | $1.78 \times 10^8$ (1) | $3.55 \times 10^8$ (1) | $2.15 \times 10^8$ (1) | $0.02 \times 10^8$ (2) | $0.34 \times 10^8$ (2) |
| 5.4 years | $2.86 \times 10^8$ (2) | $1.86 \times 10^8$ (1) | $2.24 \times 10^8$ (2) | $0.76 \times 10^8$ (2) | $0.06 \times 10^8$ (2) | $0.02 \times 10^8$ (2) |

*Number in parentheses is the number of samples.

## APPLICATION OF COLONY FORMING TECHNIQUES TO STUDIES ON LYMPHOHEMATOPOIESIS IN THE DOG

Several assays of CFU-C have been reported for dogs,[3,24,68–70] but techniques for the quantitation of dog CFU-L[44] and CFU-E[42] have been described only recently. Canine CFU-C apparently are among the more sensitive populations to radiation injury.[24] The potential importance of CFU-C to immunology is underscored because these cells are thought to be progenitors of monocyte–macrophage pathways. The increasing evidence of a major role of macrophages as accessory and effector cells for both humoral and cellular immunological processes[28] emphasizes the importance of the relation of CFU-C to immunological events. The relationship of CFU-C and immune processes has been elaborated in our study on nude mice.[23]

Assays for CFU-C and CFU-E have been extensively used in the investigation of the pathogenesis of cyclic hematopoiesis in dogs,[42] and the CFU-C assay has been used in studies of cryopreservation of bone marrow in dogs.[43]

CFU-C constantly circulate in the peripheral blood of dogs. Fliedner et al.[43] reported values of 40 to 400 CFU-C/ml of peripheral blood in normal beagles. Repeated injections of dextran sulfate (DS), a stem-cell mobilizing agent, over a course of 3 weeks increased circulating CFU-C 10-fold in those dogs. These investigators also used DS mobilization in conjunction with leukophoresis as a method for collecting large numbers of CFU-C from the peripheral blood of dogs. The mobilization and removal of CFU-C by that method was associated with a reduction of CFU-C in the bone marrow.

Fliedner et al.[43] also reported a good correlation between CFU-C levels and pluripotent stem-cell activity as reflected in the ability of peripheral blood fractions enriched for CFU-C to reconstitute lethally irradiated dogs. Further, the level of CFU-C contained in cryopreserved mononuclear cells was correlated with the ability to reconstitute lymphoid organs. In these studies early colonization of bone marrow by "niches" of erythrocytic, granulocytic, and megakaryocytic regenerative foci was related to the number of CFU-C transfused. These investigators also pointed out that the ratio of erythroid to granuloid (E/G) regenerative foci of 2 : 1 was a reversal of the 1 : 2 E/G ratio for spleen colonies in mice and suggested that the microenvironments may be different in the two species. Fliedner et al.[43] concluded that the number of CFU-C in the peripheral blood of dogs is a good indicator of stem-cell capacity for reconstitution of myeloid and lymphoid elements and therefore is an indicator of pluripotent stem-cell content.

Canine peripheral blood fractions can be enriched for hematopoietic reconstitution (stem-cell fraction) or processed for the elimination of cells responsible for graft-vs.-host disease. Ross et al.,[3] for example, used discontinuous albumin gradients to extract fractions of canine peripheral blood with improved ratios of CFU-C to PHA-reactive lymphocytes. The CFU-C enriched fraction produced highly efficient reconstitutional capacity, whereas the remaining population showed minimal capacity for reconstitution of hematopoietic elements but efficient reconstitution of lymphoid tissue associated with the induction of severe graft-vs.-host disease.

Evidence that colony stimulating factor (CSF) has a role in the physiologic regulation of granulopoiesis in dogs has recently been reported. MacVittie and Walker,[71] using the CFU-C assays, investigated the effects of endotoxin injection on canine granulopoiesis. From those studies the investigators concluded that endotoxin-induced CSF is a physiological regulator of canine granulopoiesis and that endotoxin injection stimulates CFU-C in the peripheral blood. These investigators also found that intestinal gram-negative flora influence granulopoiesis in dogs.[72] Serum CSF and marrow CFU-C were significantly reduced by the reduction of gram-negative flora. The hypothesized role of CSF as a granulopoietin in dogs is also supported by studies on CSF in dogs with cyclic hematopoiesis.[42]

We have recently extended our observations on human-bone-marrow attaching cells (stromal colonies) to studies on attaching cells from the lymph node of beagles with malignant lymphoma. Adherent cells in tissue cultures of such animals also produced mesenchymal tumors similar to human tumors when injected into nude mice (F. D. Wilson, unpublished results). The ultrastructural appearance of canine surface-adherent cells in culture and of the mesenchymal tumor in a nude mouse is shown in Figs. 6 to 8 and 17.

## APPLICATION OF CLONING TECHNIQUES TO RADIATION BIOLOGY

Although much is known about the effects of ionizing radiation on hematopoietic stem cells in mice, relatively little is known of such effects in

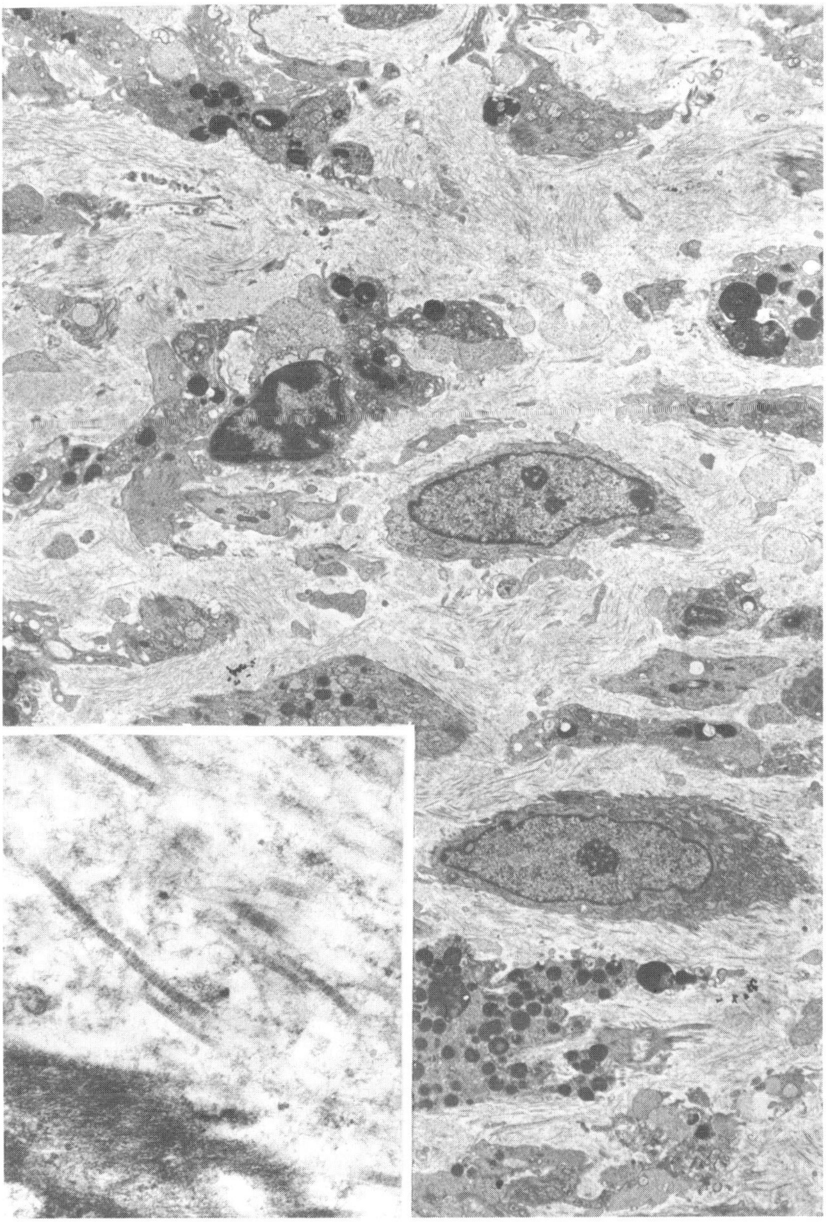

Fig. 17 Ultrastructure of cells in a mesenchymal tumor induced in a nude mouse by injection of canine "stromal" cells from culture of a lymph node of a beagle with malignant lymphoma. (Magnification, 3500 x.) Insert shows details of collagen fibers. (Magnification, 37,000 x.)

larger animals, including man. That obvious need for a long-lived animal model led us to investigations of effects of radiation on hematopoiesis in the beagle. We were interested in developing techniques for use as sensitive biodosimeters of hematopoietic injury as well as for studying changes in hematopoietic progenitors in radiation-induced leukemia in beagles.

## Effects of Continuous Irradiation on Hematopoiesis in Beagles

Continuous irradiation of the bone marrow in beagles by internal emitters results in a "hematological paradox."[70] Specifically, continuous irradiation at relatively low dose rates produces an abnormal steady state of hematopoiesis; granulopoiesis is compromised, whereas erythropoiesis is spared if not accelerated. Granulopoiesis changes, even though extramedullary sites for hematopoiesis, such as the liver and spleen, are not subject to direct radiation damage from bone-seeking nuclides, such as $^{226}$Ra and $^{90}$Sr. Thus local irradiation of bone marrow by $^{90}$Sr or bone-marrow endosteum by $^{226}$Ra induces total body changes in the normal steady-state hematopoiesis. Similar effects on lymphopoiesis have been observed in beagles inhaling insoluble forms of plutonium.[73] Those studies, in addition to providing data on the effects of continuous irradiation, are producing provocative insights into the mechanisms for the regulation of lymphohematopoiesis in both physiological and pathological conditions.

We are attempting to use specific forms of radiation (alpha, beta, and gamma) and to exploit the size advantages of the canine to develop a map for the distribution of cells at risk for radiation effects on lymphohematopoietic precursors, including their distribution in leukemia disorders. Lord, Testa, and Hendry[74] have reported a nonuniformity in the spatial distribution of hematopoietic stem and progenitor cells in the bone marrow of mice. CFU-S tend to concentrate along the endosteum, whereas CFU-C are located away from the bone/bone-marrow interface. Maloney et al.,[75] however, could not confirm those observations. That discrepancy may, in part, reflect the difficulties of retrieving specific locations of marrow within the extremely small marrow cavities of mice. Our studies in the larger dog may resolve questions about the distribution of lymphohematopoietic progenitors as assayed by our various cloning techniques. We further will use three continuous-radiation models, $^{226}$Ra (internal alpha emitter, producing endosteal radiation), $^{90}$Sr (internal beta emitter, producing total-marrow radiation), and $^{60}$Co (gamma emitter, producing total-body irradiation of both medullary and extramedullary sites of lymphohematopoiesis), to investigate the effects of radiation on the cells at risk for leukemic induction. This approach is unique for the dog model; rodents are not large enough and do not have sufficient longevity for late effects, and humans obviously cannot be exposed to radiation. Our tentative arrangement for cells at risk and the models for radiation are outlined in Fig. 18. This model was

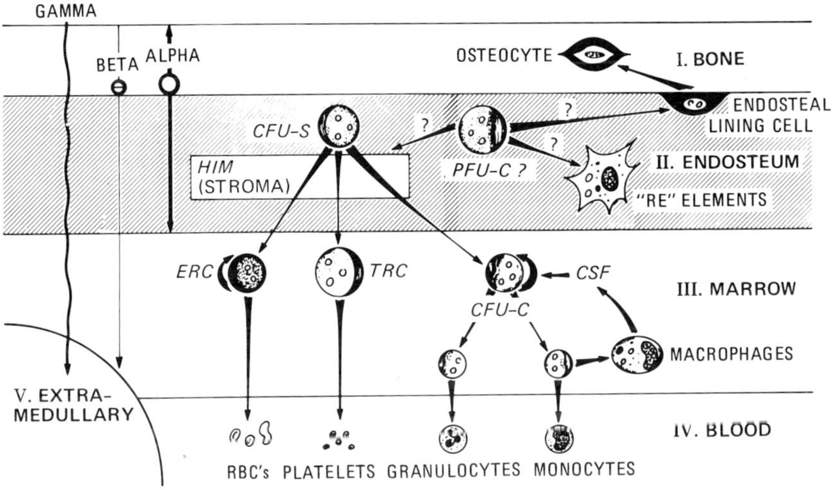


Fig. 18 Suggested location and interactions for cells at risk for radiation injury.

published previously in our efforts to relate early effects of radiation on hematopoietic parameters to the late effects of osteogenic sarcoma induction.[76]

## Studies on Progenitor-Cell Response to Acute Irradiation

The effects of acute irradiation on CFU-C and PFU-C have been investigated in weanling beagles.[24] In initial studies, groups of 2- to 3-month-old beagles were exposed to acute whole-body gamma irradiation delivered from a $^{60}$Co source. Twenty-four hours after they had been irradiated, the number of CFU-C and PFU-C per gram of bone were determined. Survival curves for the two populations indicated a D37 for CFU-C of 70 R, which is similar to D37's reported for CFU-C[19] and CFU-S[4] in mice. Populations of CFU-C are among the most sensitive to radiation injury. The D37 for the fibroblastic populations was not encompassed by the dose limits of the experiment, but the results indicated a substantial degree of radioresistance for those progenitors, which supports previous studies that suggested a D37 of about 400 R for PFU-C in mice. The results demonstrate that semisolid culture systems can be used to measure changes in CFU-C as a sensitive biodosimeter of radiation injury in the dog and simultaneously to investigate radioresistant progenitor populations—PFU-C. The effects of radiation on hematopoietic stromal elements (PFU-C) deserve more attention because of the hypothesized role of stroma in the regulation of lymphohematopoiesis.

### Radiation Effects on Lymphohematopoietic Colonies: Existence of Subpopulations of Lymphocyte Progenitors with Differing Radiosensitivities

We have previously described[44] a method for the growth of canine T lymphocyte colonies in vitro in a methylcellulose-supported culture system. We have recently exploited those techniques to study the effects of radiation on T lymphocyte progenitors in dogs. In vitro X-ray survival curves were drawn for FH enriched peripheral blood lymphocytes (PBL) that formed lymphocyte colonies in culture (CFU-L) after stimulation by PHA or Con-A. The dose–response curves for those normal T lymphocyte progenitors were best described by a biexponential expression, which suggests subpopulations of T lymphocyte progenitors with different radiosensitivities. Such observations led us to speculate that shifts in survival curves for PHA and Con-A might be predicted in dogs receiving continuous irradiation. Response curves for the in vitro, X-irradiation survival of CFU-TL of peripheral blood in dogs that had received continuous irradiation (5 R/day for 100 cumulative rads) were constructed to test this hypothesis. Shifts in survival curves (D37 about 70 R for control and about 150 R for irradiated) were indeed seen in response to continuous irradiation for Con-A- (Fig. 19) but not for PHA-induced T lymphocyte colony formation (Fig. 20). Also, continuous irradiation did not affect survival curves for canine or murine bone-marrow granulocyte–monocyte progenitors or for murine B lymphocyte progenitors in the spleen. Those results imply that PHA and Con-A stimulate different subpopulations of T lymphocyte progenitors. The results also suggest a new experimental approach for the investigation of radiation-induced injury to the immune system.

## CANINE MODELS FOR LEUKEMOGENESIS STUDIES

Although canine lymphoid neoplasias are relatively common, with an incidence of 24 per $10^5$, similar to that in humans,[77,78] myelogenous tumors are extremely rare. Myeloproliferative disorders[79–81] (MPD) are much more common for beagles subjected to continuous bone-marrow beta irradiation by $^{90}$Sr or whole-body gamma irradiation by $^{60}$Co. Those radiation-induced myeloproliferative disorders manifest as a spectrum of diseases ranging from myeloid metaplasia with myelofibrosis to overt myelogenous leukemias.[79,80] The induction of erythroleukemias also has been reported recently for beagles receiving whole-body gamma irradiation.[82] Since a high proportion of the animals that receive sufficient radiation will manifest predictably with MPD, the radiation-induced MPD model in beagles provides a truly unique experimental situation for study of the normal, preleukemic, and leukemic conditions in a single animal. Such a model provides an obvious advantage in studies directed at the etiology, pathogenesis, and therapy of leukemic disorders.

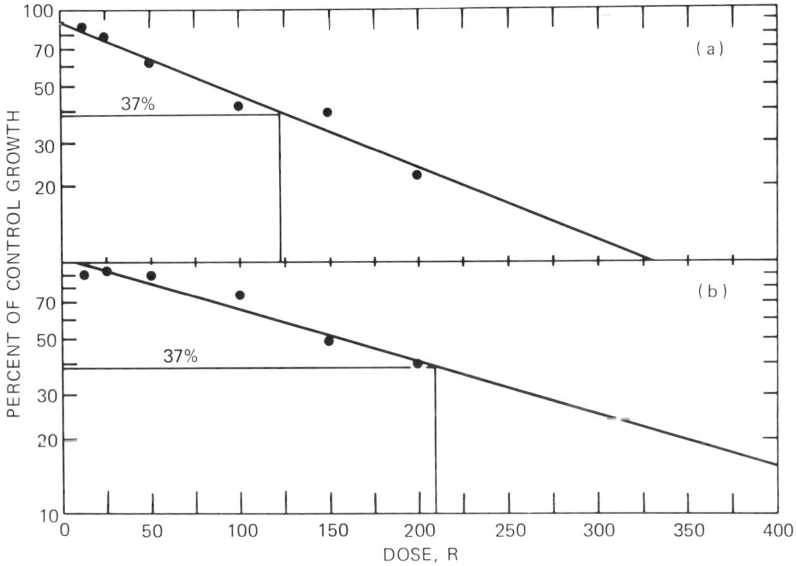


Fig. 19 X-ray dose–response curves for Con-A stimulated T lymphocyte progenitors in peripheral blood from unirradiated (a) and continuously irradiated (b) beagles.

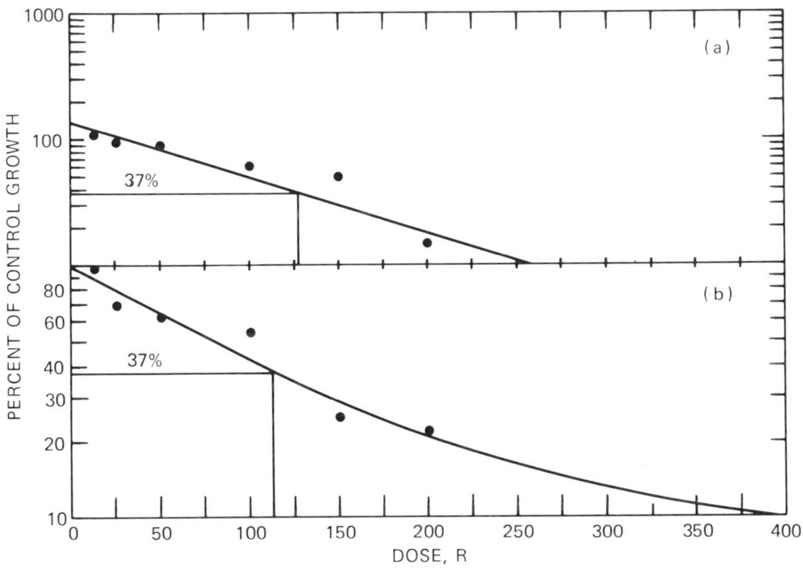


Fig. 20 X-ray dose–response curves for PHA stimulated T lymphocyte progenitors in peripheral blood from continuously irradiated (a) and unirradiated (b) beagles.

Two other unique experimental models, the in utero transplantation model and the chimeric control model, can be used with the radiation-induced MPD model. We previously described the transplantation of radiation-induced myelomonocytic leukemia in beagles by the intraperitoneal inoculation of viable leukemic cells into fetuses.[83] Also, a leukemia-associated cytogenetic marker[84] allows convenient determination of the host–recipient nature of the leukemic populations. We have also detailed the hematological values and lysozyme levels of beagles with transplantable leukemias.[85,86] This canine model was described previously and used for the transplantation of lymphosarcoma[87] and mast cell tumors.[88] Such models, in addition to producing more available leukemic animals, are invaluable for studies on leukemia transplantation phenomena, host-factor involvement in leukemogenesis, fetal immunology, and regressor–progressor factors involved in tumor growth.

Another unique model for the study of leukemogenesis is based on canine DLA types,[89] which enable the production of hematopoietic chimeras in dogs. Normal cells can be "stored" in a DLA matched chimeric host animal prior to donor irradiation. When the donor animal predictably manifests with leukemia, its own normal cells are available for comparison. Such a model is invaluable for studies on bone-marrow transplantation therapy and for defining immunologic components of leukemic disorders (i.e., a syngeneic beagle model). Also in that model, leukemic cells can be transplanted into adult histocompatible siblings. Efforts at DLA typing of beagle colonies at Davis[90] and at several other laboratories will extend the value of the chimera model.

Recently, Seed et al.[91] have reported on the preleukemic and leukemic changes in dogs receiving continuous whole-body gamma irradiation at a dose rate of 10 R/day. They recognized four preclinical phases in leukemia development: induction of severe leukopenia and thrombocytopenia, partial hematopoietic recovery, development of a subnormal steady state in hematopoiesis, and a "classical" preleukemic phase. The second phase was the critical event for subsequent development of leukemia. Approximately 50% of the dogs studied failed to enter this phase and died from aplastic anemia. Dogs destined to develop leukemia underwent partial recovery after phase I and eventually recovered after 200 to 300 days of irradiation, as evidenced by an increase in peripheral blood platelets and leukocytes. This recovery coincided with an increase in marrow cellularity, CFU-C, and marrow granulocyte reserves (the last measured by response to endotoxin). The preleukemia phase was characterized by marked oscillations in platelets, immature granulocytes in the blood, and morphological abnormalities in erythrocytes.

Cloning techniques for lymphohematopoietic progenitors in humans provide the ability to determine the capacity of leukemic cells for differentiation, the responsiveness of leukemic cells to physiological regulation by hormonal and cellular factors, and the possible production of hematopoietic regulators by leukemic cells.[5,92] In addition, in vitro experiments can determine if abnormal progenitor populations can be induced into normal growth patterns by the addition of various factors. Definite abnormal patterns of growth of progenitor populations in culture seem to be associated with leukemias. In general, leukemic cells are poor colony formers and are associated with the formation of more clusters or small colonies. Although leukemic cells are partially responsive to factors regulating normal hematopoiesis, abnormalities in regulators also have been detected in leukemia. In addition, some factors, such as PHA, induce differentiation of leukemic cells in culture.[93] Such abnormal growth patterns potentially can be used as diagnostic and prognostic aids.

In preliminary studies at our laboratory, as in similar human leukemia situations, abnormal colony size distributions in CFU-C were obtained from two beagles with radiation-induced myelogenous leukemias. One animal had been fed $^{90}$Sr (4 $\mu$Ci/day) for his entire life. Marrow from a normal dog and a nonleukemic chronically irradiated littermate was simultaneously cultured for comparison of the relative status of stem cells from dogs with radiation-induced myeloproliferative syndrome (MPS), control dogs, and chronically irradiated non-MPS dogs. The second MPS animal developed chronic granulocytic leukemia at 2.3 years of age after exposure to approximately 1000 R delivered from a $^{60}$Co source at a dose rate of 5 R/day. A clinically normal dog of the same sex and age served as a control.

Results for marrow culture (Fig. 21) show that both MPS animals exhibit abnormal colony size distribution profiles relative to controls and the chronically irradiated non-MPS animal. The number of small colonies or "aggregates" forming in the cultures from MPS animals is increased. The type of radiation used to induce MPS did not affect the shape of the colony profiles in the MPS animals.

The marked abnormalities in hematopoietic colony forming parameters (CFU-C) in MPS animals are not paralleled by changes in fibroblastic clonogenic populations (PFU-C) (Table 2). The number of PFU-C per milliliter of marrow in the $^{60}$Co-induced MPS animal is essentially the same as that in the control animal, but high PFU-C values characterize both the $^{90}$Sr-induced MPS animal and his $^{90}$Sr-irradiated non-MPS littermate. Those preliminary findings would suggest that numbers of PFU-C are not influenced by MPS, chronic irradiation with $^{90}$Sr increases these "candidate" stromal progenitor populations, and changes in PFU-C are related to animal age (compare two control animals, Table 2).

Although MPS was not associated with quantitative abnormalities in fibroblastic colonies (PFU-C), evidence for possible functional abnormalities in PFU-C was obtained in refill experiments (Table 3). The functional ability of fibroblastic colonies once derived to stimulate hematopoietic colony development was the same for controls and for the $^{90}$Sr-irradiated non-MPS animal, but such ability was dramatically increased in both animals with leukemia. Such findings suggest, as does the work of Messner, Till, and McCulloch[94] with human leukemics, abnormalities in adherent CSA-producing intramedullary populations in those MPS disorders.

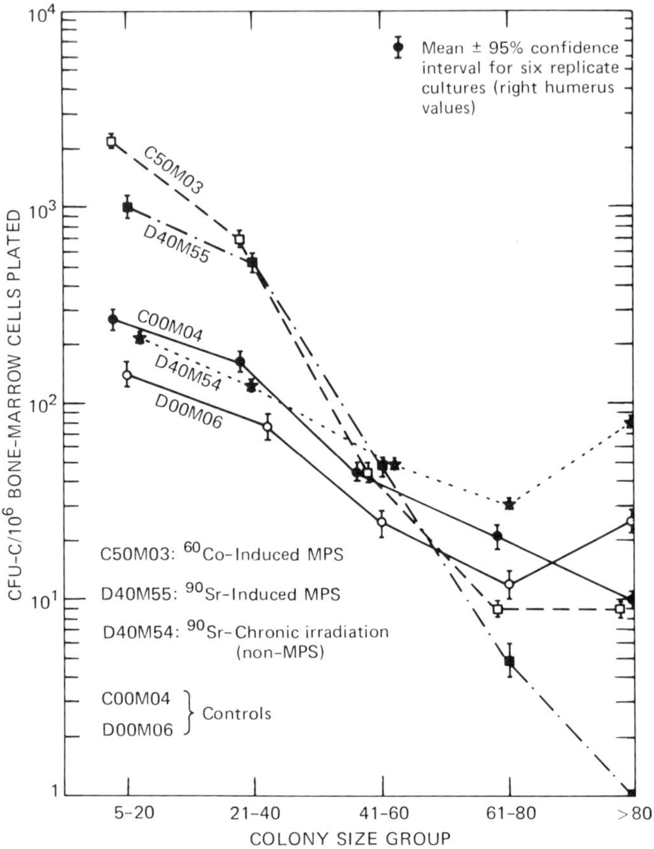


**Fig. 21** Colony size distribution for normal, $^{90}$Sr irradiated non-MPS, $^{90}$Sr-induced granulocytic leukemia, and $^{60}$Co-induced granulocytic leukemia beagles.

TABLE 2

BONE-MARROW CULTURE RESULTS FOR CONTROL BEAGLES AND DOGS WITH RADIATION-INDUCED MPS

| Treatment | CFU-C/ml bone marrow* | PFU-C/ml bone marrow* |
|---|---|---|
| Control ($^{60}$Co) | $9.2 \times 10^4$ ($\pm 6.9 \times 10^4$) | $5.2 \times 10^3$ ($\pm 3.4 \times 10^3$) |
| Control ($^{90}$Sr) | $7.5 \times 10^4$ ($\pm 2.0 \times 10^4$) | $1.8 \times 10^3$ ($\pm 0.1 \times 10^3$) |
| $^{60}$Co with MPS | $4.2 \times 10^6$ ($\pm 0.7 \times 10^6$) | $5.4 \times 10^3$ ($\pm 0.6 \times 10^3$) |
| $^{90}$Sr with MPS | $10.7 \times 10^6$ ($\pm 6.2 \times 10^6$) | $15.0 \times 10^3$ ($\pm 2.5 \times 10^3$) |
| $^{90}$Sr without MPS | $2.3 \times 10^5$ ($\pm 1.4 \times 10^5$) | $13.1 \times 10^3$ ($\pm 2.1 \times 10^3$) |

*Values = mean ± standard deviation for right and left humerus.

TABLE 3

EFFECT OF FIBROBLASTIC COLONIES ON THE GROWTH OF GRANULOCYTE–MONOCYTE COLONIES

| | CFU-C/PFU-C ratio* |
|---|---|
| Controls (2 animals, 4 samples) | 5.2 : 1 (±1.7) |
| Chronic irradiation (1 animal, 2 samples) | 5.7 : 1 |
| MPS | 12.8 : 1 (±4.3) |

*Ratio of the number of CFU-C stimulated to the number of fibroblastic colonies present in the culture. Mean ± standard deviation.

The studies on canine bone-marrow attaching cells have been extended to the granulopoietic effects of fibroblastic colonies (PFU-C) in normal humans and in patients with acute myelogenous leukemia (AML) and other myeloproliferative disorders.[55] Bone-marrow culture and refill techniques developed for dogs were modified for human cultures. Fibroblastic colonies in liquid culture from normal humans and patients with AML, preleukemia, myeloproliferative disorders (MPD), and metabolic carcinoma were overlaid with normal and leukemic marrow cells in agar. As with the canine system, increased numbers of CFU-C were found only above the PFU-C colonies that formed from normal humans. Unlike the canines with chronic myelogenous leukemia (CML), PFU-C from humans with untreated AML failed to enhance granulopoiesis. In contrast, PFU-C from patients with AML in remission stimulated CFU-C. PFU-C from patients with MPD also had little effect on granulopoiesis. The addition of CSF [placenta conditioned medium (PCM)] never augmented granulopoiesis and frequently was inhibitory. Supernatant from PFU-C cultures of normal canines and humans had little effect on colony growth. Thus the effects of PFU-C on granulopoiesis apparently are local, totally different from, and not synergistic with colony stimulating activity and are unlike the effects of CSF.

Our overall approach to the study of the role of hematopoietic stromal components in myeloproliferative disorders in humans and canines is illustrated in Fig. 22. Bone marrow obtained by aspiration is examined directly for morphological abnormalities and leukemia-associated cytogenetic marker chromosomes, such as the "Davis" chromosome in canines[84] and the $Ph^1$ chromosome in humans.[26] Liquid-phase cultures are prepared for fibroblast colonies in culture. The resultant adherent populations are subcultured repeatedly to yield morphologically and cytogenetically "pure" populations of the surface-adherent fibroblast colonies. Bone-marrow cultures also are prepared with the semisolid methylcellulose culture system for quantitation of CFU-C, CFU-L, and PFU-C. The resultant colonies are examined morphologically, and colony size distributions are determined. The response of leukemic colony

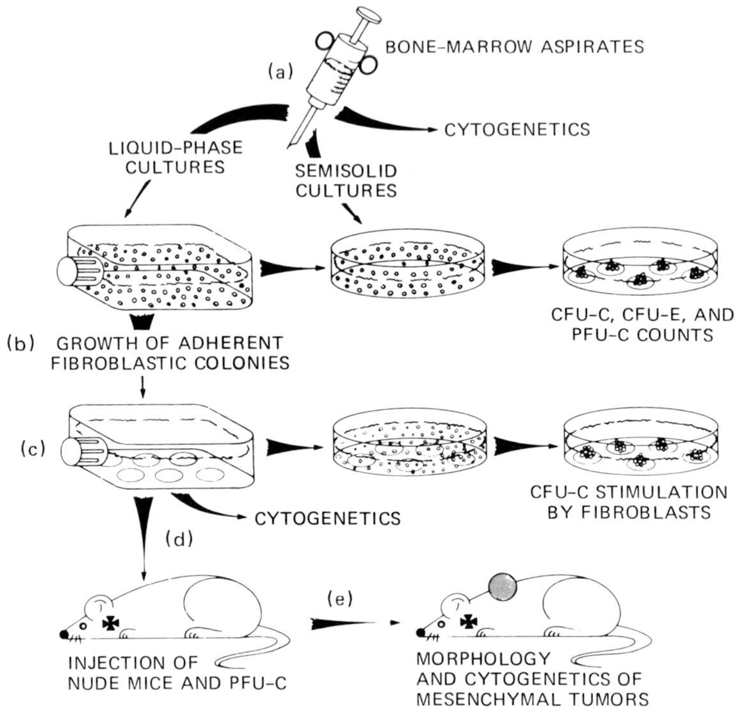


Fig. 22 Model for the investigation of stromal elements in canines. Our basic experimental approach is illustrated in this figure. Bone marrow obtained by aspiration (a) is studied directly for the presence of leukemia-associated cytogenetic marker chromosomes. Liquid-phase cultures are then prepared for obtaining fibroblastic-adherent colonies. The resultant adherent populations are subcultured repeatedly (b) until morphologically and cytogenetically "pure" populations of the surface-adherent elements are obtained (c). Fibroblastic cells are then removed by trypsinization and inoculated subcutaneously into nude mice (d). Resultant tumors are examined morphologically and cytogenetically at 1 week postinjection, and tissue cultures of tumors are prepared for additional morphological and cytogenetic studies (e).

forming cells to various humoral regulators (e.g., CSA and erythropoietin) can be investigated with this culture system. Also, the functional ability of fibroblast colonies to stimulate CFU's and the ability of the stromal cells to induce tumors in nude mice can be investigated.

## CONCLUSIONS

The relatively recent development of methods for the investigation of lymphohematopoietic stem and progenitor cells has resulted in a revolution in the advancement of experimental hematology. The coupling of those technological advances to large animal models, particularly the canine, predicts an exciting future for the unraveling of the many mysteries involved in the physiological regulation and pathological responses in lymphohematopoiesis in health and disease.

## REFERENCES


1. D. W. van Bekkum, M. J. van Noord, B. Maat, and K. A. Dicke, Attempts at Identification of Hemopoietic Stem Cell in Mouse, *Blood,* 38: 547-558 (1971).
2. K. A. Dicke, M. J. van Noord, B. Maat, U. W. Schaefer, and D. W. van Bekkum, Attempts at Morphological Identification of the Haemopoietic Stem Cell in Primates and Rodents, in *Haemopoietic Stem Cells,* G. E. W. Wolstenholme and M. O'Conner (Eds.), pp. 47-69, Ciba Foundation Symposium 13 (New Series), Elsevier, Amsterdam, 1973.
3. W. M. Ross, M. Körbling, W. Nothdurft, W. Calvo, and T. M. Fliedner, Characterization of Bone Marrow and Lymph Node Repopulating Cells by Transplanting Mononuclear Cells into Radiated Dogs, in *Experimental Hematology Today,* S. J. Baum and G. D. Ledney (Eds.), pp. 29-38, Springer-Verlag, New York, 1977.
4. J. E. Till and E. A. McCulloch, A Direct Measurement of the Radiation Sensitivity of Normal Mouse Bone Marrow Cells, *Radiat. Res.,* 14: 213-222 (1961).
5. D. Metcalf, in *Hemopoietic Colonies: In Vitro Cloning of Normal and Leukemic Cells,* G. P. Rentchnick (Ed. in Chief), Springer-Verlag, Berlin, 1977.
6. L. G. Lajtha, L. V. Pozzi, R. Schofield, and M. Fox, Kinetic Properties of Haemopoietic Stem Cells, *Cell Tissue Kinet.,* 2: 39-49 (1969).
7. J. S. Haskill, T. A. McNeill, and M. A. S. Moore, Density Distribution Analysis of In Vivo and In Vitro Colony Forming Cells in Bone Marrow, *J. Cell. Physiol.,* 75: 167-180 (1970).
8. D. J. A. Sutherland, J. E. Till, and E. A. McCulloch, A Kinetic Study of the Genetic Control of Hemopoietic Progenitor Cells Assayed in Culture and In Vivo, *J. Cell. Physiol.,* 75: 267-274 (1970).
9. M. G. Chen and J. C. Schooley, Recovery of Proliferative Capacity of Agar Colony-Forming Cells and Spleen Colony-Forming Cells Following Ionizing Radiation or Vinblastine, *J. Cell. Physiol.,* 75: 89-96 (1970).
10. K. A. Rickard, R. K. Shadduck, D. E. Howard, and F. Stohlman, Jr., A Differential Effect of Hydroxyurea on Hematopoietic Stem Cell Colonies In Vitro and In Vivo, *Proc. Soc. Exp. Biol. Med.,* 134: 152-156 (1970).
11. J. J. Trentin, Determination of Bone Marrow Stem Cell Differentiation by Stromal Hemopoietic Inductive Microenvironments (HIM), *Am. J. Pathol.,* 65: 621-628 (1971).
12. M. Tavassoli, Studies on Hemopoietic Microenvironments, *Exp. Hematol.,* 3: 213-226 (1975).
13. A. J. Friedenstein, R. K. Chailakhyan, N. V. Latsinik, A. F. Panasyuk, and I. V. Keiliss-Borok, Stromal Cells Responsible for Transferring the Microenvironment of the Hemopoietic Tissues, *Transplantation,* 17: 331-340 (1974).
14. S. H. Chan and D. Metcalf, Local Production of Colony-Stimulating Factor Within the Bone Marrow: Role of Nonhematopoietic Cells, *Blood,* 40: 646-653 (1972).
15. Y. T. Chang and R. N. Andersen, Cultivation of Mouse Bone Marrow Cells, *J. Reticuloendothel. Soc.,* 9: 568-579 (1971).
16. T. M. Dexter and M. A. S. Moore, In Vitro Duplication and "Cure" of Haemopoietic Defects in Genetically Anaemic Mice, *Nature,* 269: 412-414 (1977).

17. M. E. Gershwin, R. M. Ikeda, W. L. Kruse, F. Wilson, M. Shifrine, and W. Spangler, Age-Dependent Loss in New Zealand Mice of Morphological and Functional Characteristics of Thymic Epithelial Cells, *J. Immunol.*, 120: 971-979 (1978).
18. F. D. Wilson and L. O'Grady, Some Observations of the Hematopoietic Status In Vivo and In Vitro on Mice of Genotype $S1/S1^d$, *Blood*, 48: 601-608 (1976).
19. F. D. Wilson, L. O'Grady, C. J. McNeill, and S. L. Munn, The Formation of Bone Marrow Derived Fibroblastic Plaques In Vitro: Preliminary Results Contrasting These Populations to CFU-C, *Exp. Hematol.*, 2: 343-354 (1974).
20. A. J. Friedenstein, R. K. Chailakhjan, and K. S. Lalykina, The Development of Fibroblast Colonies in Monolayer Cultures of Guinea-Pig Bone Marrow and Spleen Cells, *Cell Tissue Kinet.*, 3: 393-403 (1970).
21. A. A. Maximow, Development of Non-Granular Leucocytes (Lymphocytes and Monocytes) into Polyblasts (Macrophages) and Fibroblasts In Vitro, *Proc. Soc. Exp. Biol. Med.*, 24: 570-572 (1926–1927).
22. F. D. Wilson, B. Greenberg, W. Spangler, M. Shifrine, M. E. Gershwin, and J. Dyck, Production of Mesenchymal Tumors in Nude Mice by $Ph^1$ Negative Fibroblasts Obtained from a $Ph^1$ Positive CML Patient: A Preliminary Report, *Exp. Hematol.*, 6: 549-557 (1978).
23. F. D. Wilson, M. E. Gershwin, M. Shifrine, and R. Graham, Increased Clonogenic (CFU-C, PFU-C) Populations from Bone Marrow and Spleen of Nude Mice, *Dev. Comp. Immunol.*, 1: 373-384 (1977).
24. F. D. Wilson et al., Quantitative Response of Bone Marrow Colony-Forming Units (CFU-C and PFU-C) in Weanling Beagles Exposed to Acute Whole-Body $\gamma$-Irradiation, *Radiat. Res.*, 74: 289-297 (1978).
25. F. D. Wilson, B. R. Greenberg, P. N. Konrad, A. K. Klein, and P. A. Walling, Cytogenetic Studies on Bone Marrow Fibroblasts from a Male–Female Hematopoietic Chimera, *Transplantation*, 25: 87-88 (1978).
26. B. R. Greenberg, F. D. Wilson, L. Woo, and H. M. Jenks, Cytogenetics of Fibroblastic Colonies in $Ph^1$-Positive Chronic Myelogenous Leukemia, *Blood*, 51: 1039-1044 (1978).
27. N. L. Petrakis, M. Davis, and S. P. Lucia, The In Vivo Differentiation of Human Leukocytes into Histiocytes, Fibroblasts and Fat Cells in Subcutaneous Diffusion Chambers, *Blood*, 17: 109-118 (1961).
28. M. J. Cline and D. W. Golde, Cellular Interactions in Haematopoiesis, *Nature*, 277: 177-181 (1979).
29. N. S. Wolf and J. J. Trentin, Hemopoietic Colony Studies. V. Effect of Hemopoietic Organ Stroma on Differentiation of Pluripotent Stem Cells, *J. Exp. Med.*, 127: 205-214 (1968).
30. S. E. Bernstein, Tissue Transplantation as an Analytic and Therapeutic Tool in Hereditary Anemias, *Am. J. Surg.*, 119: 448-451 (1970).
31. M. C. Bessis and J. Brenton-Gorius, Iron Metabolism in the Bone Marrow as Seen by Electron Microscopy: A Critical Review, *Blood*, 19: 635-663 (1962).
32. J. I. Kurland, P. W. Kincade, and M. A. S. Moore, Regulation of B-Lymphocyte Clonal Proliferation by Stimulatory and Inhibitory Macrophage-Derived Factors, *J. Exp. Med.*, 146: 1420-1435 (1977).
33. J. Kurland and M. A. S. Moore, Modulation of Hemopoiesis by Prostaglandins, *Exp. Hematol.*, 5: 357-373 (1977).
34. J. I. Kurland, The Mononuclear Phagocyte and Its Regulatory Interactions in Hemopoiesis, in *Experimental Hematology Today*, S. J. Baum and G. D. Ledney (Eds.), pp. 47-60, Springer-Verlag, New York, 1978.
35. F. W. Ruscetti and P. A. Chervenick, Regulation of the Release of Colony-Stimulating Activity from Mitogen Stimulated Lymphocytes, *J. Immunol.*, 114: 1513-1517 (1975).

36. M. J. Cline and D. W. Golde, Production of Colony-Stimulating Activity by Human Lymphocytes, *Nature*, 248: 703-704 (1974).
37. M. J. Cline and D. W. Golde, Immune Suppression of Hematopoiesis, *Am. J. Med.*, 64: 301-310 (1977).
38. W. Calvo, T. M. Fliedner, E. Herbst, E. Hügl, and C. Bruch, Regeneration of Blood-Forming Organs After Autologous Leukocyte Transfusion in Lethally Irradiated Dogs. II. Distribution and Cellularity of the Marrow in Irradiated and Transfused Animals, *Blood*, 47: 593-601 (1976).
39. A. Bøyum and R. Borgstrøm, The Concentration of Granulocytic Stem Cells in Mouse Bone Marrow, Determined with Diffusion Chamber Technique, *Scand. J. Haematol.*, 7: 294-303 (1970).
40. N. Jacobsen, H. E. Broxmeyer, E. Grossbard, and M. A. S. Moore, Colony-Forming Units in Diffusion Chambers (CFU-d) and Colony-Forming Units in Agar Culture (CFU-C) Obtained from Normal Human Bone Marrow: A Possible Parent-Progeny Relationship, *Cell Tissue Kinet.*, 12: 213-226 (1979).
41. G. R. Johnson and D. Metcalf, Pure and Mixed Erythroid Colony Formation In Vitro Stimulated by Spleen Conditioned Medium with No Detectable Erythropoietin, *Proc. Natl. Acad. Sci. U.S.A.*, 74: 3879-3882 (1977).
42. C. D. R. Dunn, J. D. Jolly, J. B. Jones, and R. D. Lange, Erythroid Colony Formation In Vitro from the Marrow of Dogs with Cyclic Hematopoiesis: Interrelationship of Progenitor Cells, *Exp. Hematol.*, 6: 701-708 (1978).
43. T. M. Fliedner et al., Hematopoietic Stem Cells in Blood: Characteristics and Potentials, in *Hematopoietic Cell Differentiation*, D. W. Golde, M. J. Cline, D. Metcalf, and C. F. Fox (Eds.), pp. 193-212, Academic Press, New York, 1978.
44. F. D. Wilson, M. Shifrine, M. E. Gershwin, W. Spangler, and J. Dyck, Growth of Canine T-Lymphocyte Colonies In Vitro, *Exp. Hematol.*, 6: 539-548 (1978).
45. C. Whaley, F. D. Wilson, G. L. Fisher, M. Shifrine, and K. McNeill, Growth of Canine Alveolar Macrophage Colonies, in *Radiobiology Laboratory Annual Report*, DOE Report UCD-472-124, pp. 47-79, University of California, Davis, California, 1977, NTIS.
46. D. Metcalf et al., Growth of B-Lymphocyte Colonies In Vitro, *J. Exp. Med.*, 142: 1534-1549 (1975).
47. D. Metcalf, J. Parker, H. M. Chester, and P. W. Kincade, Formation of Eosinophilic-Like Granulocytic Colonies by Mouse Bone Marrow Cells In Vitro, *J. Cell. Physiol.*, 84: 275-289 (1974).
48. A. Nakeff, K. A. Dicke, and M. J. van Noord, Megakaryocytes in Agar Cultures of Mouse Bone Marrow, *Ser. Haematol.*, 8: 4-21 (1975).
49. J. Shen, F. Wilson, M. Shifrine, and M. E. Gershwin, Select Growth of Human T-Lymphocytes in Single Phase Semisolid Culture, *J. Immunol.*, 119: 1299-1305 (1977).
50. F. D. Wilson, M. Shifrine, J. Dyck, C. Whaley, and S. Knox-Pillsbury, Radiation Survival Studies on Canine T-Lymphocyte Progenitors Using In Vitro Cloning Techniques: Existence of Radioresistant Subpopulations in Peripheral Blood, in *Radiology Laboratory Annual Report*, DOE Report UCD-472-124, pp. 128-130, University of California, Davis, California, 1977, NTIS.
51. J. Whang, E. Frei III, J. H. Tjio, P. P. Carbone, and G. Brecher, The Distribution of the Philadelphia Chromosome in Patients with Chronic Myelogenous Leukemia, *Blood*, 22: 664-673 (1963).
52. P. A. Chervenick, L. D. Ellis, S. F. Pan, and A. L. Lawson, Human Leukemic Cells: In Vitro Growth of Colonies Containing the Philadelphia ($Ph^1$) Chromosome, *Science*, 174: 1134-1135 (1971).

53. D. W. Golde, C. Burgaleta, R. S. Sparkes, and M. J. Cline, The Philadelphia Chromosome in Human Macrophages, *Blood,* 49: 367-370 (1977).
54. R. Ghio, G. Bianchi, B. Löwenberg, K. A. Dicke, and F. Ajmar, Effects of Fibroblasts on the Growth of Erythroid Progenitor Cells In Vitro, *Exp. Hematol.,* 5: 341-347 (1977).
55. B. Greenberg, F. Wilson, and L. Woo, Stimulation of Granulopoiesis by Fibroblastic Colonies (FC) and Relationship to Colony Stimulating Activity (CSA), *Exp. Hematol.,* 7 (Suppl. 6): 112 (1979).
56. T. M. Dexter and M. A. S. Moore, In Vitro Duplication and "Cure" of Haemopoietic Defects on Genetically Anaemic Mice, *Nature,* 269: 412-413 (1977).
57. N. L. Petrakis, M. Davis, and S. P. Lucia, The In Vivo Differentiation of Human Leukocytes into Histiocytes, Fibroblasts and Fat Cells in Subcutaneous Diffusion Chambers, *Blood,* 17: 109-118 (1961).
58. M. Seki, Hematopoietic Colony Formation in a Macrophage Layer Provided by Intraperitoneal Insertion of Cellulose Acetate Membrane, *Transplantation,* 16: 544-549 (1973).
59. F. D. Wilson, M. Tavassoli, B. R. Greenberg, and D. Hinds, Morphological Studies on "Attaching Cells" in Bone Marrow Cultures from Humans, Dogs and Mice, *Exp. Hematol.,* 7 (Suppl. 6): 19 (1979) (Abstract).
60. F. D. Wilson, B. R. Greenberg, W. L. Spangler, M. Shifrine, M. E. Gershwin, and A. K. Klein, Production of Mesenchymal Tumors in Nude Mice Using $Ph^1$ Negative "Fibroblasts" Obtained from a $Ph^1$ Positive CML Patient and Other Human Sources, in *Hematopoietic Cell Differentiation,* D. W. Golde, M. J. Cline, D. Metcalf, and C. F. Fox (Eds.), pp. 231-240, Academic Press, New York, 1978.
61. H. Wekerle, I. R. Cohen, and M. Feldman, Thymus Reticulum Cell Cultures Confer T Cell Properties on Spleen Cells from Thymus-Deprived Animals, *Eur. J. Immunol.,* 3: 745-748 (1973).
62. M. E. Gershwin, R. M. Ikeda, W. L. Kruse, F. Wilson, M. Shifrine, and W. Spangler, Age-Dependent Loss in New Zealand Mice of Morphological and Functional Characteristics of Thymic Epithelial Cells, *J. Immunol.,* 120: 971-979 (1978).
63. F. D. Wilson, C. B. Whaley, and M. Shifrine, The Formation of Human Lymphocyte Colonies In Vitro in Response to Mixed Lymphocyte Stimulation, *Exp. Hematol.,* 7 (Suppl. 6): 134 (1979).
64. D. Zucker-Franklin and G. Grusky, Ultrastructural Analysis of Hematopoietic Colonies Derived from Human Peripheral Blood, *J. Cell Biol.,* 63: 855-863 (1974).
65. M. Goldman, R. R. Pool, F. D. Wilson, and J. E. West, Nuclear Medicine Studies on the Bone and Marrow of Beagles, in *Twenty-First Gaines Veterinary Symposium,* Ames, Iowa, Oct. 20, 1971, pp. 29-34.
66. H. D. Suit, A Technique for Estimating the Bone Marrow Cellularity In Vivo Using $^{59}Fe$, *J. Clin. Pathol.,* 10: 267-269 (1957).
67. W. Calvo, T. M. Fliedner, E. W. Herbst, and I. Fache, Regeneration of Blood-Forming Organs After Autologous Leukocyte Transfusion in Lethally Irradiated Dogs. I. Distribution and Cellularity of the Bone Marrow in Normal Dogs, *Blood,* 46: 453-457 (1975).
68. K. M. Debelak-Fehir, R. Catchatourian, and R. B. Epstein, Hemopoietic Colony Forming Units in Fresh and Cryopreserved Peripheral Blood Cells of Canines and Man, *Exp. Hematol.,* 3: 109-116 (1975).
69. J. C. Marsh, M. Levitt, and A. Katzenstein, The Growth of Leukocyte Colonies In Vitro from Dog Bone Marrow, *J. Lab. Clin. Med.,* 79: 1041-1050 (1972).
70. F. D. Wilson, R. R. Pool, K. Stitzel, and M. H. Momeni, Bone—Bone Marrow Interface (Endosteum) Potential Relationship of Microenvironments in the Regulation of Response to Internal Emitters, in *The Health Effects of Plutonium and Radium,* W. S. S. Jee (Ed.), pp. 617-642, J. W. Press, Salt Lake City, Utah, 1976.

71. T. J. MacVittie and R. I. Walker, Endotoxin-Induced Alterations in Canine Granulopoiesis: Colony-Stimulating Factor, Colony-Forming Cells in Culture, and Growth of Cells in Diffusion Chambers, *Exp. Hematol.*, 6: 613-618 (1978).

72. T. J. MacVittie and R. I. Walker, Canine Granulopoiesis: Alterations Induced by Suppression of Gram-Negative Flora, *Exp. Hematol.*, 6: 639-647 (1978).

73. R. K. Jones, B. B. Boecker, C. H. Hobbs, A. J. Hulbert, and R. O. McClellan, Hematologic Effects of Inhaled $^{90}$Y, $^{91}$Y, $^{144}$Ce or $^{90}$Sr Fused Clay in Beagle Dogs, *Radiat. Res.*, 51: 470-471 (1972).

74. B. I. Lord, N. G. Testa, and J. H. Hendry, The Relative Spatial Distributions of CFU-S and CFU-C in the Normal Mouse Femur, *Blood*, 46: 65-72 (1975).

75. M. A. Maloney, R. A. Lamela, M. J. Dorie, and H. M. Patt, Concentration Gradient of Blood Stem Cells in Mouse Marrow—An Open Question, *Blood*, 51: 521-525 (1978).

76. M. Goldman, F. D. Wilson, L. S. Rosenblatt, and S. A. Book, Early Radiation Effects on Bone and Blood as "Predictors" of Osteosarcoma, in *Biological and Environmental Effects of Low-Level Radiation*, Symposium Proceedings, Chicago, 1975, pp. 335-342, STI/PUB/409 (Vol. 1), International Atomic Energy Agency, Vienna, 1976.

77. J. E. Moulton and D. L. Dungworth, Tumors of the Lymphoid and Hematopoietic Tissues, in *Tumors in Domestic Animals*, J. E. Moulton (Ed.), pp. 150-204, University of California Press, Berkeley, 1978.

78. O. W. Schalm, N. C. Jain, and E. J. Carrol, *Veterinary Hematology*, Lea & Febiger, Philadelphia, 1975.

79. D. L. Dungworth, M. Goldman, J. W. Switzer, and D. H. McKelvie, Development of a Myeloproliferative Disorder in Beagles Continuously Exposed to $^{90}$Sr, *Blood*, 34: 610-632 (1969).

80. T. E. Fritz, W. P. Norris, and D. V. Tolle, Myelogenous Leukemia and Related Myeloproliferative Disorders in Beagles Continuously Exposed to $^{60}$Co $\gamma$-Radiation, in *Unifying Concepts of Leukemia*, R. M. Dutcher and L. Chieco-Bianchi (Eds.), pp. 170-188, Karger, Basel, 1973.

81. T. M. Seed, D. V. Tolle, T. E. Fritz, R. L. Devine, C. M. Poole, and W. P. Norris, Irradiation-Induced Erythroleukemia and Myelogenous Leukemia in the Beagle Dog: Hematology and Ultrastructure, *Blood*, 50: 1061-1079 (1977).

82. D. V. Tolle, T. E. Fritz, and W. P. Norris, Radiation Induced Erythroleukemia in the Beagle Dog, *Am. J. Pathol.*, 87: 499-510 (1977).

83. M. Shifrine et al., Transplantation of Radiation-Induced Canine Myelomonocytic Leukemia, *Nature*, 232: 405-406 (1971).

84. M. Shifrine et al., Transplantation of Radiation Induced Canine Myelomonocytic Leukemia, in *Unifying Concepts of Leukemia*, R. M. Dutcher and L. Chieco-Bianchi (Eds.), pp. 158-169, Karger, Basel, 1973.

85. F. D. Wilson and M. Shifrine, Radiation-Induced Transplantable Myelomonocytic Leukemia in Beagles: Hematologic Parameters, *Exp. Hematol.*, 22: 41-44 (1972).

86. M. Shifrine, C. E. Chrisp, F. D. Wilson, and U. Hefferon, Lysozyme (Muramidase) Activity in Canine Myelogenous Leukemia, *Am. J. Vet. Res.*, 34: 695-696 (1973).

87. L. N. Owen, Transplantation of Canine Lymphocytic Leukaemia, in *Unifying Concepts of Leukemia*, R. M. Dutcher and L. Chieco-Bianchi (Eds.), pp. 139-148, Karger, Basel, 1973.

88. C. A. Bowles, W. T. Kerber, S. R. S. Rangan, R. Kwapien, W. Woods, and E. M. Jensen, Characterization of a Transplantable, Canine, Immature Mast Cell Tumor, *Cancer Res.*, 32: 1434-1441 (1972).

89. H. M. Vriesendorp, *Major Histocompatibility Complex of the Dog*, Ph.D. Thesis, Erasmus University, Rotterdam, The Netherlands, 1973.

90. K. Stitzel, H. Vriesendorp, T. P. Visser, and F. D. Wilson, Histocompatibility Studies in the Davis Beagle Population: Preliminary Report, in *Radiobiology Laboratory Annual Report,* DOE Report UCD-472-124, pp. 109-110, University of California, Davis, California, 1977, NTIS.
91. T. M. Seed, D. V. Tolle, T. E. Fritz, S. M. Cullen, L. V. Kaspar, and C. M. Poole, Haemopathological Consequences of Protracted Gamma Irradiation in the Beagle, in *Late Biological Effects of Ionizing Radiation,* Symposium Proceedings, Vienna, 1978, pp. 531-545, STI/PUB/489 (Vol. 1), International Atomic Energy Agency, Vienna, 1978.
92. D. Metcalf, In Vitro Cloning Techniques for Hemopoietic Cells: Clinical Applications, *Ann. Intern. Med.,* 87: 483-488 (1977).
93. K. A. Dicke, G. Spitzer, and M. J. Ahearn, Colony Formation In Vitro by Leukaemic Cells in Acute Myelogenous Leukaemia with Phytohaemagglutinin as Stimulating Factor, *Nature,* 259: 129-130 (1976).
94. H. T. Messner, J. E. Till, and E. A. McCulloch, Interacting Cell Populations Affecting Granulopoietic Colony Formation by Normal and Leukemic Human Marrow Cells, *Blood,* 42: 701-710 (1973).

# Canine Immunology: Current Status

3

G. S. COLGROVE
*Plum Island Animal Disease Center, Department of Agriculture, Greenport, N. Y.*
*and*
MOSHE SHIFRINE
*Laboratory for Energy-Related Health Research, University of California, Davis*

The canine species has assumed an increasingly important role as an animal model in biomedical research, particularly in the areas of transplantation immunology and immunopathology. This chapter summarizes most information relating to the immune system of the dog published since 1970; earlier studies on canine immunology are reviewed in an article by Jacoby and Griesemer.[1]

## IDENTIFICATION OF T AND B LYMPHOCYTES IN THE DOG

Immunocompetent lymphocytes are divided into two functional groups: thymus-dependent (T) cells, which are responsible for cell-mediated immune reactions, such as allograft rejection and delayed hypersensitivity; and bursa-equivalent (B) cells, which are the precursors of antibody-forming cells. The functions of the two groups are not entirely separate since T cells exert a regulatory influence on several B-cell processes.

In the dog, as in most mammalian species, it is possible to detect B and T lymphocytes by demonstration of characteristic cell surface markers. Canine B cells can be identified by the presence of immunoglobulin and/or receptors for the third component of complement on the cell membrane. Bowles, White, and Lucas[2] found that 49% of peripheral blood lymphocytes in the dog had surface immunoglobulin, as demonstrated by immunofluorescence. Zander, Boopalam, and Epstein[3] reported that 30% of cells had surface immunoglobulin and reported data on the distribution of B-cell markers in various lymphoid tissues. Both groups of workers demonstrated a receptor on canine B cells for activated complement C3, which allows formation of erythrocyte–antibody–complement (EAC) rosettes with sensitized sheep red blood cells.

Rosette formation with (unsensitized) heterologous erythrocytes (E rosettes) is used as a marker for T lymphocytes. Bach et al.[4] first reported that a percentage of lymphocytes in the peripheral blood of dogs form rosettes with human red cells, a finding that was subsequently confirmed by others.[2,3,5–8] Rosette-forming cells are found in canine thymus, spleen, lymph nodes, and bone marrow, in addition to peripheral blood.[3,5] The reaction with human erythrocytes is inhibited by anti-lymphocyte serum (ALS). Another indication of their T-cell nature is that immunoglobulin was not present on rosette-forming lymphocytes. Rosette formation between dog lymphocytes and guinea pig erythrocytes has also been described.[2,3,9] Bowles, White, and Lucas[2] reported that 36% of canine peripheral blood lymphocytes form rosettes with human erythrocytes and 32% form rosettes with guinea pig erythrocytes; however, different receptors on T-cell membranes are required for each red cell type. Zander, Boopalam, and Epstein[3] found that only a portion of the canine T-cell population is involved in rosette formation. In peripheral blood, for example, 68% of lymphocytes were positive when stained with fluorescently labeled anti-thymocyte globulin, whereas only 38% of lymphocytes formed E rosettes.

Methods for producing antisera against canine T- and B-cell receptors were described by Krakowka, Olsen, and Cockerell.[10] Erythrocyte rosette formation by T cells and EAC rosette formation by B cells could be blocked by antisera specific for T- and B-cell receptors, respectively.

Macrophage–erythrocyte rosette formation has also been described in the dog.[11] Canine alveolar macrophages form rosettes with mouse, rat, guinea pig, human, and rabbit erythrocytes, as well as with autologous and allogeneic red cells.

In Chapter 4, C. H. Miller and M. R. MacKenzie cover the subject of cell-surface receptors in greater detail.

## DEVELOPMENT OF IMMUNITY IN THE DOG

Immune reactions of vertebrate animals can be divided into two major categories: those mediated by humoral factors (antibody and complement) and those mediated by cells. Both types of responses involve elements of the lymphoreticular system, and the development of immunologic competence during fetal and neonatal life follows the differentiation and maturation of cells of the lymphoid and reticuloendothelial tissues. The histogenesis of lymphatic tissues in fetal dogs was described by Bryant and Shifrine.[12] Primordia of thymus and spleen developed at 27 to 28 days of gestation, followed by the development of primordia of larger lymph nodes (35 to 38 days) and smaller lymph nodes (46 days). The prelymphoid lymph nodes were characterized by granulocytopoiesis in paracortex and medulla. Lymphocyte development began in the thymus at 35 to 36 days of gestation and in spleen and larger lymph nodes at 52 to 53 days.

The ontogeny of humoral immune responsiveness has been studied by several groups of workers. Jacoby, Dennis, and Griesemer[13] reported that fetal dogs at the 40th day of gestation were capable of responding to bacteriophage $\phi$X-174. Reactivity to sheep red blood cells (SRBC) began at about the 48th day of gestation. Neonatal dogs were capable of responding to bovine serum albumin (BSA), but only when the antigen was incorporated into complete Freund's adjuvant. Lewis, Smith, and Garfield[14] reported that SRBC and bovine gamma globulin were immunogenic for newborn dogs but that only 19S, 2-mercapto-ethanol-sensitive antibodies were produced. Reinjection of the pups at 3 weeks of age produced a normal adult-type response composed of both 19S and 7S antibodies. Shifrine et al.[15] concluded that canine fetuses become competent for antibody production close to term. Fetuses injected at gestation days 41, 47, or 51 with keyhole limpet hemocyanine (KLH), *Mycoplasma gallisepticum*, infectious bovine rhinotracheitis virus, or chicken erythrocytes produced no antibodies. Fetuses injected at gestation day 55 with KLH and challenged with KLH 21 days after birth gave only a primary response, but 1-day-old pups injected with KLH and challenged 40 days later produced a secondary immune response. Fetuses injected with *Brucella canis* at 50 days of gestation produced agglutinating antibodies, and 1-day-old pups injected with infectious bovine rhinotracheitis virus responded with the production of neutralizing antibodies. Brown et al.[16] demonstrated that protective immunity against canine distemper virus could be induced in pups as young as 3 weeks of age provided that maternal antibodies did not interfere. Passive maternal immunity was found to interfere with immunization by either distemper virus or measles virus. The transplacental acquisition of passive maternal immunity to canine distemper has been demonstrated in colostrum-deprived pups.[17]

Cellular immune competence in fetal and neonatal dogs has been assessed by allograft rejection and lymphocyte stimulation. Dogs grafted as fetuses on the 40th or 48th day of gestation showed prolonged allograft survival (mean survival time of 42 days) when compared with dogs grafted as newborn or adults (mean survival time of 10.8 days).[18] Bryant, Shifrine, and McNeil[19] reported that cells responsive to mouse spleen cells and phytohemagglutinin (PHA) developed in the spleen, but not in the thymus, of canine fetuses by gestation day 45. The PHA-induced response of canine spleen cells was unchanged on postnatal day 90, whereas the response to mouse cells had diminished. Thymus cells remained unresponsive to mouse cells until postnatal day 50. The response of thymus cells to PHA began on gestation day 50, increased to postnatal day 3, and declined by postnatal day 90. Gerber and Brown[20] reported that lymphocytes from dogs 0 to 4 weeks of age were significantly less responsive to PHA than were lymphocytes from the same dogs at 6 to 12 weeks of age. Responses to

PHA peaked in dogs between 6 weeks and 6 months of age and then decreased with increasing age.

## COMPLEMENT

The complement system consists of a number of chemically and immunologically distinct serum proteins and functions as the primary humoral mediator of antigen–antibody reactions. Activation of the complement sequence may proceed by one of two parallel but independent mechanisms: the classical complement pathway or the alternative, or properdine, pathway. The biologic consequences of complement activation include cytolytic and cytotoxic damage, leukocyte chemotaxis, the generation of kinin, and the release of histamine from mast cells and lysosomal enzymes from phagocytes.

The first seven components of dog complement were isolated by Sargent, Johnson, and Richardson,[21] using euglobulin and ammonium sulfate precipitation and ion-exchange and exclusive chromatography. The omnipresence of the eighth component after ion-exchange chromatography and extreme lability of the ninth component prevented isolation of these last two components. Boyer, Wyde, and Bier[22] studied the relative sensitivity of canine complement components to oxidation by sodium hypochlorite. Although C1, C4, and C7 were particularly sensitive to this agent, all components of the complement system were found to decrease at high concentrations of hypochlorite.

Studies with dogs and other experimental animals to assess the role of complement in graft rejection were reviewed by Cooper and Cochrane.[23] Experiments with complement-depleted or complement-deficient animals failed to provide evidence for a major role of the complement system in allograft rejection. Xenograft rejection, however, involves simultaneous participation of multiple humoral effector systems, including the complement system.

The role of complement in the pathogenesis of chronic gingivitis was studied in dogs rendered complement deficient by treatment with carrageenan.[24] Activation of complement (by immune complexes) was not necessary for the continuous migration of neutrophils and monocytes into gingival crevices.

The endotoxin shock response of normal and decomplemented dogs was studied by Garner, Chater, and Brown.[25] The hypotension and early thrombocytopenia of endotoxin shock were abolished in dogs decomplemented with purified cobra venom factor, and the subsequent disseminated intravascular coagulation was greatly ameliorated.

## IMMUNOGLOBULINS

Immunoglobulins are glycoproteins synthesized by cells of the B lymphocyte series. Each immunoglobulin molecule contains at least one basic structural unit, or monomer, composed of two pairs of polypeptide chains. The pair containing more amino acids with a greater molecular weight is composed of two identical chains designated as the heavy (H) chains. Accordingly, the other pair contains two light (L) chains. Immunoglobulins are classified into classes and subclasses according to antigenic differences in the carboxyl terminal portion (constant or C region) of the H chains. Five classes of immunoglobulins have been described in man: IgA, IgD, IgE, IgG, and IgM. In the dog all immunoglobulin classes except IgD have been identified and at least partially characterized.

IgG is the predominant immunoglobulin class in canine serum (Table 1). Reynolds and Johnson[26] reported that purebred (foxhounds and collies) and

TABLE 1

MEAN VALUES FOR IMMUNOGLOBULINS (mg/ml) IN CANINE SERUM

| | IgA | IgM | IgG* | Ref. |
|---|---|---|---|---|
| Purebred dogs | 0.8 | 1.6 | 9.2 | 26 |
| Mongrel dogs | 0.8 | 1.4 | 14.4 | 26 |
| Mongrel dogs | 0.5 | 1.7 | 9.8 | 54 |
| Mongrel dogs | 0.6 | 2.4 | 11.3 | 29 |

*Reynolds and Johnson[26] gave IgG values for individual immunoglobulin subclasses. For comparison, values for IgG subclasses have been summed.

mongrel dogs had significantly different levels of IgG subclasses; lower concentrations of IgGa,b and IgGd were found in the purebred group.

Early colostrum is rich in IgG and IgA, whereas lesser quantities of IgM are present.[26-28] The immunoglobulin content of canine breast secretions diminishes steadily after parturition with only IgA persisting until weaning.[26,27] Secretory IgA is the predominant immunoglobulin found in canine intestinal secretions and feces, and low levels of IgM and IgG are present. IgA, IgG, and IgM are found, as well, in canine saliva and respiratory secretions, again with IgA predominating.[27,29] In progressive samples from the upper to the lower respiratory tract,[29] the concentration of IgA decreased whereas that of IgG increased. Bronchial washes contained predominantly IgG. Reynolds and Thompson[30] showed that IgG in bronchial secretions belonged to the subclass IgGd.

Local and systemic responses following respiratory immunization of dogs have been studied by several groups. Lieberman et al.[31] assayed antibody responses in tracheal secretions and in serum after tracheal pouch immunization with BSA, KLH, or SRBC. Antibody gained access to tracheal secretions by two means. Antibodies to BSA and KLH were produced systematically and transported, after a lag period, to the trachea and into secretions. Immunization with SRBC elicited both local and systemic production of IgA antibodies. IgA and IgG anti-influenza antibodies were detected in tracheal secretions following tracheal immunization of dogs with influenza virus.[32] Tracheal immunization also elicited IgM and IgG responses in the serum.[32] However, subcutaneous immunization with influenza virus was a poor method of inducing antibodies in tracheal secretions. O'Rourke et al.[33] recovered anti-KLH antibodies of the IgA, IgG, and IgM classes from canine respiratory secretions after aerosol immunization and concluded that all three immunoglobulin classes are produced locally by the respiratory tract after immunization by the respiratory route. Antibody titers found in respiratory secretions after aerosol immunization exceeded titers obtained after intravenous immunization.

Kazmierowski, Durbin, and Reynolds[34] studied the transport of radiolabeled preparations of canine IgG, IgM, and $F(ab')_2$ fragments into bronchial secretions from the circulation after intramuscular injection. Only a small portion of the passively administered IgG was recovered in respiratory specimens, an indication that transfer of IgG from the circulation is not a major process under normal conditions. Furthermore, alteration of the IgG molecule by removal of the Fc portion almost eliminated its penetration into lung secretions. IgM was not detected in respiratory specimens after passive administration. Kaltreider et al.[35] investigated the immunologic response to a particulate antigen, SRBC, instilled in dog lungs. During the primary response, antibody-forming cells were confined to pulmonary lymphoid tissue (hilar nodes and bronchoalveolar cells); none were found in systemic lymphoid tissue (spleen and peripheral lymph nodes).

The antibody response to parenterally administered antigens was studied by Mates and Hinton.[36] Precipitating antibodies to beta 2 lipoprotein were formed after primary injection of human serum. After a second dose, dogs formed precipitins to beta 2 lipoprotein, albumin, and beta globulin. The primary and secondary responses to KLH were also analyzed.

Methodology for isolating canine IgG1, IgG2ab, IgG2c, and IgM and for producing monospecific anti-immunoglobulin sera was described by Thimel.[37] Orlans[38] reported that fowl antisera to human and mouse IgG-Fc cross-precipitated with canine IgG subclasses IgG1 and IgG2. Whitacre and Lang[39] fractionated canine antilymphocyte serum by chromatography and assayed eluates for cytotoxic activity. Cytotoxicity was associated with an IgG subclass, designated IgGb according to electrophoretic mobility. Krumbacher and Schnappauf[40] reported little success in identifying allotypic markers on canine immunoglobulins. These investigators suggested that immunoglobulin allotype structures in canines may be less polymorphic than those in rabbits and mice or that they may be less immunogenic. In contrast, Saison and Colling[41] were able to demonstrate an allotypic marker associated with canine IgM.

Canine secretory IgM isolated from both colostrum and bronchial secretions reportedly contains a glycoprotein-bound secretory component.[42] This extra subunit accounted for differences in size, quaternary structure, and antigenicity between secretory IgM and serum IgM. The fact that peptide bond cleavage with cyanogen bromide did not release the bound secretory component and J chain indicated differences in the primary structure of secretory IgM and secretory IgA around the binding sites for these subunits.

Wasserman and Capra[43] determined the amino acid sequence for the Fc portion of canine IgM. When the sequence was compared with those of two human $\mu$ chains, a high degree of interspecies homology was found.

Vaerman and Heremans[44] determined that the majority of IgA in the mesenteric lymph of the dog is produced by plasma cells in the intestinal mucosa. The molecular size of IgA from mesenteric lymph was indistinguishable from that of serum IgA; the intestinal mucosa was purportedly a major source of serum IgA in the dog. Secretory IgA from canine colostrum, tracheal secretions, and tears has an additional component that is antigenically identical to free secretory component isolated from colostrum and can be distinguished from serum IgA (Ref. 45). Although canine secretory IgA from colostrum and that from fecal extracts are apparently the same molecular species, serum and secretory IgA differ in molecular weight (by approximately 75,000) and in sedimentation coefficient (10 and 11.75, respectively).[46,47] Reynolds and Johnson[46] concluded that the greater mass of secretory IgA could be accounted for by an additional, but as yet uncharacterized, component analogous to the secretory piece found in humans and rabbits. These investigators determined that the sedimentation coefficient of canine serum IgA was 115. Reaction of the immunoglobulin with an anti-human J chain antiserum implied that canine IgA contains a considerable amount of a J component.[48] A highly acidic polypeptide isolated from canine IgA had an amino acid composition corresponding closely to that reported for human J components.[48] Antiserum to human J chain cross-reacted with completely reduced and alkylated IgA from dog colostrum[49] and with J chains of canine serum IgA, secretory IgA, and IgM (Refs. 50 and 51). Thompson, Reynolds, and Waxdal[52] characterized canine IgA from serum and colostrum and free secretory component from colostrum. The reduction of colostral IgA with dithiothreitol released a secretory component which was identical to free secretory component with respect to electrophoretic migration, isoelectric focusing point, and molecular weight but which lacked some antigenic determinants. The amino acid composition and N-terminal sequence of canine free secretory component was similar to that reported for the human and cow. Virella, Slappendel, and Joudswaard[53] studied a polymeric IgA cryoglobulin isolated from a dog with multiple myeloma. The immunoglobulin contained J chain but no secretory component.

The antibacterial properties of canine IgA were investigated by Heddle and Rowley.[54] IgA antibodies showed little, if any, activity in the direct killing of

*Vibrio cholerae* in the presence of complement or in the promotion of phagocytosis. However, when mixed with a challenge inoculum of *V. cholerae* and fed to infant mice, IgA antibodies protected these mice as efficiently as IgM or IgG antibodies.

Canine reaginic antibodies (IgE) were physically separated and immunochemically differentiated from other classes of immunoglobulins by Schwartzman, Rockey, and Halliwell.[55] This distinct class of canine immunoglobulins is analogous to human IgE (Ref. 55) and cross-reacts with the human analog.[56]

IgE, found in only trace amounts in the serum, has the ability to attach to skin ("homocytotropic antibody") and to initiate reactions of immediate hypersensitivity. Both induced [anti-dinitrophenol (anti-DNP)] and spontaneous (anti-ragweed) reagins of atropic dogs have identical characteristics of heat lability, 2-mercaptoethanol sensitivity, molecular size, persistence at passive transfer sites in skin, and capacity to sensitize homologous leukocytes from normal donors for an immunospecific release of histamine.[55] Sunthoupalin, Arkins, and Fink[57] characterized homocytotropic antibodies induced with ragweed and timothy pollen extracts in nonatropic dogs. A 7S antibody had skin sensitizing and other characteristics of IgE. Matsumoto, Monkus, and Freeman[58] consistently detected homocytotropic antibody in the sera of nonatropic dogs injected intramuscularly with alum-precipitated tetanus toxoid. Less marked and consistent homocytotropic antibody responses were observed after inoculation of alum-precipitated DNP–KLH. Homocytotropic antibody to BSA in nonatropic dogs had characteristics analogous to human IgE (Ref. 59). Halliwell et al.[60] demonstrated antigen-binding activity in a dog with anti-DNP reaginic activity and identified a number of physicochemical similarities between canine and human IgE. Immunization of neonatal dogs with DNP–ovalbumin conjugates resulted in prolonged IgE antibody production against both the hapten and the carrier.[61] The anti-DNP response could be abolished by administering conjugates of DNP and a copolymer of D-glutamic acid and D-lysine. A similar immunosuppressive effect was achieved with the tolerogenic conjugate of DNP coupled to dog gamma globulin. Tada[62] reviewed current knowledge regarding regulatory mechanisms in reaginic antibody formation (in all species) with particular reference to the control of allergic disorders by suppression of IgE production.

Reaginic antibodies to cestode *(Echinococcus granulosus)* antigens were demonstrated by Williams and Perez Esandi[63] in 5 of 20 experimentally infected dogs. The antibody was thermolabile and elicited passive cutaneous anaphylactic reactions in normal canine recipients. Halliwell, Schwartzman, and Rockey[56] demonstrated homology between human and canine IgE on the basis of immunological cross-reactions. Absorption of canine reaginic serum with anti-human IgE resulted in a substantial reduction in Prausnitz–Kustner (PK) titer to ragweed. Furthermore, anti-human IgE was also effective in eliciting reversed cutaneous anaphylaxis in the dog.

Halliwell[64] used immunofluorescence to study the distribution of IgE in canine skin. The antibody was associated exclusively with mast cells and was located in the cytoplasm. In an extension of these studies, Halliwell[65] reported that all dogs in the study had immunofluorescent evidence of *Ascaris*-binding antibody on mast cells. Dogs with a high percentage of mast cells that fluoresced with anti-IgE also had large numbers of positive cells with the *Ascaris* system, which suggests that the bulk of the IgE in these animals was anti-*Ascaris*. Cells producing IgE were prominent in relatively few tissues, namely, the mucosa of the small intestine and the mediastinal and mesenteric lymph nodes. The major sites of IgE production followed the course of migrating *Toxocara canis* parasites. Healey and Gaafar[66] studied the IgE bound to mast cells in cutaneous biopsies taken from dogs with demodectic mange and from specific pathogen-free dogs. The percentage of mast cells showing specific fluorescence with anti-IgE was not different in the two groups of dogs, but the number of mast cells in the dermis of demodectic dogs was greater than that in pathogen-free dogs.

## CELL-MEDIATED IMMUNITY

Cell-mediated immune reactions result from interactions between specifically sensitized thymus-dependent (T) lymphocytes and specific antigen. The classical example of cell-mediated immunity (CMI) is the delayed skin reaction that develops in sensitized hosts after intradermal injection of specific antigen. Other immune reactions that involve CMI include acquired resistance to a variety of intracellular parasites, contact dermatitis resulting from sensitization to certain chemicals, and rejection of transplanted organs and tissues between allogeneic individuals. The bulk of recent literature on CMI in the dog stems from work relating to the major histocompatibility system (DL-A), including such reactions as mixed lymphocyte culture and cell-mediated cytotoxicity tests (see Chapter 8 for a discussion of these aspects of CMI).

Delayed cutaneous hypersensitivity reactions have been elicited in sensitized dogs by injection of specific antigen. Tuberculin skin testing after bacillus Calmette-Guérin (BCG) immunization is a reliable method of assessing cellular-immune competence in the dog,[67,68] despite occasional reports to the contrary.[69,70] Thilsted and Shifrine[68] described a test for delayed hypersensitivity involving intradermal injection of antigen [tuberculin purified protein derivative (PPD) and coccidioidin] in the pinna of the ear. Skin-test reactions, measured by change in ear thickness, correlated with the immunization histories of dogs.

Dogs sensitized with canine distemper virus exhibit delayed cutaneous hypersensitivity reactions when challenged with either distemper or measles antigen.[71] Animals sensitized with measles virus similarly exhibit delayed responses to both distemper and measles antigen, which suggests that CMI is the mechanism by which measles vaccine protects pups against canine distemper virus. Delayed cutaneous hypersensitivity to distemper vaccine was among several parameters measured by Benfield et al.[72] to evaluate the effects of immunosuppressive therapy in dogs. Absolute lymphocyte counts were the most sensitive dose-related indicators of the level of immunosuppression induced by azothioprine and corticosteroids.

Mangi et al.[73] immunized dogs with picryl guinea pig albumin and KLH and elicited delayed hypersensitivity responses to cutaneous challenge with these antigens. Mizejewski, Baron, and Poissant[74] reported that dogs with naturally occurring thyroiditis showed delayed cutaneous responses when injected with thyroglobulin or thyroid-gland extract. Delayed cutaneous hypersensitivity reactions have also been elicited in the dog by topical application of dinitrochlorobenzene[8,72] and by intradermal injection of mitogens.[75-77]

Lymphocyte stimulation assays have been used as an in vitro correlate of CMI in the dog. Schatz et al.[76] reported that canine lymphocytes reacted to the following mitogens, in decreasing order of intensity: PHA, pokeweed mitogen (PWM), concanavalin A (Con A), and sodium periodate. Angus and Yang[78] reported a fluctuating response of canine lymphocytes to PHA during sampling intervals ranging from 4 hr to 2 weeks. The variability persisted in assays of both whole blood and separated lymphocytes. Ochs et al.[67] used lymphocyte stimulation by PHA as one of several criteria for evaluating the cellular immune status of canine bone-marrow chimeras. Kaltreider and Salmon[79] used PHA responsiveness as the criterion for detecting T lymphocytes in bronchoalveolar cell populations of the dog. Results of the study indicated that T cells could be recruited into alveolar spaces after local irritation.

Substances that influence the in vitro response of canine lymphocytes to mitogens include methylprednisolone and dextran. Methylprednisolone causes a decrease in protein synthesis by PHA-stimulated cells.[80] Benjamin and Ferris[81] reported that relatively small quantities of dextran, which is frequently used in the isolation of lymphocytes from peripheral blood, markedly potentiated the response of canine lymphocytes to PHA and PWM in the final culture.

Antigen-induced transformation of dog lymphocytes was studied by Thilsted and Shifrine.[82] In dogs immunized with BCG, a positive in vitro response to PPD became evident 11 days after immunization and peaked at 18 days after immunization. Coccidioidin had a definite but variable mitogenic effect on lymphocytes from nonsensitized dogs, but higher responses were obtained with lymphocytes from a dog with coccidioidomycosis.

Virus-induced lymphocyte transformation was used by Gerber and Marron[83] to evaluate CMI in dogs vaccinated with either canine distemper virus or measles virus. Dogs vaccinated with canine distemper virus developed both CMI and neutralizing antibodies against canine distemper virus. However, dogs vaccinated with measles virus had only a cell-mediated response to canine distemper virus and did not produce antibodies against the virus.

Reif et al.[84] studied CMI to canine parainfluenza virus by lymphocyte stimulation and macrophage migration inhibition techniques. After viral infection of the respiratory tract, bronchial lymph nodes contained cells that could be stimulated by parainfluenza virus in vitro, and bronchial washings contained lymphocytes that secreted migration inhibition factor (MIF) in the presence of viral antigen. The local response to respiratory tract infection was greater than the systemic response. Splenic lymphocytes, for example, produced significantly less inhibition of macrophage migration than did bronchial lymphocytes. Migration inhibition techniques have also been used to demonstrate CMI to allogeneic cells in skin-grafted dogs.[85]

Depressed cell-mediated responses may be a feature of some diseases in the canine. In dogs infected with canine distemper virus, lymphocyte reactivity to KLH or PPD after sensitization with KLH in complete Freund's adjuvant was depressed, and lymphocyte response to the mitogens PHA and PWM was also reduced.[73] Infection with canine distemper virus also depressed delayed hypersensitivity reactions of sensitized dogs to intradermal injections of KLH and picryl guinea pig albumin. These effects were transient in dogs previously vaccinated against distemper but prolonged in susceptible dogs. Krakowka, Cockerell, and Koestner[86] reported that intradermal skin-test responses and in vitro lymphocyte responses to PHA were depressed in gnotobiotic dogs infected with canine distemper virus. Dogs with persistent loss of the skin-test response eventually died, whereas dogs that responded to PHA 14 to 21 days after viral inoculation survived the infection. Depressed CMI has also been reported in cases of demodectic mange, and serum from dogs with generalized demodectic mange reportedly suppresses the in vitro reactivity of normal dog lymphocytes to PHA (Refs. 87 and 88). In addition, treatment of dogs with anti-lymphocyte serum increased the incidence of lesions of demodicosis in dogs exposed to the mites.[89]

Studies on canine transfer factor were reported by Shifrine, Thilsted, and Pappagianis.[90] Sensitivity to *Coccidioides immitis*, as assessed by lymphocyte stimulation tests, was transferred to normal dogs by injection of transfer factor from specifically sensitized donors. Normal lymphocytes were also sensitized in vitro with PPD-specific transfer factor. Transfer factor was secreted into culture medium following the incubation of immune lymphocytes. Thymus, spleen, lymph nodes, and circulating lymphocytes of a hyperimmune dog were found to contain transfer factor.

In subsequent studies, Thilsted and Shifrine could not transfer tuberculin specific cell-mediated immunity to dogs by using dialyzable fractions or unfractionated extracts of spleen cells, lymph node cells, and peripheral blood leukocytes; such transfer could be done only with viable lymphocytes.[91] Tomar[92] used human transfer factor to transfer, and the dog developed a delayed cutaneous reaction when subsequently challenged with antigen in the same site.

## ALLERGY

Allergy can be defined as an altered state of immune reactivity, and, in the present context, the terms "allergy" and "hypersensitivity" are used interchangeably. The two basic types of hypersensitivity reactions are delayed and immediate. Both types of reactions result in tissue damage. Delayed hypersensitivity reactions, mediated by specifically sensitized thymus-dependent lymphocytes, show maximum responses between 24 and 48 hr after challenge. Examples of allergic reactions of the delayed type include contact dermatitis, which develops after repeated contact with certain chemicals, and fleabite dermatitis.[93] Krawiec and Gaafar[94] did a comparative study of the lesions of allergic contact dermatitis (delayed-type hypersensitivity) and primary irritant contact dermatitis in dogs.

Immediate hypersensitivity, mediated by antibodies of the IgE class, manifests as tissue reactions within minutes after reexposure to the sensitizing antigen. Immediate hypersensitivity is the predominant form seen clinically in the canine since dogs may develop allergies to a wide variety of environmental allergens. The most frequently encountered manifestation of immediate hypersensitivity in the dog is allergic dermatitis, usually resulting from the inhalation or ingestion of allergens. Allergic skin disorders of dogs and other animals were recently reviewed by Chamberlain,[95] Lorenz,[96] and Walton.[97] Other articles, emphasizing the diagnosis and treatment of allergic dermatitis, were authored by Austin[98] and Baker.[99] Recently published case studies of etiologic factors other than environmental or food allergens include reports on drug hypersensitivity.[100,101]

Respiratory manifestations of hypersensitivity are also seen in dogs. Kepron et al.[102] induced IgE antibody responses to hapten-carrier conjugates and elicited marked bronchoconstriction in neonatal dogs by inhalation challenge with antigen. Faith, Hessler, and Small[103] induced respiratory and cutaneous hypersensitivity to prairie sage pollen in dogs and reported that passively administered anti-sage pollen antibody greatly inhibited the respiratory response to inspired allergen. The uptake and catabolism of KLH or ragweed antigen by alveolar macrophages in dogs with respiratory hypersensitivity to the antigens[104] were identical to antigen processing in normal dogs. In a number of studies, *Ascaris*-hypersensitive dogs were used as a model of allergic asthma.[105–112] A high percentage of dogs show immediate-type cutaneous reactivity to *Ascaris suum* antigens, which is presumptively due to cross-reactivity of the *Ascaris* antigen with antigens from *Toxocara canis*. *Ascaris*-hypersensitive dogs experience bronchoconstriction when bronchially challenged with nematode antigens.

## AUTOIMMUNITY

Immune reactivity against the host's own tissue plays a dominant pathogenic role in a variety of human and animal diseases. The spectrum of autoimmune disorders ranges from organ-specific diseases (e.g., thyroiditis) to generalized systemic diseases (e.g., systemic lupus erythematosus). In the dog, at least six naturally occurring diseases are characterized by immunologically mediated injury to tissues or cells: autoimmune thyroiditis (recently reviewed by Bigazzi and Rose[113]), autoimmune hemolytic anemia, thrombocytopenia purpura, pemphigus, immune-complex glomerulonephritis, rheumatoid arthritis, and systemic lupus erythematosus. Rheumatoid arthritis is reviewed in Chapter 13, and systemic lupus erythematosus is discussed in Chapter 12.

### Autoimmune Hemolytic Anemia

Hematologic diseases of the canine were extensively reviewed by Schalm, Jain, and Carroll.[114] Dodds[115] discussed the etiology, diagnosis, and treatment of immune-mediated anemias in the dog, and Schultz[116] reviewed autoimmune disorders, including those involving erythrocytic and thrombocytic elements.

Cold-reactive red-cell agglutinins were demonstrated in a dog with Coombs-positive hemolytic anemia by Slappendal et al.[117] After exposure to cold, the dog developed cyanosis and gangrene on the ears, feet, and nose. The serum showed no distinct hemolytic properties in vitro, but it did show fixed complement to both autologous and homologous erythrocytes with agglutination of these cells at low temperatures. The agglutinin was identified as IgM. A similar case was reported by Green et al.,[118] who demonstrated cold-reactive erythrocyte autoantibodies of the IgM class in a dog with cyanosis and necrosis of the extremities.

A technique using papain-treated red cells to detect incomplete erythrocyte autoantibodies in the dog was described by Jones and Darke.[119]

### Thrombocytopenia

Schalm,[120] in a review of canine thrombocytopathies, described two cases of thrombocytopenia with a postulated autoimmune etiology. One of the dogs presented with purpura and bleeding, whereas the other had generalized lymphadenopathy and lymphatic leukemia. Schalm and Ling[121] also described clinical and hematological findings in a hemorrhagic syndrome associated with autoimmune thrombocytopenia.

Wilkins, Hurvitz, and Dodds-Laffin[122] tested 17 dogs with idiopathic thrombocytopenia purpura, characterized clinically by bleeding of the purpuric or capillary variety from, or within, the skin and mucous membranes. Twelve of the dogs with that clinical diagnosis showed antiplatelet activity in the globulin fraction of serum.[122] Antiplatelet activity was removed from the globulin fraction of six of eight dog sera by monospecific rabbit anti-canine IgG. Similarly, Joshi and Jain[123] detected anti-platelet antibody in the sera of 8 of 13 thrombocytopenic dogs and found moderate to strong immunofluorescence of megakaryocytes in bone-marrow smears stained with rabbit anti-canine globulin conjugate. Some megakaryocytes also had morphologic abnormalities, such as cytoplasmic vacuolation and nuclear fragmentation.

Joshi and Jain[124] induced immunologic thrombocytopenia experimentally in dogs by injecting anti-canine platelet serum or the IgG fraction of antiserum. Those dogs developed thrombocytopenia and signs of purpura, whereas bone-marrow samples showed attachment of anti-platelet antibody to the surface of megakaryocytes, which was associated with the appearance of morphologic abnormalities.

## Pemphigus

Stannard, Gribble, and Baker[125] and Hurvitz and Feldman[126] described a chronic, mucocutaneous disease of dogs which resembles the human disease pemphigus vulgaris. Clinically, the disease was characterized by erosions and ulcerations of the oral mucosa, various mucocutaneous junctions, and the skin. Histologically, the epithelial cells were separated one from another (acantholysis), resulting in the formation of clefts and bullae; this histologic finding was considered diagnostic for pemphigus. An immune-mediated pathogenesis was suggested by the plasmocytic–lymphocytic infiltration of lesions and a dramatic response to corticosteroid therapy. Antibodies to intracellular space substances were detected by immunofluorescent staining in the serum of four of five affected dogs.

Halliwell and Goldschmidt[127] reported a case of eruptive dermatitis with bulla formation in a dog. Autoantibodies against epidermal intercellular cement substance were demonstrated by direct and indirect immunofluorescence. Kunkle, Goldschmidt, and Halliwell[128] described a bullous eruption involving the mouth, nose, and ears of a dog. Immunofluorescent staining of lesions showed deposition of immunoglobulin and complement at the dermal–epidermal junction, and circulating anti-basement membrane zone antibodies were demonstrated by indirect immunofluorescence with canine esophagus. The condition responded to corticosteroids and further fulfilled all criteria for bullous pemphigus in man.

## Immune-Complex Glomerulonephritis

An excellent review of immune-mediated glomerulonephritis in the dog was published by Osborne and Vernier,[129] who listed the naturally occurring forms of canine immune-complex glomerulonephritis (ICG) as those seen in pyometra, *Dirofilaria immitis* infection, systemic lupus erythematosus (SLE), membranous glomerulonephritis, and chronic membranoproliferative glomerulonephritis. Muller-Peddinghaus and Trautwein,[130] in an extensive study of morphologic and immunologic characteristics of canine glomerulonephritis, examined the kidneys of 101 dogs with and without clinical signs of renal disease; 90% of these animals showed morphologic evidence of glomerulopathy. Membranous glomerulonephritis with spikes and membranoproliferative glomerulonephritis were associated with coarse subepithelial deposits of IgG and C3. Mesangial deposits of C3 were frequently found in kidneys with membranoproliferative, mesangial-proliferative, and mesangial-sclerosing glomerulonephritis. Linear immunofluorescence, which would be indicative of autoantibodies to the glomerular basement membrane (Goodpasture's syndrome), was not found in those dogs. Morphologic and immunologic aspects of canine glomerular disease have also been reviewed.[131]

In a study of 42 dogs with ICG, Murray and Wright[132] reported that more than 70% of these cases had major extrarenal lesions. By far the most common of these lesions was malignant neoplasia. Stuart, Phemister, and Thomassen[133] demonstrated ICG in 13 adult beagles that developed spontaneous proteinuria but otherwise appeared clinically normal. Granular deposits of IgG and beta IC globulins were found in glomerular capillaries and mesangium. A similar, but less intense, localization of deposits was seen in five nonproteinuric dogs. deSchepper et al.[134] and Osborne et al.[135] demonstrated ICG in dogs with nephrotic syndrome characterized by proteinuria, hypoproteinemia, hypercholesterolemia, and edema. A report of membranous glomerulonephritis associated with *D. immitis* infection in five dogs was presented by Casey and Splitter.[136] IgG was detected in glomerular capillary walls and mesangium of two of those dogs. Three of ten heartworm-infected dogs without kidney lesions had focal areas of immunoglobulin deposition in isolated glomeruli. Osborne et al.[137] described clinical and laboratory findings in a dog with SLE-associated ICG. Halliwell and Blakemore[138] found deposits of all major immunoglobulin classes (IgG, IgM, and IgA) and complement on the basement membrane of a dog with ICG. Other case studies on canine ICG include those of Kurtz et al.,[139] Murray et al.,[140] and Wright et al.[141]

The induction of ICG by complexes of antibody and canine adenovirus (CAV) has received increasing attention. Bush and Evans[142] found high titers of antibody to the virus in 15 of 20 dogs with signs of chronic renal failure. Wright et al.[143] produced a proliferative glomerulonephritis in dogs experimentally infected with CAV and demonstrated localization of IgG, C3, and viral antigen in mesangial regions. Antibody to CAV was detected in kidney eluates. Similar findings were reported in naturally occurring cases of infectious canine hepatitis.[144] CAV antigen, IgG, and C3 were detected in glomeruli, and eluates of the kidneys contained anti-CAV antibody. Evidence of circulating immune complexes in CAV-infected dogs was obtained by passively transferring canine serum to mice.[145] Mice receiving serum taken from dogs on postinfection days 8 and 9 showed mesangial localization of canine IgG. At that time antibody titers in donor dogs were low, and the deposition rate of complexes in their kidneys was maximal. Transfer of serum taken from dogs in earlier or later stages of the disease was without effect in recipient mice.

Wright[146] concluded that ICG is produced during the initial period of antibody production in dogs recovering from CAV infection; this timing distinguishes ICG from the interstitial nephritis associated with CAV infection, which probably involves a cell-mediated immune reaction.[147,148] Virus–antibody complexes are also involved in the pathogenesis of the ocular lesions of infectious canine hepatitis.[146,148–151]

A few reports have implicated immune complexes in the pathogenesis of interstitial nephritis in the dog. In studies of acute interstitial nephritis induced by *Leptospira canicola,* Morrison and Wright[152,153] demonstrated leptospiral antigen and IgG deposits in the interstitium of affected kidneys and detected anti-leptospiral antibody in renal eluates. These investigators concluded that the persistence of leptospiral antigen, alone or complexed with antibody, was responsible for the marked cellular response in acute leptospiral nephritis. Krohn et al.[154] reported immunofluorescent studies in cases of chronic interstitial nephritis. IgG and C3 were deposited in capillary walls and mesangium of the glomeruli but not in renal interstitium. Eluates from nephritic kidneys contained IgG, which reacted with autologous or homologous nephritic kidneys but not with normal tissues.

Glomerulonephritis due to anti-glomerular basement membrane autoantibodies (Goodpasture's syndrome) has not been described as a naturally occurring disease in the dog but has been produced experimentally. For example, Terman et al.[155] produced nephrotoxic glomerulonephritis in dogs by passive infusion of rabbit anti-glomerular basement membrane antibodies.

## SUMMARY

The number of publications on canine immunology, especially those dealing with clinical cases, is legion. Not all publications could be summarized in our chapter. It is obvious that our knowledge of the immunology of the dog is extensive. In fact, in some areas significant recent advances have been made, e.g., cloning of various immunohematologic populations (Chap. 2 by Wilson and Shifrine). On the other hand, there are areas where our understanding is incomplete—ability to recognize subpopulations of lymphocytes or genetic control of immune responses—in which studies are in progress to fill the gaps [see Chapters 4 (Miller and MacKenzie) and 9 (Vriesendorp and van Bekkum)]. Thus our knowledge of and techniques for studying canine immunology are more than adequate for the exploration of various clinical or pathologic disorders as models for man.

## ACKNOWLEDGMENT

This work was supported by the U. S. Department of Energy.

## REFERENCES

1. R. O. Jacoby and R. A. Griesemer, Immunologic Injury to Dogs, *Adv. Vet. Sci. Comp. Med.,* 14: 149-207 (1970).
2. C. A. Bowles, G. S. White, and D. Lucas, Rosette Formation by Canine Peripheral Blood Lymphocytes, *J. Immunol.,* 114: 399-402 (1975).
3. A. R. Zander, N. Boopalam, and R. B. Epstein, Surface Markers on Canine Lymphocytes, *Transplant. Proc.,* 7: 369-373 (1975).
4. J.-F. Bach, M. Dardenne, J. Dormont, and B. Antoine, A New In Vitro Test Evaluating Anti-Lymphocyte Serum Potency, *Transplant. Proc.,* 1: 403-407 (1969).
5. G. N. Beall, J. R. Benfield, S. R. Kruger, and P. E. Byfield, Canine Lymphocytes That Form Rosettes with Human Red Blood Cells, *J. Surg. Res.,* 17: 330-337 (1974).
6. P. Brain, J. Gordon, and W. A. Willetts, Rosette Formation by Peripheral Lymphocytes, *Clin. Exp. Immunol.,* 6: 681-688 (1970).
7. G. E. Kelly, D. C. Mears, and A. G. R. Sheil, Relationship of the Immunosuppressive Potency of Antilymphocyte Globulin to Three In Vitro Tests, *Transplantation,* 12: 443-447 (1971).
8. W. L. Joseph, F. Melewicz, and D. L. Morton, Delayed Cutaneous Hypersensitivity to Dinitrochlorobenzene as a Method of Quantitating Immunosuppressive Therapy for Organ Transplantation, *Surgery,* 69: 847-853 (1971).
9. K. Nishiya, S. Yamana, K. Yano, K. Ensako, T. Fujiwara, and T. Fuji, Studies on the Population of Human Peripheral Lymphocytes Forming Rosette with Dog Erythrocytes, *Acta. Med. Okayama,* 28: 373-375 (1974).
10. S. Krakowka, R. G. Olsen, and G. L. Cockerell, Methods for Producing T- and B-Lymphocyte Receptor-Specific Antisera, *J. Immunol. Methods,* 14: 257-265 (1977).
11. P. Lalerzari, S. L. Nehlsen, S. B. P. Sinha, M. B. Stemerman, and F. J. Veith, Spontaneous Macrophage—Erythrocyte Rosette Formation, A Species—Specific Phenomenon, *Immunology,* 27: 457-462 (1974).
12. B. J. Bryant and M. Shifrine, Histogenesis of Lymph Nodes During Development of the Dog, *J. Reticuloendothel. Soc.,* 12: 96-107 (1972).
13. R. O. Jacoby, R. A. Dennis, and R. A. Griesemer, Development of Immunity in Fetal Dogs: Humoral Responses, *Am. J. Vet. Res.,* 30: 1503-1510 (1969).
14. R. M. Lewis, C. A. Smith, and L. Garfield, Kinetics of Antibody Synthesis to Particulate and Soluble Antigen in Newborn Pups and Adult Dogs, *Am. J. Vet. Res.,* 34: 235-240 (1973).
15. M. Shifrine, J. B. Smith, M. S. Bulgin, B. J. Bryant, Y.-C. Zee, and B. I. Osburn, Response of Canine Fetuses and Neonates to Antigenic Stimulation, *J. Immunol.,* 107: 965-970 (1971).
16. A. L. Brown, J. A. Vitamvas, D. L. Merry, and W. H. Beckenhauer, Immune Response of Pups to Modified Live-Virus Canine Distemper—Measles Vaccine, *Am. J. Vet. Res.,* 33: 1447-1456 (1972).
17. S. Krakowka, D. Long, and A. Koestner, Influence of Transplacentally Acquired Antibody on Neonatal Susceptibility to Canine Distemper Virus in Gnotobiotic Dogs, *J. Infect. Dis.,* 137: 605-608 (1978).
18. R. A. Dennis, R. O. Jacoby, and R. A. Griesemer, Development of Immunity in Fetal Dogs: Skin Allograft Rejection, *Am. J. Vet. Res.,* 30: 1511-1522 (1969).

19. B. J. Bryant, M. Shifrine, and C. McNeil, Cell-Mediated Immune Response in the Developing Dog, *Int. Arch. Allergy Appl. Immunol.*, 45: 937-942 (1973).
20. J. D. Gerber and A. L. Brown, Effect of Development and Aging on the Response of Canine Lymphocytes to Phytohemagglutinin, *Infect. Immun.*, 10: 695-699 (1974).
21. A. U. Sargent, S. B. Johnson, and A. K. Richardson, The Isolation and Functional Purification of the First Seven Components of Canine Hemolytic Complement, *Immunochemistry*, 13: 823-829 (1976).
22. J. T. Boyer, P. Wyde, and M. Bier, Hypochlorite-Induced Alterations to Canine Serum Complement, *Clin. Exp. Immunol.*, 21: 345-350 (1975).
23. N. R. Cooper and C. G. Cochrane, The Participation of Complement in Renal Disease and Allograft Rejection as Assessed by Model Animal Systems, *Transplant. Proc.*, 6: 51-57 (1974).
24. R. Attström and U. Larsson, Effect of Decomplementation by Carragheenan on the Emigration of Neutrophils and Monocytes into Dog Gingival Crevices, *J. Periodontal Res.*, 9: 165-175 (1974).
25. R. Garner, B. V. Chater, and D. L. Brown, The Role of Complement in Endotoxin Shock and Disseminated Intravascular Coagulation: Experimental Observations in the Dog, *Br. J. Haematol.*, 28: 393-401 (1974).
26. H. Y. Reynolds and J. S. Johnson, Quantitation of Canine Immunoglobulins, *J. Immunol.*, 105: 698-703 (1970).
27. R. J. Heddle and D. Rowley, Dog Immunoglobulins. I. Immunochemical Characterization of Dog Serum, Parotid Saliva, Colostrum, Milk, and Small Bowel Fluid, *Immunology*, 29: 185-195 (1975).
28. H. Y. Reynolds and J. S. Johnson, Canine Immunoglobulins. III. Distribution of Immunoglobulins in Colostrum and Isolation of Secretory IgA and $7S_1$, *J. Immunol.*, 104: 1000-1008 (1970).
29. H. B. Kaltreider and M. K. L. Chan, The Class-Specific Immunoglobulin Composition of Fluids Obtained from Various Levels of the Canine Respiratory Tract, *J. Immunol.*, 116: 423-429 (1976).
30. H. Y. Reynolds and R. E. Thompson, Canine Respiratory Macrophages: Diverse Functions Plus Studies of Bacterial Phagocytosis and Killing with Serum and Bronchial Secretion Immunoglobulins, *Fed. Proc.*, 31: 936 (1972).
31. P. Lieberman, R. Patterson, V. Petersen, L. W. Chakrin, J. R. Wardell, and J. Ricks, Effect of Antigen Variation on Production of Antibody in Canine Tracheal Secretions, *J. Immunol.*, 107: 1349-1356 (1971).
32. P. Lieberman, B. S. Berlin, J. Metzger, and R. Patterson, Production of Antibody in the Canine Tracheal Pouch After Local and Systemic Influenza Immunization, *J. Immunol.*, 109: 864-869 (1972).
33. J. O'Rourke, R. Patterson, M. Roberts, and I. Suszko, The Canine Respiratory Immune Response, *Int. Arch. Allergy Appl. Immunol.*, 44: 823-832 (1973).
34. J. A. Kazmierowski, W. A. Durbin, and H. Y. Reynolds, Kinetics of Immunoglobulin Transport into Canine Bronchial Secretions, *Proc. Soc. Exp. Biol. Med.*, 152: 493-498 (1976).
35. H. B. Kaltreider, F. N. Turner, E. Adam, and M. K. Chan, Bronchoalveolar Clearance and Immunologic Reaction to an Organic Particulate Antigen Instilled into Alveolar Spaces of Dog Lungs, *Chest*, 69(2 Suppl.): 282-283 (1976).
36. A. Mates and N. A. Hinton, Immune Response in Dog. Three Antibody Formations in Dogs and Rabbits to Human Serum Proteins and Keyhole Limpet Haemocyanin, *Microbios*, 17: 175-187 (1976).
37. H. Thimel, Preparation of Monospecific Immunoglobulin Fractions (IgG, IgM) and Anti-Immunoglobulin Sera (Anti-IgG, Anti-IgM) for Quantitative Serum Immunoglobulin Measurements in the Dog, *Zentralbl. Veterinaermed., Beih.*, 22: 393-402 (1975).

38. E. Orlans, The Antigenic Interrelations of Some Mammalian IgG Subclasses Detected with Cross-Reacting Fowl Antisera to Human and Mouse IgG-Fe, *Immunology,* 28: 761-772 (1975).

39. C. Whitacre and R. W. Lang, Characterization of IgG-Containing DEAE Fractions of Canine Serum, *Transplant. Proc.,* 7: 531-536 (1975).

40. K. Krumbacher and H. P. Schnappauf, Evidence for an Allotypic Marker on Canine Immunoglobulin, *Transplant. Proc.,* 7: 545-546 (1975).

41. R. Saison and D. Colling, An Immunoglobulin (IgM) Marker in Dogs, *Transplant. Proc.,* 7: 383-385 (1975).

42. R. E. Thompson and H. Y. Reynolds, Isolation and Characterization of Canine Secretory Immunoglobulin M, *J. Immunol.,* 118: 323-329 (1977).

43. R. L. Wasserman and J. D. Capra, Amino Acid Sequence of the $F_C$ Region of a Canine Immunoglobulin M: Interspecies Homology for the IgM Class, *Science,* 200: 1159-1161 (1978).

44. J.-P. Vaerman and J. F. Heremans, Origin and Molecular Size of Immunoglobulin A in the Mesenteric Lymph of the Dog, *Immunology,* 18: 27-38 (1970).

45. J. Ricks, M. Roberts, and R. Patterson, Canine Secretory Immunoglobulins: Identification of Secretory Component, *J. Immunol.,* 105: 1327-1333 (1970).

46. H. Y. Reynolds and J. S. Johnson, Canine Copro-Immunoglobulins, *J. Infect. Dis.,* 121(Suppl.): 118-123 (1970).

47. H. Y. Reynolds and J. S. Johnson, Structural Units of Canine Serum and Secretory Immunoglobulin A., *Biochemistry,* 10: 2821-2827 (1971).

48. J. M. Kehoe, T. B. Tomasi, F. Ellouz, and J. D. Capra, Identification of "J" Chain in a Homogeneous Canine IgA Immunoglobulin, *J. Immunol.,* 109: 59-64 (1972).

49. G. C. Meinke and H. L. Spiegelberg, Antigenic Studies of J Chain, *J. Immunol.,* 112: 1401-1406 (1974).

50. K. Kobayashi, J.-P. Vaerman, H. Bazin, A.-M. Lebacq-Verheyden, and J. F. Heremans, Identification of J-Chain in Polymeric Immunoglobulins from a Variety of Species by Cross-Reaction with Antisera to Human J-Chain, *J. Immunol.,* 111: 1590-1594 (1973).

51. J.-P. Vaerman, K. Kobayashi, and J. F. Heremans, Precipitin Cross-Reactions Between Human and Animal J-Chains, *Adv. Exp. Med. Biol.,* 45: 251-255 (1974).

52. R. E. Thompson, H. Y. Reynolds, and M. J. Waxdal, Structural Composition of Canine Secretory Component and Immunoglobulin A, *Biochemistry,* 14: 2853-2860 (1975).

53. G. Virella, R. J. Slappendel, and J. Joudswaard, Multiple Myeloma, IgA Cryoglobulinemia and Serum Hyperviscosity in a Dog, *Int. Arch. Allergy Appl. Immunol.,* 55: 537-541 (1977).

54. R. J. Heddle and D. Rowley, Dog Immunoglobulins. II. The Antibacterial Properties of Dog IgA, IgM, and IgG Antibodies to *Vibrio cholerae, Immunology,* 29: 197-208 (1975).

55. R. M. Schwartzman, J. H. Rockey, and R. E. Halliwell, Canine Reaginic Antibody. Characterization of the Spontaneous Anti-Ragweed and Induced Anti-Dinitrophenyl Reaginic Antibodies of the Atopic Dog, *Clin. Exp. Immunol.,* 9: 549-569 (1971).

56. R. E. W. Halliwell, R. M. Schwartzman, and J. H. Rockey, Antigenic Relationship Between Human IgE and Canine IgE, *Clin. Exp. Immunol.,* 10: 399-407 (1972).

57. P. Sunthoupalin, J. A. Arkins, and J. N. Fink, Characterization of Induced Homocytotropic Antibody to Ragweed and Timothy Pollen in Nonatopic Dogs, *Clin. Exp. Immunol.,* 8: 825-834 (1971).

58. M. Matsumoto, T. P. Monkus, and M. J. Freeman, Induction of Canine Homocytotropic Antibody by Tetanus Toxoid and Dinitrophenylated Keyhole Limpet Hemocyanin Antigens, *Int. Arch. Allergy Appl. Immunol.,* 44: 171-180 (1973).

59. L. H. McConnell, J. A. Arkins, and J. N. Fink, Induced Homocytotropic Antibody to Bovine Serum Albumin in Nonatopic Dogs, *J. Allergy Clin. Immunol.*, 52: 47-54 (1973).

60. R. E. W. Halliwell, R. M. Schwartzman, P. C. Montgomery, and J. H. Rockey, Physicochemical Properties of Canine IgE, *Transplant. Proc.*, 7: 537-543 (1975).

61. W. Kepron, E. Rector, A. H. Sehon, and K. S. Tse, Specific Suppression of the IgE Response in Dogs, *J. Allergy Clin. Immunol.*, 57: 201 (1976).

62. T. Tada, Regulation of Reaginic Antibody Formation in Animals, *Prog. Allergy*, 19: 122-194 (1975).

63. J. F. Williams and M. V. Perez Esandi, Reaginic Antibodies in Dogs Infected with *Echinococcus granulosus*, *Immunology*, 20: 451-455 (1971).

64. R. E. W. Halliwell, The Localization of IgE in Canine Skin: An Immunofluorescent Study, *J. Immunol.*, 110: 422-430 (1973).

65. R. E. W. Halliwell, The Sites of Production and Localization of IgE in Canine Tissues, *Ann. N.Y. Acad. Sci.*, 254: 476-488 (1975).

66. M. C. Healey and S. M. Gaafar, Demonstration of Reaginic Antibody (IgE) in Canine Demodectic Mange: An Immunofluorescent Study, *Vet. Parasitol.*, 3: 107-119 (1977).

67. H. D. Ochs, et al., Immunologic Reactivity in Canine Marrow Graft Recipients, *J. Immunol.*, 113: 1039-1057 (1974).

68. J. P. Thilsted and M. Shifrine, Delayed Cutaneous Hypersensitivity in the Dog: Reaction to Tuberculin PPD and Coccidioidin, *Am. J. Vet. Res.*, 39: 1702-1705 (1978).

69. W. R. Snider, Tuberculosis in Canine and Feline Populations. Review of the Literature, *Am. Rev. Respir. Dis.*, 104: 877-887 (1971).

70. W. R. Snider, D. Cohen, J. S. Reif, S. C. Stein, and J. E. Prier, Tuberculosis in Canine and Feline Populations. Study of High Risk Populations in Pennsylvania, 1966–1968, *Am. Rev. Respir. Dis.*, 104: 866-876 (1971).

71. A. L. Brown and R. E. McCarthy, Relationship Between Measles and Canine Distemper Viruses Determined by Delayed Type Hypersensitivity Reactions in Dogs, *Nature*, 248: 344-345 (1974).

72. J. R. Benfield, E. Shors, P. Schick, P. E. Byfield, and M. O'Kita, Monitoring of Immunosuppression in Beagles, *J. Surg. Res.*, 23: 207-214 (1977).

73. R. J. Mangi, T. P. Munyer, S. Krakowka, R. O. Jacoby, and F. S. Kantor, A Canine Distemper Model of Virus-Induced Energy, *J. Infect. Dis.*, 133: 556-563 (1976).

74. G. J. Mizejewski, J. Baron, and G. Poissant, Immunologic Investigations of Naturally Occurring Canine Thyroiditis, *J. Immunol.*, 107: 1152-1160 (1971).

75. M. C. Healey and S. M. Gaafar, Immunodeficiency in Canine Demodectic Mange. II. Skin Reactions to Phytohemagglutinin and Concanavalin A, *Vet. Parasit.*, 3: 133-140 (1977).

76. M. Schatz, R. Patterson, H. M. Sommers, and K. E. Harris, Mitogen-Induced Lymphocyte and Cutaneous Reactions in Dogs, *Int. Arch. Allergy Appl. Immunol.*, 54: 121-128 (1977).

77. D. W. Scott, B. R. H. Farrow, and R. D. Schultz, Studies on the Therapeutic and Immunologic Aspects of Generalized Demodectic Mange in the Dog, *J. Am. Anim. Hosp. Assoc.*, 10: 233-244 (1974).

78. K. Angus and T. J. Yang, Lymphocyte Response to Phytohemagglutinin: Temporal Variation in Normal Dogs, *J. Immunol. Methods*, 21: 3-4 (1978).

79. H. B. Kaltreider and S. E. Salmon, Immunology of the Lower Respiratory Tract. Functional Properties of Bronchoalveolar Lymphocytes Obtained from the Normal Canine Lung, *J. Clin. Invest.*, 52: 2211-2217 (1973).

80. J. C. Rosenberg et al., In Vitro Effect of Methylprednisolone on Protein Synthesis of Activated Canine Thymus-Derived Lymphocytes, *Transplant. Proc.*, 7: 547-556 (1975).
81. S. A. Benjamin and A. C. Ferris, Dextran Potentiation of the Canine Lymphocyte Response to Plant Mitogens, *Proc. Soc. Exp. Biol. Med.*, 148: 247-250 (1975).
82. J. P. Thilsted and M. Shifrine, Lymphocyte Transformation in the Dog: Response of Lymphocytes from Normal and Immune Dogs to Phytohemagglutinin, Coccidioidin, and Purified-Protein Derivative, *Am. J. Vet. Res.*, 38: 81-87 (1977).
83. J. D. Gerber and A. E. Marron, Cell-Mediated Immunity and Age at Vaccination Associated with Measles Inoculation and Protection of Dogs Against Canine Distemper, *Am. J. Vet. Res.*, 37: 133-138 (1976).
84. J. S. Reif, F. S. Lief, and J. T. McGrath, Local and Systemic Cell-Mediated Immunity in Viral Infection of the Canine Respiratory Tract, *Transplant. Proc.*, 7: 561-565 (1975).
85. T. Kuramochi, In Vitro Detection of Transplantation Immunity: Leukocyte Migration Inhibition in the Dog, *Jpn. J. Exp. Med.*, 44: 223-226 (1974).
86. S. Krakowka, G. Cockerell, and A. Koestner, Intradermal Mitogen Response in Days: Correlation with Outcome of Infection by Canine Distemper Virus, *Am. J. Vet. Res.*, 38: 1539-1542 (1977).
87. R. Corbett, K. Banks, D. Hinricks, and T. Bell, Cellular Immune Responsiveness in Dogs with Demodectic Mange, *Transplant. Proc.*, 7: 557-559 (1975).
88. D. C. Hirsh, B. B. Baker, N. Wiger, S. G. Yaskulski, and B. I. Osburn, Suppression of In Vitro Lymphocyte Transformation by Serum from Dogs with Generalized Demodicosis, *Am. J. Vet. Res.*, 36: 1590-1595 (1975).
89. M. C. Healey and S. M. Gaafar, Immunodeficiency in Canine Demodectic Mange. I. Experimental Production of Lesions Using Antilymphocyte Serum, *Vet. Parasitol.*, 3: 121-131 (1977).
90. M. Shifrine, J. Thilsted, and D. Pappagianis, Canine Transfer Factor, in *Transfer Factor: Basic Properties and Clinical Applications*, M. S. Ascher, A. A. Gottlieb, and C. H. Kirkpatrick (Eds.), pp. 349-357, Academic Press, New York, 1976.
91. A. Khaw, C. H. Kirkpatrick, and N. O. Hill (Eds.), *Immune Regulators in Transfer Factor*, pp. 673-678, Academic Press, New York, 1979.
92. R. H. Tomar, Local Transfer of Delayed Hypersensitivity to Dogs by Human Transfer Factor, in *Transfer Factor: Basic Properties and Clinical Applications*, M. S. Ascher, A. A. Gottlieb, and C. H. Kirkpatrick (Eds.), pp. 234-236, Academic Press, New York, 1976.
93. K. P. Baker and J. O'Flanagan, Hypersensitivity of Dog Skin to Fleas—A Clinical Report, *J. Small Anim. Pract.*, 16: 317-327 (1975).
94. D. R. Krawiec and S. M. Gaafar, A Comparative Study of Allergic and Primary Irritant Contact Dermatitis with Dinitrochlorobenzene (DNCB) in Dogs, *J. Invest. Dermatol.*, 65: 248-251 (1975).
95. K. W. Chamberlain, Atopic (Allergic) Dermatitis, *Vet. Clin. N. Am.*, 4: 29-40 (1974).
96. M. D. Lorenz, Environmental Allergens in Small Animal Hypersensitivity Disorders, *Vet. Clin. N. Am.*, 4: 41-46 (1974).
97. G. S. Walton, Allergic Responses Involving the Skin of Domestic Animals, *Adv. Vet. Sci. Comp. Med.*, 15: 201-240 (1971).
98. V. H. Austin, Atopic Skin Disease. If Other Treatments Fail, Is Hyposensitization the Answer? *Mod. Vet. Pract.*, 57: 355-361 (1976).
99. E. Baker, Diagnosis and Management of Clinical Allergy, *J. Am. Vet. Med. Assoc.*, 57: 1607-1615 (1970).
100. B. B. Baker and A. A. Stannard, Drug Eruption in a Dog, *J. Am. Vet. Med. Assoc.*, 166: 500-501 (1975).

101. C. B. Carrig, Fixed Drug Eruption in a Dog, *J. Small Anim. Pract.*, 12: 185-192 (1970).
102. W. Kepron, J. M. James, B. Kirk, A. H. Sehon, and K. S. Tse, A Canine Model for Reaginic Hypersensitivity and Allergic Bronchoconstriction, *J. Allergy Clin. Immunol.*, 59: 64-69 (1977).
103. R. E. Faith, J. R. Hessler, and P. A. Small, Respiratory Allergy in the Dog: Induction by the Respiratory Route and the Effect of Passive Antibody, *Int. Arch. Allergy Appl. Immunol.*, 53: 530-544 (1977).
104. J. J. Pruzansky, I. M. Suszko, and R. Patterson, Uptake and Catabolism of Antigen by Alveolar Macrophages of Dogs with Respiratory Hypersensitivity, *Clin. Exp. Immunol.*, 25: 165-169 (1976).
105. B. H. Booth, R. Patterson, and H. Talbot, Immediate Type Hypersensitivity in Dogs: Cutaneous Anaphylactic, and Respiratory Responses to *Ascaris*, *J. Lab. Clin. Med.*, 76: 181-189 (1970).
106. A. Chiesa, D. Dain, G. L. Meyers, G.-F. Kessler, and W. M. Gold, Histamine Release During Antigen Inhalation in Experimental Asthma in Dogs, *Am. Rev. Respir. Dis.*, 11: 148-156 (1975).
107. W. M. Gold, G.-F. Kessler, D. Y. C. Yu, and O. L. Frick, Pulmonary Physiologic Abnormalities in Experimental Asthma in Dogs, *J. Appl. Physiol.*, 33: 496-501 (1972).
108. G. F. Kessler, O. L. Frick, and W. M. Gold, Immunological Allergy and Physiologic Characteristics of the Role of Reaginic Antibodies in Experimental Asthma in Dogs, *Int. Arch. Allergy Appl. Immunol.*, 47: 313-328 (1974).
109. R. D. Krell and L. W. Chakrin, Canine Airway Responses to Acetylcholine, Prostaglandin $F_{2\alpha}$, Histamine, and Serotonin After Chronic Antigen Exposure, *J. Allergy Clin. Immunol.*, 58: 664-675 (1976).
110. R. D. Krell, L. W. Chakrin, and J. R. Wardell, In Vivo Canine and Rhesus Monkey Models of Allergic Asthma, in *Immunopharmacology*, M. E. Rosenthale and H. C. Mansmann (Eds.), pp. 125-148, Spectrum Publications, Inc., New York, 1975.
111. R. D. Krell, L. W. Chakrin, and J. R. Wardell, The Effect of Cholinergic Agents on a Canine Model of Allergic Asthma, *J. Allergy Clin. Immunol.*, 58: 19-30 (1976).
112. R. Patterson and J. F. Kelley, Animal Models of the Asthmatic State, *Annu. Rev. Med.*, 25: 53-68 (1974).
113. P. E. Bigazzi and N. R. Rose, Spontaneous Autoimmune Thyroiditis in Animals as a Model of Human Disease, *Prog. Allergy*, 19: 245-274 (1975).
114. O. W. Schalm, N. C. Jain, and E. J. Carroll, *Veterinary Hematology*, Lea & Febiger, 1975.
115. W. J. Dodds, Autoimmune Hemolytic Anemia and Other Causes of Immune-Mediated Anemia: An Overview, *J. Am. Anim. Hosp. Assoc.*, 13: 437-441 (1977).
116. R. D. Schultz, Immunologic Disorders in the Dog and Cat, *Vet. Clin. N. Am.*, 4: 153-174 (1974).
117. R. J. Slappendal, C. L. G. M. VanErp, J. Joudswaard, and M. Bethlehem, Cold Hemagglutinin Disease in a Toy Pinscher Dog, *Tijdschr. Diergeneeskd.*, 100: 445-460 (1975).
118. C. E. Green, F. Kristensen, E. J. Hoff, and M. D. Wiggins, Cold Hemagglutinin Disease in a Dog, *J. Am. Vet. Med. Assoc.*, 170: 505-510 (1977).
119. D. R. E. Jones and P. G. G. Darke, Use of Papain for the Detection of Incomplete Erythrocyte Autoantibodies in Autoimmune Haemolytic Anaemia of the Dog and Cat, *J. Small Anim. Pract.*, 16: 273-279 (1975).
120. O. W. Schalm, The Blood Platelets (Thrombocytes): II. The Thrombocytopathics, *Calif. Vet.*, 25(6): 6-10 (1971).
121. O. W. Schalm and G. V. Ling, The Blood Platelets (Thrombocytes). III. The Hemorrhagic Syndrome in Thrombocytopenia, *Calif. Vet.*, 25(8): 10-15 (1971).

122. R. J. Wilkins, A. I. Hurvitz, and W. J. Dodds-Laffin, Immunologically Mediated Thrombocytopenia in the Dog, *J. Am. Vet. Med. Assoc.*, 163: 277-282 (1973).
123. B. C. Joshi and N. C. Jain, Detection of Antiplatelet Antibody in Serum and on Megakaryocytes of Dogs with Autoimmune Thrombocytopenia, *Am. J. Vet. Res.*, 37: 681-685 (1976).
124. B. C. Joshi and N. C. Jain, Experimental Immunologic Thrombocytopenia in Dogs: A Study of Thrombocytopenia and Megakaryocytopoiesis, *Res. Vet. Sci.*, 22: 11-17 (1977).
125. A. A. Stannard, D. H. Gribble, and B. B. Baker, A Mucocutaneous Disease in the Dog, Resembling Pemphigus Vulgaris in Man, *J. Am. Vet. Med. Assoc.*, 166: 575-582 (1975).
126. A. I. Hurvitz and E. Feldman, A Disease in Dogs Resembling Human Pemphigus Vulgaris: Case Reports, *J. Am. Vet. Med. Assoc.*, 166: 585-590 (1975).
127. R. E. W. Halliwell and M. H. Goldschmidt, Pemphigus Foliaceus in the Canine: A Case Report and Discussion, *J. Am. Anim. Hosp. Assoc.*, 13: 431-436 (1977).
128. G. Kunkle, M. H. Goldschmidt, and R. E. W. Halliwell, Bullous Pemphizoid in a Dog: A Case Report with Immunofluorescent Findings, *J. Am. Anim. Hosp. Assoc.*, 14: 52-57 (1978).
129. C. A. Osborne and R. L. Vernier, Glomerulonephritis in the Dog and Cat: A Comparative Review, *J. Am. Anim. Hosp. Assoc.*, 9: 101-127 (1973).
130. R. Muller-Peddinghaus and G. Trautwein, Spontaneous Glomerulonephritis in Dogs. I. Classification and Immunopathology, *Vet. Pathol.*, 14: 1-13 (1977).
131. Anon., Glomerular Disease in the Dog, *Vet. Rec.*, 95: 61-62 (1974).
132. M. Murray and N. G. Wright, A Morphologic Study of Canine Glomerulonephritis, *Lab. Invest.*, 30: 213-221 (1974).
133. B. P. Stuart, R. D. Phemister, and R. W. Thomassen, Glomerular Lesions Associated with Proteinuria in Clinically Healthy Dogs, *Vet. Pathol.*, 12: 125-144 (1975).
134. J. deSchepper, J. Hoorens, D. Mattheeuws, and J. Van Der Stock, Glomerulonephritis and Nephrotic Syndrome in a Dog, *Vet. Rec.*, 95: 434-437 (1974).
135. C. A. Osborne, R. F. Hammer, J. S. Resnick, J. B. Stevens, B. L. Yano, and R. L. Vernier, Natural Remission of Nephrotic Syndrome in a Dog with Immune-Complex Glomerular Disease, *J. Am. Vet. Med. Assoc.*, 168: 129-137 (1976).
136. H. W. Casey and G. A. Splitter, Membranous Glomerulonephritis in Dogs Infected with *Dirofilaria Immitis*, *Vet. Pathol.*, 12: 111-117 (1975).
137. C. A. Osborne, J. B. Stevens, R. McClean, and R. L. Vernier, Membranous Lupus Glomerulonephritis in a Dog, *J. Am. Anim. Hosp. Assoc.*, 9: 295-300 (1973).
138. R. E. W. Halliwell and W. F. Blakemore, A Case of Immune Complex Glomerulonephritis in a Dog, *Vet. Rec.*, 90: 275-280 (1972).
139. J. M. Kurtz, S. W. Russell, J. C. Lee, D. O. Slauson, and R. D. Schecter, Naturally Occurring Canine Glomerulonephritis, *Am. J. Pathol.*, 67: 471-482 (1972).
140. M. Murray, H. M. Pirie, H. Thompson, W. F. H. Jarrett, and A. Wiseman, Glomerulonephritis in a Dog. A Histological and Electron Microscopical Study, *Res. Vet. Sci.*, 12: 493-495 (1971).
141. N. G. Wright, E. W. Fisher, W. I. Morrison, W. B. Thomson, and A. S. Nash, Chronic Renal Failure in Dogs: A Comparative Clinical and Morphological Study of Chronic Glomerulonephritis and Chronic Interstitial Nephritis, *Vet. Rec.*, 98: 288-293 (1976).
142. B. M. Bush and J. M. Evans, Infectious Canine Hepatitis and Chronic Renal Failure, *Vet. Rec.*, 90: 33-34 (1972).
143. N. G. Wright, N. I. Morrison, H. Thompson, and H. J. C. Cornwell, Mesangial Localization of Immune Complexes in Experimental Canine Adenovirus Glomerulonephritis, *Br. J. Exp. Pathol.*, 55: 458-465 (1974).

144. W. I. Morrison, A. S. Nash, and N. G. Wright, Glomerular Deposition of Immune Complexes in Dogs Following Natural Infection with Canine Adenovirus, *Vet. Rec.*, 96: 522-524 (1975).
145. W. I. Morrison and N. G. Wright, Detection of Immune Complexes in the Serum of Dogs Infected with Canine Adenovirus, *Res. Vet. Sci.*, 21: 119-121 (1976).
146. N. G. Wright, Canine Adenovirus: Its Role in Renal and Ocular Disease: A Review, *J. Small Anim. Pract.*, 17: 25-33 (1976).
147. W. I. Morrison, N. G. Wright, and H. J. C. Cornwell, An Immunopathologic Study of Interstitial Nephritis Associated with Experimental Canine Adenovirus Infection, *J. Pathol.*, 120: 221-229 (1976).
148. G. Aguirre, L. Carmichael, and S. I. Bistner, Corneal Endothelium in Viral Induced Anterior Uneitis. Ultrastructural Changes Following Canine Adenovirus Type 1 Infection, *Arch. Ophthalmol.*, 93: 219-224 (1975).
149. S. I. Bistner, Ocular Manifestations of Systemic Disease, *Vet. Clin. N. Am.*, 3: 467-490 (1973).
150. L. E. Carmichael, B. L. S. Medic, S. I. Bistner, and G. D. Aguirre, Viral—Antibody Complexes in Canine Adenovirus Type 1 (CAV-1) Ocular Lesions: Leukocyte Chemotaxis and Enzyme Release, *Cornell Vet.*, 65: 331-351 (1975).
151. R. Curtis and K. C. Barnett, The Ocular Lesions of Infectious Canine Hepatitis. 1. Clinical Features, *J. Small Anim. Pract.*, 14: 375-389 (1973).
152. W. I. Morrison and N. G. Wright, Canine Leptospirosis: An Immunopathological Study of Interstitial Nephritis due to *Leptospira canicole*, *J. Pathol.*, 120: 83-89 (1976).
153. W. I. Morrison and N. G. Wright, Immunopathological Aspects of Canine Renal Disease, *J. Small Anim. Pract.*, 17: 139-148 (1976).
154. K. Krohn, M. Merò, A. Iksanen, and M. Sandholm, Immunologic Observations in Canine Interstitial Nephritis, *Am. J. Pathol.*, 65: 157-172 (1971).
155. D. S. Terman, D. Durante, G. Buffaloe, and R. McIntosh, Attenuation of Canine Nephrotoxic Glomerulonephritis with an Extracorporeal Immunoadsorbent, *Scand. J. Immunol.*, 6: 195-202 (1977).

# Cell-Surface Markers of Canine Lymphocytes

4

C. H. MILLER

*Departments of Pathology and Internal Medicine, Section of Hematology and Oncology, School of Medicine, University of California, Davis*

*and*

M. R. MACKENZIE

*Department of Internal Medicine, Section of Hematology and Oncology, School of Medicine, University of California, Davis*

Mammalian lymphocytes active in the immune response have been separated into two distinct populations on the basis of their unique morphologic characteristics (cell-surface receptors or markers) as well as by functional criteria. The bone marrow or bursa-derived lymphocyte (B cell) is typically identified morphologically by the presence on the cell membrane of (1) a receptor for the third complement component (C′3), (2) immunoglobulins (Ig), and (3) a receptor for the Fc portion of immunoglobulins (Fc receptor). Identification of these markers most commonly depends on rosetting procedures and fluorescent-labeling protocols.

The thymus-derived lymphocytes (T cell) in most nonmurine species can be identified by their formation of spontaneous rosettes with erythrocytes (red blood cells) from certain heterologous or, in some instances, homologous species. For example, certain mononuclear cells, presumably lymphocytes, obtained from the dog form spontaneous rosettes (E rosettes) with red blood cells obtained from humans or guinea pigs.[1-10] Unfortunately, other canine cells also form E rosettes with guinea pig or human erythrocytes,[11-14] thus complicating and confusing the identification and enumeration of canine T lymphocytes.

## ROSETTING OF CANINE CELLS WITH ERYTHROCYTES OF HETEROLOGOUS SPECIES

Red blood cells obtained from several animal species have been tested for their capacity to form rosettes with canine lymphocytes. Bowles, White, and

Lucas[6] tested red cells of human, guinea pig, rat, rabbit, sheep, bovine, horse, chicken, and pigeon origin, but only red blood cells from humans and guinea pigs formed rosettes with canine lymphocytes. The gross appearance of rosettes formed with guinea pig erythrocytes was indistinguishable from that of rosettes formed with human erythrocytes, and the percentage of E rosette-forming lymphocytes with guinea pig red blood cells (averaging 32.4%) was similar to the percentage of E rosettes formed with human red blood cells (averaging 36.3%). Further, studies by these workers[6] indicated that red cells of both origins were bound to the same canine lymphocytes. That conclusion was suggested by the observation that simultaneous incubation of canine lymphocytes with a mixture of human and guinea pig red blood cells produced a total rosetting percentage of 36.7, a number not significantly different from the mean value of 37.1% that is obtained from the use of human red cells alone.

Zander, Boopalam, and Epstein[8] studied the rosetting properties of canine lymphocytes with red blood cells of sheep, human, guinea pig, rabbit, rat, and mouse origin. Human and guinea pig red blood cells formed rosettes with 38% and 17% of the canine lymphocytes, respectively, whereas rabbit, rat, and sheep erythrocytes formed less than 1% rosettes. Similar results were obtained by Miller et al.[10] Only the red cells of humans and guinea pigs form rosettes with canine lymphocytes.

## SEPARATION OF CANINE LYMPHOCYTES FROM PERIPHERAL BLOOD

Lymphocytes have been separated from peripheral whole blood by differential sedimentation, by gradient centrifugation, by phagocytosis of iron particles, or by selectively absorbing and subsequently removing certain populations. The resulting lymphocyte preparations may contain various percentages of contaminating leukocytes, such as monocytes, eosinophils, and polymorphonuclear neutrophils.

Several of the isolation techniques use centrifugation of whole blood through a density gradient of Ficoll–Hypaque alone or as one of several steps in the isolation procedure. In the technique of Bowles, White, and Lucas,[6] Ficoll–Hypaque gradient separation alone produced a preparation containing more than 90% lymphocytes. In a study by Whitacre and Lang,[7] centrifugation through Ficoll–Hypaque was part of a three-step procedure. First, defibrinated blood was incubated with iron particles to allow phagocytosis by granulocytes and monocytes. After the removal of erythrocytes and iron-containing granulocytes with plasmagel, the lymphocyte-rich supernatant was banded by centrifugation on a Ficoll–Hypaque gradient. Residual erythrocytes, if present, were removed by hypotonic shock. Preparations obtained by this treatment contained 85 to 95% lymphocytes with 99% viability. Although those lymphocytes represented only 30% of the circulating lymphocytes, the recovered lymphocytes retained functional capacity as illustrated by their ability to undergo blast transformation when stimulated with phytohemagglutinin (PHA) or pokeweed mitogen (PWM).

A procedure that produced a mononuclear preparation containing 10 to 20% polymorphonuclear leukocytes was described by Beall et al.[4] Red cells were removed by sedimentation with plasmagel, and leukocytes were recovered by centrifugation. Contaminating red cells were lysed by 0.16*M* ammonium chloride–TRIS buffer (pH 7.2), and the lymphocytes were isolated by centrifugation on Ficoll–Hypaque.

Krakowka and Guyot[13] further defined the mononuclear cells following Ficoll–Hypaque separation by incubating these cells with latex beads (0.794-$\mu$m polystyrene). Although Ficoll–Hypaque separation yielded a population of 70 to 90% mononuclear cells, 1 to 12% of these cells ingested latex beads and were thus identified as monocytes.

In a study by Esser et al.,[14] three methods of lymphocyte separation from canine whole blood were evaluated: (1) the technique of Whitacre and Lang[7] described earlier in which blood was incubated with iron carbonyl, (2) dextran sedimentation followed by centrifugation with Ficoll–Hypaque, and (3) dextran sedimentation and subsequent incubation of the cells on a washed nylon fiber column followed by elution. The three-step iron-incubation technique of Whitacre and Lang[7] resulted in preparations that contained 96% ± 4.6% mononuclear cells, of which 21.3% ± 9.4% were latex-ingesting monocytes. Dextran sedimentation and subsequent incubation of cells on nylon fiber columns yielded variable results.

Esser et al.[14] serially sampled peripheral blood from a single dog during a 5-month period and tested the effect of two different cell-separation procedures on rosette formation. Of the 17 samples, 9 were purified with iron carbonyl treatment of defibrinated blood followed by Ficoll–Hypaque separation. The other eight samples were tested after Ficoll–Hypaque separation alone. The iron carbonyl/Ficoll–Hypaque technique gave a mean of 4.3% ± 2.0% rosette-forming cells, whereas Ficoll–Hypaque separation alone produced 36.0% ± 6.9% rosette-forming cells.

Reliable surface-marker studies depend on the quality of the isolated lymphocytes; therefore techniques used to obtain lymphocytes should be evaluated for:

1. The percentage of the total lymphocyte recovered.
2. The viability of the final preparation.
3. The techniques instituted to remove phagocytic cells (monocytes and granulocytes).
4. The morphological evaluation (by staining or other methods) of the final lymphocyte preparation and its purity.

With this information various studies performed on lymphocyte preparations can be compared directly.

## THE CANINE LYMPHOCYTE SPONTANEOUS ROSETTE: VARIATIONS IN TECHNIQUES

As with methods for separating canine lymphocytes from whole blood, the procedures for evaluating the spontaneous rosette differ greatly. Moreover, the criteria for distinguishing a rosetted from a nonrosetted cell also vary. These factors, in addition to the differences in lymphocyte separation procedures, could account for the wide range of values assigned to the normal number of T lymphocytes in canine peripheral blood and other lymphoid tissues.

In 1974, Beall et al.[4] used a rosetting procedure in which human red cells incubated with canine lymphocytes were centrifugated at 230 x *g* for 5 min. The cell suspension was counted either directly with a hemocytometer or indirectly after staining a glutaraldehyde-fixed preparation of the suspension with Wright's stain. Direct counts on 12 samples of blood from 3 dogs showed 36.7% ± 4.4% (standard error of the mean) E rosettes, whereas the stained preparation showed 22.1% ± 2.8% rosette-forming cells. If the cell suspension was incubated overnight at 4°C, the number of rosettes was not significantly increased. Shorter periods of incubation at room temperatures or at 4°C or the addition of fetal calf serum did not affect the number of rosetted cells.

Bowles, White, and Lucas[6] studied the effects of temperature (4, 22, or 37°C) and time of incubation (5, 30, or 60 min) on rosette formation by canine peripheral blood lymphocytes and human red blood cells. A cell with two or more bound erythrocytes was considered rosetted. The mean values ranged from a low of 18.3% at 4°C incubation for 30 min to a high of 39.5% at 22°C incubation for 30 min. In a second set of experiments, the mixture was incubated for 60 min at 4°C, centrifuged, and then reincubated for either 5 or 30 min at either 4, 22, or 37°C. The lowest value (19.7%) was obtained with a second incubation for 5 min at 37°C, and the highest value (31.8%) was with a second incubation for 5 min at 4°C.

Zander, Boopalam, and Epstein[8] incubated a human red blood cell–canine lymphocyte suspension with 2% fetal calf serum (absorbed with red cells) for 20 min at 37°C. After centrifugation for 5 min at 230 x *g*, the mixture was again incubated for 30 min at 4°C and resuspended by gentle shaking. The percentage of rosetted peripheral blood lymphocytes (three or more adherent red cells) from 34 adult animals averaged 38.4% ± 9.1%.

Whitacre and Lang[7] counted rosetting cells immediately after centrifugation of the human red cell–lymphocyte mixture at 200 x *g* for 5 min. The mixture contained no serum nor was it incubated. Purified samples of canine lymphocytes from whole blood samples obtained from five different adult animals showed a mean of 5.0% ± 1.3% rosette-forming cells.

Krakowka and Guyot[13] incubated neuraminidase-treated guinea pig red cells with lymphocytes in a V-bottom microliter plate. The cell mixture was incubated at 37°C for 15 min, centrifuged at 200 x *g* for 2 min, and then held undisturbed at 22°C for 1 hr. An aliquot was transferred to a hemocytometer where cells with three or more adherent red cells were counted as rosettes. The percentage of E rosette-forming lymphocytes found by this method ranged from 3.3 to 9.5%.

## CHARACTERIZATION OF THE SPONTANEOUSLY ROSETTING CANINE CELL

Although spontaneously rosetting canine lymphocytes are generally assumed to be thymus-derived lymphocytes, that assumption faces often contradictory data, including the wide range of values reported for spontaneously rosetting peripheral lymphocytes, the few spontaneously rosetting cells in the thymus, the large number of peripheral lymphocytes without markers, and the ability of certain canine nonlymphoid cells to form rosettes.

Beall et al.[4] observed the frequency, distribution, and significance of canine E rosetting cells while studying the effect of anti-lymphocyte sera on allograft rejection. The binding of the erythrocyte to the lymphocyte was thought to be a function of cell-surface proteins since incubation of the lymphocyte with 0.25% trypsin decreased the number of rosette-forming cells from 17 to 6%. By using a double-labeling system that stained with fluorescein-conjugated rabbit anti-canine gamma globulin in conjunction with spontaneous rosetting, these investigators showed that rosette-forming cells had no surface immunoglobulins, an observation consistent with the concept that E rosetted cells are not B lymphocytes.

Bowles, White, and Lucas[6] demonstrated that prior incubation of lymphocytes with human erythrocyte membrane fragments would reduce formation of rosettes formed with human red blood cells but not those formed with guinea pig red blood cells. Prior incubation with guinea pig red blood cell membrane fragments reduced the number of rosettes formed with guinea pig red blood cells but not those formed with human red blood cells. In addition, the prior incubation of the lymphocytes with either red-cell membrane fragments of human, guinea pig, or sheep origin had no effect on the numbers of cells that form erythrocyte–antibody–complement (EAC) rosettes, the procedure used to identify the third complement component. Simultaneous labeling with fluorescein-conjugated anti-canine globulin confirmed that human or guinea pig E rosetting cells did not have surface immunoglobulins.

Onions[5] used another double-labeling technique in which canine lymphocytes were simultaneously E rosetted with human red blood cells and avian EAC cells and showed that only 3% of normal canine lymph node cells formed mixed rosettes, whereas 39% formed pure EAC rosettes and 19% formed pure E rosettes.

Zander, Boopalam, and Epstein[8] showed that 90% of the lymphocytes that formed rosettes with human red cells subsequently reacted with fluorescein-labeled anti-thymocyte serum. The E rosetted cells, however, did not show the presence of surface immunoglobulins (SIg's) when double labeled with fluorescein-conjugated anti-immunoglobulins. When fluorescein-labeled anti-thymus antisera were reacted with thymus cells, 98% of the cells were positive; yet only 8% of the thymic cell population formed spontaneous rosettes.

Beall et al.[4] reported that no spontaneous rosettes formed with any cells other than typical small lymphocytes. Polymorphonuclear cells, readily recognized on stained slides, never formed rosettes. Cells that had adhered to plastic petri dishes were examined for spontaneous rosette formation to test the possibility that rosetting mononuclear cells were monocytes rather than lymphocytes. The percentage of rosette-forming cells in the adherent fraction was less than that of the original population, which suggests that most of the rosette-forming cells were nonadherent.

In contrast, Krakowka and Guyot,[13] using ingestion of latex beads as an indicator of phagocytic capacity, demonstrated that neutrophils and latex-ingesting monocytes formed rosettes as did lymphocytes. Only 22.5 to 35.4% of E rosette-forming cells could be identified as lymphocytes. They showed that natural cytophilic antibodies to guinea pig red cells in sera would explain the E rosetting capacity of the nonlymphoid cells. The sera were absorbed with guinea pig red cells to eliminate these antibodies. Use of the sera following absorption resulted in no reduction of EAC rosette-forming cells, but the number of E rosette-forming cells was reduced by 50 to 90%. This reduction was interpreted to mean that the rosettes formed were the result of naturally occurring cytophilic antibodies in the sera.

Esser et al.[14] demonstrated that mononuclear cells with phagocytic capacity have E rosetting capabilities. Iron carbonyl treatment during the cell-separation stage reduced both the number of nonlymphoid cells in the preparation and the total number of E rosetting cells. Also, 50% of the plastic adherent cells as well as some populations of polymorphonuclear leukocytes formed rosettes.

In 1976 Yang et al.[11,12] reported that the ability to form spontaneous rosettes is not limited to canine lymphoid cells. In the course of investigating cell populations that infiltrate the canine transmissible venereal sarcoma mass, they discovered that not only do the eosinophils found within the tumor mass spontaneously rosette with human red blood cells but also the tumor cells themselves have that capacity.

## VALUES FOR SPONTANEOUSLY ROSETTING CELLS IN CANINE PERIPHERAL BLOOD AND HEMATOPOIETIC TISSUES

Values reported for the percentage of E rosetting cells in canine peripheral blood (Table 1) show a wide variation, which probably reflects the differences in rosetting techniques, variation in procedures used to separate the peripheral lymphocytes, the criteria used to delineate a rosetted cell, and differences between animals.

The percentage of E rosette-forming cells in dogs varies not only between animals but also between samples from the same animal. Multiple blood samples were obtained by Beall et al.[4] from two different dogs during a 2-month period. In one animal, peripheral blood rosette-forming cells with human red blood cells ranged from 12 to 29% (mean 18%), whereas in the other animal the values ranged from 11 to 32% (mean 17%).

In another serial study,[6] eight peripheral blood samples were obtained from a single animal during a 57-day period. The percentage of spontaneously rosetting cells ranged from a low of 28% (on day 56) to a high of 51% (on day 1).

Table 2 represents a composite of values for E rosetting cells in canine thymus, lymph node, spleen, and bone marrow. Percentages of E rosettes in the thymus are consistently low, ranging from 1% (Ref. 10) to 9% (Ref. 9). The values for the dog are considerably lower than those for man and mouse; approximately 98% of thymic cells in those two species are E rosette positive. The percentages of E rosette-positive cells in the lymph node range from 6 to 55%. These disparate results as well as the wide range observed in the spleen (3 to 42%) and bone marrow (3 to 24%) could perhaps be accounted for by differences in techniques and certainly by variations in cell-separation protocols.

## USE OF ANTI-THYMOCYTE OR ANTI-LYMPHOCYTE IMMUNOGLOBULINS IN T-CELL IDENTIFICATION

Since there is a wide variation in the ability of different anti-lymphocyte sera (ALS) to immunosuppress allograft recipients, several in vitro techniques have been used to correlate the ALS with immunosuppressive abilities. One of the procedures used to evaluate immunosuppressive efficiency of ALS is the capacity of the ALS to inhibit spontaneous rosetting.[1] Kelly, Mears, and Sheil[3] in 1971 related the ability of certain anti-lymphocyte globulins to inhibit the spontaneous rosettes formed between human red blood cells and canine lymphocytes with the antisera's ability to prolong survival of full-thickness skin

TABLE 1

PERCENTAGE OF SPONTANEOUS ROSETTING CELLS IN CANINE PERIPHERAL BLOOD

| Animals tested | Criterion for rosetted cell* | Purity of preparation | Rosettes, % | Ref. |
|---|---|---|---|---|
| 3 dogs (12 samples) | ⩾4 HRBC† | 10 to 20% PMN‡ | 36.7 ± 4.4 (SEM) | 4 |
| 12 dogs (21 samples) | ⩾4 HRBC | 10 to 20% PMN | 21.7 ± 2.0 (SEM) | |
| 4 fetal siblings (pooled sample) | ⩾ 4 HRBC | Unknown | 1 | |
| 9 | ⩾2 HRBC | >90% lymphocyte | 36.3(range, 21 to 51) | 6 |
| 9 | ⩾2 GPRBC§ | >90% lymphocyte | 32.4(range, 23 to 43) | |
| 5 | ⩾3 HRBC | 85 to 94% lymphocyte | 5.0 ± 1.3 | 7 |
| 34 | ⩾3 HRBC | Unknown | 38.4 ± 9.1 | 8 |
| Unknown | ⩾3 GPRBC | Unknown | 17 | |
| 8 | Unknown | 75 to 95% mononuclear cells | 3.3 to 9.5 | 13 |
| 34 | ⩾3 HRBC | ⩾90% mononuclear cells | 13 ± 1 | 10 |
| 12 | ⩾3 HRBC | 96 ± 4.6 mononuclear cells¶ | 5.3 ± 3.3 | 14 |
| 12 | ⩾3 HRBC | 72.2 ± 15.4 mononuclear cells** | 35.5 ± 12.9 | |

*Number of adherent red blood cells.

†Human red blood cells.

‡Polymorphonuclear neutrophils.

§Guinea pig red blood cells.

¶21.9% of the mononuclear cells ingested latex beads after iron carbonyl/Ficoll–Hypaque separation.

**58% of the mononuclear cells ingested latex beads after Ficoll–Hypaque separation.

grafts between mature outbred unrelated mongrel dogs. Antisera that inhibit rosetting also were associated with prolonged graft survival.

Several commercially prepared ALS and anti-thymocyte sera were evaluated by Beall et al.[4] for their capacity to inhibit spontaneous rosetting. The effect of the anti-lymphocyte antisera was variable; some did not inhibit rosette formation. The interpretation of data from the effects of ALS on rosette

TABLE 2

VALUES FOR SPONTANEOUS ROSETTING CELLS FOUND IN CANINE HEMATOPOIETIC TISSUE

| Tissue | Animals | Criterion for rosetted cell (number of rosetted cells) | Rosettes, % | Ref. |
|---|---|---|---|---|
| Thymus | 3 pups (20 days old) | ⩾3 | 8 | 8 |
| | 2 | ⩾4 | 6 and 34 | 4 |
| | 4 fetal siblings (pooled sample) | ⩾4 | 3 to 14 (range) | |
| | 5 | ⩾4 | 9.2 ± 10.2 (SD) | 9 |
| | 16 | ⩾3 | 1 ± 0.5 (SEM) | 10 |
| | 3 pups (less than 12 weeks old) | ⩾3 | 3.8 ± 0.3 | 14 |
| Lymph node | | | 19 | 5 |
| | 3 pups (20 days old) | ⩾3 | 55 | 8 |
| | 2 | ⩾4 | 26 and 28 | 4 |
| | 4 fetal siblings (pooled sample) | ⩾4 | 32 | |
| | 8 | ⩾4 | 28.1 ± 10.5 | 9 |
| | 16 | ⩾3 | 6 ± 1.0 | 10 |
| Spleen | 3 pups (20 days old) | ⩾3 | 42 | 8 |
| | 2 | ⩾4 | 23 and 31 | 4 |
| | 4 fetal siblings (pooled sample) | ⩾4 | 3 to 15 (range) | |
| | 3 | ⩾4 | 38.1 ± 19.9 (SD) | 9 |
| | 14 | ⩾3 | 6.5 ± 1.5 (SEM) | 10 |
| Bone marrow | 2 | ⩾4 | 3 and 13 | 4 |
| | 3 pups (20 days old) | ⩾3 | 24 | 8 |
| | 12 | ⩾3 | 12 ± 2 (SEM) | 10 |

formation was further jeopardized by the procedure's lack of reproducibility as well as the observation that dead canine lymphocytes were able to form rosettes.

Anti-thymocyte globulin (ATG) was prepared by Zander, Boopalam, and Epstein[8] specifically as a reagent for characterization and/or identification of T lymphocytes. The globulins prepared from sera of rabbits that had been immunized with pup thymocytes were conjugated to fluorescein. This antisera stained 98% of the thymocytes and 68% of the peripheral lymphocytes, but only 8% of the thymus cells and 38% of the peripheral lymphocytes were E rosette positive.

## IDENTIFICATION OF T LYMPHOCYTES BY REACTION WITH MITOGEN

Certain mitogens, specifically concanavalin A (Con A) and phytohemagglutinin (PHA), are recognized as stimulating primarily the T-cell population,[15] and stimulation by these and other mitogens has been used frequently in the presumptive identification of lymphocyte subpopulations.

To identify the E rosetting cell, Beall et al.[4] separated rosette-forming cells from cells not forming rosettes with human red blood cells by centrifugation of the rosetted population on a Ficoll–Hypaque gradient and then treated both cell populations with the T-cell mitogen, PHA. The incorporation of $^{3}$H–thymidine, an indicator of stimulation, was not significantly increased in the PHA-treated rosette-forming group. Both control and PHA-treated rosetted cells incorporated more $^{3}$H–thymidine than did the nonrosetted group. Perhaps the increase in stimulation resulted from residual erythrocyte membrane on the rosette-forming cells following the separation procedure.

Whitacre and Lang,[7] with their efficient technique for obtaining pure lymphocyte populations, evaluated the functional response of lymphocytes to PHA and PWM. The cells responded to both mitogens.

Miller et al.[10] stimulated normal peripheral canine lymphocytes with Con A or PHA in either liquid culture or methylcellulose. Stimulation with Con A or PHA produced a twofold increase in the percentage of cells forming E rosettes. Furthermore, the resulting population possessed an Fc receptor, which is consistent with data of other investigators who have identified subpopulations of Fc receptor-bearing T cells in other species.[16–21]

In vitro exposure of canine thymocytes to PHA (Ref. 9) increased the percentage of E rosettes found in the thymus. Although the cells did not proliferate within 2 days after exposure to PHA, the rosetting percentages for the three animals tested increased from 4.4 to 66.1, from 2.6 to 37.0, and from 3.5 to 17.4. Apparently, stimulation of canine lymphocytes by mitogens considered specific for T cells results in an increase in cells capable of spontaneous rosette formation. The E rosette evidently identifies not the entire canine T-lymphocyte population, but a subset, namely those T cells which possess an Fc receptor.

## IDENTIFICATION OF B LYMPHOCYTES

The human and murine B lymphocyte has a variety of cell-surface markers which can be used to identify this population. The two most widely used markers are the receptor for the third component of complement (EAC3) and SIg's. The effective use of these markers requires proper preparation, including the removal of granulocytes and monocytes, both of which have surface receptors for C3 but also receptors for the heavy chain of IgG (Fc receptors). Erroneous results can occur if reagents used for identification of surface immunoglobulins contain immunoglobulin aggregates; such aggregates bind the Fc receptors, which provides positive results for SIg's. Similar considerations appear to be important in the canine system. The C3 receptor and the presence of SIg appear to be satisfactory markers for canine B cells given the constraints listed above.

## EAC3 Receptor as a Marker for B Lymphocytes

Lymphocyte receptors for EAC3 are most conveniently determined by using sheep erythrocytes that have been incubated with a subagglutinating amount of an IgM anti-sheep erythrocyte antibody and limiting amounts of complement (to minimize cell lysis). The IgM antibody is chosen instead of IgG to prevent rosette formation with cells with receptors for the Fc portion of IgG. A source of complement that is not readily lytic to the erythrocytes, such as murine or human complement, is preferred to guinea pig complement. The use of purified IgM antibody plus purified complement components C1, 4, 2, and 3 would limit any lysis or reaction with later complement components. As a practical matter, whole serum can be used if the quantity of C3 is reduced by appropriate dilutions. In human and murine systems, cells have at least two identifiable types of C3 receptors, one for C3b and one for the degradation-produced C3d. In practice, either will suffice for the identification of B cells.

If the EAC3 marker is to be used for B cells, the cell preparation should contain a minimum of nonlymphoid cell populations. After the lymphocytes have been incubated with EAC3 indicator cells, the preparations are examined, usually in a hemocytometer. At least 200 cells should be counted and those with three or more adherent erythrocytes enumerated. Controls with either uncoated or IgM-coated erythrocytes must be included to correct for nonspecific binding. Cytocentrifuge preparations of the rosette-forming cells can be stained to provide a better morphologic analysis of the cells with EAC3 receptors.

## Surface Immunoglobulins as a Marker for B Lymphocytes

Surface immunoglobulins are readily detectable by the use of fluorescent-labeled antisera. The antisera must have both high affinity and avidity and have the appropriate degree of fluorescent labeling. Nonspecific binding with receptors for the Fc portion of immunoglobulin molecules on a variety of cells, including T cells, can be eliminated by the use of either $F(ab')_2$, i.e., reagents whose Fc moiety has been removed, or antisera that have been ultracentrifuged to remove aggregates. In the human and murine systems, the principal immunoglobulins present on the cell surfaces are IgM and IgD with very small populations possessing surface IgG and IgA. Most studies in canine systems have been carried out with antisera directed against the entire gamma-globulin fraction rather than antisera specific for the individual heavy chains. Therefore no adequate information regarding the nature of the SIg on B cells in the dog is available, although the situation is presumably similar to that in man and mouse.

## VALUES FOR B CELLS IN PERIPHERAL BLOOD AND HEMATOPOIETIC TISSUES

### B Cells in Peripheral Blood

Reported values for B cells in canine peripheral blood ranged from 22% (Refs. 8 and 13) to 63% (Ref. 6) when the EAC3 receptor was used as a specific marker and from 16% (Ref. 4) to 20 to 30% (Ref. 10) or 56% (Ref. 6) when SIg's were used as a specific marker (Table 3). Because B-cell values obtained with either the EAC receptor or SIg's closely agree for each investigator, differences in lymphocyte preparation or in specific details of technique probably account for variations between studies. For the two largest series, Zander, Boopalam, and Epstein[8] and Miller et al.[10] have assayed more than 20 individual animals and report between 20 and 30% B lymphocytes in peripheral blood. Further, they both demonstrate excellent agreement between the EAC3 receptor and SIg-bearing cells. In the human system, normal B cells possess both markers under normal circumstances; the presence of either SIg alone or EAC3 marker alone is highly unusual.[22]

### B Lymphocytes in Lymph Nodes

Canine lymph nodes reportedly contain 35 to 42% cells if EAC3 receptors are used as the marker or 35 to 45% B cells if SIg's are used as the marker.[8–10] Similar values have been reported for humans.[23]

### B Lymphocytes in Spleen

Using the EAC3 marker to delineate B cells, Zander, Boopalam, and Epstein[8] reported that the spleen contained 18% B lymphocytes. Higher values, 34% B cells, were reported by Miller et al.[10] following the assay of 14 spleens, and Onions[9] reported 55% EAC receptor-positive cells (mean of 4 spleens). Using surface immunoglobulins as a criterion for B-cell identity, Zander,

TABLE 3

B-CELL MARKERS IN CANINE TISSUES

| Tissue | Marker | B cell, % | No. animals | Ref. |
|---|---|---|---|---|
| Peripheral blood | EAC3 | 22 | 17 | 8 |
| | | 35 to 63 | 3 | 6 |
| | | 36 ± 7.2 (SD) | 8 | 13 |
| | | 30 ± 1.5 (SEM) | 34 | 10 |
| | SIg | 30.5 | 20 | 8 |
| | | 35 to 56 | 3 | 6 |
| | | 18.4 ± 2.9 | 8 | 4 |
| | | 16 | 1 | 25 |
| | | 23.5 ± 2.5 (SEM) | 5 | 10 |
| Lymph nodes | EAC3 | 37 | 3(?) | 8 |
| | | 41.9 ± 13.6 (SD) | 8 | 9 |
| | | 35 ± 2.5 (SEM) | 16 | 10 |
| | SIg | 45 | 3(?) | 8 |
| | | 35 ± 3.0 (SEM) | 16 | 10 |
| Spleen | EAC3 | 18 | 3(?) | 8 |
| | | 54.6 ± 11.2 (SD) | 4 | 9 |
| | | 34 ± 2.5 (SEM) | 14 | 10 |
| | SIg | 49 | 3(?) | 8 |
| | | 41 ± 4.0 (SEM) | 14 | 10 |
| Bone marrow | EAC3 | 58 | 3(?) | 8 |
| | | 41 ± 4.0 (SEM) | 12 | 10 |
| | SIg | 71 | 3(?) | 8 |
| | | 36 ± 4.0 | 12 | 10 |
| Thymus | EAC3 | 0 | 3 | 8 |
| | | 3.5 ± 2.5 (SD) | 5 | 9 |
| | | 3 ± 0.5 (SEM) | 16 | 10 |
| | SIg | 0 | ? | 4 |
| | | 0 | 3(?) | 8 |
| | | 6.0 ± 1.0 | 16 | 10 |

Boopalam, and Epstein[8] identified 49% and Miller[10] identified 41% of the cells as B lymphocytes. These values are slightly higher than the values usually quoted for human systems.

## B Lymphocytes in Bone Marrow

In two studies of canine bone marrow, 36 to 71% of the lymphocytes were identified as B cells.[8,10] That value is considerably higher than the values reported in analogous human studies.[24]

### B Lymphocytes in the Thymus

Surface-marker studies of thymus cells have identified only 0 to 6% with surface immunoglobulins.[4,8,10] Reported values for cells bearing the C3 receptor have ranged from 0 to 3.5% (Refs. 8 to 10). The few B cells in the thymus presumably are transients rather than a permanent population.

## MITOGEN STUDIES ON B LYMPHOCYTES

In contrast to the murine B cells, no mitogens are specific for canine B cells. Miller et al.[10] studied nine different animals with lipopolysaccharide and reported no demonstrable effect. Holmberg, Manning, and Osburn[25] were unable to stimulate B-cell lymphomas with lipopolysaccharide. Dutta et al.[26] stimulated two B-cell lymphomas with either PHA, Con A, or PWM without any response. In the human, PWM apparently stimulates B cells as well as T cells. This mitogen was used by Whitacre and Lang,[7] who demonstrated some stimulation of peripheral populations, but there were no details given to define these populations as being T or B cells. Thus canine peripheral blood contains approximately 20 to 30% cells bearing B-cell markers, and between 30 and 50% of the cells in the lymph node, spleen, or bone marrow also are B cells. Differentiation of the immunoglobulin subclasses of these cells remains to be done.

Cell markers have been used to study canine disease, probably most successfully in canine lymphomas, and the results indicate that the vast majority of them are B cells (Chapter 15 of this volume).

## OTHER METHODS

Krakowka, Olsen, and Cockerell[27] used another method to identify B cells. Continuous lymphoblastoid cell lines, particularly CT45-S, express a B-cell EAC rosette marker. The media from the cell lines contain soluble lymphocyte membrane EAC3 receptors. The receptor was harvested by using a zymosan–rabbit complement–insoluble complex to extract the receptor complex from the media. The insolubilized material was washed extensively, emulsified in complete Freund's adjuvant, and injected into rabbits. The antisera formed by the rabbits to the receptor was used to inhibit the rosette formation between receptors and EAC3. This method appears promising for future development of more specific reagents that could be used for evaluating the EAC3 receptor and its relation to lymphocyte physiology.

## SUMMARY

Canine peripheral blood and lymphatic organs contain both T and B lymphocytes. The T lymphocytes are identified by rosetting with human erythrocytes. Rosetting gives variable results that could be attributed to two factors, namely, the degree of impurity of the test population and the variables in the rosetting technique itself. Because the total number of T lymphocytes in peripheral blood and the various lymphatic organs is, at best, about 36%, we believe the cells that form rosettes with human erythrocytes represent, not the entire T-cell population, but merely a subset thereof. This concept is further substantiated by evidence obtained following culture of lymphocyte with the T-cell mitogens phytohemagglutinin and concanavalin A. After culture, there was a twofold increase in an Fc receptor bearing population that rosetted with human erythrocytes.

B lymphocytes in the canine can be identified by the presence of a receptor for the third component of complement and by the presence of SIg's. Preparation of reagents specific for canine heavy and light chains will allow further identification of B-lymphocyte subpopulation and perhaps allow insight into differentiation of this lymphocyte population in the dog.

## REFERENCES

1. J. Bach, M. Dardenne, J. Dormont, and B. Antoine, A New In Vitro Test Evaluating Anti-Lymphocyte Serum Potency, *Transplant. Proc.*, 1: 403-407 (1969).
2. P. Brain, J. Gordon, and W. A. Willetts, Rosette Formation by Peripheral Lymphocytes, *Clin. Exp. Immunol.*, 6: 681-688 (1970).
3. G. E. Kelly, D. C. Mears, and A. G. R. Sheil, Relationship of the Immunosuppressive Potency of Antilymphocyte Globulin to Three In Vitro Tests: Close Correlation with Rosette Inhibition, *Transplantation*, 12: 443-447 (1971).
4. G. N. Beall, J. R. Benfield, S. R. Kruger, and P. E. Byfield, Canine Lymphocytes That Form Rosettes with Human Red Blood Cells, *J. Surg. Res.*, 17: 330-337 (1974).
5. D. E. Onions, B- and T-Cells in Canine Lymphosarcoma, *Vet. Rec.*, 97: 108 (1975).
6. C. A. Bowles, G. S. White, and D. Lucas, Rosette Formation by Canine Peripheral Blood Lymphocytes, *J. Immunol.*, 114: 399-402 (1975).
7. C. C. Whitacre and R. W. Lang, A Technique for Separation of Canine Lymphocytes and Their Use in Lymphocytotoxic, Blastogenic, and Rosette Assays, *Transfusion*, 15: 346-350 (1975).
8. A. R. Zander, N. Boopalam, and R. B. Epstein, Surface Markers on Canine Lymphocytes, *Transplant. Proc.*, 7: 369-373 (1975).
9. D. Onions, B- and T-Cell Markers on Canine Lymphosarcoma Cells, *J. Natl. Cancer Inst.*, 59: 1001-1006 (1977).
10. C. H. Miller, A. R. Carbonell, R. Peng, M. R. MacKenzie, and M. Shifrine, Cell Surface Markers on Canine Lymphocytes, *Am. J. Vet. Res.*, 39: 1191-1194 (1978).
11. T. Yang, J. R. Kennedy, and R. B. Andrews, Rosette Formation of Human Erythrocytes on Canine Transmissible Venereal Sarcoma Cells, *Am. J. Pathol.*, 83: 359-366 (1976).

12. T. J. Yang and J. R. Kennedy, Spontaneous Eosinophil–Erythrocyte Rosette Formation, *Lab. Invest.*, 35: 302-305 (1976).
13. S. Krakowka and D. J. Guyot, Rosette Formation Assays in Dogs: Lack of Specificity of E Rosettes for T Lymphocytes, *Infect. Immun.*, 17: 73-77 (1977).
14. R. E. Esser, M. Cosgrove, G. Scott, and A. B. Cosimi, Rosette Formation by Canine Leukocytes with Human Erythrocytes, *Transplantation*, 24: 223-225 (1977).
15. N. R. Ling and J. E. Kay, *Lymphocyte Stimulation*, American Elsevier Publishing Co., New York, 1975.
16. T. O. Yashida and B. Anderson, Evidence for a Receptor Recognizing Antigen Complexed Immunoglobulins on the Surface of Activated Mouse Thymus Lymphocytes, *Scand. J. Immunol.*, 1: 401-408 (1972).
17. J. A. van Boxel and D. L. Rosenstreich, Binding of Aggregated γ-Globulin to Activated T Lymphocytes in the Guinea Pig, *J. Exp. Med.*, 139: 1002-1012 (1974).
18. C. Soteria .es-Vlachos, M. I. C. Gyongyossy, and J. H. L. Playfair, Rosette Formation by Mouse Lymphocyte. III. Receptors for Immunoglobulin on Normal and Activated T Cells, *Clin. Exp. Immunol.*, 18: 187-192 (1974).
19. W. H. Fridman and P. Goldstein, Immunoglobulin-Binding Factor Present on and Produced by Thymus-Processed Lymphocytes (T Cells), *Cell Immunol.*, 11: 442-455 (1974).
20. V. Santana and J. L. Turk, Binding of Aggregated Human Immunoglobulins to Murine Thymocytes and T Cells Through Receptor for the Fc Region, *Immunology*, 28: 1173-1178 (1975).
21. A. Basten, J. F. A. P. Miller, N. L. Warner, R. Abraham, E. Chico, and J. Gainble, A Subpopulation of T Cells Bearing Fc Receptors, *J. Immunol.*, 115: 1159-1165 (1975).
22. M. R. MacKenzie and T. Paglieroni, Multiple Myeloma: An Immunologic Profile. I. Peripheral Blood Studies, *J. Immunol.*, 118: 1864-1871 (1977).
23. E. S. Jaffe, E. M. Shevach, M. M. Frank, C. W. Berard, and I. Green, Nodular Lymphoma—Evidence for Origin from Follicular B Lymphocytes, *New Engl. J. Med.*, 290: 813-819 (1974).
24. T. Paglieroni, M. R. MacKenzie, and V. Caggiano, Multiple Myeloma: An Immunologic Profile. II. Bone Marrow Studies, *J. Natl. Cancer Inst.*, 61: 943-950 (1978).
25. C. A. Holmberg, J. S. Manning, and B. I. Osburn, Canine Malignant Lymphomas: Comparison of Morphologic and Immunologic Parameters, *J. Natl. Cancer Inst.*, 56: 125-132 (1976).
26. S. K. Dutta, M. N. Novilla, M. K. Bumgardner, and A. Ingling, Lymphocyte Responsiveness to Mitogens and Quantitation of T and B Lymphocytes in Canine Malignant Lymphoma, *Am. J. Vet. Res.*, 39: 455-458 (1978).
27. S. Krakowka, R. G. Olsen, and G. L. Cockerell, Methods for Producing T- and B-Lymphocyte Receptor-Specific Antisera, *J. Immunol. Methods*, 14: 257-265 (1977).

antibody levels then declined during most of the regression sequence, which suggested immunoglobulin fixation at the tumor site. Histologically, regression was associated with neutrophilic infiltrates and focal necrosis in tumors. Levels of virus-neutralizing antibody paralleled those of cytotoxic antibody; their relationship in the circulation was not clear, but the activity of each was directed toward virus-specified antigens.

Ryan and Durick[27] reported that antiserum to normal dog kidney cells stimulated cell division in a transformed dog kidney cell line. The stimulation was obtained with the gamma-globulin fraction of the antiserum, and the effect could be removed by absorption with either the transformed dog kidney cells or the primary dog kidney cell cultures used to raise the antiserum. Apparently, the growth-enhancing activity of the antiserum was directed against normal tissue (histocompatibility) antigens.

## CELL-MEDIATED IMMUNITY TO CANINE TUMORS

Immunological rejection of most solid, nonhematologic neoplasms, like the rejection of normal tissue allografts, depends on the activity of the cellular immune system. One of the most intriguing phenomena in tumor immunology is the inhibition or blocking of potentially effective cellular response to tumor antigens by factors in the sera of cancer-bearing individuals.[28] In vitro studies have shown that small numbers of sensitized lymphocytes accelerate tumor growth, whereas large numbers of specifically immune lymphocytes inhibit tumor growth.[29-31] That finding prompted Prehn[30] and Prehn and Lappe[31] to hypothesize that the normal immune response to neoplasia may have a dual role; in the early stages the cell-mediated response might directly stimulate rather than inhibit tumor growth.

Cell-mediated immunity of dogs to their own tumors has been demonstrated by a variety of methods. Tsoi, Weiden, and Storb[32] studied lymphocyte reactivity to autochthonous tumor cells in dogs with lymphosarcomas or nonhematologic solid tumors. In approximately 62% of the cases, lymphocytes from tumor-bearing dogs were stimulated by their own neoplastic cells in vitro. The effect of autologous serum on the blastogenic response was inconsistent, and blocking activity was not conclusively demonstrated. In another report[33] the same investigators tested KCl extracts of lymphomas and other canine neoplasms for their ability to induce cell-mediated reactions. Lymphocytes from most tumor-bearing dogs underwent blastogenesis in response to autochthonous tumor extracts. Autologous serum failed to block the in vitro blastogenic response. Delayed hypersensitivity responses to intradermal injections of tumor extracts were an uncommon and unpredictable means of demonstrating cellular immunity.

Cell-mediated reactivity to canine mammary tumors was demonstrated by the leukocyte migration technique.[34] Mixing autochthonous tumor cells with peripheral blood leukocytes produced moderate to marked inhibition of migration.

Fidler et al.[35,36] and Brodey, Fidler, and Bech-Nielsen[37] reported that the addition of large numbers of lymphocytes from tumor-bearing dogs inhibited the growth of autochthonous tumor cells in vitro, whereas the addition of only a few sensitized lymphocytes stimulated tumor growth. These effects were independent of the source of lymphocytes; cells from regional (draining) lymph nodes, distant nodes, or peripheral blood of tumor-bearing dogs were equally effective in inhibiting or stimulating autochthonous tumors. Autologous serum blocked the inhibitory effect of large additions of lymphocytes. Also, sera from dogs with the same histologic type of tumor blocked allogeneic lymphocyte cytotoxicity, whereas normal dog serum or sera from dogs with a different type of neoplasm did not abrogate the cell-mediated response. The stimulation of tumor growth by a few lymphocytes was potentiated by autologous serum from dogs with progressively growing neoplasms.

Similar results were obtained in a study of canine oral melanomas.[38] The cultured tumor cells were destroyed by large numbers of autochthonous lymphocytes, whereas tumor cell growth was stimulated by a few lymphocytes. Autologous serum blocking activity was also demonstrated, and sera from dogs with the same histologic type of tumor cross-reacted to block cell-mediated immunity. In some cases the stimulation of tumor growth by small additions of lymphocytes was significantly enhanced by autologous serum.

Hess et al.[39] investigated cell-mediated responses to canine venereal tumors by assaying lymphocyte reactivity in mixed lymphocyte–tumor cultures. Venereal tumor cells stimulated both lymphocytes from normal dogs and lymphocytes from dogs with regressing tumors. The in vitro response of lymphocytes from dogs with rapidly progressing tumors, however, was significantly depressed by tumor cells, and autologous progressor serum further inhibited that response. The degree of lymphocyte reactivity during the progressive phase of tumor growth correlated with the clinical outcome of the neoplastic process. Tumors in animals that were significantly stimulated by tumor cells eventually regressed, whereas tumors in animals that were nonreactive in mixed lymphocyte–tumor cultures metastasized.

Yang, Roberts, and Jones[40] investigated the relationship between cellular infiltration of transplanted canine venereal sarcomas and growth or regression of the tumors. The percent viable lymphoreticular cells in tumor cell suspensions showed an inverse correlation with the tumor mass, which suggested that the degree of cellular infiltration might be a measure of the host immune response. However, the lymphoreticular cell mass in a tumor was a function of its surface area. Thus the relationship between lymphoreticular cell numbers and tumor mass did not necessarily indicate that the infiltrates were active immune effector cells.

Warren et al.[41] used lymphocyte-mediated cytotoxicity and indirect leukocyte migration–inhibition assays to study cell-mediated immunity in dogs with lymphomas or solid nonhematologic tumors. Lymphocytes from 8 of 14 dogs with lymphomas were cytotoxic to autochthonous tumor cells. In four of five animals tested, autologous serum was able to block lymphocyte-mediated cytotoxicity. Lymphoma cells from different dogs cross-reacted in 8 of 20 instances; lymphocytes from tumor-bearing dogs were cytotoxic to lymphoma cells from another dog. In half the cases, lymphocytes from tumor-bearing dogs produced migration–inhibition factor when cultured with autochthonous tumor cells.

Bech-Nielsen, Reif, and Brodey[42] correlated the clinical course of osteosarcoma with serum blocking activity after amputation in 9 of 11 dogs. Six of eight dogs with metastatic dissemination of tumor cells after amputation showed detectable serum blocking activity 1 to 2 months before overt metastasis. In two of three dogs that remained clinically free of osteosarcoma after surgery, serum blocking activity was never demonstrable. In the third animal, blocking activity in the serum disappeared 3 months after amputation.

## T- AND B-CELL MARKERS ON CANINE LYMPHOMA CELLS

Studies on the origin of canine lymphoid tumors indicate that B-cell neoplasias probably predominate. In two separate studies, Holmberg et al.[43,44] noted that the majority of tumors showed B-cell characteristics (surface immunoglobulin or complement receptors). Lymphomas of a T-cell nature were not found. A similar preponderance of B-cell malignancies was reported by Onions,[45,46] although T-cell lymphomas and tumors with mixtures of T and B cells were also identified. In four cases of thymic lymphosarcoma, two possessed rosette-forming (T) cells and two were devoid of surface receptors. Two cases of lymphoblastic leukemia consisted of cells devoid of surface markers. Results of another study[47] support the conclusion that B-cell lymphomas predominate in the dog. In two cases B cells constituted an average of 80% of the peripheral blood lymphocytes and 82% of the lymph node lymphocytes.

## IMMUNOLOGIC ABNORMALITIES ASSOCIATED WITH NEOPLASIA IN THE DOG

Abnormal responses of the humoral and/or cellular immune systems have been described in certain canine malignancies. Predictably, immunologic defects are most commonly associated with tumors of the lymphoid cell series. The most extensively documented abnormality is the excessive production of immunoglobulins in plasma cell malignancies. Canine myeloma has been associated with hyperproteinemia from IgA (Refs. 48 to 54) and IgM (Refs. 48, 49, and 55) and Bence Jones proteinemia and proteinuria.[48,49,56-58] IgA and IgM hyperproteinemias have also been reported in dogs with lymphocytic leukemia.[59]

Depressed immune responses have been described in tumor-bearing dogs. Weiden et al.[60] reported that dogs with lymphosarcoma had low serum levels of IgG and significant impairment of antibody responses to sheep erythrocytes and bacteriophage. In contrast, dogs with solid, nonhematologic tumors had normal levels of IgG and responded normally to both of those antigens. Reticuloendothelial system function, as assessed by phage clearance, was impaired in dogs with lymphosarcoma but not in dogs with solid tumors.

Owen, Bostock, and Halliwell[61] measured antibody responses to a dinitrophenyl (DNP)/human serum albumin conjugate in dogs with lymphosarcoma. The secondary (but not the primary) anti-DNP response in dogs with lymphosarcoma was significantly lower than that in normal dogs.

A deficiency in antibody production in dogs with mastocytomas has been described.[62] Measurement of early agglutinating antibody titers following inoculation of *Brucella abortus* vaccine indicated that the deficiency was most pronounced soon after antigen stimulation, which suggested a defect in IgM production.

Depressed cell-mediated responses in dogs with lymphosarcomas and in dogs with solid, nonhematologic tumors were described.[60] In vitro lymphocyte responses to phytohemagglutinin (PHA) and allogeneic lymphocytes were deficient in both groups of dogs, and response to tuberculin challenge after sensitization with BCG showed impaired delayed-type hypersensitivity reactions. Survival of allogeneic skin grafts in dogs with solid tumors did not differ from that in normal dogs, but survival of both first- and second-set skin grafts was prolonged in dogs with lymphosarcoma. In contrast, Owen, Bostock, and Halliwell[61] reported that rejection of skin grafts in 10 dogs with lymphosarcoma was essentially normal.

Bowles et al.[63] reported that lymphocytes from dogs bearing spontaneous lymphomas showed depressed in vitro proliferative responses to PHA, pokeweed mitogen, and streptolysin-O. Lymphocyte responses in mixed leukocyte cultures were also lower in dogs with lymphomas than in normal dogs. Those responses were further depressed by combination chemotherapy. Lymphocyte responsiveness began to return 2 to 3 days after chemotherapy but did not reach levels of normal, untreated dogs.

Various cell-mediated responses in dogs receiving a carcinogenic treatment of 4-aminobiphenol were monitored by Rippe et al.[64] Before the development of overt neoplastic disease (bladder cancer), those dogs showed no change in cellular immune status as measured by leukocyte migration inhibition, cytotoxicity tests, or phagocytosis. However, a trend toward lowered blastogenic responses was noted.

## IMMUNOTHERAPY

Some basic principles underlying cancer immunotherapy, with particular reference to the dog, were discussed by Brodey,[65] MacEwen,[66,67] Owen,[68] and Theilen.[69] More comprehensive reviews of cancer immunotherapy are available in several recently published monographs and texts.[70–73]

In the dog, as in other species, most approaches to the immunologic treatment of cancer have involved the following manipulations either alone or in combination: (1) nonspecific stimulation of the immune system with adjuvants such as bacillus Calmette-Guérin (BCG); (2) specific, active immunization with the patient's own tumor cells or antigenic components of the cells; or (3) passive immunotherapy with antiserum produced in another animal.

Hess et al.[74] reported that intralesional BCG therapy of canine venereal tumors was highly effective in causing regression of the tumors. Test animals received intralesional BCG in one of two growing tumors, whereas lesions of control animals received saline. Both injected and noninjected lesions of BCG-treated dogs regressed at a significantly faster rate than did lesions of control dogs. Tumor regression was correlated with lymphocyte responses in serial in vitro assays. Lymphocytes from BCG-treated dogs were significantly more responsive to tumor cells (in mixed lymphocyte–tumor cultures) than were those from control dogs. BCG therapy did not affect lymphocyte activation by PHA or numbers of T and B lymphocytes in peripheral blood.

Dogs with osteosarcomas treated with intravenous BCG after amputation of limb survived longer than untreated dogs.[75] In six dogs receiving intravenous injections of BCG or BCG plus X-irradiated autologous tumor cells after amputation of affected limbs, the appearance of metastasis was slightly delayed, and four of the six BCG recipients lived approximately 1 year after the limb had been amputated. The average survival time for 10 dogs with amputated limbs but not given BCG was only 18 weeks (range, 6 to 49 weeks).

Bech-Nielsen et al.[76] also studied the effects of BCG on the clinical course of osteosarcoma and correlated those effects with in vitro immune reactivity. The median survival time for dogs treated with intradermal injections of BCG after amputation of limbs was 40 weeks, which was significantly longer than the 13-week survival time for dogs whose limbs had been amputated for osteosarcoma but not given BCG. The cellular response of lymphocytes to tumor cells was essentially the same in BCG-treated and untreated dogs. Serum blocking activity decreased significantly soon after surgery in both groups of dogs, but 2 to 3 months after surgery, serum blocking activity was slightly lower in BCG-treated dogs than in controls. Eventually, the postsurgical serum blocking activity in BCG recipients rose to about the same level as that in the controls, correlating in time with radiographic appearance of metastasis.

A cutaneous melanosarcoma in one dog regressed after intralesional injection of BCG (Ref. 77). The dog subsequently showed delayed cutaneous reactivity to intradermal injections of BCG.

Efforts to treat canine malignancies with other nonspecific immunologic stimulants were reported by MacEwen.[66,67] In two dogs with lymphosarcoma, treatment with levamisole and thiabendazole was associated with a reduction in lymph node size. *Corynebacterium parvum* was administered by intralesional and intravenous routes to a dog with lymphosarcoma. All lesions regressed within a 2-month period, and no evidence of lymphosarcoma was found in skin or lymph nodes when the dog was later examined by necropsy.

Treatment of canine malignancies by active immunization has been attempted by several groups of workers. Sedlacek et al.[78–80] reported regression of canine mammary tumors following immunization with enzymatically modified, autochthonous tumor cells. Tumor cells were treated with neuraminidase, which removes cell surface sialic acid and thus alters the antigenicity of the cell membrane. Neuraminidase also has adjuvant activity. Subcutaneous reinjection of $2 \times 10^7$ treated cells caused a striking reduction in volume of the residual mammary tumor, but greater cell numbers ($1 \times 10^8$) were associated with accelerated tumor growth and metastasis. Tumor cells that had not been modified by neuraminidase treatment had no effect on tumor growth when injected back into the host.

Juillard, Boyer, and Snow[81] treated dogs with spontaneous lymphomas with neuraminidase-modified, irradiated autochthonous tumor cells repeatedly injected into the lymphatic vessels of the distal extremities. The intralymphatic route was chosen on the supposition that lymph nodes contain populations of unstimulated but potentially immunoreactive cells, which may require a certain threshold of antigen concentration before generating an effective response. All dogs so treated showed a reduction in volume of measurable peripheral lymph nodes. In subsequent experiments, Juillard et al.[82] injected irradiated but unmodified autochthonous lymphoma cells by the intralymphatic route and monitored the effect on distant lymphadenopathies by radiography. Intralymphatic infusions were associated with a marked shrinkage of the para-aortic nodes, but the reductive effect was temporary.

Benjamini et al.,[83] Crow,[84] and Theilen, Worley, and Benjamini[69] reported that dogs with lymphosarcoma that were treated first with chemotherapy and then with immunotherapy had longer survival times than dogs receiving chemotherapy alone. Immunotherapy consisted of a vaccine of chemically-modified (acetoacetylated) extracts of autochthonous lymphosarcoma cells. That approach to immunotherapy was based on the finding that certain chemical modifications of protein antigens lead to the formation of an immunogen that preferentially induced a cell-mediated response to the native antigen. By that theory chemically modified tumor preparations might selectively induce cell-mediated immunity to the tumor without evoking a humoral (blocking) response.

Passive cancer immune therapy involves the transfer of antibodies produced by one animal into the tumor-bearing animal. That tack proved unsuccessful for treating canine tumors in one series of experiments. Epstein et al.[85] reported unsuccessful attempts to treat canine tumors by passive immunization in cross-circulation experiments. Cross circulation was carried out in five cases between a tumor-bearing dog and a normal partner. Before cross circulation, the normal dog received intradermal and subcutaneous inoculations of tumor tissue from the partner. Four dogs had lymphosarcoma and one had renal cell carcinoma. The initial period of cross circulation was associated with an anaphylactic reaction in the dog with the tumor, and death in anaphylaxis occurred in three dogs. Transient and incomplete regression of tumors was noted in two dogs, but both eventually died with extensive tumor involvement.

Passive treatment of lymphoid neoplasias with normal blood constituents has been reported by MacEwen,[66] who states that the blood or plasma of normal dogs contains an antileukemic factor, which may be a component of the complement system. A dog with lymphoblastic leukemia confined to the bone marrow was given repeated infusions of whole blood and fresh plasma over a 2-week period. That treatment reportedly produced a remission, as evidenced by a progressive decrease in lymphoblasts and lymphocytes in bone marrow aspirated. The dog was still alive 20 weeks after diagnosis, a survival period considerably longer than that usually associated with the malignancy. MacEwen[67] further tested blood or plasma therapy on five dogs with lymphosarcoma; a complete remission was observed in one of those dogs. (That dog may have been the same animal described in the previous report,[66] but relevant data are lacking.)

## CONCLUSIONS

Investigations with canine neoplasms clearly demonstrate that tumors in the dog, like those in other species, possess antigens, which evoke a specific and multicomponent immune response. That response involves both humoral and cellular elements, which interact in a complex and sometimes opposing manner. Cell-mediated immunity plays a major role in tumor rejection; the in vitro killing of neoplastic cells by specifically sensitized lymphocytes has been demonstrated in several tumor systems. Serum-mediated immunity may play a lesser role in immune defense against cancer, but the demonstration of blocking factors in the sera of cancer-bearing individuals has made the study of humoral immune responses an equally important aspect of tumor immunology. If, as Heppner[28] states, the total immune response in vivo depends on a "balance" of humoral and cellular elements, the cancer immunologist must learn to tip that balance in favor of the host by manipulating each of the elements independently. Tumor–host relationships defined thus far in the dog indicate that the canine species provides a suitable model for pursuit of that goal.

## ACKNOWLEDGMENT

This work was supported by the U. S. Department of Energy.

## REFERENCES

1. E. J. Foley, Antigenic Properties of Methylcholanthrene-Induced Tumors in Mice of the Strain of Origin, *Cancer Res.,* 13: 835-837 (1953).
2. G. Moller and E. Moller, The Concept of Immunological Surveillance Against Neoplasia, *Transplant. Rev.,* 28: 3-16 (1976).
3. J. Rygaard and C. O. Povlsen, Is Immunological Surveillance Not a Cell-Mediated Immune Function? *Transplantation,* 17: 135-136 (1974).
4. J. Rygaard and C. O. Povlsen, The Nude Mouse vs. the Hypothesis of Immunological Surveillance, *Transplant. Rev.,* 28: 43-61 (1976).
5. O. Haller, M. Hansson, R. Kiessling, and H. Wizzell, Role of Nonconventional Natural Killer Cells in Resistance Against Syngeneic Tumour Cells In Vivo, *Nature,* 270: 609-611 (1977).
6. R. O. Jacoby and R. A. Griesemer, Immunologic Injury to Dogs, *Adv. Vet. Sci. Comp. Med.,* 14: 149-207 (1978).
7. F. M. Burnet, Immunological Surveillance in Neoplasia, *Transplant. Rev.,* 7: 3-25 (1971).
8. T. E. Starzl, I. Penn, C. W. Putnam, C. G. Growth, and C. G. Halgrimson, Iatrogenic Alterations of Immunologic Surveillance in Man and Their Influence on Malignancy, *Transplant. Rev.,* 7: 112-145 (1971).
9. W. L. Joseph, F. Melewicz, and D. L. Morton, Spontaneous Development of Mammary Adenocarcinoma Following Prolonged Immunosuppression in the Dog, *Cancer Res.,* 30: 2606-2608 (1970).
10. J. R. Benfield, E. C. Shors, M. Okita, K. Matsumura, and A. H. Cohen, Effects of Localized Chemical Carcinogenesis and Immunosuppression upon Bronchial Preneoplasia in the Dog, *J. Thoracic Cardiovasc. Surg.,* 74: 752-760 (1977).
11. D. Cohen, The Biological Behavior of the Transmissible Venereal Tumour in Immunosuppressed Dogs, *Eur. J. Cancer,* 9: 253-258 (1973).
12. G. R. Betton and L. N. Owen, Proceedings: The Growth of Spontaneous Melanoma Allografts in Immunosuppressed Neonatal Dogs, *Br. J. Cancer,* 30: 183-184 (1974).
13. G. R. Betton and L. N. Owen, Allogeneic Grafts of Spontaneous Canine Melanomas and Their Cell Culture Strains in Neonatal Immunosuppressed Dogs, *Br. J. Cancer,* 34: 374-380 (1976).
14. C. A. Bowles, W. T. Kerber, S. R. S. Rangan, W. A. Woods, and E. M. Jensen, Studies of a Transplantable Canine Sarcoma, in *Unifying Concepts of Leukemia,* R. M. Dutcher and L. Chieco-Bianchi (Eds.), Bibl. Haemat. No. 39, pp. 149-157, S. Karger, Basel, 1973.
15. L. N. Owen, Transplantation of Canine Lymphocytic Leukemia, in *Unifying Concepts of Leukemia,* R. M. Dutcher and L. Chieco-Bianchi (Eds.), Bibl. Haemat. No. 39, pp. 139-148, S. Karger, Basel, 1973.
16. M. Shifrine et al., Transplantation of Radiation-Induced Canine Myelomonocytic Leukemia, in *Unifying Concepts of Leukemia,* R. M. Dutcher and L. Chieco-Bianchi (Eds.), Bibl. Haemat. No. 39, pp. 158-169, S. Karger, Basel, 1973.
17. C. A. Bowles, W. Hagen, J. Ditmore, W. T. Kerber, W. A. Woods, and E. M. Jensen, Immunofluorescent Studies of Cultured Canine Tumor Cells, *Int. J. Cancer,* 10: 28-35 (1972).

18. H. D. Snow and W. D. Winters, Humoral Antibodies to Tumor-Associated Antigens, *Calif. Vet.*, 28: 12-15 (1974).
19. W. D. Winters and H. D. Snow, Immunodiffusion Studies of Canine Lymphosarcoma Antigens, *J. Natl. Cancer Inst.*, 53: 1027-1032 (1974).
20. W. D. Winters and H. D. Snow, Precipitating Antibodies to Canine Lymphosarcoma-Associated Antigens, *Am. J. Vet. Res.*, 35: 1333-1335 (1974).
21. D. Cohen, Detection of Humoral Antibody to the Transmissible Venereal Tumor of the Dog, *Int. J. Cancer*, 10: 207-212 (1972).
22. R. B. Epstein and B. T. Bennett, Histocompatibility Typing and Course of Canine Venereal Tumors Transplanted into Unmodified Random Dogs, *Cancer Res.*, 34: 788-793 (1974).
23. B. T. Bennett, Y. Taylor, and R. Epstein, Segregation of the Clinical Course of Transmissible Venereal Tumor with DL-A Haplotypes in Canine Families, *Transplant. Proc.*, 7: 503-505 (1975).
24. B. T. Bennett, K. M. Debelak-Fehir, and R. B. Epstein, Tumor-Blocking and -Inhibitory Serum Factors in the Clinical Course of Canine Venereal Tumor, *Cancer Res.*, 35: 2942-2947 (1975).
25. W. E. Beschorner, A. D. Hess, S. T. Nerenberg, and R. B. Epstein, Characterization of Serum Blocking and Inhibition of Canine Venereal Tumors, *Proc. Am. Assoc. Cancer Res.*, 18: 187 (1977).
26. D. O. Slauson, B. I. Osburn, M. Shifrine, and D. L. Dungworth, Regression of Feline Sarcoma Virus Induced Sarcomas in Dogs. II. Immunologic Investigations, *J. Natl. Cancer Inst.*, 54: 371-377 (1975).
27. W. L. Ryan and M. A. Durick, Antibody Stimulation of Tumor Cells, *Proc. Soc. Exp. Biol. Med.*, 155: 420-423 (1977).
28. G. H. Heppner, Blocking Antibodies and Enhancement, *Ser. Haematol.*, 5: 41-66 (1972).
29. R. T. Prehn, The Immune Reaction as a Stimulation of Tumor Growth, *Science*, 176: 170-171 (1972).
30. R. T. Prehn, Do Tumors Grow Because of the Immune Response of the Host? *Transplant. Rev.*, 28: 34-42 (1976).
31. R. T. Prehn and M. A. Lappe, An Immunostimulation Theory of Tumor Development, *Transplant. Rev.*, 7: 26-54 (1971).
32. M. S. Tsoi, P. L. Weiden, and R. Storb, Lymphocyte Reactivity to Autochthonous Tumor Cells in Dogs with Spontaneous Malignancies, *Cell Immunol.*, 13: 431-439 (1974).
33. M. S. Tsoi, P. L. Weiden, and R. Storb, Immunologic Reactivity to Soluble Autochthonous Tumor Extracts in Dogs with Spontaneous Malignancies, *J. Immunol.*, 116: 1134-1139 (1976).
34. M. J. Ulvund, Cellular Immunity to Canine Mammary Tumor Cells Demonstrated by the Leukocyte Migration Technique, *Acta Vet. Scand.*, 16: 95-114 (1975).
35. I. J. Fidler, R. S. Brodey, and S. Bech-Nielsen, In Vitro Immune Stimulation-Inhibition to Spontaneous Canine Tumors of Various Histologic Types, *J. Immunol.*, 112: 1051-1060 (1974).
36. I. J. Fidler, R. W. McWilliams, S. Bech-Nielsen, and M. B. Budman, Role of the Regional Lymph Node in Neoplasia: Cellular Mediated Reactivity In Vitro by Autologous Regional or Distant Lymph Nodes or Peripheral Blood Lymphocytes of Dogs with Spontaneous Neoplasma, *Immunol. Commun.*, 4: 325-335 (1975).
37. R. S. Brodey, I. J. Fidler, and S. Bech-Nielsen, Correlation of In Vitro Immune Response with Clinical Course of Malignant Neoplasia in Dogs, *Am. J. Vet. Res.*, 36: 75-80 (1975).

38. I. J. Fidler and R. W. McWilliams, Immune Reactivity In Vitro to Spontaneous Canine Oral Tumors, *J. Med.*, 6: 323-336 (1975).

39. A. Hess, B. Cunningham, B. Taylor-Bennett, and R. Epstein, In Vitro Correlates of the In Vivo Course of the Canine Transmissible Venereal Tumor Studied by Mixed Lymphocyte-Tumor Cultures, *Transplant. Proc.*, 7: 507-511 (1975).

40. T.-J. Yang, R. S. Roberts, and J. B. Jones, Quantitative Study of Lymphoreticular Infiltration into Canine Transmissible Venereal Sarcoma, *Virchows Arch. B, Cell Path.*, 20: 197-204 (1976).

41. R. P. Warren, M. S. Tsoi, B. H. Henderson, P. L. Weiden, and R. Storb, Demonstration of Tumor-Associated Antigens in Dogs, *Transplant. Proc.*, 7: 481-484 (1975).

42. S. Bech-Nielsen, J. S. Reif, and R. S. Brodey, Pre- and Post-Operative Studies of In Vitro Cell-Mediated Reactivity in Canine Osteosarcoma, *Am. J. Vet. Res.*, 39: 87-93 (1978).

43. C. A. Holmberg, J. S. Manning, and B. I. Osburn, Canine Malignant Lymphomas: Comparison of Morphologic and Immunologic Parameters, *J. Natl. Cancer Inst.*, 56: 125-135 (1976).

44. C. A. Holmberg, J. S. Manning, and B. I. Osburn, Malignant Lymphoma with B-Lymphocyte Characteristics in Dogs, *Am. J. Vet. Res.*, 38: 1877-1879 (1977).

45. D. E. Onions, B- and T-Cells in Canine Lymphosarcoma, *Vet. Rec.*, 97: 108 (1975).

46. D. Onions, B- and T-Cell Markers on Canine Lymphoma Cells, *J. Natl. Cancer Inst.*, 59: 1001-1006 (1977).

47. S. K. Dutta, M. N. Novilla, M. K. Bumgardner, and A. Ingling, Lymphocyte Responsiveness to Mitogen and Quantitation of T and B Lymphocytes in Canine Malignant Lymphoma, *Am. J. Vet. Res.*, 39: 455-458 (1978).

48. A. I. Hurvitz, J. M. Kehoe, and J. D. Capra, Characterization of Three Homogeneous Canine Immunoglobulins, *J. Immunol.*, 107: 648-654 (1971).

49. A. I. Hurvitz and R. Prata, Immunoproliferative Disorders with Monoclonal Gammopathies in the Dog, *J. Am. Vet. Med. Assoc.*, 156: 1246-1247 (1970).

50. W. Medway, W. T. Weber, J. A. O'Brien, and L. J. Krawitz, Multiple Myeloma in a Dog, *J. Am. Vet. Med. Assoc.*, 150: 386-395 (1967).

51. J. H. Rockey and R. M. Schwartzman, Skin Sensitizing Antibodies: A Comparative Study of Canine and Human PK and PCA Antibodies and a Canine Myeloma Protein, *J. Immunol.*, 98: 1143-1151 (1967).

52. V. J. Shepard, W. J. Dodds-Laffin, and R. J. Laffin, Gamma A Myeloma in a Dog with Defective Hemostasis, *J. Am. Vet. Med. Assoc.*, 160: 1121-1127 (1972).

53. R. M. Shull et al., Serum Hyperviscosity Syndrome Associated with IgA Multiple Myeloma in Two Dogs, *J. Am. Anim. Hosp. Assoc.*, 14: 58-70 (1978).

54. G. Virella, R. J. Slappendel, and J. Goudswaard, Multiple Myeloma, IgA Cryoglobulinemia and Serum Hyperviscosity in a Dog, *Int. Arch. Allergy Appl. Immunol.*, 55: 537-541 (1977).

55. J. D. Capra and A. I. Hurvitz, Characterization of a Monoclonal Canine Macroglobulin, *J. Immunol.*, 105: 949-955 (1970).

56. A. I. Hurvitz, J. M. Kehoe, J. D. Capra, and R. Prata, Bence Jones Proteinemia and Proteinuria in a Dog, *J. Am. Vet. Med. Assoc.*, 159: 1112-1116 (1971).

57. C. A. Osborne, V. Perman, J. H. Sauttner, J. B. Stevens, and G. F. Hanlon, Multiple Myeloma in the Dog, *J. Am. Vet. Med. Assoc.*, 153: 1300-1319 (1968).

58. N. L. Warner, M. Potter, and D. Metcalf, Multiple Myeloma and Related Immunoglobulin-Producing Neoplasma, in *International Union Against Cancer*, Vol. 13, Springer Verlag, 1974.

59. E. G. MacEwen, A. I. Hurvitz, and A. Haynes, Hyperviscosity Syndrome Associated with Lymphocytic Leukemia in Three Dogs, *J. Am. Vet. Med. Assoc.*,170: 1309-1312 (1977).

60. P. L. Weiden et al., Immune Reactivity in Dogs with Spontaneous Malignancy, *J. Natl. Cancer Inst.*, 53: 1049-1056 (1974).

61. L. N. Owen, D. E. Bostock, and R. E. W. Halliwell, Cell-Mediated and Humoral Immunity in Dogs with Spontaneous Lymphosarcoma, *Eur. J. Cancer,* 11: 187-191 (1975).
62. E. B. Howard, Immunologic Defect in Mastocytoma-Bearing Dogs, *J. Am. Vet. Med. Assoc.,* 151: 1308-1310 (1967).
63. C. A. Bowles, D. Lucas, M. I. Bull, and R. G. Graw, Longitudinal Study of Cellular Immune Reactivity in Dogs with Spontaneous Lymphoma Following Combined Modality Therapy, *Am. Assoc. Cancer Res. Proc.,* 16: 146 (1975).
64. D. F. Rippe, G. Ortiz-Muniz, N. L. Block, C. M. Lynne, R. B. Shouger, and M. M. Sigel, Urinary Bladder Carcinogenesis in the Dog: Preliminary Studies on Cellular Immunity, *Transplant. Proc.,* 7: 495-501 (1975).
65. R. S. Brodey, Immunological Therapy of Canine Neoplasma, in *Proceedings of the 22nd Annual Gaines Veterinary Symposium,* pp. 16-19, 1973.
66. E. G. MacEwen, General Concepts of Immunotherapy of Tumors, *J. Am. Anim. Hosp. Assoc.,* 12: 363-373 (1976).
67. E. G. MacEwen, An Immunologic Approach to the Treatment of Cancer, *Vet. Clin. North Am.,* 7: 65-75 (1977).
68. L. N. Owen, Therapy of Neoplasia in Domestic Animals, *Adv. Vet. Sci. Comp. Med.,* 20: 224-252 (1976).
69. G. H. Theilen, M. Worley, and E. Benjamini, Chemoimmunotherapy for Canine Lymphosarcoma, *J. Am. Vet. Med. Assoc.,* 170: 607-610 (1977).
70. J. E. Harris and J. G. Sinkovics, *The Immunology of Malignant Disease,* C. V. Mosby Co., Saint Louis, 1976.
71. G. Lamoureux, R. Turcotte, and V. Portelance (Eds.), *BCG in Cancer Immunotherapy,* Grune & Stratton, New York, 1976.
72. W. D. Terry and D. Windhorst, Immunotherapy of Cancer: Present Status of Trials in Man, in *Progress in Cancer Research and Therapy,* Vol. 6, Raven Press, New York, 1978.
73. J. Wybran and M. J. Staquet, Clinical Tumor Immunology, in *Symposium of Clinical Tumor Immunology,* Brussels, 1975, Pergamon Press, New York, 1976.
74. A. D. Hess, R. Catchatourian, A. R. Zander, and R. B. Epstein, Intralesional Bacillus Calmette-Guérin Immunotherapy of Canine Venereal Tumors, *Cancer Res.,* 37: 3990-3994 (1977).
75. L. N. Owen and D. E. Bostock, Effects of Intravenous BCG in Normal Dogs and in Dogs with Spontaneous Osteosarcoma, *Eur. J. Cancer,* 10: 775-780 (1974).
76. S. Bech-Nielsen, R. S. Brodey, I. J. Fidler, D. A. Abt, and J. S. Rief, The Effect of BCG on In Vitro Immune Reactivity and Clinical Course in Dogs Treated Surgically for Osteosarcoma, *Europ. J. Cancer,* 13: 33-41 (1977).
77. R. L. Grier, C. O. Thoen, and D. L. Harris, Regression of Cutaneous Melanosarcoma Following Intralesional Mycobacterium Bovis BCG Injection: A Case Report, *J. Am. Anim. Hosp. Assoc.,* 14: 76-81 (1978).
78. H. H. Sedlacek, H. Meesmann, and F. R. Seiler, Regression of Spontaneous Mammary Tumors of Dogs After Injection of Neuraminidase-Treated Tumor Cells, *Behring Inst. Mitteilungen,* 55: 349-355 (1974).
79. H. H. Sedlacek, H. Meesmann, and F. R. Seiler, Regression of Spontaneous Mammary Tumors in Dogs After Injection of Neuraminidase-Treated Tumor Cells, *Int. J. Cancer,* 15: 409-416 (1975).
80. H. H. Sedlacek, F. R. Seiler, and H. G. Schwick, Neuraminidase and Tumor Immunotherapy, *Klin. Wochenschr.,* 55: 199-214 (1977).
81. G. J. F. Juillard, P. J. J. Boyer, and H. D. Snow, Intralymphatic Infusion of Autochthonous Tumor Cells in Canine Lymphoma, *Int. J. Radiat. Oncol. Biol. Phys.,* 1: 479-503 (1976).

82. G. J. F. Juillard et al., Regional Intralymphatic Infusion (ILI) of Irradiated Tumor Cells with Evidence of Distant Effects, *Cancer,* 39: 126-130 (1977).
83. E. Benjamini, G. H. Theilen, M. Torten, S. Fong, S. Crow, and A. M. Henness, Tumor Vaccines for Immunotherapy of Canine Lymphosarcoma, *Ann. N.Y. Acad. Sci.,* 277: 305-312 (1976).
84. S. E. Crow, Chemo-Immunotherapy of Canine Lymphosarcoma, *Am. Assoc. Cancer Res. Proc.,* 17: 212 (1976).
85. R. B. Epstein, T. C. Graham, R. Storb, and E. D. Thomas, Studies of Marrow Transplantation, Chemotherapy and Cross-Circulation in Canine Lymphosarcoma, *Blood,* 37: 349-359 (1971).

# Natural Cell-Mediated Cytotoxicity in the Canine

# 6

G. R. BETTON*
*Laboratory for Experimental Surgery, Erasmus University, Rotterdam, The Netherlands*

The canine provides a unique model for the study of neoplasia in man since that species exists as a large outbred population living in man's environment to an advanced age and suffering spontaneous neoplasms of a diversity paralleling that in man. Cases are presented for treatment with facilities similar to those in use in human medicine. Material from tumor cases can be obtained through stages of surgery, metastasis, radiotherapy, and chemotherapy and/or immunotherapy in a species where experimentation is possible.

Human tumor immunology, after a promising upsurge in the 1960's, has become a complex subject with in vitro results frequently showing a lack of specificity on more careful examination. The nature of the putative tumor specific antigen advanced in the immune surveillance theory[1] remains elusive.[2]

## SPECIFICITY OF CELL-MEDIATED IMMUNITY TO CANINE TUMORS

The area of canine tumor immunology (see Chapter 5) has been slow in developing. Evidence for antigen expression on canine tumor cells has been derived from the results of leukocyte migration inhibition tests,[3,4] lymphocyte blastogenesis,[5] colony inhibition,[6] and chromium release cytotoxicity[3] assays.

---

*Present address: ICI Pharmaceuticals Division, Mereside, Alderley Park, Macclesfield, Cheshire, United Kingdom.

Other studies failed to demonstrate histologic-type specificity of immune responses in tumor-bearing dogs.[7-9] The transmissible venereal tumor (see Chapter 19) represents an allograft in which canine transplantation antigens (DLA) rather than tumor antigens generate an immune response.

## NATURAL KILLER (NK) CELL ACTIVITY

### Discovery

During the search for histologic-type specific antigens expressed on allogeneic tumor cells in vitro, objective analysis of the data from cell-mediated cytotoxicity assays (48- to 72-hr microtest) showed a dominant nonselective component[10] in reactions previously interpreted as being tumor specific in early poorly controlled studies. Fractionation studies showed the effector "N" cell[11] to lack T or B lymphocyte markers. Workers using the chromium ($^{51}Cr$) release assay of cell-mediated immunity have used the term natural killer (NK) or spontaneous lymphocyte-mediated cytotoxicity to describe the nonspecific cell-mediated cytotoxicity observed in both tumor bearers and healthy controls.

### Species Distribution of NK Cell Activity

Recent studies on NK cell activity have revealed a wide species distribution, including the mouse,[12,13] rat,[14,15] guinea pig,[16,17] hamster,[18] and dog[7] as well as humans.[19,20]

### Variation of NK Cell Activity with Time

Natural cytotoxicity develops in the first few weeks of life[15,21,22] and declines in old age. Healthy human donors showed fairly stable individual levels of NK cell activity under carefully controlled sequential experiments.[23]

Sequential testing of a series of dogs for NK and mitogen-induced cell-mediated cytoxicity (MICC) showed considerable week-to-week variations (Fig. 1). The initiation of NK cell activity in the dog has not been determined.

### NK Cell Activity in Cancer Patients

The discovery of cell-mediated cytotoxicity against tumor target cells in both cancer patients and healthy controls has naturally raised the question of the in vivo significance of NK cell activity[19] in relation to classical concepts of specific immunological surveillance against tumor cells. The athymic nude mouse, which lacks a T lymphocyte immune response but which has good NK

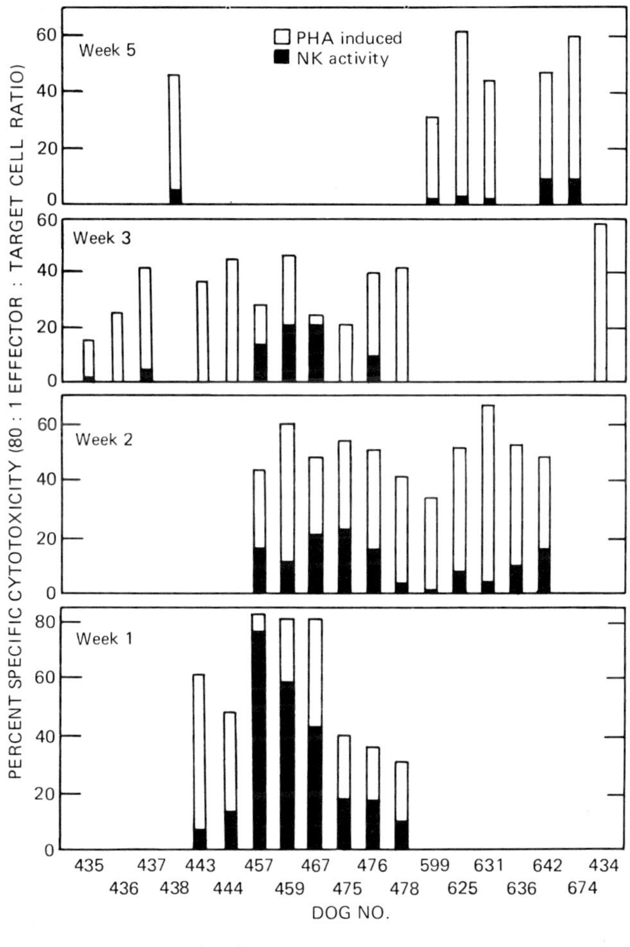


**Fig. 1** Variation of natural killer (NK) cell activity and PHA mitogen-induced cell-mediated cytotoxicity with time in individual dogs.

cell activity, does not show an increased frequency of neoplasia.[24] Also, NK cell activity in vitro has been correlated with tumor resistance in vivo; the transfer of bone-marrow cells containing NK cell precursors from high NK cell activity strains to recipients with low NK cell activity and low tumor resistance reconstitutes tumor resistance.[25]

A decline in NK cell activity with tumor progression has been reported in man[26,27] and in a small series of melanoma cases in the dog.[7]

## Pharmacological Suppression of NK Cell Activity

The anti-macrophage agents silica and carrageenan depress NK cell activity,[28] but whether the effect is direct or indirect is unknown. Cyclophosphamide and cortisone also depress NK cell activity.[28-30] A panel of cancer chemotherapeutic agents varied in ability to suppress NK cell activity in the mouse.[31] Radioactive strontium, a chronic bone-marrow suppressant, also depressed NK cell activity.[32]

### Pharmacological Enhancement of NK Cell Activity

A wide range of agents generate NK cell activity in vivo. Poly-ionosinic acid–cytosinic acid (poly IC), an interferon inducer, elevates NK cell activity.[30] In vitro, interferon directly stimulated NK cell activity.[33]

Tumor cells liberate prostaglandins capable of inhibiting NK cell activity, and agents that block prostaglandin synthetase (indomethacin, fenclozic acid, acetylsalicylic acid, and 2-6 xylenol) enhance NK cell activity against tumor cell targets.[34]

### Effects of Biological Agents on NK Cell Activity

Viral agents, such as lymphocytic choriomeningitis virus (LCMV) of mice, enhance NK cell activity;[35,36] nonspecific cytotoxicity precedes the appearance of T lymphocytes specifically cytotoxic against LCMV.[37] Viruses may affect NK cell activity via interferon induction.

Bacterial agents, such as bacillus Calmette-Guérin (BCG) and *Corynebacterium parvum*, have found clinical applications in cancer immunotherapy[38] and also elevate NK cell activity in the mouse,[39-41] rat,[36] dog,[42] and man.[43] Because BCG immunotherapy has been used in the dog to treat osteosarcoma,[44,45] mammary carcinoma,[46] and transmissible venereal tumor,[47] mediation of its therapeutic efficacy by NK cell activation merits further study.[40]

Antigenic challenge with normal or neoplastic allogeneic cells also elevates NK cell activity in vitro[39] and protects against tumor challenge.[48]

## THE MECHANISM OF NK CELL ACTIVITY

### In Vitro Assays for NK Cell Cytotoxicity

In the area of tumor immunology, two principal assay systems have been used. The microtest assay, with high effector : target cell ratios over a 48- to 72-hr period, is scored optically (visual or computer) for the relative survival of adherent target cells. The $^{51}Cr$ release assay is normally performed for a short incubation period (4 to 8 hr) at lower effector : target cell ratios, and cytotoxicity is quantitated by measuring the release of $^{51}Cr$-labeled material from lysed target cells into the medium in a gamma scintillation counter. A comparison of available assays is discussed by Bean et al.[49]

Most in vitro tests have used established cell lines derived from normal or malignant tissue. Such lines yield reproducible results, but prolonged growth in vitro renders the designation of lines as "normal" rather questionable.

A range of canine tumor and normal lines is available, and these lines have performed satisfactorily as target cells[7] in the $^{51}$Cr release assay (Fig. 2). Established cell cultures susceptible to nonspecific NK cell cytotoxicity include melanomas, mammary carcinomas, osteosarcoma, normal thymus fibroblasts (Fig. 3), and normal kidney epithelial cultures. In contrast to early cell cultures of primary tumors in the dog,[7] a melanoma line transplanted to an allogeneic immunosuppressed recipient was also susceptible to NK lysis when reestablished in vitro for a short period. The CT45 canine thymocyte line (supplied by Dr. D. E. Onions, University of Glasgow) was always resistant to lysis.

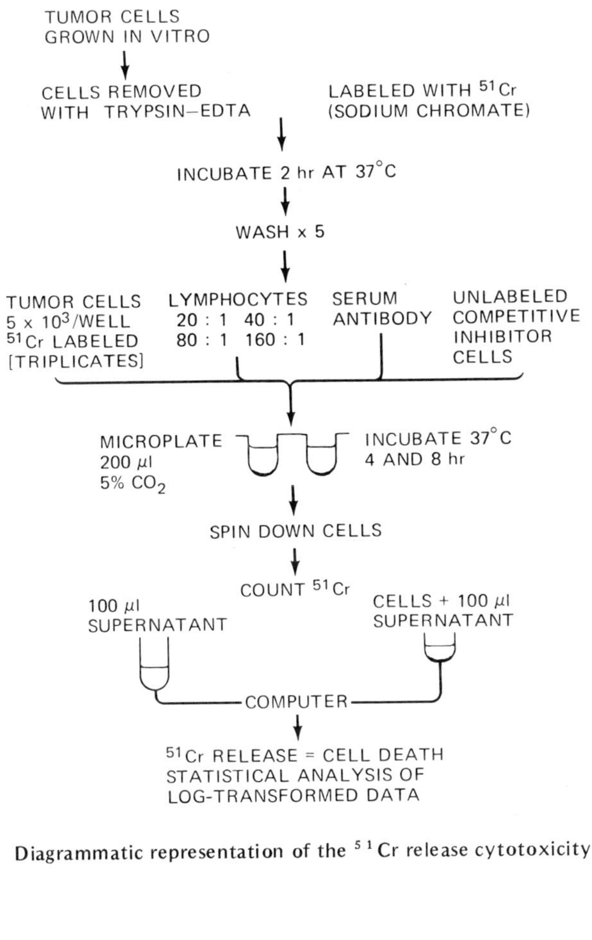


**Fig. 2 Diagrammatic representation of the $^{51}$Cr release cytotoxicity assay.**

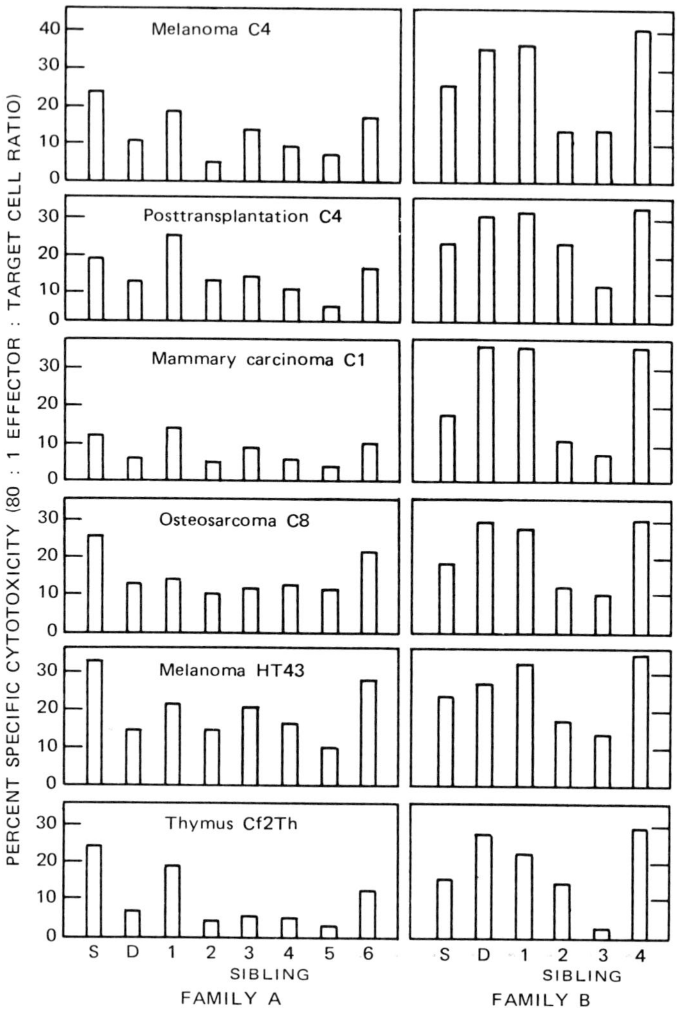


Fig. 3 Level of natural killer cell activity against a variety of target cells for dogs from two families.

The cell lines used in the $^{51}Cr$ release assay showed stable morphological and cultural characteristics. Tumor cells showed classical loss of contact inhibition (Figs. 4a to 4d), whereas kidney epithelium cultures ceased proliferation once confluent (Fig. 4e). A proportion of fresh tumor explants failed to maintain in vitro growth; only senescent fibroblasts were left (Fig. 4f). Tumorigenicity in vivo[50] and membrane characteristics of transformed cells[51] have been previously reported for the tumor lines used. Maintenance of *Mycoplasma*-free cultures is an essential prerequisite for target cell application.

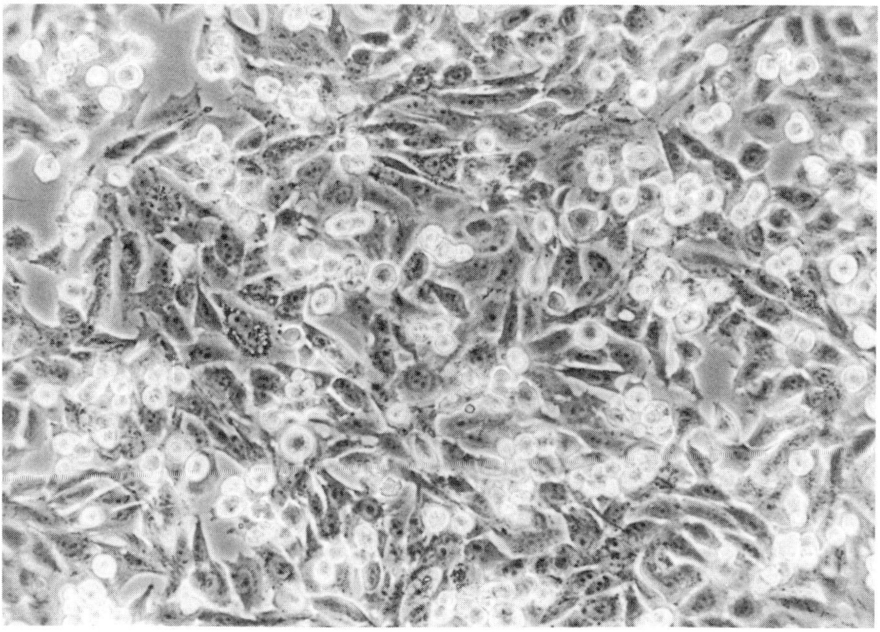

(a)

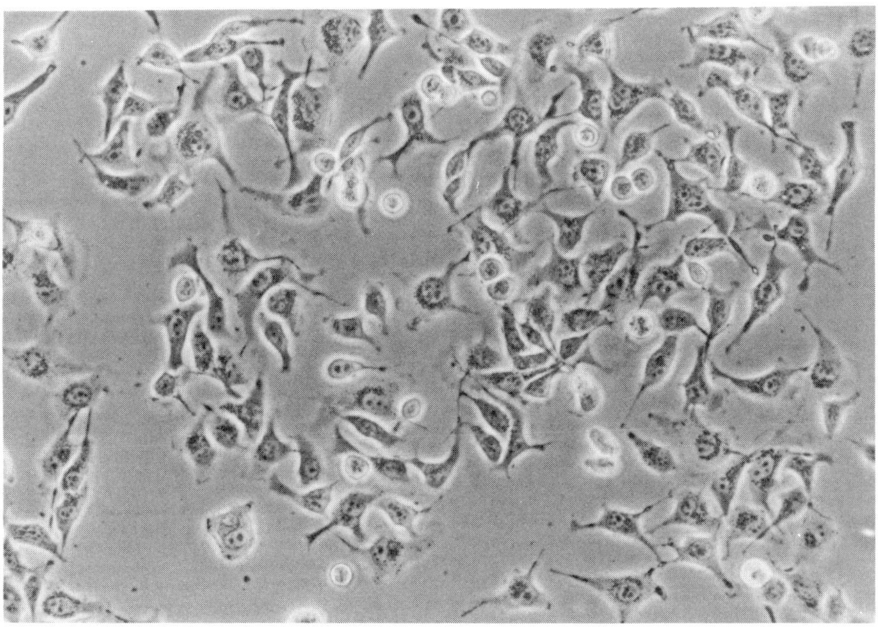

(b)

Fig. 4 (See page 107 for legend.)

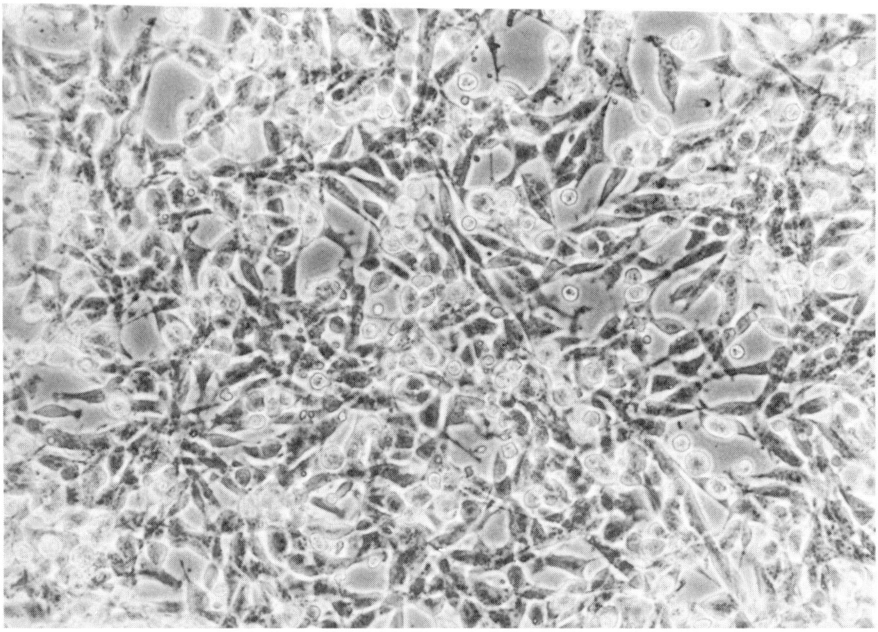

(c)

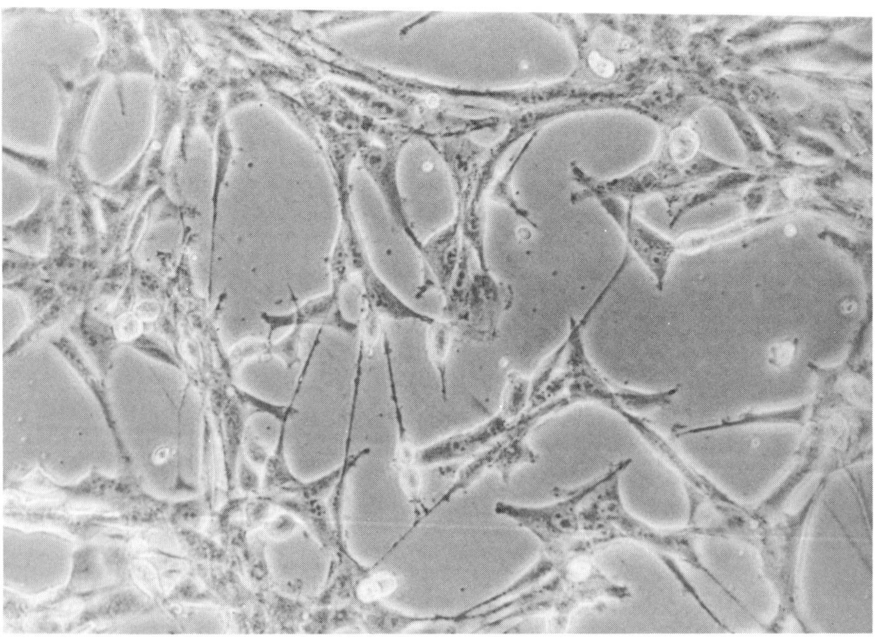

(d)

Fig. 4 (See following page for legend.)

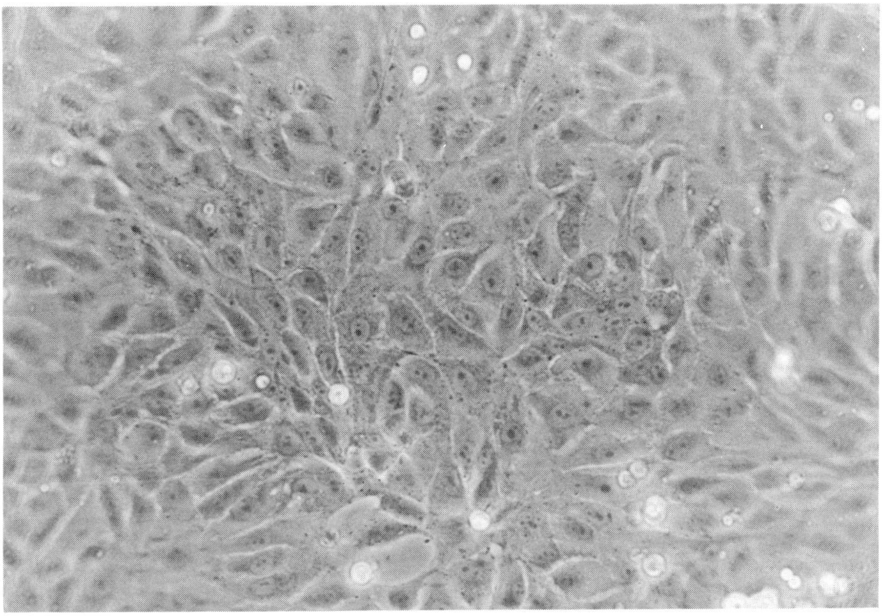

(e)

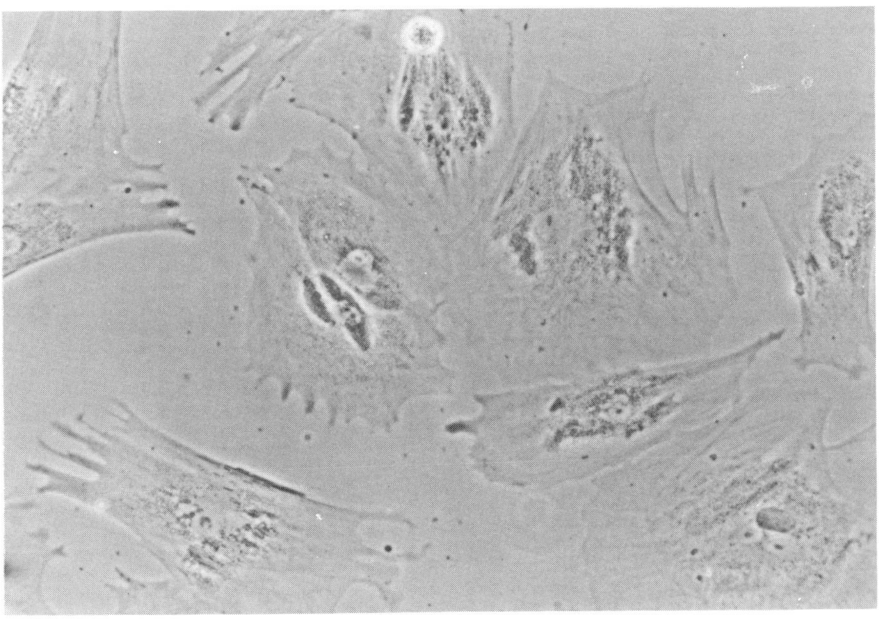

(f)

Fig. 4 Morphology of canine cell cultures used as target cells in cytotoxicity tests for NK activity. (a) Mammary carcinoma C1 (H73-2295). (b) Melanoma C4 (RVC347). (c) Melanoma HT43. (d) Melanoma C7 (VI). (e) Kidney epithelium (MDCK). Note contact inhibition. (f) Senescent tumor culture containing only fibroblasts (HT69, third passage).

## Specificity of NK Cell Activity

Nk cell activity is characterized by nonspecificity, presence in unimmunized animals, and widespread cytotoxic capability against allogeneic tumor target cells. Selective and nonselective components have been separated mathematically in an analysis of total cytotoxic activity.[10]

Since target cell variables, such as sensitivity to lysis by NK cells, may make direct comparisons difficult, the relative affinity of NK cells for target cells of given types can be determined with unlabeled competitive inhibitor cells.[52] The application of the competitive inhibition technique to the dog NK model showed cross-inhibition as well as cross-cytotoxicity (Fig. 3). Cross-inhibition by the addition of unlabeled inhibitor cells was also observed in BCG-stimulated nonspecific cytotoxicity.[42] No group selective antigenicity was detected on canine target cell lines studied in $^{51}Cr$ release tests. In contrast, a selective component was identified by inhibition tests using the microtest system.[52] Differential sensitivity of T and B lymphocyte lines to NK cell activity[53–55] has not been examined in the dog.

Nonspecificity of NK cell activity has been further demonstrated by adsorption studies. Effector cells are incubated overnight on an adherent cell monolayer, and the nonbound effector cells are tested for NK cell activity against the adsorbing and unrelated target cell lines. Nonadherent effector cells showed a reduction in NK cell activity against specific adsorbing target cells and unrelated target cells. Nonspecific PHA-induced cytotoxicity was unaffected by adsorption (Table 1). Results indicate that canine NK cells are not a pool of specifically committed effector cells directed at a wide range of antigens.

If the NK cell system provides an in vivo surveillance mechanism against neoplasia, selectivity of NK cells for malignant as opposed to normal host cells must operate. Alternatively, in vivo arming, inhibitory, sensitivity, or suppressor factors specific for malignant tumor clones may serve to direct the nonselective NK cell activity observed in vitro into an effective means of selectively eliminating potential malignancies. Selectivity for malignant cells has not been observed in vitro,[13] and alternative modifiers of in vivo activity still have not been identified.

The presence of endogenous Retrovirus-coded antigens in murine target cells led to the proposition that C-type viruses were the true target for NK cell activity.[56] However, Retrovirus antigens were found not to be the target of NK cell activity in the mouse,[57] and other factors, such as xenogeneic activity, render that ubiquitous antigen an unlikely target for NK activity.[19]

The suggestion that antigenic "HM" determinants from fetal calf serum in cell culture medium[58] were the target for NK activity has been discounted.[59,60] Canine data on that aspect are not available.

## Effect of Enzyme Treatment on NK Cell Activity

Protease treatments have been investigated to identify the target receptor moiety on NK cells. Trypsin depressed NK activity,[61–63] but activity was restored by incubation of effector cells in human serum. Preadsorption of the serum with a given target cell selectively depressed NK activity to that target cell.[62] Attempts to specifically rearm NK cells with antibody to given human HLA antigen specificities were thwarted by nonspecific cytophilic antibodies in the anti-HLA sera.[64] Protease treatment therefore indicated that NK cell activity in man, assayed in the microtest, is a function of cytophilic antibody gained in vivo.

TABLE 1

EFFECT OF PREADSORPTION OF EFFECTOR CELLS* ON CROSS-REACTIVITY AND SPECIFICITY OF NK CELL ACTIVITY

| | Percent specific lysis at 8 hr at 80 : 1 effector : target cell ratio | | | | | |
|---|---|---|---|---|---|---|
| | Mammary carcinoma (H73-2295) | | Melanoma (RVC347) | | Melanoma + PHA (RVC347-MICC) | |
| Donor | — | + | — | + | — | + |
| 1 | 2.4 | | | | 57.1 | 66.1 |
| 2 | | | 2.0 | | 15.4 | 11.4 |
| 3 | 1.9 | | | | 25.3 | 26.2 |
| 4 | 3.6 | | 5.1 | | 42.4 | 27.0 |
| 5 | 5.4 | | | | 37.0 | 29.9 |
| 6 | 1.9 | 1.6 | | | 44.7 | 47.9 |
| 7 | 12.2 | 2.4 | 13.4 | 9.2 | 29.0 | 24.2 |
| 8 | 10.3 | 8.1 | 21.0 | 18.9 | 46.5 | 53.0 |
| 9 | 18.9 | 4.8 | 20.9 | 5.9 | 24.0 | 36.9 |
| 10 | 2.1 | | | | 20.3 | ND |
| 11 | 8.8 | | 9.1 | | 39.4 | 27.2 |
| 12 | 1.7 | | | | 41.8 | 38.8 |

*Lymphocytes from donors were tested for NK and MICC activity after overnight incubation on plastic (— preadsorption) or on a monolayer of viable H73-2295 cells (+ preadsorption).

The canine NK cell is resistant to treatment with 0.05% trypsin plus ethylenediaminetetraacetic acid for 10 min at 20°C (Fig. 5) but is somewhat sensitive to 0.25% trypsin for 15 min at 37°C (Table 2). The sensitivity of canine NK cells to trypsinization was much less than that of human NK cells, and xenogeneic human NK cell activity to canine target cells, which was high before trypsinization, could not be restored by adding 10% pooled dog serum to the test medium (Table 2). Without evidence for reconstitution of NK cell activity by antibody, the mode of NK cell suppression by protease in the dog cannot be evaluated.

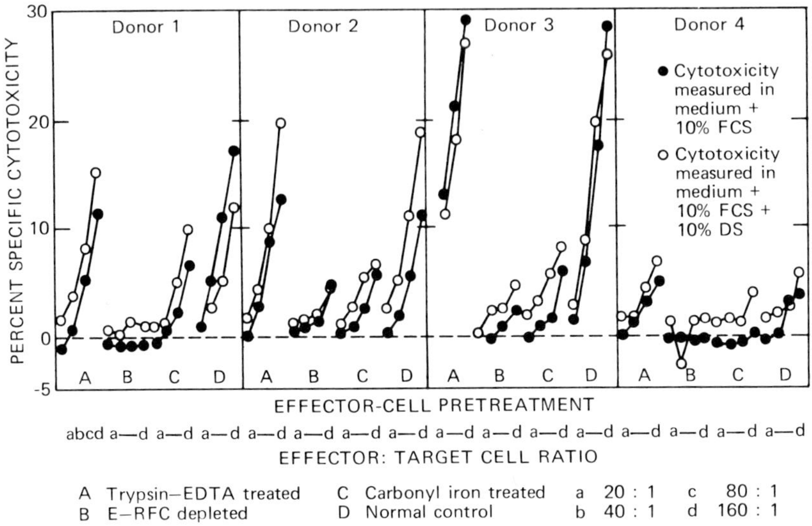


Fig. 5 Effect of effector cell pretreatment on cytotoxicity against mammary carcinoma C1 in medium with or without added dog serum (DS). Cytotoxicity was measured in medium plus 10% FCS (●) or medium plus 10% FCS plus 10% DS (○).

## CHARACTERIZATION OF THE NK EFFECTOR CELL

### Subclasses of Lymphocytes

The discoveries which demonstrated that lymphocytes are not a homogenous population of cells initially identified two classes. The thymus-dependent T lymphocyte forms spontaneous erythrocyte rosette forming cells (E-RFC) with red cells of specific species, is nonadherent to nylon wool, lacks membrane immunoglobulin, and bears specific antigenic markers, such as $\theta$, Thy, and TL. Bone-marrow-derived (or bursa derived in birds) B lymphocytes are progenitors of antibody forming cells, bear membrane immunoglobulin, have receptors for complement C′3, and adhere to nylon wool. By deduction, a poorly characterized "null" (neither T nor B) lymphocyte class completed the nonphagocytic mononuclear population of peripheral blood and lymphoid tissues.

### Erythrocyte Rosette Forming Cells (E-RFC)

T lymphocyte E-RFC populations can be separated from non-T cells by isopycnic density gradient centrifugation, and purified populations subsequently can be assayed for NK cell activity. Most investigators have found that NK cell

TABLE 2

EFFECTS OF TRYPSIN PRETREATMENT OF CANINE AND HUMAN NK EFFECTOR CELLS ON CYTOTOXICITY (WITH AND WITHOUT ADDITION OF SERUM) AGAINST HUMAN AND CANINE TARGET CELLS

| | | Percent $^{51}Cr$ release at 8 hr with 80 : 1 effector : target cell ratio | | |
|---|---|---|---|---|
| | | Target cells | | |
| | | Human myeloid | Canine melanoma | |
| Donors | Trypsin pretreatment | K562 (no serum added) | C4 (no serum added) | C4 (10% dog serum added) |
| Canine | | | | |
| 457 | — | 26 | 38 | 50 |
| | + | 16 | 32 | 31 |
| 459 | — | 18 | 40 | 39 |
| | + | 20 | 27 | 28 |
| 625 | — | 19 | 36 | 33 |
| | + | 18 | 26 | 28 |
| 434 | — | 16 | 30 | 31 |
| | + | 17 | 28 | 27 |
| Human | | | | |
| 1 | — | 75 | 70 | 77 |
| | + | 27 | 25 | 28 |
| 2 | — | 80 | 80 | 79 |
| | + | 59 | 46 | 49 |
| Medium only | | 18.5 | 25 | 28 |

activity in man resides in non-T-cell fractions,[11,65–69] but one group of researchers has shown that NK effector cells express low-affinity receptors for erythrocytes (low-affinity E-RFC).[70] In other research, both E-RFC-positive T cells and non-E-RFC populations were reported to be NK effectors.[71,72]

NK effector cells were found to be non-T cells in the mouse[21,73] and in the rat.[14] A comprehensive review of markers on NK cells is given by Herberman and Holden.[19]

Our studies of NK cell activity in the dog have included fractionation procedures aimed at depletion of E-RFC to examine the role of T cells in NK cell functions in the $^{51}Cr$ release assay. Canine lymphocytes form spontaneous rosettes with both human and guinea pig erythrocytes, and E-RFC depletion resulted in depression of NK cell activity of healthy canine donors (Fig. 5). BCG-stimulated cytotoxicity was similarly depressed by E-RFC depletion of

T cells.[42] A control rosetting procedure with a nonrosetting erythrocyte (sheep) did not affect NK cell activity or mitogen-induced cytotoxicity (Table 3).

Rosette formation with human and guinea pig erythrocytes is purportedly a specific canine T lymphocyte marker,[74,75] but thymocytes do not form rosettes, and some neoplastic lymphocytes were reported to be double labeled for E-RFC and complement C′3 rosettes, a B-cell marker.[76] Furthermore, canine

TABLE 3

EFFECT OF FRACTIONATION OF PERIPHERAL BLOOD LYMPHOCYTES ON NK AND MICC ACTIVITY ON MELANOMA C4 TARGET CELLS

| | Percent specific cytotoxicity at 8 hr at 80 : 1 effector : target cell ratio | | |
|---|---|---|---|
| | Effector cell preparation | | |
| Donor | Normal Ficoll–Hypaque centrifugation | Minus guinea pig E-RFC | Minus sheep E-RFC |
| | NK Activity | | |
| 1 | 3.1 | | 4.3 |
| 2 | 5.5 | 5.5 | 9.2 |
| 3 | 12.9 | 3.1 | 15.3 |
| 4 | 8.0 | 3.1 | 4.3 |
| | MICC Activity | | |
| 1 | 28.3 | 17.0 | 28.3 |
| 2 | 49.7 | 48.4 | 52.2 |
| 3 | 37.1 | 27.0 | 52.2 |
| 4 | 44.7 | 32.1 | 38.3 |

mononuclear phagocytes also form E rosettes,[77,78] probably via Fc receptors for cytophilic antibody on the erythrocyte. The apparent T-cell nature of the canine NK cell may therefore be based on a spurious erythrocyte marker for canine T cells, and the anomalous finding of canine NK cell residing in an E-RFC fraction could be explained by an affinity for the Fc region of natural antibody on human and guinea pig erythrocytes. Until T-cell markers in the dog are better defined, the E-RFC identity of the canine NK cell must be treated with caution.

## Receptors for Immunoglobulin Fc on NK Cells

Effector cells of NK activity in man were depleted by rosetting with antibody coated erythrocytes (EA-RFC), presumably via receptors for the Fc region of immunoglobulin.[11,67,69,70] In other work, Fc receptors were expressed only on some NK cells and were not a prerequisite for NK cell activity.[72] Murine NK cells show weaker Fc affinity.[79] In most cases, NK cell affinities for bound immunoglobulin in the form of IgG, IgM, or immune complexes or aggregated immunoglobulin[80] have not been measured separately. Pinpointing the immunoglobulin source of NK cell activity could help to further identify the nature of the NK effector cell[81,82] and may explain, for example, the rise in human NK cell activity after incubation in vitro at 37°C.[83,84]

## Relationship to Killer (K) Cells and Antibody Dependent Cytotoxicity

The expression of receptors for bound immunoglobulin on NK cells and the evidence of a role for cytophilic antibody in NK cell activity suggest a similarity between NK and K cells. Like NK cells, K cells are not T or B cells but have a low-affinity E rosette forming capacity.[85] Although there are some differences between NK and K cells, both may represent a pre-T-lymphocyte bone-marrow-derived cell.[19] NK and K cell functions have not been compared in the dog. The canine K cell is a nonadherent non-B cell (S. Livnat, personal communication); E-RFC activity and other markers are not known.

## Relationship to the Effector Cells of Mitogen-Induced Cell-Mediated Cytotoxicity (MICC)

The presence of T lymphocyte mitogenic lectins, such as phytohemagglutinin (PHA) or concanavalin A (Con-A), in a lymphocyte/target cell mixture induces rapid nonspecific cytotoxicity (MICC). B lymphocyte mitogens (LPS and pokeweed mitogen) are inactive in MICC assays.[86] Blastogenesis of lymphocytes in response to the mitogens is not a prerequisite, and cytotoxicity, probably induced by lectin cross-linking of effector and target cells, is high after 4 hr of incubation.

MICC against erythrocyte target cells is mediated by mononuclear phagocytes, whereas nucleated tumor target cells are lysed by T lymphocytes.[87] Human MICC effector cells are found in both E-RFC-positive and -negative fractions, but all bear Fc receptors.[88]

In the dog, the addition of PHA to a cytotoxicity test generates a powerful nonspecific cell-mediated cytotoxic reaction. Simultaneous testing for MICC and NK activity against the same target cell shows high MICC and a lower level of NK activity (Fig. 1), and both NK activity and MICC show chronological variation for individual donors. Correlation of total cytotoxicity (MICC plus NK) vs. NK activity for one experiment gave a 0.63 correlation coefficient, but PHA-induced MICC alone (total cytotoxicity minus NK) was unrelated to the level of NK activity (Fig. 6). Effector cell pretreatment by E-RFC depletion or

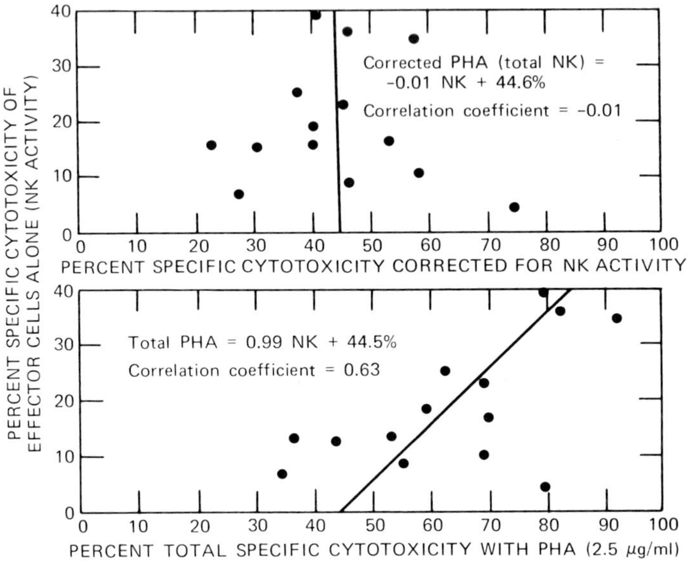


**Fig. 6 Regression analysis of natural killer (NK) vs. PHA mitogen-induced cell-mediated cytotoxicity (MICC). Target cell, melanoma C4.**

carbonyl iron treatment depressed NK activity and the NK component of total MICC activity (Fig. 7).

Preabsorption of effector cells on target cell monolayers similarly reduced NK activity, whereas MICC was intact (Table 1).

Both PHA and staphylococcal protein A (SpA) are powerful mitogens for canine lymphocytes,[89] but only PHA generates an immediate MICC reaction (Table 4). SpA binds specifically to Fc regions of immunoglobulin, and the failure of SpA to generate MICC may reside in a non-T-cell specificity, a failure to cross-link effector to target cell, or a blocking of Fc sites required for MICC function.

## GENERATION OF NK CELL ACTIVITY IN VITRO

The in vitro response of lymphocytes to antigenic stimulation has been widely used for the generation of specifically cytotoxic T lymphocytes.

Stimulation of lymphocytes with allogeneic cells in mixed lymphocyte cultures (MLR) results in the generation of nonspecific "anomalous" cytotoxicity around the fifth day of culture.[90–92] Stimulation with PPD antigen also produces nonspecific cytotoxic activity[93] by production of a mitogenic factor. An NK cell activating factor (NAF) has been found in MLR supernatants.[94]

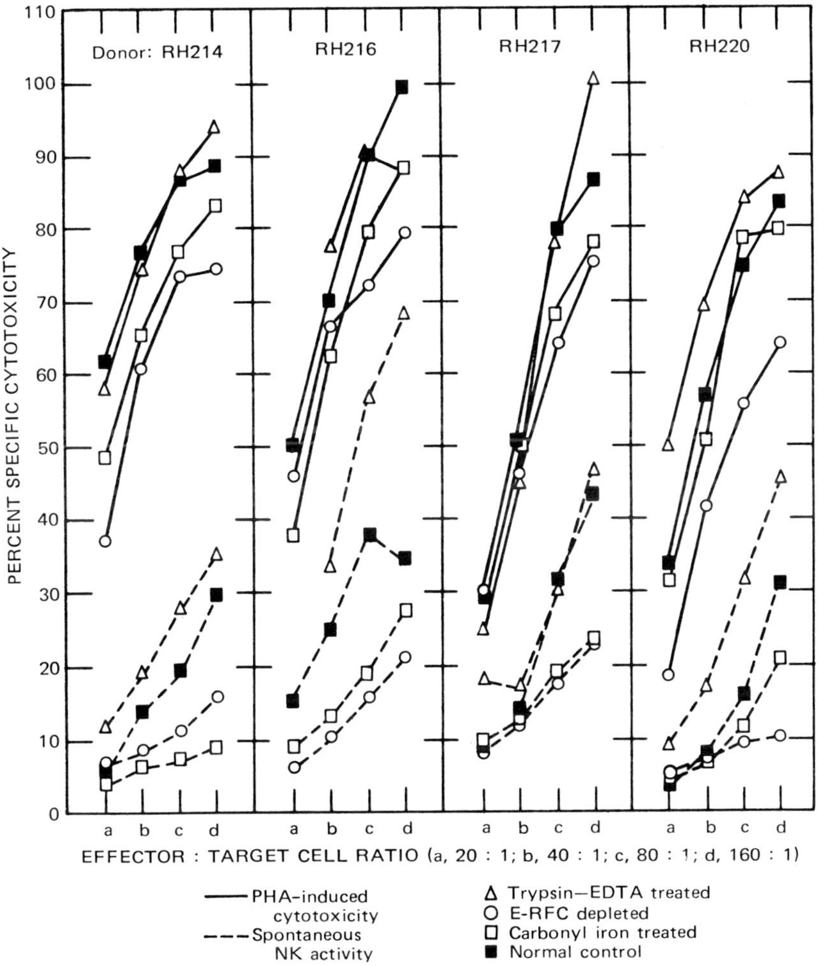


Fig. 7 Effects of effector cell pretreatments on NK and PHA mitogen-induced cell-mediated cytotoxicity (MICC, not corrected for incorporated NK activity).

Interferon,[95] or fetal calf serum,[96] also activated NK cells in vitro. The development of spontaneous cytotoxicity for macrophage-mediated erythrocyte targets has been attributed to loss of suppressor cells.[97] Whether these in vitro findings represent the mechanisms for in vivo stimulation of NK activity requires further examination.

Long-term in vitro culture of canine lymphocytes by SpA stimulation[89] has shown high NK cell activity after 7 days in preliminary experiments, but 6-day MLR cells failed to show NK activity, perhaps for technical reasons.

TABLE 4

EFFECT OF PHA AND STAPHYLOCOCCAL PROTEIN A (SpA) MITOGENS ON NK ACTIVITY ($^{51}Cr$ RELEASE ASSAY) AGAINST MELANOMA C4 TARGET CELLS

| | Percent $^{51}Cr$ release at 8 hr at 160 : 1 effector : target cell ratio | | |
|---|---|---|---|
| Donor | Cells alone (NK) | Cells plus PHA (MICC) | Cells plus SpA |
| 1 | 52 | 72 | 51 |
| 2 | 49 | 76 | 48 |
| 3 | 36 | 74 | 45 |
| 4 | 39 | 75 | 43 |
| 5 | 50 | 84 | 60 |
| 6 | 47 | 74 | 51 |
| 7 | 56 | 77 | 53 |
| 8 | 24 | 58 | 29 |
| 9 | 45 | 84 | 54 |
| 10 | 44 | 76 | 51 |
| 11 | 29 | 69 | 30 |
| 12 | 26 | 87 | 31 |
| Medium control | 19 | 18 | 19 |

# IMMUNOGENETIC ASPECTS OF NK CELL ACTIVITY

## Lack of Compatibility Requirement for NK Cell Cytotoxicity

Products of the major histocompatibility complex (MHC) serve not only as antigens in allograft rejection (see Chapter 8) but also play a key role in cell cooperation during the generation of an immune response[98] and in the recognition of target antigens by cytotoxic T lymphocytes.[99] The NK phenomenon, however, was identified with allogeneic target cells, which NK cells would not recognize or share MHC products (DLA antigens) with except by chance. No requirement for MHC compatibility was observed in mouse NK cytotoxicity[57] or in the rat.

For the dog, all susceptible target cell lines were similarly lysed by lymphocytes from a panel of donors (Fig. 3); those results would not be expected if recognition of MHC products were required for expression of cytotoxicity in the outbred canine species. Differing levels of NK cytotoxicity in haplotype-identical siblings would also be incompatible with a DLA recognition requirement for NK cell expression.

## NK Activity Against Xenogeneic Target Cells

As with syngeneic and allogeneic cells, nonspecific NK activity extends to xenogeneic target cells, albeit sometimes at a lower level than with target cells of the donor species. Xenogeneic activity has been observed for human and mouse,[100,101] human and rat,[102] and human and guinea pig cell combinations.[17] Interspecies cross-reactivity between MHC antigeneic determinants[103] may, nevertheless, permit specific recognition of target antigens in xenogeneic combinations.

Our preliminary studies with canine and human effector and target cell combinations have demonstrated lysis of human and canine target cells by human trypsin-sensitive effector cells (Table 2). The converse activity of canine NK cells was of a lower order of magnitude against either canine or human target cells. A cytophilic antibody explanation of NK cell activity would bypass a xenogeneic specificity problem since interspecies cross-reactivity of "natural" antibodies and Fc receptor specificity are known.

## Genetic Determination of NK Cell Activity

Different levels of NK cell activity among inbred strains of mice are genetically determined by several loci,[104] including MHC-linked genes.[105] Bone-marrow reconstitution in syngeneic mice results in the marrow donor's NK cell level replacing the recipient's genetically determined NK cell activity level,[106] which demonstrates the bone-marrow origin of (or regulation of) NK cells. NK cell activity in the rat was not linked with the MHC complex.[107] The human HLA-A3, B7 MHC haplotype has been associated with low NK cell activity.[108]

The allogeneic resistance phenomenon in bone-marrow grafting (see Chapter 9) is genetically controlled by Hh genes and may represent an in vivo manifestation of NK cell activity[28,109,110] with some discrepancies, such as hydrocortisone sensitivity.[29]

Studies of NK cell activity within canine families of known DLA serological and lymphocyte-determined antigen composition have been carried out. DLA gene products were identified with internationally recognized reagents at Rotterdam, The Netherlands, and NK cell activity against a reference melanoma target cell was measured by the $^{51}Cr$ release assay. Levels of cytotoxicity within families showed large variations that were not segregated with the chromosome bearing the DLA complex (Figs. 8a to 8c). DLA-identical siblings, which were also negative in mixed-lymphocyte reactions, showed major differences in level of NK activity. No evidence for a non-MHC-linked gene locus regulating NK level could be obtained from the available family studies. Nongenetic effects, such as temporal variation of NK levels in the same donor, apparently dominate the level of NK activity in the dog.

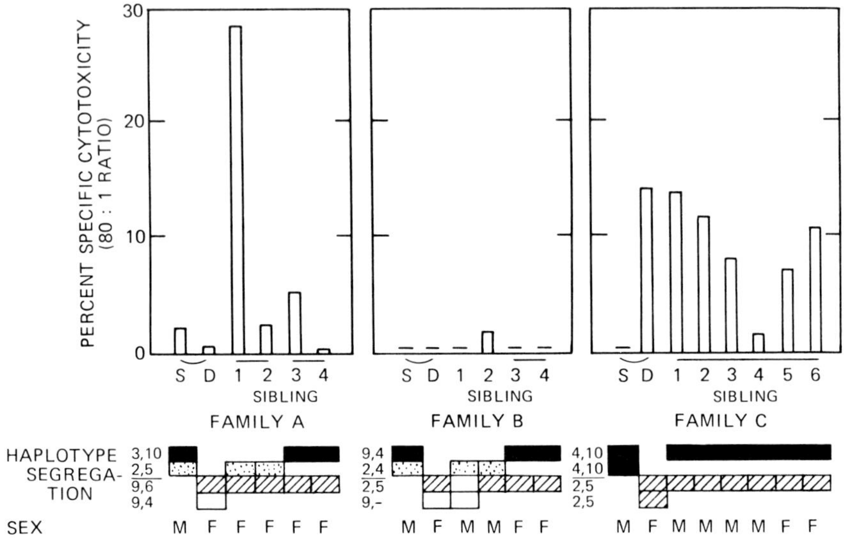
PERCENT SPECIFIC CYTOTOXICITY (80 : 1 RATIO)
0
10
20
30
S D 1 2 3 4
SIBLING
FAMILY A
S D 1 2 3 4
SIBLING
FAMILY B
S D 1 2 3 4 5 6
SIBLING
FAMILY C
HAPLOTYPE SEGREGATION
3,10
2,5
9,6
9,4
9,4
2,4
2,5
9,-
4,10
4,10
2,5
2,5
SEX
M F F F F F
M F M M F F
M F M M M M F F

(a)

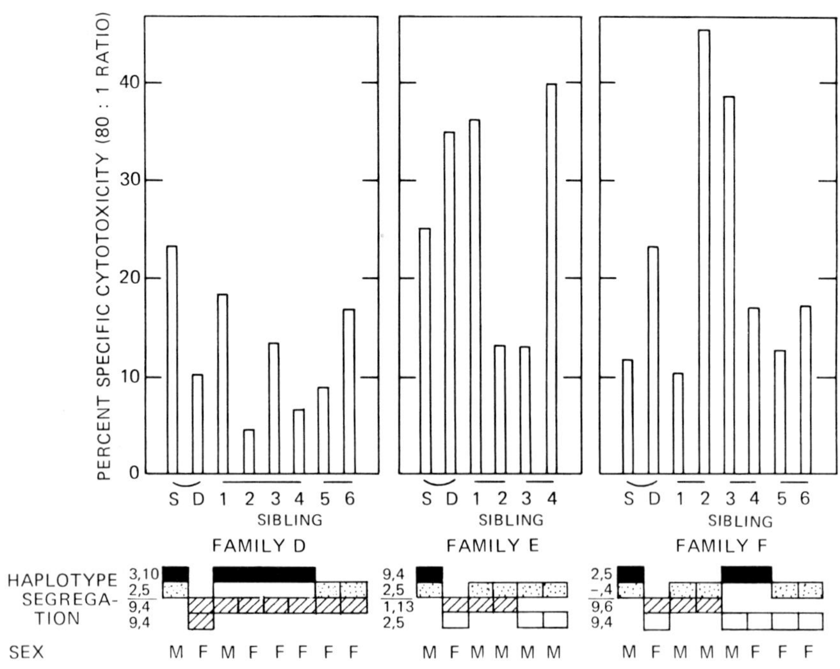
PERCENT SPECIFIC CYTOTOXICITY (80 : 1 RATIO)
0
10
20
30
40
S D 1 2 3 4 5 6
SIBLING
FAMILY D
S D 1 2 3 4
SIBLING
FAMILY E
S D 1 2 3 4 5 6
SIBLING
FAMILY F
HAPLOTYPE SEGREGATION
3,10
2,5
9,4
9,4
9,4
2,5
1,13
2,5
2,5
-,4
9,6
9,4
SEX
M F M F F F F F
M F M M M M
M F M M M F F F

(b)

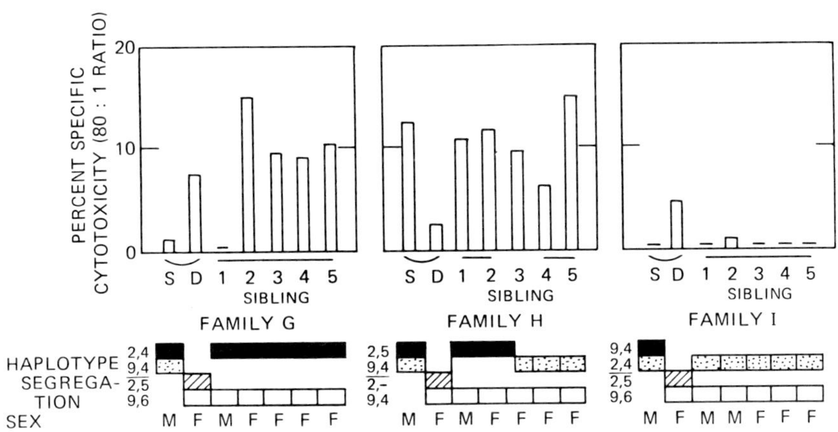


(c)

Fig. 8 Lack of association of level of NK activity with the segregation of DLA histocompatibility haplotypes in nine families (A to I). The segregation of sire (S) and dam (D) DLA-bearing chromosomes (■, ▒, paternal; ▨, □, maternal) to siblings 1, 2, 3 . . . is shown by the bar coding. The DLA-A and DLA-B serologically defined specificities on each chromosome (e.g., 3,10) are listed for each family according to international nomenclature.[109] ——, haplotype identical siblings.

## CONCLUSIONS ABOUT NK CELL ACTIVITY IN THE DOG

Studies of tumor immunology in the dog, as in man, have led to the discovery of a powerful NK cell cytotoxic activity. The dog therefore shows similarities with other laboratory species but provides a model closer to the clinical situation in man for studying tumor immunology at all stages of disease and therapy. The depression of NK cell cytotoxicity in metastatic cases and NK stimulation by BCG immunotherapy parallel studies in man. The dog provides the best model in that area of cancer research.

A range of canine target cells, both malignant and nontransformed, has been tested in the $^{51}$Cr release cytotoxicity assay and has demonstrated a lack of selectivity of the NK cell activity both as direct cytotoxicity and in inhibition studies. The absence of a selective element in the canine studies reported may result from the use of the $^{51}$Cr release assay because the microtest assay of NK cell activity in man did demonstrate a selective element.

The depletion of NK cell activity in the dog by E rosette formation of effector cells, a property of T lymphocytes, is contrary to results obtained in laboratory rodents and, to some degree, in man. Verification of the apparent T-cell identity of the canine NK cell is still required, however, since E rosetting with human or guinea pig erythrocytes may not be a specific marker for T lymphocytes in the dog. Data on Fc receptor expression on NK cells and their relationship to K-cell function are also required.

Mitogen-induced cell-mediated cytotoxicity is not related to the level of NK cell activity. Xenogeneic activity of human NK cells against canine target cells was much stronger than in the converse situation.

Immunogenetic studies showed no evidence for genetic control or DLA histocompatibility complex linkage of NK cell activity in the dog families studied, although a genetic basis with environmental modulation of NK cell activity could be operative.

The NK cell system in the dog provides a useful model for human cell-mediated immune reactions to neoplasia. Nonspecific, cell-mediated reactions, which may regulate tumor development, could be amenable to manipulation for therapy and have prognostic value. Other benefits in the areas of transplantation, autoimmune diseases, and infectious disease, may become evident in the future.

## ACKNOWLEDGMENTS

I wish to thank Dr. H. M. Vriesendorp for cooperation in immunogenetics studies, D. R. Morgan, Amelie Dekkers-Bijma, and Ineke Punselie-Wevers for competent technical assistance, and J. Nicklin for typing the manuscript.

Results reported are derived from studies at the Department of Clinical Veterinary Medicine, University of Cambridge, England, and at the Laboratory for Experimental Surgery, Erasmus University, Rotterdam, The Netherlands, and were supported by the Wellcome Trust, the Cancer Research Campaign, and the Medical Research Council of Great Britain. These studies were made possible by Dr. L. N. Owen (Cambridge) and Dr. D. L. Westbroek (Rotterdam).

## REFERENCES

1. F. M. Burnet, Immunological Surveillance in Neoplasia, *Transplant. Rev.*, 7: 3-25 (1971).
2. R. B. Herberman, Immunogenicity of Tumor Antigens, *Biochim. Biophys. Acta*, 473: 93-119 (1977).
3. R. P. Warren, M-S. Tsoi, B. H. Henderson, P. L. Weiden, and R. Storb, Demonstration of Tumor Associated Antigens in Dogs, *Transplant. Proc.*, 7: 481-484 (1975).
4. M. J. Ulvand, Cellular Immunity to Canine Mammary Tumor Cells Demonstrated by the Leukocyte Migration Technique, *Acta Vet. Scand.*, 16: 95-114 (1975).
5. M-S. Tsoi, P. L. Weiden, and R. Storb, Immunologic Reactivity to Soluble Autochthonous Tumor Extracts in Dogs with Spontaneous Malignancies, *J. Immunol.*, 116: 1134-1139 (1976).
6. I. J. Fidler, R. S. Brodey, and S. Bech-Nielsen, In Vitro Immune Stimulation Inhibition to Spontaneous Canine Tumors of Various Histologic Types, *J. Immunol.*, 112: 1051-1060 (1974).
7. G. P Betton and N. T. Gorman, Cell-Mediated Responses in Dogs with Spontaneous Neoplasms. I. Detection of Cell-Mediated Cytotoxicity by the Chromium-51 Release Assay, *J. Natl. Cancer Inst.*, 61: 1085-1093 (1978).

8. N. T. Gorman and G. R. Betton, Cell Mediated Responses in Dogs with Spontaneous Neoplasms. II. Detection by the Direct Leukocyte Migration Inhibition Technique, *J. Natl. Cancer Inst.*, 61: 1095-1100 (1978).
9. D. E. Onions, L. N. Owen, and D. E. Bostock, Leukocyte Migration Inhibition Responses in Canine Lymphosarcoma, *Int. J. Cancer*, 22: 503-507 (1978).
10. M. Takasugi and M. R. Mickey, Interaction Analysis of Selective and Non-Selective Cell-Mediated Cytotoxicity, *J. Natl. Cancer Inst.*, 57: 255-261 (1976).
11. M. Kiuchi and M. Takasugi, The Non-Selective Cytotoxic Cell (N Cell), *J. Natl. Cancer Inst.*, 56: 575-582 (1976).
12. E. Ojo and H. Wigzell, Natural Killer Cells May Be the Only Cells in Normal Mouse Lymphoid Cell Populations Endowed with Cytolytic Ability for Antibody Coated Tumor Target Cells, *Scand. J. Immunol.*, 7: 297-306 (1978).
13. M. E. Nunn, R. B. Herberman, and H. T. Holden, Natural Cell-Mediated Cytotoxicity in Mice Against Non-Lymphoid Tumor Cells and Some Normal Cells, *Int. J. Cancer*, 20: 381-387 (1977).
14. G. R. Shellam, Gross Virus-Induced Lymphoma in the Rat. V. Natural Cytotoxic Cells Are Non-T Cells, *Int. J. Cancer*, 19: 225-235 (1977).
15. M. E. Nunn, J. Y. Djeu, M. Glaser, D. H. Lavrin, and R. B. Herberman, Natural Cytotoxic Reactivity of Rat Lymphocytes Against Syngeneic Gross Virus-Induced Lymphoma, *J. Natl. Cancer Inst.*, 56: 393-399 (1976).
16. F. Arnaud-Battandier, B. M. Bundy, and D. L. Nelson, Natural Killer Cells in Guinea Pig Spleen Bear Fc Receptors, *Eur. J. Immunol.*, 8: 400-406 (1978).
17. A. Altman and H. J. Rapp, Natural Cell Mediated Cytotoxicity in Guinea Pigs: Properties and Specificity of Natural Killer Cells, *J. Immunol.*, 121: 2244-2252 (1978).
18. S. K. Datta, J. J. Trentin, and J. G. Sinkovics, Lymphoma Cytotoxic Natural Killer Cells in Hamsters, *Proc. Am. Assoc. Cancer Res.*, 19: 193 (1978).
19. R. B. Herberman and H. T. Holden, Natural Cell Mediated Immunity, *Adv. Cancer Res.*, 27: 305-377 (1978).
20. E. Lotzova and K. B. McCredie, Natural Killer Cells in Mice and Man and Their Possible Biological Significance, *Cancer Immunol. Immunother.*, 4: 215-221 (1978).
21. R. Kiessling, E. Klein, H. Pross, and H. Wigzell, "Natural" Killer Cells in the Mouse. II. Cytotoxic Cells with Specificity for Mouse Moloney Leukemia Cells. Characteristics of the Killer Cell, *Eur. J. Immunol.*, 5: 117-121 (1975).
22. G. R. Shellam and N. Hogg, Gross Virus-Induced Lymphoma in the Rat. IV. Cytotoxic Cells in Normal Rats, *Int. J. Cancer*, 19: 212-224 (1977).
23. A. C. Levin, R. J. Massey, and F. Deinhardt, Spontaneous Human Mononuclear Cell Cytotoxicity to Culture Tumor Cells: Reproducibility of Serial Measurements with the Use of the Chromium-51 Release Microcytotoxicity Assay, *J. Natl. Cancer Inst.*, 60: 1283-1294 (1978).
24. O. Stutman, Tumor Development After 3-Methylcholanthrene in Immunologically Deficient Athymic Mice, *Science*, 183: 534-536 (1974).
25. O. Haller, M. Hansson, R. Kiessling, and H. Wigzell, Role of Nonconventional Natural Killer Cells in Resistance Against Syngeneic Tumor Cells In Vivo, *Nature*, 270: 609-611 (1977).
26. M. Takasugi, A. Ramseyer, and J. Takasugi, Decline in Natural Nonselective Cell Mediated Cytotoxicity in Patients with Tumor Progression, *Cancer Res.*, 37: 413-418 (1977).
27. H. F. Pross and M. G. Baines, Spontaneous Human Lymphocyte Mediated Cytotoxicity Against Tumor Target Cells. I. The Effect of Malignant Disease, *Int. J. Cancer*, 18: 593-604 (1976).
28. E. Lotzova and C. A. Savary, Possible Involvement of Natural Killer Cells in Bone Marrow Graft Rejection, *Biomedicine*, 27: 341-344 (1977).

29. P. S. Hochman and G. Cudkowitz, Different Sensitivities to Hydrocortisone of Natural Killer Cell Activity and Hybrid Resistance to Parental Marrow Grafts, *J. Immunol.*, 119: 2013-2015 (1977).
30. J. R. Oehler and R. B. Herberman, Natural Cell-Mediated Cytotoxicity in Rats. III. Effects of Immunopharmacologic Treatments in Natural Reactivity and on Reactivity Augmented by Polyionosinic-Polycytidilic Acid, *Int. J. Cancer*, 21: 221-229 (1978).
31. A. Mantovani, W. Luini, G. Peri, A. Vecchi, and F. Spreafico, Effect of Chemotherapeutic Agents on Natural Cell-Mediated Cytotoxicity in Mice, *J. Natl. Cancer Inst.*, 61: 1255-1261 (1978).
32. O. Haller and H. Wigzell, Suppression of Natural Killer Cell Activity with Radioactive Strontium: Effector Cells Are Marrow Dependent, *J. Immunol.*, 118: 1503-1506 (1977).
33. G. Trinchieri and D. Santoli, Anti-Viral Activity Induced by Culturing Lymphocytes with Tumor Derived or Virus Transformed Cells. Enhancement of Human Natural Killer Cell Activity by Interferon and Antagonistic Inhibition of Susceptibility of Target Cells to Lysis, *J. Exp. Med.*, 147: 1314-1333 (1978).
34. M. J. Droller, P. Perlmann, and M. U. Schneider, Enhancement of Natural and Antibody Dependent Lymphocyte Cytotoxicity by Drugs Which Inhibit Prostaglandin Production by Tumor Target Cells, *Cell. Immunol.*, 39: 154-164 (1978).
35. R. M. Welsh, Cytotoxic Cells During Lymphocytic Choriomeningitis Virus Infection of Mice. I. Characterization of Natural Killer Cell Induction, *J. Exp. Med.*, 148: 163-181 (1978).
36. J. R. Oehler, L. R. Lindsay, M. E. Nunn, H. T. Holden, and R. B. Herberman, Natural Cell-Mediated Cytotoxicity in Rats. II. In Vivo Augmentation of NK Cell Activity, *Int. J. Cancer*, 21: 210-220 (1978).
37. K. Pfizenmaier, H. Trostmann, M. Rollinghoff, and H. Wagner, Temporary Presence of Self-Reactive Cytotoxic T Lymphocytes During Lymphocytic Choriomeningitis, *Nature*, 258: 238-240 (1975).
38. E. M. Hersh, J. U. Gutterman, and G. M. Mavligit, BCG as Adjuvant Immunotherapy for Neoplasia, *Annu. Rev. Med.*, 28: 489-515 (1977).
39. R. B. Herberman, M. E. Nunn, H. T. Holden, S. Staal, and J. Y. Djeu, Augmentation of Natural Cytotoxic Reactivity of Mouse Lymphoid Cells Against Syngeneic and Allogeneic Target Cells, *Int. J. Cancer*, 19: 555-564 (1977).
40. A. Ghaffar and R. T. Cullen, In Vitro Antitumor Effect of Lymphoid Cells from *Corynebacterium parvum* Treated Mice: Effect of Route of *C. parvum* Administration, *J. Natl. Cancer Inst.*, 58: 717-720 (1977).
41. E. Ojo, O. Haller, A. Kimura, and H. Wigzell, An Analysis of Conditions Allowing *Corynebacterium parvum* to Cause Either Augmentation or Inhibition of Natural Killer Cell Activity Against Tumor Cells in Mice, *Int. J. Cancer*, 21: 444-452 (1978).
42. G. R. Betton, N. T. Gorman, and L. N. Owen, Cell Mediated Cytotoxicity in Dogs Following Systemic or Local BCG Treatment Alone or in Combination with Allogeneic Tumor Cell Lines, *Eur. J. Cancer*, 15: 745-754 (1979).
43. N. Thatcher and D. Crowther, Changes in Non-Specific Lymphoid (NK, K, T Cell) Cytotoxicity Following BCG Immunization of Healthy Subjects, *Cancer Immunol. Immunother.*, 5: 105-107 (1978).
44. L. N. Owen and D. E. Bostock, Effects of Intravenous BCG in Normal Dogs and in Dogs with Spontaneous Osteosarcoma, *Eur. J. Cancer*, 10: 775-780 (1974).
45. S. Bech-Nielsen, R. S. Brodey, I. J. Fidler, D. A. Abt, and J. S. Reif, The Effect of BCG on In Vitro Immune Reactivity and Clinical Course in Dogs Treated Surgically for Osteosarcoma, *Eur. J. Cancer*, 13: 33-41 (1977).
46. D. E. Bostock and N. T. Gorman, Intravenous BCG Therapy of Canine Mammary Carcinoma in Bitches After Surgical Excision of the Primary Tumour, *Eur. J. Cancer*, 14: 879-883 (1978).

47. A. D. Hess, R. Catchatourian, A. R. Zander, and R. B. Epstein, Intra-lesional Bacillus Calmette-Guérin Immunotherapy of Canine Venereal Tumors, *Cancer Res.*, 37: 3990-3994 (1977).
48. R. H. Bear, O. A. Reholt, and D. Pressman, Syngeneic Tumor Rejection Induced by Immunization with Normal Allogeneic Tissue, *Immunol. Commun.*, 6: 217-225 (1977).
49. M. A. Bean et al., Evaluation of In Vitro Methods for Assaying Tumor Immunity, in *In Vitro Methods in Cell Mediated and Tumor Immunity*, B. R. Bloom and J. R. David (Eds.), pp. 27-65, Academic Press, New York, 1976.
50. G. R. Betton and L. N. Owen, Allogeneic Grafts of Spontaneous Canine Melanomas and Their Cell Culture Strains in Allogeneic Immunosuppressed Dogs, *Brit. J. Cancer*, 34: 374-380 (1976).
51. G. R. Betton, Agglutination Reactions of Spontaneous Canine Tumor Cells Induced by Concanavalin A Demonstrated by an Isotopic Assay, *Int. J. Cancer*, 18: 687-696 (1976).
52. M. Takasugi, Y. Koide, D. Akira, and A. Ramseyer, Specificities in Natural Cell-Mediated Cytotoxicity by the Cross-Competition Assays, *Int. J. Cancer*, 19: 291-297 (1977).
53. D. M. Callewaert, J. Kaplan, W. D. Peterson, and J. J. Lightbody, Spontaneous Cytotoxicity of Human Lymphoblast Cell Lines Mediated by Normal Peripheral Blood Lymphocytes. I. Differential Susceptibility of T- Versus B-Cell Lines, *Cell. Immunol.*, 33: 11-19 (1977).
54. A. Ono, D. B. Amos, and H. S. Koren, Selective Cellular Natural Killing Against Human Leukaemic T Cells and Thymus, *Nature*, 266: 546-548 (1977).
55. Y. Koide and M. Takasugi, Specificities in Natural Cell-Mediated Cytotoxicity Against Lymphoblastoid Cell Lines. I. Selective Inhibition by Cross Competition, *Eur. J. Immunol.*, 8: 818-822 (1978).
56. R. B. Herberman, M. E. Nunn, and D. H. Lavrin, Natural Cytotoxic Reactivity of Mouse Lymphoid Cells Against Syngeneic and Allogeneic Tumors. I. Distribution of Reactivity and Specificity, *Int. J. Cancer*, 16: 216-229 (1975).
57. S. Becker, E. M. Fenyo, and E. Klein, The "Natural Killer" Cell in the Mouse Does Not Require H-2 Homology and Is Not Directed Against Type or Group Specific Antigens of Murine C Viral Proteins, *Eur. J. Immunol.*, 6: 882-885 (1976).
58. R. F. Irie, K. Irie, and D. L. Morton, Characteristics of Heterologous Membrane Antigen on Cultured Human Cells, *J. Natl. Cancer Inst.*, 53: 1545-1551 (1974).
59. H. F. Pross, S. S. Luk, and M. G. Baines, Spontaneous Human Lymphocyte Mediated Cytotoxicity Against Tumor Target Cells. V. The Role of Serum Derived Heterologous Membrane Antigens, *Int. J. Cancer*, 21: 291-298 (1978).
60. O. Haller, Mouse Natural Cell-Mediated Cytotoxicity to Human Leukemia: Independence of Serum Source Used for Target Cell Culturing and Cytotoxicity Assay, *J. Natl. Cancer Inst.*, 60: 1433-1438 (1978).
61. D. Akira and M. Takasugi, Loss of Specific Natural Cell Mediated Cytotoxicity with Absorption of Natural Antibodies from Serum, *Int. J. Cancer*, 19: 747-755 (1977).
62. Y. Koide and M. Takasugi, Determination of Specificity in Natural Cell-Mediated Cytotoxicity by Natural Antibodies, *J. Natl. Cancer Inst.*, 59: 1099-1105 (1977).
63. O. Eremin, J. Ashby, and J. P. Stephens, Human Natural Cytotoxicity in the Blood and Lymphoid Organs of Healthy Donors and Patients with Malignant Disease, *Int. J. Cancer*, 21: 35-41 (1978).
64. J. Takasugi, Y. Koide, and M. Takasugi, Reconstitution of Natural Cell-Mediated Cytotoxicity with Specific Antibodies, *Eur. J. Immunol.*, 7: 887-892 (1977).
65. P. Hersey, A. Edwards, J. Edwards, E. Adams, G. W. Milton, and D. S. Nelson, Specificity of Cell-Mediated Cytotoxicity Against Human Melanoma Lines: Evidence for "Non-Specific" Killing by Activated T-Cells, *Int. J. Cancer*, 16: 173-183 (1975).

66. B. M. Vose, F. Vanky, S. Argov, and E. Klein, Natural Cytotoxicity in Man; Activity of Lymph Node and Tumor Infiltrating Lymphocytes, *Eur. J. Immunol.*, 7: 753-757 (1977).
67. O. Eremin, R. R. A. Coombs, D. Plumb, and J. Ashby, Characterization of the Human Natural Killer (NK) Cell in Blood and Lymphoid Organs, *Int. J. Cancer*, 21: 42-50 (1978).
68. U. Galili and M. Schlesinger, Regulation of the Cytotoxic Effect of Human "Normal Killer Cells" on Tumor Cell Lines by Neuraminidase Treated T Lymphocytes, *Cancer Immunol. Immunother.*, 4: 33-39 (1978).
69. T. Bakacs, P. Gergely, and E. Klein, Characterization of Cytotoxic Human Lymphocyte Subpopulations: The Role of Fc Receptor Carrying Cells, *Cell. Immunol.*, 32: 317-328 (1977).
70. W. H. West, G. B. Cannon, H. D. Kay, G. D. Bonnard, and R. B. Herberman, Natural Cytotoxic Reactivity of Human Lymphocytes Against a Myeloid Cell Line: Characterization of Effector Cells, *J. Immunol.*, 118: 355-361 (1977).
71. M. A. Kall and H. S. Koren, Heterogeneity of Human Natural Killer Cell Populations, *Cell. Immunol.*, 40: 58-68 (1978).
72. R. L. H. Bolhuis, H. R. E. Schuit, A. M. Nooyen, and C. P. M. Ronteltap, Characterization of Natural Killer (NK) Cells and Killer (K) Cells in Human Blood: Discrimination Between NK and K Cell Activities, *Eur. J. Immunol.*, 8: 731-740 (1978).
73. R. B. Herberman, M. E. Nunn, H. T. Holden, and D. H. Lavrin, Natural Cytotoxic Reactivity of Mouse Lymphoid Cells Against Syngeneic and Allogeneic Tumors. II. Characterization of Effector Cells, *Int. J. Cancer*, 16: 230-239 (1975).
74. C. A. Bowles, G. S. White, and D. Lucas, Rosette Formation by Canine Peripheral Blood Lymphocytes, *J. Immunol.*, 114: 399-402 (1975).
75. C. H. Miller, A. R. Carbonell, R. Peng, M. R. MacKenzie, and M. Shifrine, Cell Surface Markers on Canine Lymphocytes, *Am. J. Vet. Res.*, 39: 1191-1194 (1978).
76. D. E. Onions, B- and T-Cell Markers on Canine Lymphosarcoma Cells, *J. Natl. Cancer Inst.*, 59: 1001-1006 (1977).
77. P. Lalezari, S. L. Nehlson, S. B. P. Sinha, M. B. Stemerman, and F. J. Veith, Spontaneous Macrophage Erythrocyte Rosette Formation: A Species Specific Phenomenon, *Immunology*, 27: 457-462 (1974).
78. S. Krakowka and D. J. Guyot, Rosette Formation Assays in Dogs: Lack of Specificity of E Rosettes for T Lymphocytes, *Infect. Immun.*, 17: 73-77 (1977).
79. R. B. Herberman, S. Bartram, J. S. Haskill, M. Nunn, H. T. Holden, and W. H. West, Fc Receptors on Mouse Effector Cells Mediating Natural Cytotoxicity Against Tumor Cells, *J. Immunol.*, 119: 322-326 (1977).
80. S. Targan and M. Jondal, Monolayer Immune Complex (MIC) Fractionation of Fc Receptor Bearing Human Spontaneous Killer Cells, *J. Immunol. Methods*, 22: 123-129 (1978).
81. J. B. Winfield, P. I. Lobo, and M. E. Hamilton, Fc Receptor Heterogeneity: Immunofluorescent Studies of B, T and "Third Population" Lymphocytes in Human Blood with Rabbit IgG b4/Anti-b4 Complexes, *J. Immunol.*, 119: 1778-1784 (1977).
82. S. S. Froland, J. B. Natvig, and T. B. Michaelson, Binding of Aggregated IgG by Human B Lymphocytes Independent of Fc Receptors, *Scand. J. Immunol.*, 3: 375-380 (1974).
83. I. McConnell and C. M. Hurd, Lymphocyte Receptors. II. Receptors for Rabbit IgM on Human T Lymphocytes, *Immunology*, 30: 835-839 (1976).
84. L. Moretta, M. Ferrarini, M. L. Durante, and M. C. Mingari, Expression of a Receptor for Ig by Human T Cells In Vitro, *Eur. J. Immunol.*, 5: 565-569 (1975).
85. W. H. West, R. B. Boozer, and R. B. Herberman, Low Affinity E-Rosette Formation by the Human K Cell, *J. Immunol.*, 120: 90-95 (1978).

86. B. Bonavida and T. P. Bradley, Studies on the Induction and Expression of T Cell-Mediated Immunity. V. Lectin Induced Non-Specific Cell-Mediated Cytotoxicity by Alloimmune Lymphocytes, *Transplantation*, 21: 94-102 (1976).
87. D. L. Nelson, B. M. Bundy, H. E. Pitchon, R. M. Blaese, and W. Strober, The Effector Cells in Human Peripheral Blood Mediating Mitogen Induced Cellular Cytotoxicity and Antibody Dependent Cellular Cytotoxicity, *J. Immunol.*, 117: 1472-1481 (1976).
88. B. Bonavida, A. Robins, and A. Saxon, Lectin Dependent Cellular Cytotoxicity in Man, *Transplantation*, 23: 261-270 (1977).
89. G. R. Betton, A. M. Dekkers-Bijma, and P. M. Lansdorp, Mitogenic Responses of Canine Peripheral Blood Lymphocytes to Staphylococcal Protein A, *J. Immunol. Methods*, 32: 157-166 (1980).
90. D. M. Callewaert, J. J. Lightbody, J. Kaplan, J. Jaroszewski, W. D. Peterson, and J. C. Rosenberg, Cytotoxicity of Human Peripheral Blood Lymphocytes in Cell Mediated Lympholysis; Antibody Dependent Cell Mediated Lympholysis and Natural Cytotoxicity Assays After Mixed Lymphocyte Culture, *J. Immunol.*, 121: 81-85 (1978).
91. J. K. Seeley and S. H. Golub, Studies on Cytotoxicity Generated in Human Mixed Lymphocyte Cultures. I. Time Course Study and Target Spectrum of Several Distinct Concomitant Cytotoxic Activities, *J. Immunol.*, 120: 1415-1422 (1978).
92. A. E. Butterworth and D. Franks, Non-Specific Cytotoxic Effects of Antigen Transformed Lymphocytes. III. Relationship to Specific Cytotoxicity, *Cell. Immunol.*, 16: 74-81 (1975).
93. A. E. Butterworth, Non-Specific Cytotoxic Effects of Antigen Transformed Lymphocytes: Kinetics, Cell Requirements in the Role of Recruitment, *Cell. Immunol.*, 7: 357-369 (1973).
94. Y. Koide and M. Takasugi, Augmentation of Human Natural Cell-Mediated Cytotoxicity by a Soluble Factor. I. Production of N-Cell Activating Factor (NAF), *J. Immunol.*, 121: 872-878 (1978).
95. S. Einhorn, H. Blomgren, and H. Strander, Interferon and Spontaneous Cytotoxicity of Peripheral Lymphocytes by Human Leukocyte Interferon, *Int. J. Cancer*, 22: 405-412 (1978).
96. J. R. Ortaldo, G. D. Bonnard, and R. B. Herberman, Cytotoxic Reactivity of Human Lymphocytes Cultured In Vitro, *J. Immunol.*, 119: 1351-1357 (1977).
97. A. V. Muchmore, J. M. Decker, and R. M. Blaese, Spontaneous Cytotoxicity of Human Peripheral Blood Mononuclear Cells Toward Red Blood Cell Targets. II. Time Dependent Loss of Suppressor Cell Activity, *J. Immunol.*, 119: 1686-1689 (1977).
98. A. Munro and S. Bright, Products of the Major Histocompatibility Complex and Their Relationship to the Immune Response, *Nature*, 264: 145-152 (1976).
99. G. M. Shearer and A. M. Schmitt-Verhulst, Major Histocompatibility Complex Restricted Cell-Mediated Immunity, *Adv. Immunol.*, 25: 55-92 (1977).
100. O. Haller, R. Kiessling, A. Orn, K. Karre, K. Nilsson, and H. Wigzell, Natural Cytotoxicity to Human Leukemia Mediated by Mouse Non-T Cells, *Int. J. Cancer*, 20: 93-103 (1977).
101. M. Hansson, K. Karre, T. Bakacs, R. Kiessling, and G. Klein, Intra- and Inter-Species Reactivity of Humans and Mouse Natural Killer (NK) Cells, *J. Immunol.*, 121: 6-12 (1978).
102. M. R. Potter and M. Moore, Organ Distribution of Natural Cytotoxicity in the Rat, *Clin. Exp. Immunol.*, 34: 78-86 (1978).
103. J. Martinis and F. H. Bach, Human LD Antigens Are Present on Xenogeneic Cells, *Nature*, 266: 540-542 (1977).
104. G. G. Petranyi, R. Kiessling, and G. Klein, Genetic Control of "Natural" Killer Lymphocytes in the Mouse, *Immunogenetics (N. Y.)*, 2: 53-61 (1975).

105. G. G. Petranyi, R. Kiessling, S. Povey, G. Klein, L. Herzenberg, and H. Wigzell, The Genetic Control of Natural Killer Cell Activity and Its Association with In Vivo Resistance Against Moloney Lymphoma Isografts, *Immunogenetics (N. Y.)*, 3: 15-28 (1976).
106. O. Haller, R. Kiessling, A. Orn, and H. Wigzell, Generation of Natural Killer Cells: An Autonomous Function of the Bone Marrow, *J. Exp. Med.*, 145: 1411-1416 (1977).
107. J. R. Oehler, L. R. Lindsay, M. E. Nunn, and R. B. Herberman, Natural Cell-Mediated Cytotoxicity in Rats. I. Tissue and Strain Distribution and Demonstration of a Membrane Receptor for the Fc Portion of IgG, *Int. J. Cancer*, 21: 204-209 (1978).
108. D. Santoli, G. Trinchieri, C. M. Zmijewski, and H. Koprowski, HLA-Related Control of Spontaneous and Antibody Dependent Cell Mediated Cytotoxic Activity in Humans, *J. Immunol.*, 117: 765-770 (1976).
109. R. Kiessling, P. S. Hochman, O. Haller, G. M. Shearer, H. Wigzell, and G. Cudkowitz, Evidence for a Similar or Common Mechanism for Natural Killer Cell Activity and Resistance to Hemopoietic Grafts, *Eur. J. Immunol.*, 7: 655-663 (1977).
110. H. M. Vriesendorp, H. Grosse-Wilde, and M. E. Dorf, The Major Histocompatibility Complex of the Dog, in *Major Histocompatibility System in Man and Animals*, E. Gotze (Ed.), pp. 129-163, Springer/Verlag, New York, 1977.

# Canine Blood Groups

# 7

C. STORMONT
*and*
Y. SUZUKI
*Department of Reproduction, School of Veterinary Medicine, University of California, Davis*

During early plasmaphoresis experiments with dogs,[1] some pairs of animals showed serologic incompatibility. Incompatibility was manifested in vivo by clinical symptoms associated with transfusion reactions and in vitro by serological reactions between the red cells of one animal and the antibodies in the serum of the other animal in the experimental pair. In the first systematic studies of such experimentally produced antibodies, Young and coworkers[2-4] identified eight alloantigenic determinants or blood factors and laid the foundation for the contemporary knowledge of dog blood groups.

## NOMENCLATURE

Originally, blood factors were given alphabetical designations, A through G, in the order of their discovery. Two categories of A-positive red cells were recognized; a strongly reacting subtype $A_1$ was distinguished from the weakly reacting intergrades lumped under the designation A′, later renamed $A_2$.[3,4] Factors E and G, which were "lost,"[5] are not considered further. The addition[6,7] of factors Tr and He completes the roster of recognized dog blood factors (see Table 1).

In 1973, the *First International Workshop on Canine Immunogenetics*[7] adopted a new nomenclature; each factor was labeled with CEA (canine erythrocyte antigen) and a number in the sequence 1 to 8. In 1976, DEA (dog erythrocyte antigen) was substituted for CEA.[8] In 1978, another nomenclature was proposed to overcome the failure of the CEA and DEA nomenclatures to consider genetically independent systems of blood groups.[9] In that proposal,

TABLE 1

HISTORICAL BLOOD-FACTOR DESIGNATIONS IN DOG BLOOD GROUPS

| Original | 1973 | 1976 | 1978 proposal |
|---|---|---|---|
| A | CEA 1 | DEA 1 | Aa |
| $A_1$ | CEA 2 | DEA 2 | $Aa_1$ |
| B | CEA 3 | DEA 3 | Ba |
| C | CEA 4 | DEA 4 | Ca |
| D | CEA 5 | DEA 5 | Da |
| F | CEA 6 | DEA 6 | Fa |
| Tr | CEA 7 | DEA 7 | Tra |
| He | CEA 8 | DEA 8 | * |

*No designation.

which has not yet been officially adopted, each genetic system is represented by a capital letter and the blood factor by a small letter. For example, DEA-1 and DEA-2, which are subtypes and belong in the same system, are represented by Aa and $Aa_1$. When a new blood factor is identified and placed in the A system, it will be designated Ab etc. (see Table 2).

## SEROLOGY OF DOG BLOOD GROUPS

Alloimmune antisera are used as blood-typing antibodies or reagents for the recognized blood factors. Fresh autogenous serum is used with anti-A, whereas the other reagents are used as saline agglutinins.[2–4] The fresh autogenous serum enhances the agglutination of A-positive red cells but has no effect with other specificities. The complement in the fresh serum also produces rapid hemolysis of some of the A-positive red cells, which are designated $A_1$. Cells with weakly reactive subtypes of A, known as $A_2$, are not hemolyzed by anti-A and complement. Some of the $A_2$ subtypes are difficult to detect in ordinary saline agglutination and require the use of rabbit anti-dog globulin to bring out the reactions.[3]

A problem long recognized by investigators of dog blood groups is the spontaneous, progressive hemolysis of dog red cells in the presence of any normal dog serum, including autogenous serum.[1,2] That effect, although poorly understood, is not mediated by complement and is not alleviated by heating the serum at 56°C. Observations of the hemolytic activity of any dog alloantibodies in the presence of dog complement must therefore be limited to those hemolysins, invariably of specificity anti-$A_1$, which are powerful enough to effect hemolysis within 15 to 20 min after the beginning of the tests. The spontaneous hemolysis also creates problems in evaluating agglutination reactions when the blood-typing antisera are of low titer and weak avidity.

TABLE 2

INCIDENCE OF BLOOD FACTORS IN RANDOM DOGS

| Blood factor | | United States[4-6] | | Netherlands[7] | |
|---|---|---|---|---|---|
| Original name | 1976 designation | No. of dogs tested | Percent of dogs with factor | No. of dogs tested | Percent of dogs with factor |
| A | DEA 1 | 954 | 63.6* | 100 | 66 |
| $A_1$ | DEA 2 | 332 | 44.6 | 100 | 51 |
| B | DEA 3 | 867 | 5.5 | 100 | 0 |
| C | DEA 4 | 947 | 98.4 | 100 | 97 |
| D | DEA 5 | 764 | 22.3 | 100 | 17 |
| F | DEA 6 | 165 | 99.4 | 100 | 97 |
| Tr | DEA 7 | 197 | 53.8 | 100 | 47 |
| He | DEA 8 | † | 40.0 | 100 | 39 |

*Incidence of factor $A_2$ can be obtained by subtracting incidence of $A_1$ from incidence of A.

†Not given.

As a practical solution to the problem of spontaneous hemolysis, we perform two serial 20-min absorptions on undiluted, heated (30 min at 56°C) antiserum with a ratio of thrice-washed red cells to serum of 1 : 2 per absorption. Although the red cells of any dog are effective in removing the lytic property from the serum, we use the red cells of the dog that produced the antiserum so that none of the alloantibodies are removed. With such absorbed antisera, hemolytic and agglutination tests can stand for several hours after their setup and thereby permit full evaluation of the reactions.[10]

For our hemolysis tests, fresh-frozen rabbit serum is the source of complement. Rabbit complement must be absorbed with dog red cells at temperatures in the range of 1 to 4°C to remove heteroantibodies, mainly of Forssman specificity, which all rabbits have for dog red cells.

The use of antiglobulin tests will lead to recognition of certain dog alloantibodies that otherwise would remain cryptic.[11] Nevertheless, saline agglutination tests will continue to be useful. In our laboratory we find that the "microagglutination"[12] tests increase the sensitivity of the agglutination reactions and thereby permit higher dilutions of the antibody reagents. Sometimes the sensitivity of agglutination tests is enhanced by the addition of dextran and bovine serum albumin.[13]

## THE Tr SYSTEM

In 1971, Bowdler et al.[6] discovered a dog blood factor, designated Tr, that was serologically related to A of people. Tr is also similar to "A-like" specificities recognized in other species (e.g., J of cattle, R of sheep, and A of pigs) in that it is not an intrinsic property of the red cells. Rather, Tr is a soluble substance, produced in tissues, that is adsorbed by the red cells from the blood plasma.[14] Tr is closely associated with canine secretory antigen, designated CSA-A,[15,16] which is also serologically related to human A.[7]

A new antibody, designated anti-O, which was discovered recently, permits the distinction of three phenotypes in the Tr system: Tr, O, and "–," or the absence of both Tr and O.[11]

## INCIDENCE AND INHERITANCE OF BLOOD FACTORS

Each of the eight dog blood factors shows similar incidences in two widely separated, random dog populations (Table 2). The incidence data shown in Table 2 are the only information available on "He."

Limited family data[4,17,18] show that each blood factor is inherited as a simple autosomal dominant. Types $A_1$ and $A_2$ are presumably controlled by allelic genes.[3] Segregation data have shown that blood factor A is independent of B and C and that B is independent of D.[4,17,18] With no other segregation studies, the blood factors, other than $A_1$ and $A_2$, have been assigned to separate genetic systems largely on the basis of population data alone.

## OTHER CANINE BLOOD FACTORS

In 1968, Rubenstein and coworkers[13] detected 11 canine blood factors with alloimmune antisera, which reacted as agglutinins in the presence of dextran and bovine serum albumin. All but one of the alloimmune antisera were raised by skin allografts or the injection of lymph node cells. One agglutinin, designated anti-6, was found regularly as a naturally occurring agglutinin in the serum of dogs lacking factor 6. Blood factor 6 was serologically related to human A, which indicates an earlier recognition of what could have been the Tr factor. Segregation data (20 families with 109 offspring) indicated that each of the 11 factors was inherited independently of the others.

In 1975, Suzuki and coworkers[10] identified four new factors, numbered 4, 5, 6, and 7. Two of those factors, 6 and 7, formed a closed system akin to the MN types of people and the FV types of cattle. No dogs in that study lacked both factors. Thus the three phenotypes were 6+7–, 6–7+, and 6+7+. Recent population data in conjunction with limited family data indicate that 6 and 7 are probably coded by codominant, autosomal alleles.

In 1978, Colling and Saison[11] produced alloimmune antisera that required the use of antiglobulin and, with that antisera, identified five blood factors that apparently were controlled by nonallelic genes.

## BLOOD TRANSFUSIONS AND NEONATAL ISOERYTHROLYSIS (NI)

Sera from 145 random dogs were tested for naturally occurring alloantibodies against blood factors A, B, C, D, and F.[4] Alloagglutinins found included 3 sera with anti-B, 14 with anti-D, and 3 with antibodies against unknown blood factors. The low incidence of antibodies (15%) and their low titers (not exceeding 1/8) and avidity could account for the absence of reports of transfusion reactions involving randomly paired unimmunized dogs. Transfusions of $A_1$-positive blood produce severe reactions in dogs possessing immune anti-$A_1$ as a result of previous injections or transfusions of $A_1$-positive cells.[2,4] Those experimentally induced transfusion reactions, all involving blood factor A, are characterized by hemoglobinuria and high rates of red-cell destruction, but the reactions are usually not fatal. Significantly, immune alloantibodies against blood factors other than $A_1$ were not implicated as a cause of transfusion reactions.

In the transfusing of dogs, especially those with multiple transfusions over a period of several weeks, the clinician should consider the high alloimmunogenicity of blood factor $A_1$ and the risk, estimated at 25%, of randomly pairing an $A_1$-positive donor with an A-negative recipient.[4] Clearly, the most effective way of avoiding the risk of severe reactions from repeated transfusions is to use known A-negative dogs as donors.

Hemolytic disease of the newborn does not occur naturally in dogs. Experimental hemolytic disease, however, has been produced in pups[19] that ingested colostrum from females that had produced alloantibodies as a result of blood transfusions. As in the case of transfusion reactions, all cases of experimentally produced NI traced to blood factor $A_1$ and the alloantibodies elicited by that blood factor.[2,4] Such antibodies, unlike anti-B, -C, or -D, acted as strong hemolysins not only in vitro but also in vivo.

An unexpected case of NI in a litter of pups should alert the clinician to probable cause: the mother of the pups, presumed to be A-negative, has produced anti-A as a result of a blood transfusion or as a result of immunization with a vaccine prepared from dog blood cells or tissues; she had been bred to a male that possessed blood factor $A_1$. The specificity of the offending antibodies is almost certain to be anti-A if they act as in vitro hemolysins.

## IMPORTANCE OF RESEARCH ON DOG BLOOD FACTORS

Since the early developments in the studies of dog blood groups,[1–3] relatively little progress has been made. For example, all but two of the eight blood factors listed in the 1976 DEA nomenclature were described in those early reports. However, considerably more antigens mark dog red cells than the nomenclature would indicate.[9-11]

Perhaps the greatest need in dog blood group research is for international comparisons akin to those which have been so helpful in expanding knowledge about blood groups in such species as cattle, sheep, pigs, and horses. In such comparisons each laboratory would test a representative panel of dog red cells consigned by one of the participating laboratories. The "duty" laboratory would compile the results from all laboratories and provide each participating laboratory with a copy of the overall comparisons. In that manner each laboratory could compare its results with several other parallel tests on the same panel of blood cells. That procedure would ensure accurate identification of all specificities that are truly different from one another and could also lead to the identification of supertypic and subtypic relationships among the various blood factors. Such procedure, although not discouraging the exchange of blood-typing reagents, offers an alternative to avoid the problems of false positive and false negative reactions which are so often associated with the use of reagents by workers who have had no previous experience with those reagents.

The number of inquiries we receive concerning the use of blood-typing tests to resolve problems of questionable paternity arising in the breeding of dogs suggests a compelling reason for encouraging such international comparison tests; such work might facilitate the development of tests that could be used efficaciously and reliably in the solution of paternity problems.[20] There are other reasons for the development of canine blood-typing tests that are capable of detecting the vast majority of dog erythrocyte antigens. For example, such tests could provide a reliable record of identification and could be used in determining the success or failure of bone-marrow allografts.

In humans, mice, pigs, chickens, and horses, there is evidence that at least one locus controlling red-blood-cell alloantigenic markers is within the major histocompatibility complex.[21-24] Such a locus has not yet been identified in dogs.

The above need, together with the importance of the dog as a biomedical model, calls for renewed interest and support for studies of the blood groups that would parallel and augment the efforts devoted to the immunogenetic analysis of the MHC in this species.

## REFERENCES

1. A. Wright, Isohemolysins and Isoagglutinins Occurring in Dogs, *Proc. Soc. Exp. Biol. Med.*, 34: 440-443 (1936).
2. L. E. Young, D. M. Ervin, and E. L. Yuile, Hemolytic Reactions Produced in Dogs by Transfusion of Incompatible Dog Blood and Plasma. I. Serologic and Hematologic Aspects, *Blood*, 4: 1218-1231 (1949).
3. R. M. Christian, D. M. Ervin, and L. E. Young, Observations on the In Vitro Behavior of Dog Isoantibodies, *J. Immunol.*, 66: 37-50 (1951).

4. S. N. Swisher and L. E. Young, The Blood Grouping Systems of Dogs, *Phys. Rev.*, 41: 495-520 (1961).
5. S. N. Swisher, R. Bull, and J. Bowdler, Canine Erythrocyte Antigens, *Tissue Antigens*, 3: 164-165 (1973) (abstract).
6. S. J. Bowdler, R. W. Bull, R. Slating, and S. N. Swisher, Tr: A Canine Red Cell Antigen Related to the A-Antigen of Human Red Cells, *Vox Sang.*, 20: 542-554 (1971).
7. Joint Report of First International Workshop on Canine Immunogenetics, *Tissue Antigens*, 3: 145-162 (1973).
8. R. W. Bull, R. Saison, and S. N. Swisher, Current Status of the Dog Erythrocyte Antigen (DEA) System, *Anim. Blood Groups Biochem. Genet.*, 8 (Suppl. 1): 12 (1977) (abstract).
9. R. Saison and D. Colling, A Suggested Nomenclature for Canine Red Cell Blood Groups. Presented at the *XVIIth International Conference of Animal Blood Groups and Biochemical Polymorphism*, Leningrad, Aug. 14–18, 1978.
10. Y. Suzuki, C. Stormont, B. G. Morris, M. Shifrine, and R. Dobrucki, New Antibodies in Dog Blood Groups, *Transplant. Proc.*, 7: 365-367 (1975).
11. D. Colling and R. Saison, A Report of Five New Canine Red Cell Groups and a New Red Cell Allele in the Tr System. Presented at the *XVIIth International Conference of Animal Blood Groups and Biochemical Polymorphism*, Leningrad, Aug. 14–18, 1978.
12. T. G. Wegmann and O. Smithies, A Simple Hemagglutination System Requiring Small Amounts of Red Cells and Antibodies, *Transfusion*, 6: 67-73 (1965).
13. R. Rubinstein, M. Morigado, D. A. Blumenstock, and J. W. Ferrebee, Isohemagglutinins and Histocompatibility in the Dog, *Transplantation*, 6: 961-969 (1968).
14. R. W. Bull, H. M. Vriesendorp, A. Zweibaum, and S. N. Swisher, The Inapplicability of DEA-7 as a Canine Bone Marrow Transplantation Marker, *Transplant. Proc.*, 7: 575-577 (1975).
15. R. Oriol et al., Determination of the Structure of the Immunodominant Groups of the Canine Secretory Alloantigens (CSA) by Radioimmunoassay, *Transplant. Proc.*, 7: 523-528 (1975).
16. A. Zweibaum, R. Oriol, N. Feingold, E. Dussaulx, M. Rousset, and G. Chevalier, Studies on Canine Secretory Alloantigens (CSA), *Tissue Antigens*, 4: 115-129 (1974).
17. C. Cohen and J. L. Fuller, The Inheritance of Blood Types in the Dog, *J. Hered.*, 44: 225-228 (1953).
18. S. N. Swisher, L. E. Young, and N. Trabold, In Vitro and In Vivo Studies of the Behavior of Canine Erythrocyte Isoantibody Systems, *Ann. N. Y. Acad. Sci.*, 97: 15-25 (1962).
19. L. E. Young et al., Hemolytic Disease in Newborn Dogs, *Blood*, 4: 291-313 (1951).
20. N. M. Vriesendorp, B. Duyzer-Den Hartog, B. M. J. Smid-Mercx, and D. L. Westbroek, Immunogenetic Markers in Canine Paternity Cases, *J. Small Anim. Pract.*, 15: 693-699 (1974).
21. G. D. Snell, J. Dausset, and S. Nathenson, *Histocompatibility*, pp. 118 and 216, Academic Press, New York, 1976.
22. V. Hruban, M. Simon, J. Hradecky, and F. Jilek, Linkage of the Pig Main Histocompatibility Complex and the J Blood Group System, *Tissue Antigens*, 7: 267 (1976).
23. J. R. L. Pink, W. Droege, K. Hala, V. C. Miggiano, and A. Ziegler, A Three-Locus Model for the Chicken Major Histocompatibility Complex, *Immunogenetics (N. Y.)*, 5: 203 (1977).
24. E. Bailey, C. Stormont, Y. Suzuki, and A. Trommershausen-Smith, Linkage of Loci Controlling Alloantigens on Red Blood Cells and Lymphocytes in the Horse, *Science*, 204: 1317 (1979).

# 8 Canine Histocompatibility Testing

H. M. VRIESENDORP
*Radiobiological Institute TNO, The Netherlands*

## GENETIC CONTROL OF ALLOGRAFT SURVIVAL

Under normal circumstances a healthy mammalian applies secure autocratic principles. Intruding microorganisms are recognized as foreign and are removed or inactivated by the immune system. Likewise the unnatural act of organ transplantation is not tolerated by the recipient, who will reject the introduced foreign material.

Rejection is the stumbling block as well as the intellectual challenge of organ transplantation. The transplantation of a normal organ is not the easy solution for life-threatening organ failure because the recipient's immune system, unconvinced of the desirability of accepting the new organ, will reject it. Two fundamentally different approaches were developed to fool the recipient into accepting a new organ. The first approach to prevent rejection is to abrogate the effector mechanisms of the host immune system by immunosuppressive agents. Immunosuppression in dogs was recently reviewed elsewhere.[1] The second approach involves the recognition mechanism of the host. If the newly introduced organ is histocompatible, i.e., similar to the recipient tissues (*histos,* in Greek), the host will not mount an immune response against that organ. Histocompatibility is determined by cell membrane antigens. This chapter is a brief review of the current state of knowledge of dog histocompatibility membrane determinants.

Table 1 lists the generally accepted nomenclature of the possible different donor–recipient combinations. The immunological specificity of rejection is

TABLE 1

DEFINITION OF DONOR–RECIPIENT PAIRS

| Label | Donor origin |
|---|---|
| Autologous | Recipient |
| Isologous | Same inbred strain as recipient (mouse)<br>Monozygotic twin of recipient (man) |
| Allogeneic | Same species as recipient, not isologous |
| Xenogeneic | Different species than recipient |

demonstrated in experimental animals by the unlimited survival of autologous or isologous transplants. In unmodified recipients allogeneic transplants are usually rejected after an initial period of acceptance. Xenografts are always rejected, usually within a few hours, except when the species of donor and recipient are phylogenetically close, e.g., fox, wolf, and dog.[2] Then survival times approach those of allogeneic combinations.

## MAJOR AND MINOR HISTOCOMPATIBILITY SYSTEMS

Without immunosuppression, allograft survival is usually short in unrelated donor–recipient pairs; prolonged survival is found only in those extremely rare cases when donor and recipient happen to be histocompatible. The low frequency of histocompatibility matching between unselected, unrelated donor–recipient pairs can be explained by either one extremely polymorphic histocompatibility system or several systems, which could be less polymorphic. With a single polymorphic histocompatibility system, 25% of the grafts in random littermate donor–recipient pairs would survive indefinitely.[3] A summary of selected kidney allograft experiments in dogs (Fig. 1) argues against the single-system assumption. However, allograft survival in random littermate combinations is obviously different from allograft survival in random unrelated combinations, and the influence of a single genetic system can be seen in the littermate combinations. Survival in those donor–recipient pairs segregates in three groups. In one group of littermate pairs (relative frequency, approximately 25%), allograft survival is as short as that in unrelated donor combinations. Of the remaining 75%, about two-thirds (50% of the total) (group 2) survive between 12 and 35 days; the final 25% of animals (group 3) survive for even longer periods after transplantation. The relative frequency of these groups can be explained by simple Mendelian genetics for one system. A complete (or so-called 2 haplotype) difference for this system between littermates (a chance

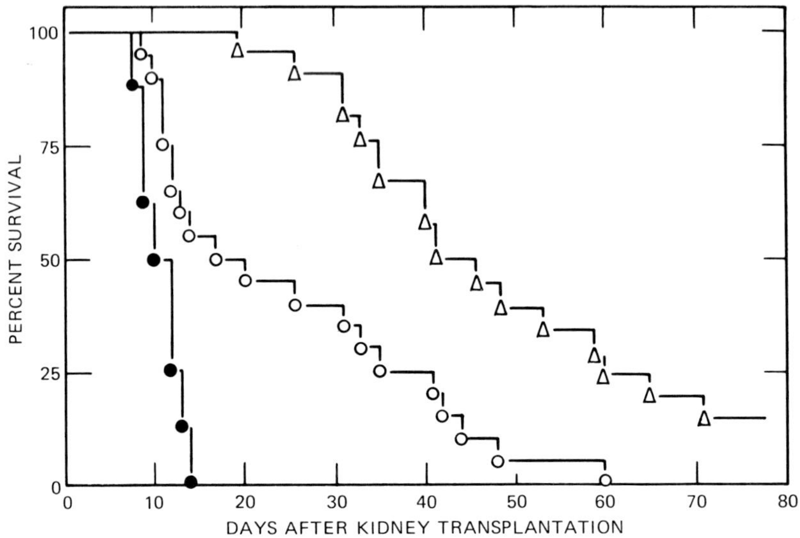


**Fig. 1 Dog kidney allograft survival from random unrelated and littermate donors. ●, unrelated donor (n = 8).[10a] ○, random littermate (n = 20).[10] △, DLA-identical littermate (n = 21).[10,13]**

of 25% by Mendelian genetics) is associated with short survival (group 1). Partial identity (1 haplotype difference) is associated with group 2 animals, and complete identity (identical haplotypes) will be present for group 3. There has been no indefinite survival in group 3 dogs. Thus several histocompatibility systems appear to exist, but one of the systems is more important than the others. In dogs, that major histocompatibility system was labeled DLA.[4] Later, when it was determined that DLA consisted of more than one locus, the word "system" was exchanged for "complex," in analogy to the human histocompatibility nomenclature.[5,6] The selection of dog littermate pairs for identity for antigens of that strong histocompatibility system (DLA) yields consistent long-term survival, but indefinite survival is significantly below 25%, which indicates the presence of more than one additional histocompatibility system. The dog allograft experience mirrors the experience in other experimental animal models as well as in man, where one strong (or major) and several weak (or minor) histocompatibility systems have been identified.[7] The separation of genetic control into one major histocompatibility complex (MHC) and several minor systems and the differences between the two kinds of systems (Table 2) have implications for the definition of the most realistic goal for histocompatibility studies. The number of different histocompatibility systems will not allow complete matching between unrelated individuals or even siblings. Minor histocompatibility differences are indeed of minor importance, difficult to identify prospectively, and relatively easy to suppress immunologically. The

TABLE 2

DIFFERENCES BETWEEN MAJOR AND MINOR HISTOCOMPATIBILITY SYSTEMS

| No. of different systems | Influence on grafted survival | Inactivation by immuno-suppression | Prospective identification of antigens | Histology of rejection[9-11] |
|---|---|---|---|---|
| | | Major Histocompatibility System | | |
| 1 | Pronounced[8] | Diffucult | Easy[8] | Acute graft ischemia: severe cellular infiltrate |
| | | Minor Histocompatibility System | | |
| Many | Weak; only demonstrated when identity present for the major histocompatibility complex; sometimes sensitization required[14] | Easy[12,13] | Difficult[14] | Chronic graft ischemia due to endarteritis; mild sometimes disap-pearing cellular infiltrate |

place of histocompatibility studies could thus be in the combination of partial matching (donor selection for major histocompatibility antigens) and immuno-suppression for the remaining minor histocompatibility differences. That principle, if properly applied, should afford prolonged graft survival.

## BIOLOGICAL SIGNIFICANCE OF HISTOCOMPATIBILITY

In addition to the search for the most effective and practical way to identify a histocompatible organ donor, investigators have turned their attention to the possible biological significance of histocompatibility genes. In the not too distant past, every genetic variation within a species was thought to be of biological significance.[15,16] More-recent evidence suggests that the majority of genetic polymorphisms have probably evolved by chance and not by pressures of natural selection.[17] Therefore the extreme polymorphism of the MHC is not acceptable as a priori evidence of the biological importance of the MHC. However, in man as well as in experimental animals, suggestive evidence for a biological significance of the MHC includes associations between MHC antigens and diseases[18] as well as between MHC antigens and parameters of biological "fitness," such as immune response genes.[19] In addition, population studies of MHC genes have indicated that present-day MHC phenotype frequencies and, in particular, the occurrence of so-called linkage disequilibrium are best explained if natural selection acts on MHC-coded structures.[20,21] Several hypotheses on the nature of the biological significance of MHC genes have been put forward.[19,22-24] All suggest some form of a regulatory role for the MHC in the development or control of tissue differentiation and immune responsiveness. Those hypotheses are now being tested in many laboratories, but the complexity of the problem as well as the models proposed for the biological significance of histocompatibility genes have impeded progress. The definition of that biological significance would greatly help the development of further applications of histocompatibility studies outside the field of transplantation.

## THE MAJOR HISTOCOMPATIBILITY COMPLEX DLA

### Nomenclature

Uniform and simple nomenclature is a prerequisite for proper communications, a fortiori for a complicated genetic area such as the MHC. The recombination frequencies between the different loci in the DLA area indicate that still many (>100) unidentified loci are within the DLA complex as it is currently defined.[25] Probably not all the loci (defined or undefined) will have a histocompatibility function. Until otherwise resolved, all loci within the DLA region will be considered as belonging to the MHC. The MHC structures known for the longest time are the serologically defined (SD) antigens[26] and the lymphocyte defined (LD) antigens responsible for mixed leukocyte culture (MLC) reactivity.[27]

In 1974, a nomenclature subcommittee was established for dog histocompatibility determinants by the International Union of Immunological Studies. This subcommittee adheres to a nomenclature policy formulated by Swisher[28] in 1963 for biomedical notation systems and to a maximum resemblance to human HLA nomenclature decisions. A report of the subcommittee,[29] published in 1977, recommends labeling all loci within the DLA complex with consecutive letters, starting with A. The loci of the SD series of multiple alleles are labeled DLA-A, DLA-B, and DLA-C. Their alleles receive the prefix DLA, followed by a dash, the letter of the locus, and their allelic number. DLA alleles are numbered from 1 upward in sequence of their discovery. The highest assigned number is 13. Alleles are not numbered per series; thus the DLA-A, DLA-B, and DLA-C series show gaps in their allele numbers.

The genetic control of MLC reactivity is polygenic in the dog. The assignment of MLC alleles to one of the different MLC loci (sometimes labeled DLA-D and DLA-E, although additional MLC loci might exist) is incomplete. Therefore the use of the prefix DLA and a consecutive number beginning with 50 is recommended for LD alleles to allow for an easy distinction from alleles of the SD series. Letters indicating the MLC locus for a given allele will be inserted later after clarification of the MLC system. Investigators are advised to use a local provisional nomenclature for new alleles of old loci and for new loci and their alleles until an international comparison has been performed. Thereafter an official designation will be considered in a meeting of the Nomenclature Committee.

The DLA complex includes loci other than the DLA-A to -E loci. Primed lymphocyte testing (PLT) defines PLT antigens in sensitized (initial) and secondary lymphocyte donors. PLT antigens probably overlap to some degree with DLA-D and -E antigens, but differences between PLT and DLA-D or -E antigens have been noted.[30] Presently no available assay modification allows for a separate recognition of MLC and PLT antigens.

R, which represents resistance, is a genetic locus (or loci) controlling a major part of the host vs. graft reaction in bone-marrow (BM) transplantation (see Chapter 9). An in vitro assay identifying alleles of R remains to be defined.[31]

IrGT, IrGA, and IrGL are three similar loci controlling the immune response to different copolymers of amino acids. GA stands for the L-glutamic acid and L-alanine polymer, GL for the L-glutamic acid and L-lysine polymer, and GT for the L-glutamic acid and L-tyrosine polymer.[32] PGM-3 is an abbreviation for phosphoglucomutase 3. That enzyme in leukocytes has variant forms differing in electrophoretic mobility.[33]

## Methods

**Microlymphocytotoxicity.** Antigens of the MHC reside in high density on cell membranes of blood lymphocytes and are thus readily available for testing. The bottleneck of any serological technique is the specificity of the antiserum used in the test. For a proper definition of the specificity of an antiserum against antigens of the MHC, large numbers of unrelated individuals as well as complete families need to be tested. The details of such a specificity analysis, which are similar for all species, have been published previously for dogs.[25,32,34–36] The technique involves finding sera with similar reaction patterns. In contrast to red-cell serology, more than one serum is required for the recognition of a single leukocyte antigen because each anti-leukocyte serum will have "false"-positive or "false"-negative results. The leukocyte antigen can be found only by accepting the most common reaction in a group of similarly reacting antisera. Then genetic models are built on the distribution of leukocyte antigens in families and in populations of unrelated individuals. That process should define the monofactorial nature of a given leukocyte group and the relationships between leukocyte groups. The possible relations between two groups are (1) alleles of the same locus, (2) alleles of different linked loci, and (3) alleles of different unlinked loci.

A microtechnique for histocompatibility testing sera extends available antisera, whereas wasteful 0.05-ml-droplet methods will consume most of the antisera during the specificity analysis and only allow the testing of a few individuals. Lymphocytes are isolated from venous peripheral blood by Ficoll–Isopaque gradients. Two different modifications have been in use in dog microlymphocytotoxicity testing. In the first, a one-stage test, the cells are suspended in complement and are added to an equal amount of antiserum for 30 min at 37°C. The second modification is a two-stage test performed at room temperature; equal volumes of cells and antiserum are incubated for 30 min, followed by a second incubation period with complement. The original descriptions of those tests can be found in the manual of techniques of the first workshop on canine immunogenetics[37] and in other publications.[6,34,36,38] Currently both technique modifications are performed in Falcon trays and fixed with formalin to allow reading of the reactions at a convenient time. Reproducibility of both tests has been between 90 and 95%. Increasing the amount of complement in the test (four parts complement to one part each of cells and serum) has given higher reproducibility (H. M. Vriesendorp, unpublished observations). The selection of the test procedure depends on the antisera. Antisera can be raised by immunizations with grafts (skin) or leukocyte injections or by pregnancy. One-stage testing was most useful for antisera produced by skin grafting.[4] Two-stage testing gives too many extra positive results with such sera and is more appropriate for pregnancy-induced antisera, which are more reactive in two-stage testing.[32] In addition to the sensitivity difference between the two tests, some antibodies are reactive in only one of the two test modifications.[39] Consequently antisera can only be trusted in the test modification that was used for the specificity analysis. Changes in test procedures (as in sera exchanges between laboratories) will often lead to changes in sera specificities.

Some of the genetic variation in the SD loci remains undefined, so-called blank alleles (Table 3). Since most of the currently available antisera were raised in beagles, blanks have the lowest frequency in that dog population, with less than one-third of the genetic variation for DLA-A, DLA-B, and DLA-C undefined. Mongrel dogs show a higher percentage (~50%) of blanks. Dog breeds other than beagles will show higher or lower blank frequencies than mongrel dogs, depending on the influence of genetic drift on the currently recognized antigens.[15] Obviously, blank antigens can be uncovered by the production and analysis of new antisera. The ideal donor–recipient pairs for sera production are parent–child combinations or child–child combinations with a single DLA haplotype difference.[26,40]

**Mixed Lymphocyte Cultures.** After a few days of culture, a mixture of lymphocytes from two unrelated dogs will show an enhanced DNA synthesis. Several successful culture methods[27,41,42] can be used with different growth media (MB 752/1, RPMI 1640, and MEM-s buffered with $NaHCO_3$ or Hepes). Cultures need 10 to 20% normal dog serum, preferably from unimmunized male dogs, to exclude the possibility of pregnancy-induced anti-DLA antisera.

TABLE 3

LOCI, ALLELES, AND METHODS OF DETECTION OF THE DLA COMPLEX

| Locus | Alleles | Method of detection |
|---|---|---|
| DLA-A | 1, 2, 3, 7, 8, 9, 10, blank* | Antisera in one- or two-stage microlymphocytotoxicity |
| DLA-B | 4, 5, 6, 13, blank | Antisera in one- or two-stage microlymphocytotoxicity |
| DLA-C | 11, 12, R15, blank | Antisera in one- or two-stage microlymphocytotoxicity |
| DLA-D and DLA-E† | 50, 51, 52, 53, 54, 55, 56, 57, 58, R1, R2, R3, R4, blank | Homozygous typing cells as stimulators in unilateral mixed lymphocyte cultures (MLC) |
| PLT | Undefined | Secondary MLC |
| R | Undefined | Hemopoietic regeneration after standardized total-body irradiation and bone-marrow transplantation |
| IrGT | IrGT | Specific antibody formation |
| IrGA | IrGA | after low-dose antigen inocula- |
| IrGL | IrGL | tion in complete Freund's adjuvant |
| PGM-3 | 3.1, 3.2, 3.3 | Electrophoresis of leukocyte lysate on cellulose acetate gel |

*Blank indicates that additional allele(s) remain to be defined.
†Allocation of alleles to D or E not possible yet.

The MLC can be modified to enable the expression of DNA synthesis by only one of the participants in the mixture (the so-called responder). For this purpose, the cells of the other animal (stimulator) are irradiated with 2000 rads or treated with mitomycin-C. Stimulator antigens absent in the responder initiate the increased deoxyribonucleic acid (DNA) synthesis. Labeled thymidine ($^{3}$H or $^{14}$C) is generally used to quantify the increased DNA synthesis in the mixture. The incorporation of labeled thymidine usually peaks between day 6 and day 7 of culture for mixtures from related animals and slightly earlier for mixtures from unrelated animals.[43,44] The ratio of stimulator to responder cells also influences MLC results. Ratios below one lead more often to negative results.[43,44] Many variables influence the final net increase in DNA synthesis measured at the end of the culture; an MLC is therefore a qualitative rather than a quantitative assay. No evidence for a gene dose effect was obtained in dogs.[43,44]

A given MLC antigen can be identified with stimulator cells that are homozygous for that antigen.[45] Homozygous typing cells can be made in the dog by mating DLA-identical littermates. One-half of their offspring will be homozygous for one of the two DLA haplotypes of the parents. (A haplotype is

TABLE 4

PRINCIPLES OF HOMOZYGOUS TYPING CELLS[44]

| Test | MLC mixture | | MLC response |
|---|---|---|---|
| | Responder | Stimulator* | |
| 1.A | AB | AAx | – |
| 1.B | AA | ABx | + |
| 2 | AB | CCx | + |
| 3 | CD | AAx | + |
| 4 | CD | CCx | – |

Conclusion: Individual AB has MLC antigen "A" and does not possess MLC antigen "C."

*x, irradiated with 2000 rads.

the DLA information carried on one chromosome.) Table 4 illustrates the principles of identification of such a homozygote in MLC. In a so-called unilateral one-way reaction, the homozygote AA will not stimulate AB, but AB will stimulate AA. In necessary control reactions, AB is able to respond and AA is able to stimulate. Finally, AB does not carry antigens of another homozygous typing cell CC (test 2, Table 4).

Several arguments propose the genetic control of MLC reactivity by more than one locus. First, some DLA haplotypes have more than one MLC/LD antigen.[20] Also, recombination within the DLA complex cannot be explained by one locus.[46,47] Finally, some experiments have shown a high frequency of positive MLC's between unrelated individuals with two different MLC/LD antigens in common.[46]

The effects of a multiple loci model on the homozygous typing cell method have not been analyzed yet through lack of appropriate methodology. One of the possibilities is that typing cells are homozygous for several antigens, and some antigens will be shared by different typing cells. The effect of partial identity between stimulating homozygous typing cells and the cells of the responder is unknown. Further population and family studies with a large panel of typing cells will be required to define optimal use of the MLC as a genetic tool. Such testing should also lead to the assignment of LD antigens to one of the MLC-controlling loci.

**Primed Lymphocyte Testing.** The PLT test described for human lymphocytes can be performed successfully with dog lymphocytes too.[30,48] After 10 days of primary unilateral MLC, cells can be counted and set up in secondary cultures with fresh stimulators. The processing and harvesting of cells after 2 days of additional culture is similar to that of normal MLC's with the exceptions that in PLT fewer cells can be used and that the secondary stimulation cells do not require prior irradiation or mitomycin treatment in such a short culture period.

Positive PLT results are strongly, but not absolutely, correlated with similar LD differences between the first stimulator and responder and the second stimulator and responder. However, about one-third of the positive PLT occurs when first and second stimulators have no LD antigen in common. In addition, about 10% of the PLT is negative in secondary stimulation with the original stimulation cell when the primary MLC is clearly positive. Those data suggest differences between the genetic control of MLC and PLT.[30] A possible explanation is either polygenic control for both phenomena, with some overlap, or a completely separate genetic control, with a high degree of linkage disequilibrium between alleles of the different genetic systems. The available evidence indicates that PLT reactivity is genetically influenced from within the DLA complex only.[30]

**Resistance.** The resistance to or the rejection (R) of allogeneic BM cells after total-body irradiation is under polymorphic genetic control, distinct from LD and SD loci (see Chapter 9). The number of BM takes in random, unrelated donor–recipient pairs indicates that at least 1 locus with 6 alleles (95% confidence limits: 3 to 17 alleles) of equal frequency and strength must control this phenomenon (see Chapter 9, Table 11). That estimate is based on mathematical formulas published by Simonson[3] on immunogenetic systems. The assumption of equal frequency and strength of alleles is introduced to facilitate computations. The data can also be explained by one locus with more alleles of unequal frequency or by more loci with a lesser degree of polymorphism. Those speculations can be verified only with the development of a reliable in vitro method that measures R alleles prospectively. Without such a test, BM transplantation and the in vivo evaluation of BM engraftment retrospectively will be the only way to assign differences for R between BM donor and recipient.[31]

**Immune-Response Genes and PGM-3.** The methodology for immunization and the measurement of specific antibody responses for immune-response (Ir) genes in dogs have been published previously.[32] In brief, 100 $\mu$g of GA and 50 $\mu$g of DNP-GL or 100 $\mu$g of GT in complete Freund's adjuvant are injected deep intramuscularly into the hind legs. Intradermal tests of 10 and 50 $\mu$g for each antigen follow 3 weeks later. Serum samples are collected on days 0, 21, and 28 after the start of the immunization procedure and assayed for antibody by a Farr assay.[49] A major problem of Ir studies in dogs is that a fair number (±25%) of animals with a specific antibody response showed only a small positive response. Further experimentation is required to define conditions for producing a wider difference in antibody response between responder and nonresponder animals.

PGM-3 variants can be determined by the techniques published by Meera Khan et al.,[33] who have demonstrated the linkage between PGM-3 and DLA.[50]

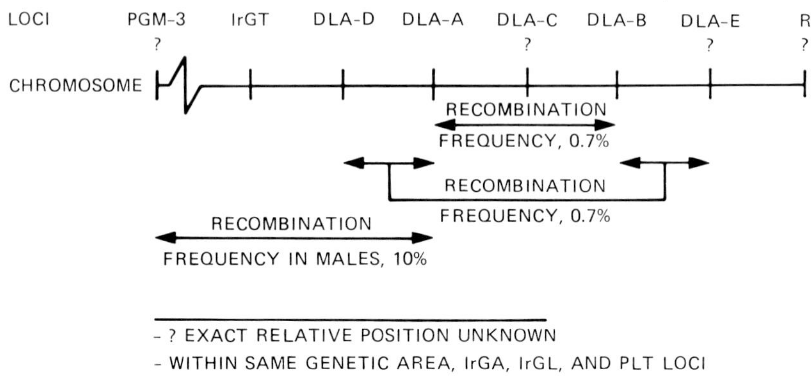


Fig. 2 Current model of the genetic organization of the DLA complex.

## Genetic Organization

Theoretically gene products of the same genetic locus can be measured in different ways. Therefore gene products that are measured differently do not necessarily belong to different loci. For example, SD and LD antigens could be identical antigens, which are measured one time serologically and another time by MLC. The various loci identified so far within the DLA region are separate loci because their polymorphic gene products show different distribution patterns in populations of unrelated animals. In families gene products of linked loci will usually be inherited en bloc or in repulsion. Exceptions will be produced by crossing over and recombination. Studies of such recombinant families allow determination of the order within a group of linked loci. Recombinants between some of the loci on the DLA chromosome have been identified[32] (Fig. 2). Informative recombinants for PGM-3 and DLA-D have not been found, and the PGM-3 position is assumed to be analogous to man. PGM-3–MHC linkage can be demonstrated only in male segregation patterns.[50] The found DLA recombinants are documented or referenced in a recent review.[32] Positions of DLA-C, DLA-E, and R are still unverified by recombinants.[8,32]

Discrimination between alleles of linked loci is more difficult in dogs than in other outbred species because of a high degree of linkage disequilibrium in dogs. Normally linkage equilibrium describes the independence of alleles of linked loci after they have reached an equilibrium state in the random population.[51] For example, the presence of allele *a* of locus A in an individual does not influence the chance that another allele *b* of linked locus B is present in that individual. Linkage disequilibrium[52] is the reverse situation; *a* and *b* are often found in the same individual, and consequently the frequency of the phenotype ab is much higher than the product of the population frequencies of *a* and *b*. Linkage disequilibrium can result from several conditions: when a new allele originating from a recent mutation has not yet reached equilibrium in the population, after a so-called "founder" or bottleneck effect (i.e., populations in which a few ancestors contributed most of the genes to much more numerous next generations), or in the case of natural selection favoring certain combinations of linked genes over others.

The dog shows a high degree of linkage disequilibrium between SD alleles of different loci as well as between SD and LD alleles and between LD alleles of different loci.[20,53] The higher degree of linkage disequilibrium in the DLA complex than in HLA in man[54] or RhLA in rhesus monkeys[55] can be explained by a different population structure in dogs causing a more pronounced founder effect.[6] The founder effect should not be equated with inbreeding. In general, dog breeders will not breed family members, and even dogs apparently show preference for unrelated mating partners.[56] Dog breeders will breed preferentially with a few "champion" animals (founders). Genes of those animals and thus also their DLA haplotypes will be overrepresented in next generations. Inbreeding would increase the frequency of homozygosity for polymorphic genetic systems, but dog genetic systems show Hardy Weinberg equilibrium,[35,57] which indicates that the observed frequency of homozygosity is not significantly different from the frequency expected from random mating.[52] Thus the founder effect will lead to some loss of genetic variation through "gene drift" with gene loss or gene fixation,[15] but the effect is not the same as inbreeding.

Suggestive evidence for linkage disequilibrium caused by a natural selection process has been found for the MHC determinants of man and dog.

## Correlations with Graft Survival

The most practical solution to graft rejection is to match donor and recipient for major histocompatibility differences and suppress the inevitable remaining minor histocompatibility differences by immunosuppression. The genetic complexity of the major histocompatibility complex interferes with a general application of that approach.

For siblings, some donor–recipient pairs will be identical for the whole MHC. For histocompatibility purposes, the more distantly located PGM-3 information can probably be disregarded. The recombination frequency between the remaining loci is so low that the majority of littermate donor and recipient pairs will be identical for the whole MHC regardless of the locus and the technique used for donor selection. The chances for a MHC-identical donor among littermates are $1-(3/4)^n$, in which n is the number of siblings.

If no MHC-identical sibling is available, the selection of an unrelated donor is hampered by the ignorance of which loci within the MHC are really coding for histocompatibility antigens. LD antigens have some histocompatibility effect but do not fully explain the results of MHC matching in littermate donor–recipient pairs.[43] The recombination frequency within the DLA complex, which is similar to that of the human HLA system, indicates the presence of some 100 additional, presently unrecognized loci in the MHC region.[25] At least some of those loci will be polymorphic and might contribute to the control of allograft survival. In analogy to the murine situation, probably several loci within the MHC, including non-LD loci, control graft survival.[8,43,58] Since not all the histocompatibility loci within the MHC have been identified, the principle of partial donor selection and immunosuppression cannot be applied in unrelated donor–recipient pairs.

In unrelated donor–recipient pairs, almost every kind of MHC matching will have some advantage through the effects of linkage disequilibrium. Matching unrelated donor–recipient pairs for a system without a histocompatibility effect will produce significant graft prolongation over mismatched pairs through accidental matching for alleles of a linked, true, histocompatibility system. That matching will often be incomplete since linkage disequilibrium, which is the cause of that accidental matching, is never 100%. Therefore indirect matching will never be as good as matching through a direct identification of the true histocompatibility determinants. Further studies in experimental animals will be required for complete identification of all histocompatibility loci within the MHC. Even then the cumulative polymorphism of these loci might be so high that donor–recipient matching between unrelated would remain an unattainable dream.

In littermate donor–recipient pairs, DLA matching will remain the most effective tool for prolonging survival of every conceivable, transplantable organ (including frozen nerves).[8]

## MINOR HISTOCOMPATIBILITY SYSTEMS

In human transplantation, identity for the major histocompatibility complex will not always safeguard against rejection problems. Over 25% of recipients of HLA-identical bone-marrow cells die from graft-vs.-host disease (GvH), and, despite maximal immunosuppression, approximately 25% of HLA-identical kidneys are rejected within 5 years.[59,60] Those complications are caused by histocompatibility differences controlled by minor histocompatibility systems. Solutions to these problems require further experimental animal work. The dog model has the advantage of providing sibling donor–recipient pairs, which show the same immunogenetic relation as HLA-identical siblings.

### Methods of Analysis for Minor Histocompatibility Systems

Studies with inbred mice have defined several principles for the analysis of minor histocompatibility systems.[61] With slight modifications the same methodology can be applied to dogs.[14]

Two differences with major histocompatibility antigens complicate analyses of minor systems. First, multiple minor antigens, belonging to systems on many different chromosomes, will simultaneously influence graft survival. Second, a prospective identification of minor antigens is difficult. The problems caused by the nature of minor histocompatibility antigens can be solved in several ways. For example, if a genetic marker has been linked to a minor histocompatibility system, prospective identification of minor antigens can be attempted with the appropriate animals. Another example is the use of third-party donors, different from either graft donor or recipient, for sensitization. Only those minor histocompatibility antigens which are shared by sensitization donor and graft donor will be enhanced; thus the number of active minor systems will be limited, and the confusion created by the simultaneous activity of multiple systems will decrease. A last example is the possibility to distinguish a minor histocompatibility system on the Y chromosome from a system on the X chromosome or an autosomal chromosome. All male-into-female combinations will be incompatible for a Y chromosome locus, and grafts in such combinations should always be rejected. The percentage of incompatible pairs (or rejection) for an autosomal locus is dependent on the polymorphism of such a locus but in contrast to an X-linked locus independent of the genders of donor and recipient. Simplified mathematical models for the predictions involved in such an analysis, with application in dog kidney and bone-marrow transplantation, have tentatively identified X- and Y-linked histocompatibility systems in dogs.[14] A minor histocompatibility locus on the canine X chromosome influences kidney graft survival in sensitized DLA-identical littermate pairs but has no influence on GvH disease in those donor–recipient pairs.[14] The set of minor histocompatibility loci that influences the allograft survival of one organ might be different from the set that influences survival of another organ. By recent estimate, only a few (±3) minor histocompatibility loci influence kidney graft survival in dogs, but many more loci (>20) influence skin-graft survival in mice.[30,43,58]

## Minor Histocompatibility Systems in Dogs

**Blood Groups.** Blood groups can be considered minor histocompatibility systems if they influence the survival of allogeneic blood cells after transfusion. Immunizations across differences for the dog erythrocyte antisera (DEA) systems A and D have led to shortened red-cell survival with repeated transfusions of blood donors with the same phenotype as the sensitizing donor.[62]

**Canine Secretory Allo-Antigen System.** The human ABO blood-group system has been shown to have some histocompatibility effect.[63] An immunogenetic system in dogs with some similarities to the human ABO and Lewis systems is the canine secretory allo-antigen (CSA) system.[64,65] The equivalent system in rabbits was of some importance for skin allograft survival.[66] In dogs CSA incompatibility apparently offered some protection against GvH reactions in DLA-identical donor–recipient pairs.[67] The mechanism of that protective effect is unknown. Systematic studies of possible histocompatibility effects of CSA structures under different circumstances in BM transplantation or transplants of other organs have not been performed yet.

**Phosphoglucomutase 2 (PGM-2).** PGM-2 is a red-cell enzyme that shows electrophoretic variation in dogs. PGM-2 disparity was significantly more frequent in DLA-identical littermate donor–recipient combinations with GvH than in combinations without GvH.[67] That finding is compatible with the existence of a minor locus controlling GvH reactions linked to the PGM-2 locus. Two similar linkage groups in the dog are PGM-2 and a minor GvH locus and PGM-3 and a major GvH locus. Attempts to raise typing reagents for alleles of the hypothetical PGM-2-linked GvH locus by repeated skin grafts between five DLA-identical, PGM-2 different pairs did not yield antibodies against erythrocytes or lymphocytes of the skin donor. Blood transfusions might be a more-effective immunization procedure (H. M. Vriesendorp, unpublished observations).

**Sex Chromosomes.** Ohno[68] has suggested homology between sex chromosomes of all mammals and as a consequence a conservation of linkage groups on mammalian sex chromosomes. By that hypothesis the minor histocompatibility locus on the X and the Y chromosome in mice[69,70] should have a counterpart in histocompatibility loci on those chromosomes in dogs.[14] Serological or other methods for a prospective determination of antigens controlled by the X-chromosome histocompatibility locus are not available. The dog Y antigen, however, was demonstrated to be similar to the corresponding antigen in mice and men.

## ROLE OF HISTOCOMPATIBILITY STUDIES

Histocompatibility is an important variable in transplantation experiments. Defined histocompatibilities of donor and recipient are required not only in studies of allograft survival but also in studies of graft histology, function, and effect of immunosuppression. In dogs, histocompatibility can be achieved by controlling DLA and sex differences in littermate pairs. For studies involving donor–recipient combinations that are different for major and minor histocompatibility antigens, some researchers choose dissimilar mongrel dogs. That approach is not reliable. Animals from the same litter can appear quite different.[71] Thus chance histocompatibility matching will not be excluded by selecting for anatomical dissimilarity. The use of littermates different for two

DLA haplotypes is a much better way of introducing a standard histoincompatibility. For most experimental purposes, all the required histocompatibility differences can be obtained with DLA-typed littermate donor–recipient pairs.

Histocompatibility research should be continued for a definition of better selection procedures in unrelated donor–recipient pairs. Dog studies could make a major contribution because of the large average litter size in the species. DLA recombinant families in combination with the availability of DLA-identical littermate pairs offer excellent starting points for a further analysis.

In addition to their use in transplantation research, major histocompatibility antigens can find application in other fields of research. The genetic structure of a dog population or interrelationships between dog populations can be defined by those polymorphic markers. Disputed paternity can be checked by the same markers.[72] In human histocompatibility testing correlations have been found between certain diseases and histocompatibility antigens.[18] The use of dogs for such studies is difficult since the usually weak correlations necessitate the availability of large populations of diseased and normal dogs.[53] As an additional problem, a considerable portion of non-beagle dogs show undefined genetic variation within DLA systems (blanks). That situation reduces the chance for significant results from such DLA-disease susceptibility studies. The most economical way to check for correlations between DLA determinants and specific diseases or other biological phenomena is to study dog families with a segregation of the phenomenon (e.g., some animals have the disease whereas others do not). In particular, if three-generation families are available, only a few animals will be required to test the hypothesis of a connection between DLA and that particular phenomenon. Positive results in that area would offer great possibilities for a further, direct analysis of the biological significance of MHC determinants.

## REFERENCES


1. H. M. Vriesendorp, Applications of Transplantation Immunology in the Dog, *Adv. Vet. Sci. Comp. Med.,* 23: 229 (1979).
2. C. Hammer, C. Chaussy, J. von Scheel, H. Pongratz, E. Roscher, and W. Brendel, Survival Times of Skin and Kidney Grafts Within Different Canine Species in Relation to Their Genetic Markers, *Transplant. Proc.,* 7: 439 (1975).
3. M. Simonson, Immunogenetic Speculations on Human Kidney Transplants, *Transplantation,* 4: 354 (1966).
4. H. M. Vriesendorp, C. Rothengatter, E. Bos, D. L. Westbroek, and J. J. van Rood, The Production and Evaluation of Dog Allolymphocytotoxins for Donor Selection in Transplantation Experiments, *Transplantation,* 11: 440 (1971).
5. Joint Report of the Fourth International Histocompatibility Workshop, in *Histocompatibility Testing 1970,* P. I. Terasaki (Ed.), p. 17, Munksgaard, Copenhagen, 1970.
6. H. M. Vriesendorp, *The Major Histocompatibility Complex of the Dog,* Thesis, Erasmus University, Rotterdam, The Netherlands.

7. D. Gotze, *The Major Histocompatibility Systems in Man and Animals,* Springer-Verlag New York, Inc., New York, 1977.
8. H. M. Vriesendorp, A. B. Bijnen, D. L. Westbroek, and D. W. van Bekkum, Genetics and Transplantation Immunology of the DLA Complex, *Transplant. Proc.,* 9: 293 (1977).
9. D. L. Westbroek, C. Rothengatter, H. M. Vriesendorp, J. J. van Rood, R. G. J. Willighagen, and M. J. de Vries, Histocompatibility and Allograft Rejection in Canine Small Bowel Transplants, Evidence for the Existence of a Major Histocompatibility Locus in the Dog, *Transplant. Proc.,* 3: 157 (1971).
10. D. L. Westbroek et al., The Influence of DLA Histocompatibility on the Function and Pathohistological Changes in Unmodified Canine Renal Allografts, *Transplantation,* 14: 52 (1972).
10a. H. Obertop, A. B. Bijnen, H. M. Vriesendorp, and D. L. Westbroek, Prolongation of Renal Allograft Survival in DLA Tissue Typed Beagles After Third Party Blood Transfusions and Immunosuppressive Treatment, *Transplantation,* 26: 255 (1978).
11. O. C. K. M. Penn, I. MacDicken, and E. Bos, Histopathology of Rejection in DL-A Identical Canine Cardiac Allografts, *Transplantation,* 22: 313 (1976).
12. R. Storb et al., Marrow Grafts Between DL-A Matched Canine Littermates, *Transplantation,* 15: 92 (1973).
13. J. de Gruyl et al., Influence of DLA Matching, ALS and 24 Hour Preservation on Isolated Pancreas Allograft Survival, *Transplant. Proc.,* 3: 755 (1973).
14. H. M. Vriesendorp, A. B. Bijnen, A. C. M. van Kessel, H. Obertop, and D. L. Westbroek, Minor Histocompatibility Systems in Dogs, in *Experimental Hematology Today 1978,* S. J. Baum and G. S. Ledney (Eds.), p. 109, Springer-Verlag New York, Inc., New York, 1978.
15. D. S. Falconer, *Introduction to Qualitative Genetics,* Oliver & Boyd, Edinburgh, 1964.
16. E. B. Ford, *Genetic Polymorphism,* Faber & Faber, London, 1965.
17. H. Harris, Polymorphism and Protein Evolution. The Neutral Mutation Random Drift Hypothesis, *J. Med. Gen.,* 8: 444 (1921).
18. J. Dausset and A. Svejgaard, *HLA and Disease,* Munksgaard, Copenhagen, 1977.
19. H. O. McDevitt and W. F. Bodmer, HLA, Immune Response Genes and Disease, *Lancet,* 1: 1269 (1974).
20. H. Grosse-Wilde et al., Immunogenetics of Seven LD Alleles of the DL-A Complex in Mongrels, Beagles and Labradors, *Transplant. Proc.,* 7 (Suppl. 1): 159 (1975).
21. W. F. Bodmer, H. Cann, and A. Piazza, Differential Genetic Variability Among Polymorphisms as an Indicator of Natural Selection, in *Histocompatibility Testing 1972,* J. Dausset and J. Colombani (Eds.), p. 753, Munksgaard, Copenhagen, 1973.
22. W. F. Bodmer, Evolutionary Significance of the HLA System, *Nature,* 237: 139 (1972).
23. P. C. Doherty and R. M. Zinkernagel, A Biological Role for the Major Histocompatibility Antigens, *Lancet,* 1: 1406 (1975).
24. N. K. Jerne, The Somatic Generation of Immune Recognition, *Eur. J. Immunol.,* 1: 1 (1971).
25. W. F. Bodmer, J. G. Bodmer, and M. V. Tripp, Recombination Between the LA and 4 Loci of the HLA System, in *Histocompatibility Testing 1970,* P. I. Terasaki (Ed.), p. 187, Munksgaard, Copenhagen, 1970.
26. H. M. Vriesendorp, R. B. Epstein, J. D'Amaro, D. L. Westbroek, and J. J. van Rood, Polymorphism of the DL-A System, *Transplantation,* 14: 299 (1972).
27. H. Grosse-Wilde et al., Identification of Four MLC Specificities in the Dog, *Tissue Antigens,* 4: 229 (1974).
28. S. N. Swisher, Biomedical Notation Systems, *J. Am. Med. Assoc.,* 185: 21 (1963).
29. Report of the IUIS Committee on Nomenclature of DLA Determinants, *Transplant. Proc.,* 9: 1909 (1977).

30. P. Landsdorp, H. M. Vriesendorp, G. R. Betton, and A. B. Bijnen, 1980, Secondary Responses of Alloantigen Primed Dog Lymphocytes, *Tissue Antigens,* 15: 40 (1980).
31. H. M. Vriesendorp, B. Löwenberg, T. P. Visser, S. Knaan, and D. W. van Bekkum, The Influence of Genetic Resistance and Silica Particles on Survival After Bone Marrow Transplantation, *Transplant. Proc.,* 8: 483 (1976).
32. H. M. Vriesendorp, H. Grosse-Wilde, and M. E. Dorf, The Major Histocompatibility System of the Dog, in *The Major Histocompatibility System in Man and Animals,* D. Gotze (Ed.), p. 129, Springer-Verlag New York, Inc., New York, 1977.
33. P. Meera Khan, W. R. T. Los, J. A. van der Does, and R. B. Epstein, Isoenzyme Markers in Dog Blood Cells, *Transplantation,* 15: 624 (1973).
34. E. D. Albert et al., Serology and Genetics of the DL-A System. I. Establishment of Specificities, *Tissue Antigens,* 3: 417 (1973).
35. Joint Report of 1st International Workshop on Canine Immunogenetics, *Tissue Antigens,* 3: 145 (1973).
36. R. Saison and E. Doble, The Reproducibility of Reactions Obtained with Cytotoxic Antisera Directed Against Various DL-A Specificities, *Transplant. Proc.,* 7: 353 (1975).
37. D. L. Westbroek (Ed.), *Manual of Techniques Used in the First International Workshop on Canine Immunogenetics,* Medical Faculty, Erasmus University, Rotterdam, The Netherlands, 1972.
38. B. M. J. Smid-Mercx, B. Duyzer-den Hartog, T. P. Visser, and H. M. Vriesendorp, Serological Studies of Canine Histocompatibility Antigens, *Transplant. Proc.,* 7: 361 (1975).
39. R. W. Bull, H. M. Vriesendorp, H. Obertop, A. B. Bijnen, J. Jeekel, J. de Gruyl, and D. L. Westbroek, Effect of Prior Third-Party Blood Transfusions on Canine Renal Allograft Survival, *Transplantation,* 26: 249 (1978).
40. J. A. van der Does, J. J. van Rood, W. Walker, and R. B. Epstein, Consequent Intrafamilial Immunization for DL-A Haplotyping in Canines, *J. Exp. Med.,* 137: 494 (1973).
41. E. Gluckman, M. Parr, E. Mickelson, M. L. Schroeder, and R. Storb, Mixed Leukocyte Cultures in Dogs. A Microtechnique Using Gradient Separated Cells, *Transplantation,* 15: 642 (1973).
42. S. F. Goldmann and H. D. Flad, Histocompatibility Testing in Dogs. I. A Semimicromized Lymphocyte Culture Technique for Histocompatibility Matching in Dogs, *Tissue Antigens,* 5: 145 (1975).
43. A. B. Bijnen, *Donor Selection for Renal Transplantation,* Thesis, Erasmus University, Rotterdam, The Netherlands, 1978.
44. A. B. Bijnen, A. M. Dekkers-Bijma, H. M. Vriesendorp, and D. L. Westbroek, Kinetics of Mixed Lymphocyte Reactions in Dogs, *Immunogenetics,* 8: 287 (1979).
45. H. Grosse-Wilde et al., One Way Nonstimulation in MLC-Related to DL-A Homozygosity, *Transplant. Proc.,* 5: 1567 (1973).
46. A. B. Bijnen, H. M. Vriesendorp, H. Grosse-Wilde, and D. L. Westbroek, Polygenic Control of Mixed Lymphocyte Reactions in Dogs, *Tissue Antigens,* 9: 187 (1977).
47. S. F. Goldmann, K. Krumbacher, H. Schnappauf, R. P. Huget, and H. D. Flad, Histocompatibility Testing in Dogs. II. Leukocyte Typing in Relation to the Mixed Lymphocyte Culture (MHC) Reactivity, *Tissue Antigens,* 5: 155 (1975).
48. M. J. Sheehy and R. H. Bach, Primed LD Typing (PLT). Technical Considerations, *Tissue Antigens,* 8: 157 (1976).
49. M. E. Dorf, E. K. Dunham, J. P. Johnson, and B. Benacerraf, Genetic Control of the Immune Response. The Effect of Non-H-2 Linked Genes on Antibody Production, *J. Immunol.,* 112: 1329 (1973).
50. P. Meera Khan, H. M. Vriesendorp, R. Saison, W. S. Volkers, W. R. T. Los, and B. A. Dopert, Linkage Between PGM3 and the Genes Determining the Major Histocompati-

bility Complex (MHC) in *Canis familiaris* (the Domestic Dog), *Cytogenet. Cell Genet.*, 22: 585 (1978).

51. C. C. Li, *Human Genetics*, McGraw-Hill Book Company, New York, 1961.
52. L. L. Cavalli Sforza and W. F. Bodmer, *The Genetics of Human Populations*, W. H. Freeman & Co., San Francisco, 1970.
53. H. M. Vriesendorp, J. D'Amaro, J. A. van der Does, D. L. Westbroek, and R. B. Epstein, Analysis of the DL-A System in Families and Populations of Healthy and Diseased Individuals, *Transplant. Proc.*, 5: 311 (1973).
54. P. L. Mattiuz, D. Ihde, A. Piazia, R. Cepellini, and W. F. Bodmer, New Approaches to the Population Genetic and Segregation Analysis of the HLA System, in *Histocompatibility Testing 1970*, P. Terasaki (Ed.), p. 193, Munksgaard, Copenhagen, 1970.
55. H. Balner and W. van Vreeswijk, The Major Histocompatibility Complex of Rhesus Monkeys (RhL-A). V. Attempts at Serological Identification of MLR Determinants and Postulation of a I Region in the RhLA Complex, *Transplant. Proc.*, 7: 13 (1975).
56. F. A. Beach, Coital Behavior in Dogs. VIII. Social Affinity, Dominance, and Sexual Preference in the Bitch, *Int. J. Comp. Etiology*, 36: 131 (1970).
57. Joint Report of the 2nd International Workshop on Canine Immunogenetics, *Transplant. Proc.*, 8: 289 (1976).
58. J. Klein, *Biology of the Mouse Histocompatibility-2 Complex*, Springer-Verlag New York, Inc., New York, 1975.
59. Advisory Committee to the Renal Transplant Registry. Thirteenth Report of the Human Renal Transplant Registry, *Transplant. Proc.*, 9: 9 (1977).
60. E. D. Thomas, A. Fefer, C. D. Buckner, and R. Storb, Critical Review: Current Status of Bone Marrow Transplantation for Aplastic Anemia and Acute Leukemia, *Blood*, 49: 671 (1977).
61. W. H. Hildemann, Components and Concepts of Antigenic Strength, *Transplant. Rev.*, 3: 5 (1970).
62. S. N. Swisher and E. Young, The Blood Grouping Systems of Dogs, *Physiol. Rev.*, 14: 495 (1961).
63. R. Cepellini, P. L. Mattuiz, G. Cudeller, and M. Visetti, Experimental Allotransplantation in Man. II. The Role of A1, A2, and B Antigens, *Transplant. Proc.*, 1: 390 (1969).
64. A. Zweibaum, R. Oriol, N. Feingold, E. Dussaulx, M. Rousset, and G. Chevalier, Studies on Canine Secretory Alloantigens (CSA), *Tissue Antigens*, 4: 115 (1974).
65. R. Oriol et al., Determination of the Structure of the Immunodominant Groups of the Canine Secretory Alloantigens (CSA) by Radioimmunoassay, *Transplant. Proc.*, 7: 523 (1975).
66. A. Zweibaum and E. Bouhou, Studies on Digestive Groups. II. Influence of the Digestive Group A System on Skin Allografts in Rabbits, *Transplantation*, 15: 294 (1973).
67. H. M. Vriesendorp, A. B. Bijnen, C. Zurcher, and D. W. van Bekkum, Donor Selection and Bone Marrow Transplantation in Dogs, in *Histocompatibility Testing 1975*, F. Kissmeyer-Nielsen (Ed.), p. 963, Munksgaard, Copenhagen, 1975.
68. S. Ohno, *Monographs on Endocrinology. Sex Chromosomes and Sex Linked Genes*, Springer-Verlag, Heidelberg, 1967.
69. D. W. Bailey, Histocompatibility Associated with the X-Chromosome in Mice, *Transplantation*, 1: 70 (1963).
70. E. Y. Eichwald and L. R. Slimzer, Note, *Transplant. Bull.*, 2: 148 (1955).
71. C. R. Stockard, *The Genetic and Endocrine Basis for Differences in Form and Behaviour as Elucidated by Studies of Contrasted Pure-Line Dog Breeds and Hybrids*, American Anatomical Memoirs, Wistar Institute of Anatomy and Biology, Philadelphia, 1941.
72. H. M. Vriesendorp, B. Duyzer-den Hartog, B. M. J. Smid-Mercx, and D. L. Westbroek, Immunogenetic Markers in Canine Paternity Cases, *J. Small Anim. Pract.*, 15: 693 (1974).

# Bone-Marrow Transplantation in the Canine

# 9

H. M. VRIESENDORP
*and*
D. W. VAN BEKKUM
*Radiobiological Institute TNO, The Netherlands*

## GENERAL ASPECTS OF MARROW TRANSPLANTATION

Several models of hemopoiesis, a self-renewal system,[1] have been proposed in the past, but currently a "monophyletic" model is generally accepted (Fig. 1). The monophyletic model postulates one hemopoietic stem cell (HSC) for all the different hemopoietic cell lines—the erythrocyte, granulocyte, platelet, and lymphocyte lines. For each differentiation line, a separate group of "committed progenitor" cells will differentiate and finally deliver the specific end cells. The model is based on the observations of several investigators[2–5] that properties of "marked" HSC's can be traced (after differentiation) in end cells from all the hemopoietic cell lines. The multipotential HSC has two important tasks, the production of self copies and the initiation of differentiation into one of the four differentiation lines.

The monophyletic model carries a number of important consequences for experimental and clinical medicine. The HSC is the top of the hemopoietic development pyramid. The crowded cellular conditions in active bone marrow (BM) are caused by cells that are descendents of the HSC; the HSC is the least-frequent cell. In mouse BM, for example, it is estimated that approximately 0.5% of the cells are HSC's.[6] The paucity hinders the morphological identification of HSC's. A morphological identification of the HSC is possible only in cell suspensions where the concentration of the HSC is increased to about 10%.[7] For mice and rats,[6,7] such concentrated suspensions have been produced, and a candidate HSC was identified morphologically by electron microscopy. Similar cells were identified in human, subhuman primate and dog BM cell suspensions.[8,9]

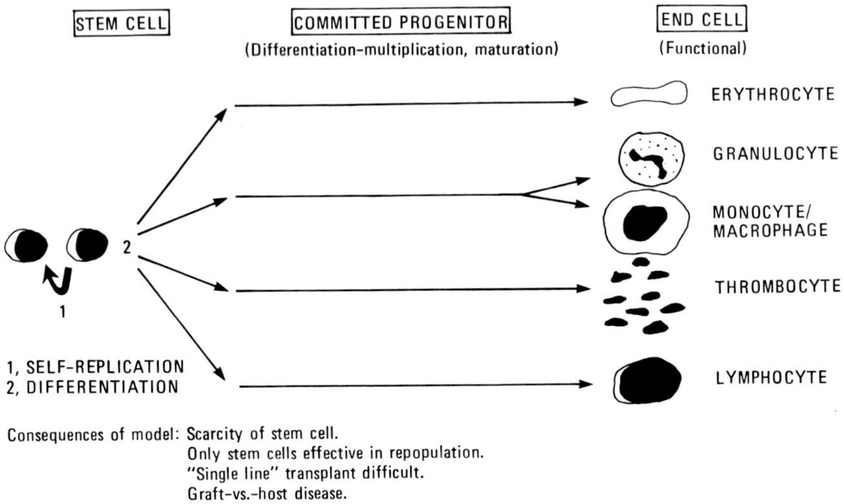


Fig. 1 Monophyletic model of hemopoiesis.

Another consequence of the monophyletic model is that a lasting replacement of the hemopoietic system can be achieved only through an HSC transplant. With a BM transplant, most of the cells injected are not HSC's. Since those other cells lack the property of self-replication, they will gradually die out in the host. Only the injected HSC's and their self-copies will be able to effect lasting hematological repopulation of the host. A priori, each successful BM transplantation will lead to the generation of donor blood cells in all the lines indicated in Fig. 1. This model does not permit a single-line transplant (e.g., only red cells in the case of a pure red-cell aplasia). However, single-line transplants have been observed infrequently in experimental animals as well as in man.[10–13] The monophyletic model, although compatible with most of the currently available information, may not be correct, or, alternatively, under still undefined conditions; HSC's of different genetic origins can coexist within the same host and each develop in a separate number of hemopoietic lines.

Another consequence of monophyletic hemopoiesis is that under all circumstances immunocompetent cells will develop from the injected HSC, not an inevitable event if hemopoiesis were "polyphyletic" and a separate stem cell produced lymphocytes. The transplantation of donor lymphocytes entails the serious complication that, if important antigenic differences exist between donor and recipient, donor cells will reject the recipient. This complication, unique to the transplantation of immunocompetent cells, has been labeled the reverse transplantation reaction or graft vs. host (GvH) reaction.

This chapter on BM transplantation in the dog illustrates the remaining limitations of the application of this approach in human patients as well as the advantages and disadvantages of the dog as a preclinical model.

## Relevance of Dog Model

The purpose of a BM transplant in man is to replace a defective or malignant hemopoietic system, or part of it, with a normal hemopoietic system. Bone-marrow transplantation in experimental animals can be used to study the difficulties encountered or anticipated in human patients (preclinical models) or as an experimental device of its own right in fundamental studies of hemopoiesis or immune reactivity.

Obvious economic considerations have led to the predominant use of small rodents in hemopoiesis experimentation, whereas the dog has mainly been used in preclinical experiments. The dog offers some advantages not available in other animal models. The size and temperament of dogs allow for intensive individual care and analysis. The large litter size and relatively short gestation and generation time facilitate genetic studies and, in particular, the use of littermate donor–recipient pairs, which have the same immunogenetic relationships as human sibling donor–recipient pairs. Moreover, the HSC concentration in the BM in the dog apparently is similar to that in man.[14]

Currently, BM transplantation is used in human patients suffering from severe combined immune deficiencies, aplastic anemia, or acute leukemia.[15–17] Questions specifically related to these diseases cannot be studied in the dog because of the lack of appropriate models. Leukemia models have been defined in small rodents[18,19] and might become available in dogs after further studies of the myeloproliferative disorders induced by continuous, low-dose-rate, total-body irradiation (TBI) (see Chapter 15). For aplastic anemia, no animal model resembles the human situation sufficiently to justify further study. Again, dogs receiving continuous, low-dose-rate TBI might provide such a model in the future. A severe combined immunodeficiency disease has been identified in horses but not in other larger animal species.[20] Thus currently the role of the dog model in BM transplantation research is limited to a preclinical analysis of the peculiarities of the procedure.

## Chronological Order of Events in Bone-Marrow Transplantation

Bone-marrow transplantation can be divided into a number of phases (Fig. 2), each with characteristic problems. The first phase is the preparation, or "conditioning," of the recipient. The next step is the actual transplant, the intravenous infusion of a cell suspension containing sufficient HSC's to produce a timely recurrence of functional end cells in the peripheral blood. Then, if host vs. graft (HvG) reactivity (host resistance) is sufficiently suppressed to avoid rejection of the allogeneic HSC's, these cells will start to deliver end cells between days 10 and 20 after transplantation. The transition period, when host blood cells have disappeared and donor blood cells have not produced in sufficient number, can last for several weeks. In this phase frequent infectious

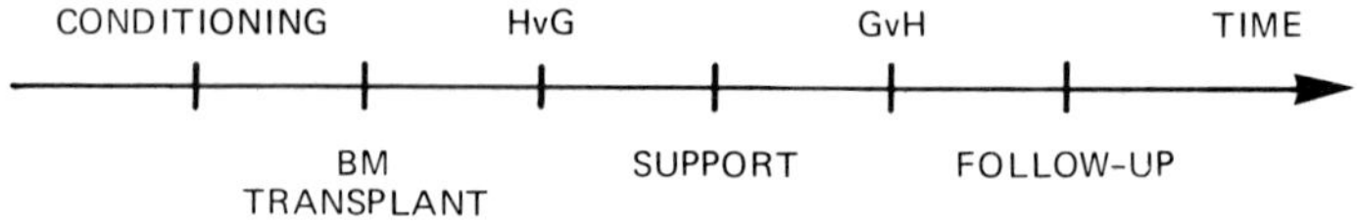


Fig. 2 Chronological order of events in bone-marrow transplantation.

complications and spontaneous hemorrhages require prophylactic as well as symptomatic supportive care. As soon as donor hemopoiesis can be demonstrated and sometimes even before that time, the attention of the clinician or experimentalist will have to be focused on the development and possible complications of GvH reactivity. The final phase of BM transplantation is the evaluation of end results, determined by the balance between benefits of a new hemopoietic system and side effects (early as well as late).

## CONDITIONING

### Aims of Conditioning

Severe combined immune deficiency, aplastic anemia, and acute leukemia, the most common indications for human BM transplantation, can illustrate the principles of conditioning (Table 1). In the body hemopoiesis is limited to specialized tissues. Replication and differentiation of donor HSC's can proceed only if the proper "space" is available in those tissues. Although detailed information on the nature or anatomical location of space is lacking, the first aim of conditioning is to create sufficient space for the incoming HSC's. Histological examination of BM biopsies indicates that space is not available for

TABLE 1

CONDITIONING STRATEGIES FOR BONE-MARROW TRANSPLANTATION

| | Severe combined immune deficiency | Aplastic anemia | Acute leukemia |
|---|---|---|---|
| Space for HSC | No | Yes | No |
| Resistance to HSC | No | Yes | Yes |
| Tumor cell elimination | No | No | Yes |
| Aims of conditioning | Space? | Decrease resistance | Increase space, decrease resistance, and eliminate tumor cells |

allogeneic HSC's in patients with severe combined immune deficiency (SCID) disease except perhaps in the thymus. In patients without conditioning, T lymphocytes carrying donor-type markers emerge after BM transplantation, whereas other hemopoietic end cells continue to show recipient types.[11] Patients with aplastic anemia (AA) apparently have sufficient space in the BM, whereas leukemic patients generally do not.

Lasting functional activity of the donor HSC's depends on prevention of rejection of these cells by the recipient. The tendency of the recipient to inactivate allogeneic HSC's has been designated "resistance,"[21] although the term "rejection," or HvG, offers a better description of the event. In this chapter the terms "resistance" and "HvG" are used for historical reasons. Both terms indicate the same phenomenon. Resistance does not occur in SCID patients, who lack a functioning immune system. Resistance is manifested in AA and leukemic patients. Sometimes it is increased as a result of prior sensitization against HSC antigens through cross-reactive antigens in blood transfusions.

A third aim of conditioning is relevant only in patients with leukemia. The replacement of the malignant hemopoietic system by a new normal hemopoietic system will be of lasting usefulness only if all clonogenic leukemic cells in the host can be eliminated. Therefore conditioning agents should also lead to a pronounced and, if possible, definitive eradication of the leukemic cells of the patients.

## Conditioning with Single Agents

**Increasing Space.** Animal death, after high-dose chemotherapy or radiotherapy, with signs of hematological insufficiencies can sometimes be prevented by an infusion of autologous BM cells harvested before the insult. The survival of platelets and leukocytes is so short that a single infusion will not contribute to improved survival through reduced thrombopenia or leukopenia. If lethality can be prevented by a BM cell transfusion, the dose-limiting toxicity of the investigated agent is at the HSC level, and, moreover, sufficient space for the transfused HSC's is available. Space can be measured independently of resistance by comparing the results of autologous rescue experiments with those of similar tests with allogeneic BM. A summary of the results of autologous BM rescue experiments after high-dose chemotherapy or radiotherapy in dogs (Table 2) shows that all agents tested appear to create sufficient space. In another experiment, lethal toxicity of intravenous nitrogen mustard within a week after drug administration could not be prevented by autologous BM infusion, which demonstrates that the hemopoietic system is not the dose-limiting tissue for this drug in dogs.[28] If dose-limiting lethal toxicity of an agent can be prevented by autologous BM transplantation, nonhematological toxicity will become dose limiting with a further increase in agent dose. Gastrointestinal toxicity is usually the limiting factor after hematological toxicity (Table 2). L-asparaginase (L-Asp)

TABLE 2

RESCUE WITH AUTOLOGOUS BONE-MARROW CELLS AFTER SUPRALETHAL CHEMOTHERAPY OR RADIOTHERAPY

| Agent* | Dose | $10^8$ Autologous BM cells/kg | Fraction of surviving dogs | Dose-limiting toxicity | Reference |
|---|---|---|---|---|---|
| Thio–TEPA | 6 mg/kg (iv) | ?† | 3/3 | GI‡ | 22 |
| CCNU | 20 mg/kg (iv) | 3 | 4/4 | GI | 23 |
| DMM | 10 mg/kg (iv) | ±5 | 7/7 | GI | 24 |
| CY | 100 mg/kg (iv) | ±3† | 8/10 | GI | 25 |
| CY | 120 g /kg (iv) (days –3, –1, and –1) | 7 | 9/13 | Bladder | 26 |
| TBI | 750 rads | 0.5 | 4/4 | GI | 27 |

*Thio–TEPA, triethylenethiophosphoramide; CCNU, 1-(2-chloroethyl)-3-cyclohexyl-1-nitrosourea; DMM, dimethylmyleran; CY, cyclophosphamide; and TBI, total-body irradiation, 300-kV X rays, 16 rads/min, 3.0 mm Cu half-value layer.

†Cryopreserved BM.

‡GI, gastrointestinal.

and anti-lymphocyte serum (ALS), conditioning agents with other dose-limiting toxicities, have not received much attention as single agents in conditioning, presumably because they do not create sufficient space. However, differences in dose-limiting toxicity of conditioning agents can decrease the toxicity of multiple-agent conditioning protocols. For example, several combinations of ALS or L-Asp and the agents listed in Table 2 have been shown to be effective in preparing recipients for allogeneic BM cells.

**Reducing Resistance.** Unfortunately, no prospective in vitro test to measure resistance[27] in unimmunized recipients is presently available. Resistance can only be quantified in vivo by determining the ratio between the number of allogeneic and autologous BM cells required for hemopoietic recovery in 50% of the nonsensitized animals (Table 3). After supralethal TBI, considerable resistance remains against BM cells of an allogeneic donor identical for the major histocompatibility complex (MHC); four times as many BM cells are required from an MHC-identical donor as from an autologous or isologous source. Strong resistance against MHC-mismatched BM cells remains under those conditions since 40 to 60 times as many BM cells are needed for rescue. In experiments of single-agent conditioning in dogs with dimethylmyleran (DMM), cyclophosphamide (CY), and TBI (Table 4), TBI was the most effective conditioning agent in MHC-identical pairs, with complete chimerism obtained in almost all animals. In contrast, conditioning with DMM and CY led to incomplete chimerism in MHC-identical donor–recipient pairs, possibly because of the relatively minor

TABLE 3

RELATIVE BONE-MARROW DOSE REQUIRED FOR 50% RADIOPROTECTION AFTER SUPRALETHAL TOTAL-BODY IRRADIATION*[14,27,29-31]

| Donor | Mice | Dogs | Rhesus monkeys |
|---|---|---|---|
| Autologous/isologous | 1 | 1 | 1 |
| MHC-identical† | 4 | 4 | No data |
| MHC-mismatched† | 60 | ±60 | ±40 |

*800 rads X ray.
†MHC, major histocompatibility complex.

TABLE 4

ALLOGENEIC RESISTANCE AFTER SINGLE-AGENT CONDITIONING

| | | MHC-mismatched donor | | MHC-identical donor | | | |
|---|---|---|---|---|---|---|---|
| Agent* | Dose | Source of transplant | Fraction of dogs with take | Source of transplant | Fraction of dogs with take | Chimerism | Reference |
| DMM | 10 mg/kg (iv) | BM + PBL† | 0/4 | BM + PBL | 8/16 | Incomplete | 24, 32 |
| CY | 120 mg/kg (iv) (days −3, −2, and −1) | BM | 1/4 | BM | 6/9 | Incomplete | 33 |
| | | BM + PBL | 3/6 | BM + PBL | 9/9 | Incomplete | |
| TBI | 750 rads | BM | 0/6 | BM | 24/24 | Complete | 34 |
| | | BM + PBL | 9/11 | BM | 10/11 | Complete | 35 |

*DMM, dimethylmyleran; CY, cyclophosphamide; and TBI, total-body irradiation, 300-kV X rays, 16 rads/min, 3.0 mm Cu half-value layer.
†BM, bone marrow, 4 to 7 x $10^8$ $kg^{-1}$ (iv); PBL, peripheral blood leukocytes, 1.3 to 3.3 x $10^9$ $kg^{-1}$ (iv).

effects of these agents on host resistance. The addition of peripheral blood lymphocytes (PBL) to BM cells, which decreases allogeneic resistance, yielded lasting "takes" in all MHC-identical combinations after CY but not after DMM conditioning. In MHC-mismatched donor–recipient pairs, takes can be obtained if very high numbers of BM cells are given or if moderate numbers of BM cells are given in combination with PBL. Again, TBI appears to be the most effective single conditioning agent with moderate activity of CY and no activity of DMM.

**Tumor-Cell Eradication.** Conditioning agents differ in their efficacy to eradicate tumor cells. TBI has the advantage of killing noncycling tumor cells as efficiently as cycling cells. Pharmacological sanctuaries do not exist for that agent. A disadvantage of TBI is that the number of decades of tumor cells that will be killed by the maximum acute TBI dose (± 800 rads, 300-kV X rays or its equivalent) is limited to approximately four when the $D_0$'s (70 to 300 rads) found for various tumor-cell lines are taken into account.[36] In this respect CY and DMM might be more effective than TBI. For example, CY and myleran effectiveness in a rat acute myeloid leukemia model is about seven decades (Ref. 37 and A. Hagenbeek, Radiobiological Institute, Rijswijk, The Netherlands, unpublished observations). Since dog leukemia models are not available (if they were, they would be extremely expensive), further quantitative studies in small-animal models for predicting responses in human cancer patients will be required for the selection of conditioning agents with maximum tumoricidal activity.

## Role of Total-Body Irradiation

**History.** The observation that started research in BM transplantation was that the injection of normal BM cells after accidental TBI could prevent death as a consequence of BM aplasia.[38] This procedure also led to the establishment of a new donor-type hemopoietic system. Thereafter deliberate irradiation of animals was used to investigate the peculiarities and possible benefits of BM transplantation. Later, radiomimetic drugs were introduced as alternatives to TBI.[14] None of the currently known agents is capable of fulfilling all the goals of conditioning as a single agent. TBI appears to be most useful in suppressing resistance. Further development of multi-agent conditioning is needed.

**Physics.** The uncontrolled exposure of animals to ionizing radiation will lead to nonhomogeneous dose distributions, inexplicable results, and differences in results between laboratories. Although the selection of a TBI procedure will depend largely on the facilities available, several variables must be controlled for a regulated, reproducible exposure. The European Late Effects Project (EULEP) group of the European Atomic Energy Community has published a detailed protocol for X-ray dosimetry. Appendix 2 of that report summarizes the recommendations and is reprinted in part below:

> A description of the exposure conditions for x-irradiations has to comprise the following parameters:
>
> – the quality of the radiation: high voltage, the current of the x-ray machine and the HVL (half value layer).
> – the absorbed dose at the volume of interest or the exposure at that position and the way the absorbed dose was derived.
> – the temporal distribution of absorbed dose (dose rate, fractionation, timing of successive exposures, etc.).
> – the geometrical conditions of irradiation (distance from focus to center of animal, field size, scattering materials, etc.).
> – the ratio between maximum and minimum absorbed dose in the volume of interest.

It is recommended that in the majority of radiobiological experiments class A irradiation should be applied (ICRU report 10e), i.e. a ratio of less than 1.15 between maximum and minimum absorbed dose in the volume of interest. To comply with this condition, the following procedures are suggested:

- exposure variations over the irradiation field used (in the absence of biological objects) should not exceed ±3% of the average exposure. In some cases rotation of the objects during exposure can give an improvement of the homogeneity of the dose distribution.
- maximal source-to-object distances should be used in order to diminish the effect of the inverse square law.
- a sufficient filtration of the radiation should be used (to diminish the effect of the absorption) e.g. an HVL of at least 2.5 mm Cu.

For dogs, a bilateral exposure with at least 300-kV X rays is recommended. Unilateral exposure or lower energy irradiation will lead to nonhomogeneous dose distributions. Correction factors of relative biological effectiveness (RBE) must be applied to high-energy X-ray, gamma-ray, or fission neutron radiation to obtain a dose equivalent to 300-kV X rays. For TBI, the RBE for high-energy X-ray and gamma radiation is usually given as 0.8; for fission neutrons, it is approximately 2.[31,39] In addition to the type of radiation, the dose and dose rate determine the RBE of a radiation procedure.[40,41] Also, the size and age of the dog influence the radiation response;[42-44] so experimental animals should be of similar age and size with minor differences corrected with published factors.[43,44]

**TBI Dose.** The syndromes resulting from TBI in untreated recipients have been described previously for many different species.[1,30] Briefly, the three major syndromes are the BM syndrome, the gastrointestinal (GI) syndrome, and the central nervous system (CNS) syndrome (Table 5). With an effective BM transplantation regimen, GI toxicity becomes dose limiting. Routinely the highest possible dose of TBI, the dose just under the one leading to a lethal GI syndrome, has been given to experimental animals under the assumption that maximum doses provide the required maximum suppression of host resistance. However, recent results in dogs have shown that such a dose is a considerable overdose in allogeneic MHC-identical donor–recipient pairs[14] (Table 6). A dose of 500 rads (300-kV X rays) was sufficient to obtain a complete radiation chimera, and the lower TBI dose is less toxic. After 500 rads of TBI, no GI toxicity is produced, and thrombocytopenia and endogenous infections occur significantly later (1 and 2 days, respectively) than after 750 rads of TBI (H. M. Vriesendorp, unpublished observations). Monkeys suffer a lower lens cataract incidence and no skeletal growth inhibition after 500 rads.[44a,44b] Radiation pneumonitis is not observed after that radiation dose in man.[45] A disadvantage of a lower TBI dose in cancer patients would be a decrease in tumor-cell kill by one to two decades. However, TBI alone is not sufficient for complete tumor eradication, and the necessary conditioning with multiple agents can only be

TABLE 5

MORTALITY AFTER TOTAL-BODY IRRADIATION

| | Syndrome | | |
|---|---|---|---|
| | Bone marrow | Gastrointestinal | Central nervous system |
| TBI dose,* rads | 500 to 1000 | 750 to 10,000 | 10,000 |
| Cause of death | BM aplasia<br>Infections<br>Hemorrhages | Denudation of GI mucosa<br>Fluid, electrolyte loss<br>Infections (infrequent) | Apnea<br>Cardiac arrest |
| Day of death after TBI | 7 to 30 | 3 to 7 | 0 to 1 |
| Therapy | BM transplant<br>Hematological support | Fluid, electrolyte replacement<br>Gastrointestinal decontamination | None |
| Efficacy of radioprotective therapy | Good | Only at lower TBI dose range | None |

*Dose differs for each species. Important species differences exist in radiation sensitivity. See text for possible mechanisms.

TABLE 6

TBI DOSE AND MHC-IDENTICAL BONE-MARROW TRANSPLANTS*

| TBI dose,* rads | Number of dogs tested | Percent complete chimeras 100 days after TBI† |
|---|---|---|
| 750 (supralethal) | 20 | 20/20 |
| 500 ($\pm LD_{100}$) | 5 | 5/5 |
| 350 ($\pm LD_{50}$) | 5 | 0/5 |

*300-kV X rays, 10 mA, half-value layer 3.0 mm Cu; dose rate, 16 to 18 rads/min.

†Dogs received $4 \times 10^8$ BM cells/kg 6 to 24 hr after TBI.

given in combination with radiation if the additional toxicity will not kill the recipient. The addition of other agents to conditioning with TBI might also give a better suppression of resistance against MHC-mismatched BM cells.

As a final point in a discussion of the optimal TBI dose, larger species apparently have a lower $LD_{50}$ and $LD_{100}$ for TBI because of a negative correlation between body weight and HSC concentration.[14] Thus larger species develop a BM syndrome after a lower TBI dose and require the transplantation of more BM cells to survive than smaller species.

**TBI Fractionation and Dose Rate.** Variations in TBI dose rate or fractionation of TBI will influence the biological effect of the procedure. TBI should be given at a dose rate and in a fractionation scheme that will create sufficient space and have a maximum effect on host resistance and tumor cells. Simultaneously, the chosen set of TBI variables should not induce severe side effects on normal tissues. The TBI dose rate and fractionation, incorporating all the endpoints of space, resistance, tumor kill, and normal tissue response, have not been studied. In one study with dogs, an important therapeutic gain was obtained (decreased resistance, no extra toxicity) by lowering the dose rate to ±8 rads/min,[46] but other investigators have been unable to reproduce those results.[34,35] Experiments in rats have shown that a higher dose rate of approximately 20 rads/min yields best results for that species.[47] A drawback of a low dose rate is the inconvenience to the animal and experimentalist of long exposure times. Therefore the present incomplete knowledge of the effects of dose rate argues for the use of the highest possible dose rate. A short interval between the start of the BM insult (conditioning) and the start of the protection procedure (transfusion of HSC) is advantageous. A long interval will cause a prolonged period of BM aplasia requiring complicated supportive measures that are not always successful. Preliminary results[48] suggest that the use of a few low-dose TBI fractions before a supralethal dose of TBI might provide an effective suppression of host resistance against DLA-mismatched BM. In that study, three fractions of approximately 350 rads, given on alternate days, before a final dose of 750 rads were most effective. The current recommendation is to use unfractionated TBI, pending further confirmation of the possible therapeutic gains of fractionated TBI.

Partial irradiation with shielding of dose-limiting or irrelevant tissues has been proposed for treatment of patients with no malignant disorders. In tumor patients all tissues are relevant, and shielding procedures might be counterproductive if tumor cells are disseminated in the shielded areas. Shielding procedures also usually involve shielding of BM and therefore diminish the creation of space and the abrogation of resistance in the host. A special kind of partial-body irradiation investigated for use in small rodents[12] is an equivalent of the type of irradiation used for human patients with stages II and III of Hodgkin's disease, the so-called total lymphoid irradiation (TLI). The majority of tissues containing lymphocytes is irradiated during this multifractionated procedure. Hematological toxicity is acceptable because large portions of the hematopoietic system are not irradiated. After the completion of this procedure, resistance apparently is decreased, and, in a proportion of the mice, allogeneic marrow cells are accepted. Incomplete chimerism (i.e., a combination of donor and recipient type hemopoiesis) develops and seems to preclude GvH disease. The application of TLI could be advantageous in certain patients, such as those with aplastic anemia, with a small space requirement and not needing eradication of malignant cells. The limited experience with TLI as conditioning for BM transplantation in larger experimental animals, such as dogs, and in man requires further studies for definition of optimal conditions.

## Conditioning with Two Agents

**Rationale for Multiple-Agent Conditioning.** Single-agent conditioning fails in the case of BM transplantation across an MHC barrier because of insufficient suppression of resistance and also fails in patients with disseminated cancer because a proportion of clonogenic tumor cells will survive the conditioning. The addition of a second agent to the conditioning regimen can improve the success rate in both cases. Unfortunately, quantitative tumor models for enumerating the fraction of tumor cells killed by an experimental agent or procedure have not been defined for dogs.

**Selection of Agents.** The efficacy of TBI on resistance against MHC-identical BM cells and the advantageous properties of this agent for tumor-cell kill make TBI a logical choice as a first agent in multiple-agent conditioning. The second agent can be selected by using various criteria. An important consideration is the dose-limiting toxicity of the agent. In the setting of BM transplantation, hematological toxicity can be neglected if the second agent is no longer active at the time of the BM transplant. Nonhematological toxicity, however, should be different from that of the first agent (TBI). Another consideration is the mechanism of action of the agent. For example, the function of most irradiated cells is unaffected unless cell division is required for that function. Resistance might be suppressed more effectively if TBI is combined with other agents that do affect cell function regardless of cell division. The principle of combining agents with different mechanisms of action and nonoverlapping dose-limiting toxicities has been developed in human oncology.[49] The application of the principle to BM transplantation will induce maximum anti-tumor effects as well as optimal abrogation of resistance.

**Current Experience in Conditioning with Two Agents.** In dogs, conditioning with a second agent, in addition to TBI, sometimes reduces allogeneic resistance (Table 7). In other cases, such as with CY, which is effective as a single-

TABLE 7

ALLOGENEIC RESISTANCE TO BONE-MARROW GRAFTS FROM MHC-MISMATCHED DONOR AFTER CONDITIONING WITH TWO AGENTS

| Agent 1* | Agent 2* | Agent 1 dose (on day 0) | Agent 2 dose (day) | $10^8$ BM cells/kg (day 1) | Fraction of dogs with takes | Reference |
|---|---|---|---|---|---|---|
| TBI† | CY | 750 rads | 25 mg/kg (−3 and −2) | 4 | 0/3 | Vriesendorp (unpublished observations) |
| | 6MP‡ | 600 rads | 12.5 mg/kg (−7, −6, −5, and −4) | 10 | 5/7 | 50 |
| | MTX | 860 rads | 5 mg/kg (0, 2, 4, 6, and 8) | 20 | 8/10 | 51 |
| | $SiO_2$ | 750 rads | 100 mg/kg (1) | 4 | 25/29 | 27, 52 |
| | | 750 rads | 100 mg/kg (1) | 4 | 2/8 | 35 |
| | ALS | 450 rads (2 times) (day 0 and +1) | 2 ml/kg (sc) (3 times per week from day 7) | 13 | 8/8 | 53 |
| | | 750 rads | 0.5 ml/kg (sc) (−9 to −4) | 4 | 1/3 | 35 |
| | | 750 rads | 2 ml/kg (sc) (−7, −5, −3, and −2) | 4 | 0/2 | 34 |
| | L-Asp | 750 rads | 500 IU/kg (1 to 10) | 4 | 2/5 | 54 |
| | CBP | 750 rads | 10 mg/kg (−7) | 4<br>4 | 2/5<br>0/4 | 34<br>55 |
| DMM | ALS | 10 mg/kg (iv) | 0.5 ml/kg (sc) (−6 to −1) | 14§ | 2/7 | 24<br>32 |

*TBI, total-body irradiation; CY, cyclophosphamide; 6MP, 6-mercaptopurine; MTX, methotrexate; $SiO_2$, silica particles, <5 μm; ALS, anti-lymphocyte serum; L-Asp, L-asparaginase; CBP, *Corynebacterium parvum;* and DMM, dimethylmyleran.

†In rads, normalized to 300-kV X rays, 16 rads/min., 3.0 mm Cu half-value layer.

‡Four animals received, in addition, urethane before TBI.

§In addition, 1.3 x $10^9$ donor peripheral blood leukocytes/kg were given. In MHC-identical pairs, most animals showed takes with incomplete chimerism.

conditioning agent in rats, man, and dogs, the results in two-agent protocols are disappointing. CY has a good record as an anti-tumor agent in man and, in mouse spleen assays, gives considerable suppression of resistance,[56,57] but the experience in dogs with combinations of TBI and CY is negative, primarily because of overlapping toxicity. The tolerance of dogs for CY is low with massive GI toxicity and side effects on bladder mucosa and possibly the heart.[56]

The anti-metabolites 6-mercaptopurine (6-MP) and methotrexate (MTX) in combination with TBI were effective in suppressing resistance as previously reported.[50,51] Both drugs are immunosuppressive, active against a number of human tumors,[58] and lead to GI toxicity. In addition, liver toxicity can result from 6-MP administration.[56] One attempt of leucovorine "rescue" after high-dose MTX was inconclusive because the toxicity of the chosen MTX–leucovorine rescue protocol to normal tissues remained high.[59] The possible therapeutic gain of the use of those two drugs because of their differential effects on resistance, tumor-cell kill, and GI toxicity remains to be determined. Further studies incorporating anti-metabolites with TBI in conditioning regimens appear to be warranted.

Silica particles were tested as conditioning agents in dogs because studies in mice suggested a significant effect on resistance.[57] Different investigators[27,35] have found a varying degree of efficacy of this additional conditioning agent, but its immediate toxicity (diffuse intravascular coagulation, thrombopenia, and arteritis) and its potential late effects (fibrosis and carcinogenesis) prohibit clinical application.

Anti-lymphocyte serum (ALS) is a xenogenic (usually horse or rabbit) antiserum against lymphocytes. Anti-tumor activity of ALS is limited to tumor cells sharing antigens with the lymphocytes used for immunization, but its immunosuppressive potency, its ability to simultaneously suppress HvG as well as GvH reactions, and the difference in toxicity between ALS and other conditioning agents have led to preliminary studies of the effectiveness of ALS as a second conditioning agent. In dogs, ALS treatment before TBI did not suppress resistance, but treatment with ALS after TBI was effective in one study.[53] One drawback of this agent is that its efficacy can be evaluated only by elaborate in vivo tests, e.g., by determination of the prolongation of skin allograft survival; reliable prospective in vitro tests for its efficacy remain to be defined. The dose-limiting toxicities of ALS in conditioning for BM transplantation are allergy to foreign proteins after repeated exposure,[60] an increased susceptibility to viral and fungal infections,[60,61] and toxicity to the transplanted HSC as well as to erythrocytes and thrombocytes.[60,62] Some sera are contaminated with anti-glomerulus membrane antibodies, which causes renal toxicity.[63] In combination with DMM, ALS was not, or at best marginally, effective in conditioning recipients for BM and PBL of an MHC-mismatched donor.[32] Further tests with optimally defined ALS are required before its true value in conditioning can be evaluated.

L-asparaginase is a relatively nontoxic immunosuppressive agent in dogs[64] and also has some activity against dog lymphoma.[65] The dose-limiting toxicity of L-Asp in dogs is not known, but in man liver and pancreas toxicity and allergy to the antigens of the microorganism used for L-Asp production limit its use.[56] A drawback of L-Asp in dogs is that its activity against resistance is reproducible but small.[34,54]

*Corynebacterium parvum,* effective in mouse spleen assays, had no effect on resistance in the few dogs tested.

The experience with conditioning with two agents indicates that optimal multiagent conditioning protocols have not yet been defined. Combinations of TBI with anti-metabolites or steroids might be rewarding in future studies. Anti-metabolites have proved most effective additions to TBI in the past. Steroids, although not yet tested, fulfill most of the criteria for including an agent in a multiple-agent conditioning regimen.

## BONE-MARROW TRANSPLANT

### Homing

Hemopoietic stem cells are known for their migratory behavior under physiological conditions. They travel via blood vessels and are scattered among peripheral blood nucleated cells, albeit in low concentration.[3] When HSC's are infused intravenously, they will follow physiological pathways. After a stay of unknown duration in the peripheral blood, a percentage of HSC's will cross the blood–marrow barrier and "home" into the BM space.[66] However, the mechanisms of homing remain ill understood. For mice that have received supralethal TBI, approximately 30% of the injected HSC's actually reach the sites of hemopoiesis (marrow and spleen).[67] The reasons for the loss of the remaining HSC's remain obscure. After the iv injection, some HSC's may be trapped in the lung capillaries for at least 48 hr.[68]

#### Route of Administration of HSC's

The homing behavior of HSC's precludes the necessity of finding ways to deposit HSC's directly in the BM cavity. Intra-bone medullary injection of BM cells is complicated and painful and has no advantages over an iv injection.[69] When compared with the iv route, fewer HSC's home after intraperitoneal (ip) and subcutaneous (sc) injections (efficiencies of 2 and 0%, respectively, of the iv route in mice).[30] Conversely, in monkeys the ip route was more effective (30% of the iv route).[70] The cause of that difference of results in mice and monkeys is unknown. The possibility that a proportion of iv-injected HSC's is trapped in the lung capillaries could be tested by comparing iv HSC administration with intra-aorta HSC administration.

### Bone-Marrow Cell Dose

The quantity of BM cells used for transplantation is important. If insufficient cells are given, the conditioned animal will die from the complications of BM aplasia. The optimum number of BM cells is the one that will give a fast hemopoietic recovery yet not increase the risks of GvH disease. The incidence and severity of GvH disease are related to the number of transplanted allogeneic T lymphocytes. Thus an increase in the BM cell dose also increases the chances of GvH complications.[30] However, the T lymphocyte content of normal dog BM cell suspensions is small, and GvH disease amplification by an overdose of BM cells is more a problem in primate BM transplantation,[71,72] where BM contains more T lymphocytes.

The minimum BM cell dose required for radioprotection differs by species. Later species have a lower $LD_{50}$ for TBI and require more BM cells per kilogram of body weight for rescue than do smaller species, probably because of the lower HSC concentration in larger species.[14] In dogs 4 x $10^7$ autologous BM cells/kg apparently is a safe rescue dose.[14,73] For allogeneic donors identical for the MHC, about 4 times as many (i.e., 1.5 to 2 x $10^8$/kg) BM cells are required; if a donor differs for the MHC with the recipient, 60 times as many BM cells (i.e., ±2.5 x $10^9$/kg) will be needed for primary radiation protection.[51]

In experimental protocols that require the use of cryopreserved BM cells, those cells can be quantitatively substituted for fresh BM. In the best currently available cryopreservation programs, more than 70% of HSC's can be recovered after thawing (Ref. 74 and H. M. Vriesendorp, unpublished observations). Various techniques for cryopreservation of dog BM cells have been described.[73–75] Optimal hemopoietic recovery after cryopreservation requires controlled freezing programs, storage in liquid nitrogen, rapid thawing and stepwise dilution, and in vitro monitoring by CFU-C assays.[74,76,77] The CFU-C assay does not measure HSC's but a progenitor cell close to the HSC's. In the absence of an in vitro HSC assay, CFU-C measurement before and after cryopreservation is the best way of correcting for HSC loss in the cryopreservation procedure.

## Other Sources of Hemopoietic Stem Cells

The ideal source of allogeneic HSC's has a high concentration of HSC's for establishment of donor hemopoiesis and a low concentration or no postthymic T lymphocytes because those cells will cause acute GvH disease. The relative concentration of HSC's and T lymphocytes in tissues has been determined for the mouse (Table 8), but data for other species are less complete. As in the mouse, dog BM does not contain appreciable numbers of postthymic T lymphocytes, but primate BM does contain such cells.[71,72]

The limited canine experience with transplantation of sources of HSC's other than BM is summarized in Table 9. With a single exception, canine fetal liver cells have not given rise to permanent hemopoietic grafts. Extrapolation of the relative HSC concentration of mouse fetal livers to the dog and the increased cell numbers required from MHC-mismatched donors indicates that insufficient cells

TABLE 8

CONCENTRATION OF HEMOPOIETIC STEM CELLS AND POSTTHYMIC T LYMPHOCYTES IN VARIOUS TISSUES OF THE MOUSE*

| Tissue | Relative HSC concentration | Percent postthymic T lymphocytes |
|---|---|---|
| Bone marrow | 100† | 0 |
| Fetal liver | 15 | 0 |
| Spleen | 5 | 30 |
| Peripheral blood leukocytes | 1 | 40 |
| Lymph node cells | 0 | 60 |

*Data are from Refs. 3 and 78.

†In mouse bone marrow, approximately 5 HSC per $10^3$ bone-marrow cells.[6]

probably are the cause of the failures of dog fetal liver transplantation. The single "take" might be explained by the increased chances for a good histocompatibility match as a consequence of common ancestors of the recipient and the liver donors in that case.[81]

The potential for success of liver HSC transplants is limited by the logistics of obtaining liver cells. Most of the previous experiments have pooled yields of several livers, and obtaining more dog fetal liver cells than used in those experiments would be complicated. Even if the administration of more cells were successful, duplication of those techniques for human (MHC-mismatched) recipients would be impossible. Without new developments, priorities for additional research in this field in dogs appear to be low.

Low doses of dog fetal spleen cells did not lead to hemopoietic recovery in supralethally irradiated dogs.[80] The concentration of HSC's in fetal spleen is probably even lower than that in fetal liver, and transplantation of fetal spleen cells suffers at least the same drawbacks inherent in the use of fetal liver cells in dogs. Thus the lower GvH incidence after transplantation of allogeneic murine fetal tissues cannot be easily verified in the dog model[86] and presently does not appear to be applicable in the treatment of human patients.

Spleen cells, from adult or fetal donors, added to BM cells apparently increase the percentages of "takes,"[82] probably by abrogation of host resistance by allogeneic lymphocytes in the spleen cell suspensions. The same explanation might hold for the observation that BM takes can be obtained with relatively low numbers of PBL's. The number of granulocyte/macrophage colonies formed in culture (CFU-C) by dog PBL's is only about 1/100 the number formed from equal numbers of dog BM cells (H. M. Vriesendorp, unpublished observations). Even with those few HSC's, the high percentage of lymphocytes among PBL's

TABLE 9

TRANSPLANTATION OF CANINE HEMOPOIETIC STEM CELLS FROM TISSUES OTHER THAN BONE MARROW

| Dose, $10^8$ cells/kg | TBI, rads (300-kV X rays on day 0) | Donor–recipient combination | Fraction of takes | Reference |
|---|---|---|---|---|
| **Fetal liver HSC*** | | | | |
| ? plus fetal spleen | 450 | Unrelated | 4?‡/16 | 79 |
| 0.6 to 20 | 850 | Unrelated | 0/4 | 80 |
| 22 plus thymic cells (0.7) | 650 | Related | 1/4 | 81 |
| 3 | 750 plus 100 mg $SiO_2$/kg (iv) | Unrelated | 0/2 | H. M. Vriesendorp (unpublished observations) |
| **Fetal spleen HSC*** | | | | |
| 0.3 to 0.6 | 850 | Unrelated | 0/2 | 80 |
| **Peripheral blood leukocyte HSC** | | | | |
| Cross circulation (days +1 to +9) | 750 | Unrelated | 10/10 | 82 |
| 50 (days +1 and +2) | 750 | Unrelated | 10/10 | 83 |
| 15† | 750 | Autologous | 4/4 | 84 |
| | | MHC-identical littermate | 12/12 | 84 |
| **Thoracic duct cell HSC** | | | | |
| 15 | 750 | Unrelated | 0/5 | 85 |

*Multiple fetuses at near term were used for single recipient in most cases. Only Vriesendorp used single donors with pregnancy duration of approximately 7 weeks.

†Cryopreserved cells. Initially obtained with a continuous-flow cell centrifuge after CFU-C mobilization with dextran sulfate.

‡One animal showed temporary engraftment by genetic markers. In the other three, no markers were available. Takes of long-term survival were not observed in animals receiving higher doses of TBI.

effectively supplements the decrease in host resistance from TBI and might permit a lasting establishment of a low number of allogeneic HSC's. Unfortunately, the success of such transplants is short lived for the same reason; the high numbers of allogeneic lymphocytes will lead to severe, acute GvH reactions. Another explanation for the unexpected efficacy of low numbers of dog peripheral blood nucleated cells could be that dogs have a higher HSC concentration in blood that is not correlated with an increase in CFU-C numbers. Transplantation of $1 \times 10^9$ autologous mononuclear cells/kg from the peripheral blood led to hemopoietic recovery after CY conditioning.[87]

Quantitative comparisons between blood and BM in HSC rescue studies are needed to clarify this issue of relative HSC concentration in blood and BM.

A low dose of thoracic duct cells contains insufficient HSC's for hemopoietic recovery after supralethal TBI.[85]

## HOST VS. GRAFT

### Definition

Lasting activity of the transplanted HSC's is the goal of BM transplantation. The chances for realizing that goal are determined by (1) the efficacy of the conditioning regimen, (2) the donor–recipient relationship, (3) the number of transplanted HSC's, (4) immunizations of the recipient preceding the BM transplant, and (5) "contaminating" lymphocytes in the transplant.

The HvG reaction is the tendency of the host to prevent the establishment of allogeneic HSC's. The severity of the HvG reaction is reflected in the hemopoietic recovery after conditioning. If hemopoietic recovery is not manifest in the peripheral blood within 3 weeks after transplantation, the host has rejected the graft. In solid organ transplantation, immunosuppression can be successfully applied after signs of organ rejection (e.g., serum creatinine rise after a kidney allograft). In BM transplantation, however, the diagnosis of rejection is made when it can no longer be reversed. An in vitro assay that would predict such an outcome of BM transplantation would be extremely helpful by preventing useless transplants and speeding investigations of factors influencing HvG. In vitro tests of HvG reactions in mice include spleen colony formation or spleen IUdR uptake assays.[57] Those assays cannot be applied to nonrodent species, and even in rodents they are unreliable for the prediction of the number of BM cells needed for transplantation.[27] Without prospective in vitro tests, predictions of HvG must be based on in vivo experience from experiments in which the factors affecting HvG are controlled.

In animals immunized with allogeneic tissues before BM transplantation, the HvG reaction is sometimes increased when effective immunization is induced against HSC antigens by cross-reactive antigens from those tissues. This increase in HvG appears to be predictable in vitro by cocultures of donor BM and recipient lymphocytes. Lymphocytes of unsensitized recipients stimulate colony formation of erythrocyte progenitor cells (CFU-E) in donor BM. Sensitization of the recipient effectively prevents that stimulation of CFU-E.[88]

The experience with blood transfusions preceding BM transplants is summarized[89,90] in Table 10. Transfusion of donor blood prior to the BM transplant reduces the percentage of successful transplants. Repeated transfusions with BM donor blood decrease the take incidence to 0, whereas repeated third-party (i.e., different from BM donor or recipient) transfusions are followed

TABLE 10

FREQUENCY OF TAKES IN DOGS GIVEN BLOOD TRANSFUSIONS PRIOR TO BONE-MARROW TRANSPLANTATION[89,90]

| BM donor–recipient | No blood | One-time-donor blood (day −10) | Three-time-donor blood | Three-time third-party blood | One-time-donor blood | | | Three-time-donor tissue (days −24, −17, and −10) | | |
|---|---|---|---|---|---|---|---|---|---|---|
| | | | | | Day −1 | Day −10 | Day −90 | Epithelial cells | Platelets | Red cells |
| MHC-identical | 55/55 | 2/7 | 0/19 | 8/11 | Not done | Not done | Not done | 0/6 | 8/15 | 7/12 |
| MHC-mismatched | 13/15 | 3/12 | Not done | 3/9 | 1/7 | 3/12 | 8/16 | Not done | Not done | Not done |
| Experimental question | | BM donor or third-party blood | | | Duration of sensitization | | | Tissue distribution of sensitizing antigens | | |

by higher (~50%) BM take incidence. The incidence of BM takes in MHC-identical and MHC-mismatched donor–recipient pairs appears to be similar after BM donor as well as third-party blood transfusions, which indicates that minor histocompatibility differences and MHC differences are equally effective in sensitizing the future BM host. The onset of sensitization is almost immediate and lasts for at least 3 months. Sensitizing antigens occur, probably, on epithelial cells as well as on blood leukocytes but to a lesser extent or not at all on platelets or red cells. The decreased take incidence in recipients of prior platelet or red-cell transfusions can be explained by sensitization induced by leukocytes surviving the purification procedure.[89]

## Genetic Control of HvG

Bone-marrow cells and HSC's, in particular, are no exceptions to the rule that their survival in allogeneic hosts is curtailed by histoincompatibility with the host. The first results[91] indicating a histocompatibility influence on HvG in dogs were reported in 1968. In BM transplantation, histoincompatibilities between the donor and the recipient can be overcome by increasing the cell dose. A larger increase of cells will be required for a major histocompatibility difference; a smaller increase of cells will be needed for a minor histocompatibility difference, which induces only mild HvG reactions. The number of BM cells transplanted significantly correlated with the peripheral leukocyte values 13 days after supralethal TBI in MHC-identical donor–recipient pairs (Fig. 3b) as well as in autologous transplants (Fig. 3a). The correlation between BM cell dose and peripheral leukocyte levels remains significant until approximately day 21 after TBI. About three to four times as many BM cells are required from an MHC-identical donor as from an autologous donor to yield the same leukocyte level after 13 days. Often hemopoietic recovery is not observed in dogs transplanted with less than $2.5 \times 10^7$ autologous BM cells/kg or less than $1 \times 10^8$ MHC-identical BM cells/kg after supralethal TBI (H. M. Vriesendorp, unpublished observations).

The MHC consists of many closely linked loci, of which only a minority has been identified (see Chapter 8). The loci within the canine MHC that can be identified, the DLA subregions designated as LD and SD, were correlated with HvG response to allogeneic BM transplants.[52,92,93] LD identity between a donor and a host is a negative unilateral MLC reaction in which host lymphocytes are not stimulated by the irradiated donor lymphocytes. SD identity indicates that a donor does not possess any of the known antigens of the first, second, or third serologically defined series of multiple alleles (officially called DLA-A, DLA-B, and DLA-C) that are absent in the host. Allogeneic BM transplants usually take if the donor and recipient are DLA-identical siblings but not if they are DLA-, LD-, and SD-different siblings (Table 11). The single exception (no take of an LD- or SD-identical sibling)

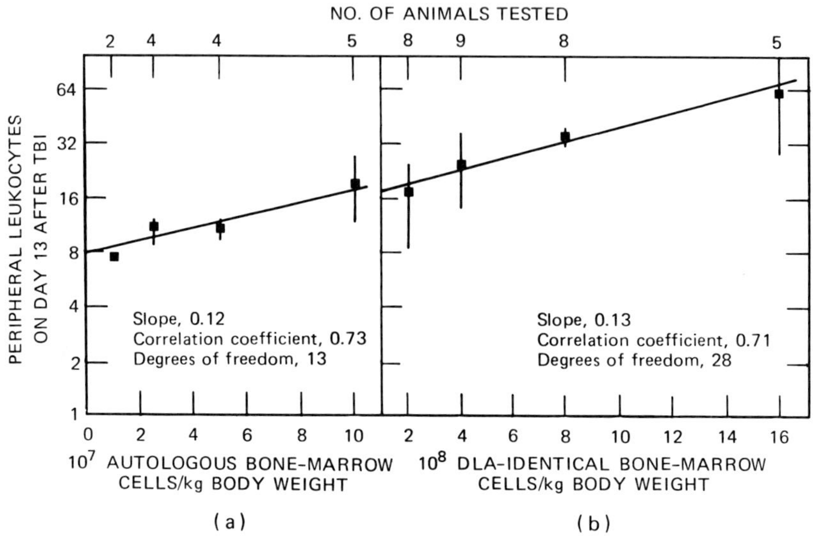


Fig. 3 Correlation between bone-marrow cell dose and peripheral blood leukocytes on day 13 after irradiation in autologous (a) and MHC-identical (b) receptors.

TABLE 11

GENETICS OF HvG AGAINST ALLOGENEIC DOG BONE MARROW*

| DLA differences between donor and host† | | Relation between donor and host | Proportion of persisting takes |
|---|---|---|---|
| LD | SD | | |
| ≠ | ≠ | Siblings | 0/6 |
| = | = | Siblings | 39/40 |
| = | ≠ | Unrelated | 0/6 |
| ≠ | = | Unrelated | 0/3 |
| ≠ | ≠ | Unrelated | 2/25 |
| = | = | Unrelated | 8/8 |

*Standard protocol: 750 rads TBI, 300-kV X rays, 16 to 18 rads/min, 3.0 mm Cu half-value layer, followed by 4 x $10^8$ BM cells/kg (iv) 6 to 24 hr later.

†MHC-mismatched BM cells are represented by ≠, and MHC-identical BM cells, by =.

indicates that HvG is controlled by a locus or loci different from LD and SD. Those results in siblings are corroborated by the results for nonrelated animals, where 2 out of 25 unrelated combinations produced a take notwithstanding LD and SD differences. In contrast, when both LD and SD were matched between unrelated animals, lasting takes occurred in eight out of eight animals. These data indicate that the genes controlling HvG, although different from LD and SD, show a high degree of linkage disequilibrium with LD and SD genes (see Chapter 8 for discussion of the phenomenon of linkage disequilibrium). The most reliable way to secure identity for HvG-controlling genes between donor and recipient is with littermates. The low frequency of recombination between LD/SD genes and HvG-controlling genes will almost guarantee identity for all MHC genes (including HvG-controlling genes) in littermate donor–recipient pairs that are SD- and LD-identical. The high degree of linkage disequilibrium in the random dog population between alleles of the DLA complex will cause identity for HvG-controlling genes in donor–recipient pairs that are matched for other (LD and SD) genes of the MHC. In littermate as well as in unrelated combinations, a positive, direct identification of HvG-controlling genes would be preferable over the present indirect practice of relying on the low frequency of recombination within the MHC or the high degree of linkage disequilibrium between alleles of different loci within the MHC. Obviously, such an approach cannot be applied until development of an in vitro test that identifies directly the gene products of HvG-controlling genes.

## Abrogation

**Conditioning.** Earlier in this chapter the effect of conditioning on resistance (HvG) in the dog was summarized. No combination of drugs, biological agents, (as ALS or L-Asp) and irradiation, with the possible exception of high-dose fractionated TBI,[48] will abrogate resistance sufficiently for low numbers of allogeneic MHC-mismatched BM cells. This problem deserves considerable attention in view of the absence of an MHC-identical donor in approximately two-thirds of the human patients who otherwise could benefit from a BM transplantation.

In recipients of MHC-identical BM, sensitization by preceding donor blood transfusions can be abrogated by a regimen of procarbazine, anti-lymphocyte serum, and TBI. TBI with procarbazine or anti-lymphocyte serum alone was ineffective.[94] The same protocol was not effective, or at best was only marginally effective, in conditioning unimmunized recipients for an MHC-mismatched bone-marrow transplant,[35] possibly because minor histoincompatibilities[95] are easier to suppress by immunosuppressive measures than are major histoincompatibilities (see Chapter 8).

**Timing of BM Infusion.** Results of studies in mice suggested that the effector cells in HvG reactions would die within 100 hr after TBI.[57] Thereafter HvG

TABLE 12

CONDITIONING REGIMENS FOR ABROGATION OF RESISTANCE IN DOGS

| Regimen† | | Proportion of takes |
|---|---|---|
| Higher TBI (900 rads) | | 0/1‡ |
| Split TBI (200 rads on day −7 and 700 rads on day 0) | | 0/3 |
| Delayed BM cell transfusion | | |
| Day +3 | | 1/4 |
| Day +5 | | 0/6 |
| CY, 25 mg/kg (iv), days −2 and −1 | +750 rads | 0/3§ |
| L-Asp, 500 IU/kg (iv) (6 or 10 times), days −2 to +10 | TBI on day 0 | 4/13 |
| $SiO_2$ (<5 μm), 100 mg/kg (iv), day +1 | | 25/29 |
| CBP, 0.10 mg/kg (iv), day −7 | | 0/4 |

*From Ref. 55 and H. M. Vriesendorp (unpublished observations).

†TBI, total-body irradiation, 300-kV X rays, 16 rads/min, 3.0 mm Cu half-value layer; CY, cyclophosphamide; L-Asp, L-asparaginase; $SiO_2$, silica particles; CBP, *Corynebacterium parvum.*

‡Three additional animals died from GI toxicity.

§One additional animal died from GI toxicity.

would be diminished considerably. In dogs, that concept is incorrect. Hemopoietic recovery was not observed in three out of four dogs receiving $4 \times 10^8$ MHC-mismatched BM cells/kg 3 days after TBI or six out of six dogs receiving MHC-mismatched cells 5 days after TBI (Table 12). The single case of hemopoietic recovery might be explained by a chance match for HvG-controlling genes. Those animals showed the same "abortive" rise of peripheral blood leukocytes[1] as animals receiving their transplant within 30 hr after TBI. This abortive rise was evident 8 to 10 days after BM cell infusion regardless of the day of TBI (H. M. Vriesendorp, unpublished observations), which indicates that the abortive rise is caused by donor BM cells (HSC's) and not by host HSC's surviving the TBI. The signs of donor BM reactivity do not continue because the donor BM cells are rejected.

**Donor Leukocytes.** The rejection of the donor HSC's can be prevented by adding donor leukocytes to the BM cells.[96] A dose of MHC-mismatched BM cells, which alone would not lead to hemopoietic recovery after supralethal TBI, can do so if donor leukocytes from the peripheral blood are transfused too. Two doses of $3 \times 10^8$ leukocytes/kg on days 2 and 3 after TBI effectively abrogated resistance.[35] The active principle in the donor leukocytes is probably the dividing lymphocyte since irradiated leukocytes were ineffective.[35] The minimum number of donor lymphocytes required for abrogation of resistance has not been determined. The mechanism of improvement of the take of DLA-mismatched BM cells by donor lymphocytes after TBI is probably by killing host lymphocytes since one of the targets of GvH reactions is host lymphocytes. Killing host lymphocytes will give additional suppression of host resistance.

## Resistance vs. Rejection

The studies in mice[57] that led to the postulation of resistance as a mechanism for BM allograft rejection, distinct from classical solid organ allograft rejection, indicated three important differences between this phenomenon and normal organ allograft rejection. First, the ontological development of resistance would occur later (± day 22 after birth) than the capability to reject a skin graft (± day 7 after birth). Resistance would be effectuated within 48 hr after the injection of the cells, whereas rejection takes at least 10 days. Also, special conditioning agents, given in addition to supralethal TBI, would effectively abrogate resistance but would not prolong skin allograft survival. The ontology of resistance vs. rejection has not been tested in dogs so far, but the other differences between the two phenomena have not been found in experiments with the dog. DLA-mismatched BM cells disappear in approximately 10 days after TBI. The rejection time of allogeneic dog BM is similar to that of other allografts in unmodified recipients.[93] The activity of DLA-identical BM disappears in approximately 20 days after 350 rads of TBI (Fig. 4). That time is similar to the rejection time of DLA-identical skin grafts in unmodified recipients.[93] After day 20, host hemopoiesis gradually recovers in most of the animals given 350 rads of TBI. The dips in the leukocyte recovery curves after 750 and 500 rads(± day 20) can be explained as a reflection of the ongoing reactions between donor and recipient hemopoiesis, each trying to "kill" the other. Genetic marker studies verify that the second, continuing rise of leukocyte levels after day 23 indicates that the donor wins the war after 500 or 750 rads of TBI.[14] Conditioning by TBI will not completely remove host hemopoiesis. The fraction of HSC's that survives TBI[14] (after 750 rads, approximately $10^{-5}$) remains to be removed by the donor. The limited effect of TBI on BM allograft survival is comparable to the meager effect of TBI on kidney allograft survival (H. M. Vriesendorp, unpublished observations).

Measures that were effective in boosting murine splenic hemopoiesis after TBI and allogeneic BM transplantation do not stimulate hemopoietic recovery in the dog with the notable exception of $SiO_2$ particles (Table 12). Another exception is that L-Asp apparently improves the take of MHC-mismatched BM in dogs but has no effect on murine splenic hemopoiesis (CFU-S assay; H. M. Vriesendorp, unpublished observations).

The differences between the mouse and the dog can probably be accounted for by the difference in endpoint assays, spleen colonies, or IUdR uptake in the mouse vs. primary radiation survival and BM histology in the dog. In dog studies no compelling evidence indicates a distinct rejection mechanism (resistance) for

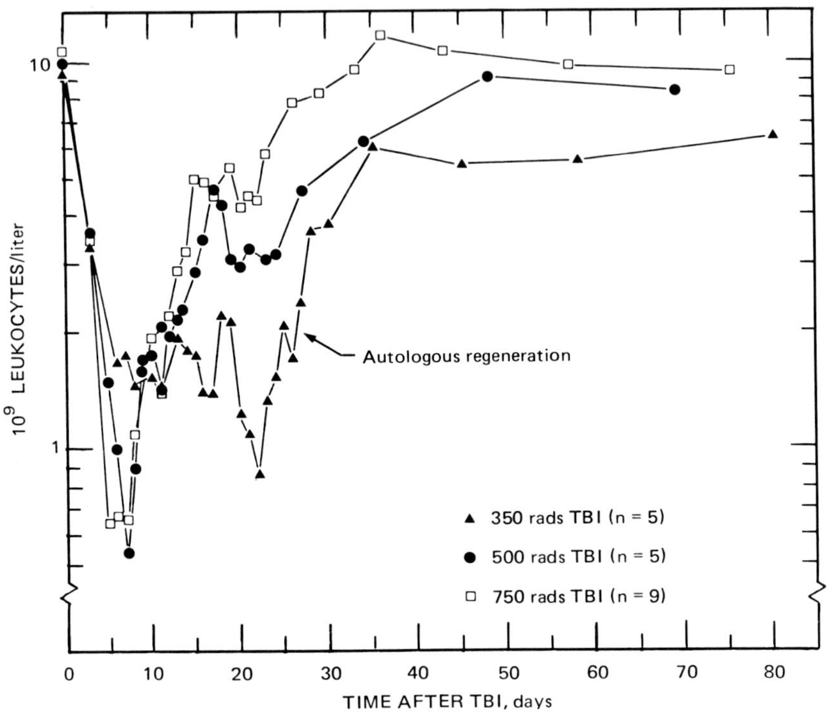


**Fig. 4** Leukocyte recovery after TBI and bone-marrow transplantation of DLA-identical donors (4 x $10^8$ cells per kilogram of body weight).

allogeneic BM cells. In the future the term "resistance" should be limited to descriptions of the murine phenomenon for which it was introduced. The continuing use of that term or its equivalent Hh (hemopoietic histocompatibility) in describing the rejection of BM cells is not supported by the available data and is therefore confusing. "Rejection" and "HvG" are terms that describe the phenomenon of failure of hemopoietic engraftment adequately and do not have the additional, incorrect, connotations of the term "resistance."

## SUPPORT

### Preventive Measures Before Conditioning

The introduction of specific preventive measures before BM transplantation in dogs will depend on the nature of the planned experiments. The obvious aim is to prevent or limit the seriousness of posttransplant complications. In addition to the normal, general measures taken in dog experimentation (e.g., vaccination and quarantine), animals should be screened for potential troublesome foci of infection or bleeding. The suggested procedure entails an inspection of the oral cavity and removal of dental calculus, the digital emptying of anal glands, and a parasitological and bacteriological analysis of fecal material. Before conditioning, feces should be free of parasites. Parasites in the GI tract will cause local damage to the GI mucosa, creating spots from which blood loss can occur during severe, additional GI toxicity (as in the GI syndrome after TBI) or during periods of thrombopenia. Asymptomatic *Pseudomonas aeruginosa* carrier dogs should not be used for irradiation experiments in view of the severity of endogenous infections caused by that organism in granulopenic animals. A prophylactic course of oral colistin or polymyxine B will usually be sufficient to decontaminate the animal of that microorganism. At sites for vena punctures, infusions, and subcutaneous fluid administration, the coat should be shaved before irradiation to prevent the possible complications of that procedure (hemorrhages and infections) at a later stage of the experiment. In addition, baseline clinical data that will be needed in subsequent animal care and result evaluation should be obtained before TBI.

## Gastrointestinal Toxicity

In animals receiving TBI with a dose of 700 rads or higher (300-kV X ray; dose rate, 16 to 18 rads/min), GI symptoms start approximately 60 hr after TBI. Symptoms include vomiting and watery diarrhea and a compulsive drinking habit followed by more vomiting. Without treatment those symptoms usually become rapidly more severe. The animals develop lethargy, become comatose, and die within 7 days after TBI. Postmortem blood cultures are usually negative. Occasionally at autopsy blood is found within the lumen and wall of the GI tract. However, cause of death is usually extreme fluid and electrolyte loss caused by diarrhea and vomiting. That course of events can be prevented by stopping oral fluid and food intake on the third day after irradiation and switching to twice-daily sc administration of Ringer's lactate to a total daily amount of roughly 75 ml per kilogram of body weight. From day 6 onward, normal fluid and food intake can be resumed. If animals lose blood through the rectum in the first week of the TBI, immediate blood transfusions are indicated without awaiting the development of further clinical signs. Regardless of the volume of blood lost, such an event indicates a massive GI blood loss due to radiation damage to the GI tract. In contrast, rectal blood loss in a later phase of the experiment indicates periods of thrombopenia, and whole blood or thrombocytes need only be given if hematocrit or peripheral blood thrombocyte counts are dangerously low.

If during an episode of GI toxicity the hematocrit increases over 55 notwithstanding sc fluid support, iv infusion of low-molecular-weight dextran solutions or Ringer's lactate is sometimes beneficial in a volume of 100 $cm^3$ per dog of 10 kg repeated two to three times a day, depending on urine output.

The $LD_{50}$ for the GI syndrome in dogs after TBI is 850 rads. If the GI tract is decontaminated of bacteria, yeasts, and fungi with oral unabsorbable antibiotics prior to TBI and if the dog is housed under laminar airflow conditions, the $LD_{50}$ is increased to 1000 rads.[97] That increase in resistance to the GI syndrome indicates that in conventional animals infections contribute to the lethality of the GI syndrome at the lowest TBI dose causing the GI syndrome. The $LD_{50}$ for the GI syndrome in dogs is lower than that for rodents and that for rhesus monkeys. That difference could be a consequence of the fact that the length of the GI tract is shorter in dogs (carnivores) than in rodents (herbivores) or monkeys (omnivores). This suggestion of an inverse correlation between length of the GI tract and susceptibility to the GI syndrome is based on the assumption that the number of surviving GI crypt stem cells[1] determines the severity of the GI syndrome. A shorter GI tract will have fewer stem cells, and the system is therefore more susceptible to stem-cell loss from TBI. That explanation is compatible with the observed efficacy of shielding only a small part of the ilium during TBI in the prevention of the GI syndrome.[98]

## Hematological Support

After supralethal TBI and BM transplantation, most end cells of the host hemopoietic system will gradually disappear from the peripheral blood. Only lymphocytes die immediately after irradiation without undergoing cell division (so-called interphase death). The disappearance of other end cells from the peripheral blood is determined by the marrow transit time of their progenitor compartment and the life span of the end cell.[1] The average duration of the nadir for each type of hemopoietic end cell, with the exception of erythrocytes, is shown in Table 13. The life span of erythrocytes is so long that donor hemopoiesis will deliver sufficient red cells before host erythrocytes die out. A clinically significant shortage of red cells will occur only if additional erythrocytes are lost through hemorrhages caused by thrombopenia or as a result of hemolytic anemia caused by GvH reactions. In contrast, the much shorter survival times of platelets and leukocytes cause clinically significant problems. Transfusions can be given as prophylaxis or in response to clinical symptoms. All transfusions during the aplastic phase after conditioning should be irradiated to prevent acute GvH reactions from allogeneic lymphocytes in the transfusion. A dose of 1500 rads will inactivate approximately 10 decades of lymphocytes ($D_0 \approx 70$ rads) yet not influence the function of platelets,[99] red cells,[21] or granulocytes.[101] The risks of acute GvH reactions also prohibit a priori the use of nonirradiated lymphocytes in supportive care. In the future, however, conditions might be defined for safe transfusion of lymphocytes for the purpose of achieving more-complete or more-rapid immune reconstitution. Seven months after TBI and BM transplantation, the infusion of BM donor lymphocytes did not induce acute GvH disease in DLA-identical donor–recipient combinations.[102]

TABLE 13

HEMOPOIETIC RECOVERY AND SUPPORT PHASE AFTER TOTAL-BODY IRRADIATION

| Cell type | Low cell count, days after TBI | Currently available supportive measures |
|---|---|---|
| Granulocytes | 4 to 14 | Antibiotics and granulocyte transfusions |
| Reticulocytes | 3 to 14 | Packed red cells |
| Thrombocytes | 8 to 20 | Platelet transfusions |
| Monocytes | 4 to 20 | None |
| Lymphocytes | 1 to 20 to 100 | None |

The indications for transfusion differ among investigators. Usually thrombocytes are transfused as prophylaxis if their number falls below $20 \times 10^9$ per liter; red cells will be given if the hematocrit falls below 20%. No role for monocyte transfusions has been defined. Granulocyte transfusions were beneficial in model studies of pseudomonas or candida infections.[101,103] The high numbers of granulocytes required for transfusion and the difficulties encountered in obtaining those numbers from single donors indicate that granulocyte transfusion techniques are presently too complicated for routine application in dogs.

## Infectious Complications

At the end of the first week after TBI and BM transplantation, bacteria from the GI tract usually cause endogenous infections. An antibiogram of GI flora, determined immediately before or after TBI, will indicate which antibiotics to use. The recommendation is to give parenterally two different, effective, bactericidal antibiotics without BM toxicity. Usually within a week after the start of such an antibiotic treatment, a second rise in body temperature signals the emergence of microorganisms resistant to the antibiotic regimen. A repeated antibiogram of fecal flora can be used to corroborate that impression and to select new antibiotics. At all times antibacterial agents should be given in conjunction with an oral anti-fungal agent to prevent overgrowth with fungi. Pimaricin (by mouth, 160 mg twice a day) is an effective treatment. Nystatin (by mouth) is less appropriate in dogs since its administration is often followed by almost rebound vomiting.

At later stages of the experiment, beyond day 20 after TBI, fevers are not always caused by endogenous infections from the GI tract. Pneumonia is a frequent complication, and empiric parenteral treatment with ampicillin is sometimes useful. Hemopoietic recovery should enable the host to handle most infections without the help of antibiotics at this stage unless GvH lesions prevent the clearance of infections.

Antibiotics are continued for 5 days after normalization of the body temperature, and the anti-fungal agents are given for an additional 3 days. Preliminary studies on the effects of total or selective prophylactic decontamination of the GI tract with oral, nonabsorbable antibiotics on endogenous infections as well as on GvH disease have indicated that total decontamination, although difficult to maintain, effectively prevents endogenous infections. Selective removal of aerobic gram-negative microorganisms, yeasts, and fungi does decrease the severity of GvH disease but not the incidence of endogenous infections.[104]

## GRAFT VS. HOST DISEASE

### Definition

Graft vs. host disease has been defined primarily in experimental animals other than the dog, but the principles that emerged from those small-rodent studies[30,71,72,105] can be safely extrapolated to other species, including the dog. GvH disease apparently is caused by long-living postthymic T-cells of the donor.[105] Indirect confirmation of that concept in the dog model is the reported decrease in GvH severity after in vitro BM incubation with anti-lymphocyte serum[106] or the reported increase in GvH disease when donor lymphocytes are added to the BM graft. The postthymic T-cells, which initiate the GvH disease, are immunocompetent cells with anti-host specificity. The subsequent development of GvH disease will also involve other donor and host cells in a nonspecific way.[105]

GvH disease can be caused via one of two pathways, each leading to a separate type of GvH and requiring a different approach for treatment.[67] The first type, the so-called acute GvH disease, is caused by the postthymic T-cells of the donor being infused concurrently with BM cells. That form of GvH disease usually causes death within 4 weeks after transplantation. The second type is the "delayed" GvH disease, which develops later, usually between 30 and 100 days after transplantation. The cells inducing the delayed syndrome are T lymphocytes, which develop through the processes of proliferation and differentiation out of the donor HSC's. That form of GvH disease carries a lower mortality and often resolves spontaneously or remains in an indolent form. In man a third type of GvH disease, "chronic" GvH disease, occurs later or persists longer than acute and delayed GvH diseases.[107,108] That form shows some similarities with autoimmune diseases such as scleroderma and chronic active hepatitis. An effective form of treatment has not yet been found. An experimental animal model for chronic GvH disease is currently not available.

## Diagnosis

In dogs, GvH disease can be diagnosed in several ways, but any mode of diagnosis must include controls with autologous BM transplants, which, by definition, are not affected by GvH disease. That approach will exclude the possibility that an incorrect diagnosis of GvH disease may be made on the basis of signs or symptoms that are caused, not by allogeneic lymphocytes, but rather by other factors, such as conditioning, supportive care, preexisting disease, or infections.

A diagnosis of GvH disease can be based on survival, histological signs, clinical symptoms, and biochemical or hematological tests. Survival can be a misleading endpoint for GvH disease because animals sometimes die from causes other than GvH disease, or conversely GvH reactions gradually decrease in severity and do not kill the recipient, particularly in dogs transplanted with BM from an MHC-identical donor. Therefore survival studies used as a definition of GvH disease should be accompanied by a histological analysis.

Histological evaluation of rectal biopsies or skin biopsies can detect GvH disease. For skin biopsies, the lip is the most appropriate site if skin as well as mucosa is included in the biopsy. Skin from other areas of the body is often not involved in GvH lesions (C. Zurcher, Institute for Experimental Gerontology, Rijswijk, The Netherlands, personal communication). A rectal biopsy technique has been described for dogs.[109]

Analyses of autopsy material of dogs with GvH disease have emphasized histological signs of GvH disease in skin, lymph nodes, liver, and GI tract.[110,111] In addition to those "classical" GvH target tissues, histological GvH lesions have been noted in lung, trachea, uroepithelium, prostate, salivary duct, and pancreatic duct (Table 14). The characteristic GvH lesions in lymph nodes (fibrinoid deposition in the center of the lymphatic follicles)[30] were found only in those dogs which received donor lymph node cells in addition to BM cells.

Daily clinical inspection will yield a preliminary diagnosis of GvH disease. Skin lesions will manifest as hyperkeratosis, erythema, hair loss, and subsequent skin atrophy. Sometimes mucositis affects the nasal and oral cavity and eyes. Diarrhea, blood-stained urine, and icterus are other clinical signs sometimes caused by GvH disease. In a retrospective study (H. M. Vriesendorp and C. Zurcher, unpublished observations), an almost complete correlation was found between clinical symptoms and histological signs of liver GvH disease. Highly significant correlations were also found between histological signs of skin or GI tract GvH disease and clinical observations. More skin lesions were found clinically than histologically, which indicates that the experimentalists were either too eager to find GvH disease or GvH disease had disappeared before autopsy. Frequently lip biopsies performed at regular intervals before death show such regression. In contrast, GI GvH disease was found more often by histology than by clinical observation, probably because of the usual confusion regarding the cause of

TABLE 14

GRAFT-VS.-HOST LESIONS IN VARIOUS TISSUES IN DOGS AFTER BONE-MARROW TRANSPLANTATION*

| BM transplant | | | Mean survival, | Percentage of dogs† with lesions in | | | | | | | | | | | | | |
|---|---|---|---|---|---|---|---|---|---|---|---|---|---|---|---|---|---|
| Type | No. | Take | days | Skin | Lip | Oropharynx | Stomach | Ileum | Colon | Liver | Lung | Trachea | Uro-epithelium | Prostate | Epididymis | Salivary duct | Pancreatic duct |
| Allogeneic | 33 | + | 29 (11 to 108)‡ | 48 | 42 | 45 | 15 | 73 | 36 | 67 | 60 | 21 | 30 | 19 | 12 | 15 | 3 |
| Allogeneic | 25 | + | 33 (11 to 149)‡ | 0 | 4 | 0 | 0 | 0 | 0 | 4 | 4 | 0 | 4 | 0 | 0 | 4 | 0 |
| Allogeneic | 12 | — | 15 (12 to 21)‡ | 0 | 0 | 0 | 0 | 0 | 0 | 0 | 0 | 0 | 0 | 0 | 0 | 0 | 0 |
| Autologous | 12 | + | 33 (11 to 48)‡ | 0 | 0 | 0 | 0 | 0 | 0 | 0 | 0 | 0 | 0 | 0 | 0 | 0 | 0 |

*From unpublished data of C. Zurcher.

†Only dogs surviving for 11 days or longer were used for calculation; in the autologous group, animals were killed to obtain histological controls with comparable survival times after TBI.

‡Numbers in parentheses are ranges.

diarrhea in the setting of BM transplantation. For example, antibiotics and radiation may also cause diarrhea. Clinical mucositis was found in cases both with and without histological signs of GvH disease in the inflamed tissues.

A biochemical serum analysis can reliably indicate liver GvH disease. Serum levels of alkaline phosphatase and glutamic pyruvate transferase (alanine aminotransferase) increase if dogs go into hypovolemic or septic shock, but liver GvH disease will induce yet higher elevations of those values (more than five times normal). A daily determination of peripheral blood cells has also been informative in GvH diagnosis. Generally, granulocytosis and thrombocytopenia accompany severe GvH reactions. Those blood-cell changes may be related to the GvH reactions per se or to complicating secondary infections. Another "hematological" sign of GvH disease in dogs is Coombs' positive hemolytic anemia. If the animal survives, the anemia will gradually disappear as the host erythrocytes are replaced by donor red cells. Because of the various modes of GvH disease expression, comparisons between different laboratories or between different experimental groups within one laboratory can be made only if the same methods of GvH diagnosis are used.

## Modification of GvH Disease

Several factors influence the incidence and the severity of GvH disease. These factors can be conveniently divided into genetic and nongenetic factors. Genetic factors, the histocompatibility differences between donor and recipient, influence the number of donor postthymic T-cells with specific anti-host activity. Nongenetic factors affect mainly the aspecific amplification of the anti-host activity of donor T-cells.

**Genetic Factors.** Clinical or histological signs of GvH disease do not result from autologous BM transplantation in dogs. The most important genetic factor controlling GvH disease in allogeneic donor–recipient combinations is the MHC or DLA in the dog (see Chapter 8). The first report[112] indicating the important influence of DLA on GvH disease in 1971 has been confirmed subsequently in various laboratories.[26,92,111] Survival after BM transplantation from a DLA-identical sibling is determined by several modifying factors. Long, uncomplicated survival is the rule when BM cells only are used for transplantation.[34] The occasional GvH reactions are of the delayed type and are usually transient. Lethal GvH disease is infrequent. The absence of acute GvH disease in such donor–recipient combinations can be explained by the low concentration of postthymic T-cells in dog BM, which is also responsible for the negative PHA or Con A responses of dog BM cell suspensions (H. M. Vriesendorp, unpublished observations). The incidence and the severity of GvH disease can be increased in DLA-identical donor–recipient combinations by adding lymphocytes to the transplant.[95] Under those conditions acute GvH disease can develop and the

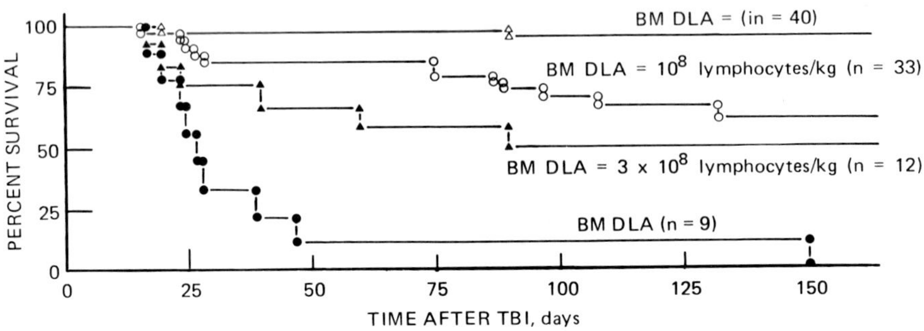


**Fig. 5** Influence of donor lymphocytes on survival of recipients of DLA-identical bone marrow.

percentage of animals developing GvH disease or dying from GvH disease is related to the number of lymphocytes added (see Fig. 5). If, instead of survival, other criteria for the diagnosis of GvH disease, such as skin biopsies, clinical inspection, and biochemical and hematological determinations, are used, the incidence of GvH disease in the different groups of Fig. 5 is 25% for DLA-identical BM only, 60% for DLA-identical BM plus 1 x $10^8$ lymph node cells/kg, and almost 100% for DLA-identical BM plus 3 x $10^8$ lymph node cells/kg.

Clinical symptoms and histological signs of GvH disease are similar in all dogs that die before day 30 after TBI regardless of MHC identity in the donor–recipient combination. The most severe GvH disease is observed in animals receiving both HSC's and lymphocytes from peripheral blood of MHC-identical donors[84] because peripheral blood has the most unfavorable (i.e., high) lymphocyte/HSC ratio.

GvH disease in DLA-identical BM transplants is under the genetic control of minor histocompatibility systems. The observed correlation between GvH disease and the number of lymphocytes transplanted is explained best by the assumption that reactive lymphocytes are required to induce GvH disease over a given histocompatibility difference. If histocompatibility systems differ in their requirement for lymphocytes, more minor histocompatibility systems will become apparent clinically when more lymphocytes are added. Such a model is compatible with the observation that in unimmunized animals a higher percentage of lymphocytes are reactive against allogeneic major histocompatibility antigens than against allogeneic minor histocompatibility antigens. The model also fits tne experience with solid organ allografts. Sometimes the influence of minor histocompatibility antigens can be demonstrated only after prior immunization of the recipient. Such an immunization can be considered as another method of increasing the number of lymphocytes reactive against a given minor histocompatibility antigen.[95]

In contrast, if, with CY instead of TBI conditioning, buffy coat cells from a DLA-identical donor are added to BM cells, the incidence of persisting chimerism after BM transplantation increases, but severe GvH reactions remain infrequent.[33] A possible explanation for the difference in effect of donor leukocytes after CY or TBI conditioning is that CY is a less effective conditioning agent than TBI. The addition of donor leukocytes after CY conditioning will kill host lymphocytes and thereby decrease the HvG reaction and incomplete chimerism. Donor leukocytes are also killed during the reaction with host leukocytes (so-called allergic death), and insufficient numbers of donor leukocytes survive to initiate acute GvH disease. After TBI conditioning fewer host lymphocytes remain, and relatively more donor leukocytes will be available for the initiation of acute GvH disease. Quantitative information that would support that possibility is lacking. A similar experience has been obtained in human BM transplantation, where the addition of donor leukocytes after CY conditioning only slightly increases GvH reactions.[113]

A serological identification of dog minor histocompatibility antigens is not yet possible, but a genetic approach shows that a locus on the Y chromosome, a locus close to the autosomal locus for phosphoglucomutase-2, and at least one additional autosomal locus probably control minor histocompatibility antigens capable of triggering GvH disease.[92,95] The most practical solution to the problem of GvH disease in DLA-identical BM transplantation (or MHC-identical transplants in other species) is to prevent lymphocyte contamination in the BM cell suspension or to remove lymphocytes from such suspensions. In contrast, attempts to match MHC-identical donor–recipient pairs for minor histocompatibility systems are bound to fail because the number of polymorphic, independently inherited histocompatibility systems is large and the number of available donors (identical for at least the MHC) is small.

DLA identity is easily verified between littermates. The recombination frequency within the DLA complex is so low that identity in a littermate for one of the polymorphic systems in the complex usually indicates that the same parental chromosomal areas coding for DLA have been inherited. In unrelated donor–recipient pairs, complete DLA identity is not necessarily concomitant with identity for a single DLA locus. Control of GvH disease by more than one locus creates difficulties in the selection of donors because the low phenotype frequency of a given set of GvH-controlling DLA antigens in a single recipient requires large donor pools.

If the results of studies with mice can be extrapolated, multilocus control of GvH disease within the MHC is expected in the dog too.[114] The known LD and SD loci within the DLA complex do not have an important influence on GvH disease[26,92,115] (Fig. 6), which suggests that one or more presently unknown loci within DLA participate in GvH disease control. Those loci are distinct from LD and SD loci, and their gene products are presently not recognized, which complicates the selection of a donor for recipients lacking an MHC-identical

| Donor–recipient combination | Loci of the DLA complex LD | SD1 (DLA-A) | SD2 (DLA-B) | Number tested |
|---|---|---|---|---|
| Related 0 DLA haplotype difference | = | = | = | 7 |
| Related 1 DLA haplotype difference | ≠ | ≠ | ≠ | 11 |
| Related 2 DLA haplotype difference | ≠ | ≠ | ≠ | 3 |
| Unrelated | ≉ | = | = | 3 |
| Unrelated | ≈ | ≠ | = | 3 |
| Unrelated | ≠ | ≠ | ≠ | 7 |

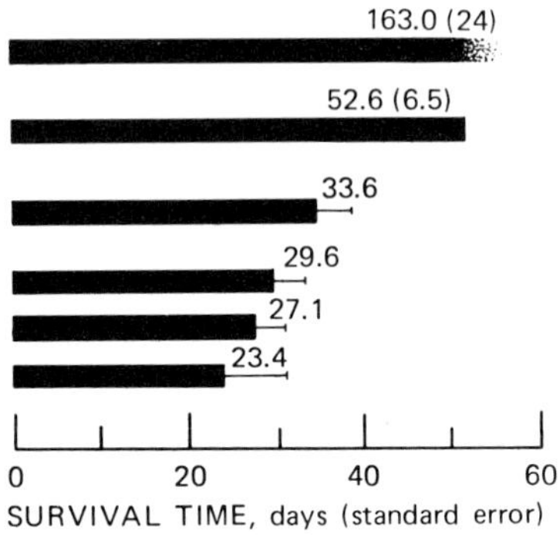


Fig. 6 Influence of host–donor matching on survival of radiation chimeras. All animals received 750 rads of TBI followed by 100 mg of $SiO_2$ per kilogram of body weight (iv) and $4 \times 10^8$ BM cells per kilogram of body weight.

sibling donor. Half-matched related BM donors (one DLA haplotype different) apparently offer better survival chances than do LD- or SD-matched unrelated donors.[93,116] The use of family donors different by one MHC haplotype deserves further study since for almost every human patient such a donor is available. However, a much better prevention of HvG as well as GvH reactions will be required before those donor–recipient pairs can be recommended for general use in dogs or man.

**Nongenetic Factors.** Small numbers of DLA-identical BM cells induce rapid hematopoietic recovery after TBI. Hematological support is usually not required. In one study irradiated blood transfusions given on days 7, 14, 21, and 28 after TBI aspecifically increased the severity of the GvH reactions but did not influence the incidence of GvH disease.[34] The relation between blood transfusion and GvH severity needs to be confirmed in a larger series.

The severity of canine GvH disease in DLA-identical donor–recipient pairs transplanted with BM and lymph node cells can also be influenced by removing *Enterobacteriaceae,* pseudomonas, fungi, and yeasts from the GI tract of the recipient with daily oral nonabsorbable antibiotics. Such treatment does not remove anaerobic and gram-positive aerobic microorganisms. The mechanisms of this GvH mitigating procedure are still obscure.[104] Studies in mice suggest that bacterial invasion of the GvH lesions of the intestinal mucosa increases the severity of such lesions and, in addition, increases the severity of GvH lesions elsewhere in the body, possibly through immunogenic bacterial antigens, which cross-react with tissue antigens of the BM recipient.[117]

TABLE 15

LETHAL GRAFT-VS.-HOST REACTIONS IN DLA-IDENTICAL DONOR–RECIPIENT PAIRS*[95,104]

| Gender | | | | |
|---|---|---|---|---|
| Donor | Recipient | Lethality | | |
| Male | Male | 2/10 | 4/18 | $p < 0.01$ |
| Male | Female | 2/8 | | |
| Female | Male | 6/13 | 13/27 | |
| Female | Female | 7/14 | | |

*750 rads TBI, 300-kV X rays, 16 rads/min, 3.0 mm Cu half-value layer.

†2 x $10^8$ BM cells/kg, 1 or 3 x $10^8$ lymph node cells/kg, equally distributed over the different groups.

Another influence on GvH disease is the gender of the donor (Table 15). Female BM donors significantly increase the lethality of GvH disease. In human BM transplantation, the effect of donor gender is similar.[15] The genotype of the recipient apparently does not modify the "killing instinct" of female lymphocytes. Perhaps female lymphocytes are more aggressive than male lymphocytes, or females have more T lymphocytes than males. The higher incidence of autoimmune diseases in females than in males (dog or man) might be another reflection of the same phenomenon.

## Abrogation of GvH Disease

**Prophylaxis.** When GvH disease is expected, either a "prophylactic" or a "therapeutic" approach can be adopted. Prophylaxis involves prevention of the complications of GvH beforehand, whereas therapy involves deferred treatment after GvH complications become apparent. The low efficacy of the currently available agents against established GvH disease argues for a prophylactic approach, which keeps the therapeutic agents in reserve if GvH disease develops notwithstanding prior prophylaxis.

Prophylactic treatment must take into account the modifying factors of donor selection, lymphocyte contamination of transplanted cells, post-transplantation transfusion care, GI microflora, and sex of BM donor. Ideal prophylaxis conditions would be to use BM cells from a DLA-identical male donor with no contaminating lymphocytes, no posttransplantation transfusions, and selective removal of GI microorganisms.

Lymphocytes can be removed from a BM cell suspension by purification of the HSC's on a discontinuous albumin gradient,[9,118] incubation of the BM cells with anti-lymphocyte serum (ALS),[106,119] or pretreatment of the recipient with ALS.[120,121] All procedures carry the risk of decreasing the number of active HSC's, which would impair the chances for a rapid hemopoietic recovery. In addition, ALS is a biological product with unpredictable efficacy and toxicity. A gradient procedure can be better defined and more easily transferred from one species to another. Further experimentation with that technique should define the best lymphocyte removal method.

The microflora of the GI tract can be influenced by either total or selective decontamination. A total decontamination procedure will remove all bacterial and fungal microorganisms from the host. That procedure has the advantage of also preventing endogenous infections during the aplastic phase of BM transplantation. The disadvantages of total decontamination are high costs, difficult technology, and the high risk and serious complications of breaks in the isolation procedure. The totally decontaminated host will be rapidly colonized by an inadvertently introduced contaminant, which is likely to cause a life-threatening infection. The selective removal of *Enterobacteriaceae,* pseudomonas, and fungi from the GI tract results from giving the animal oral antibiotics with a limited spectrum of activity (e.g., a combination of polymyxine B, neomycin, and pimafucin).[104] Selective treatment will not disturb the aerobic gram-positive bacteria and anaerobic microbes of the GI tract. That procedure also is easier to apply, the risks of failure are lower than in the case of total decontamination, and the incidence of exogenous infections is decreased. A drawback is that selective decontamination does not prevent endogenous infections.[104] However, those endogenous infections during selective decontamination are less severe than in conventional animals and respond well to appropriate, parenterally administered antibiotics.

An additional method for GvH prophylaxis is posttransplantation methotrexate (MTX) treatment. The experience in dogs shows that MTX is extremely effective in suppressing delayed-type GvH disease in MHC-identical donor–recipient pairs.[111] In MHC-mismatched combinations, survival is prolonged with MTX,[122] but almost all recipients of MHC-mismatched BM die notwithstanding MTX treatment. A disadvantage of MTX (and most other posttransplantation chemotherapy) is hematological and GI toxicity. The difference in efficacy of MTX prophylaxis on GvH disease between MHC-identical and MHC-mismatched donor–recipient pairs can be explained if that procedure only decisively affects delayed-type GvH disease. For maximum GvH prophylaxis, a more effective and less toxic treatment of acute GvH disease is needed. A combination of a lymphocyte depletion method and selective GI decontamination potentially offers a better GvH disease prophylaxis than MTX, but it needs to be evaluated accurately.

Cyclophosphamide has been effective against GvH disease in mice and monkeys[72] but has proved ineffective in trials with DLA-mismatched canine combinations in schedules similar to the effective MTX protocol.[123] Experience with the new agent Cyclosporin A has not yet been obtained in the dog.

**Treatment of Established GvH Disease.** The treatment of established GvH disease has not been well studied in dogs. ALS treatment doubles the median survival time; prednisone has no effect.[124] A major problem with the application of ALS is HSC toxicity to the extent that treatment animals either do not obtain a functioning BM graft or lose it after obtaining it.[53,121,124] CY, MTX, cytosine arabinoside, L-Asp, 6-mercaptopurine, and procarbazine have no effect on GvH disease.[125]

## FOLLOW-UP

### Aims of Transplant

Usually the principal aim of BM transplantation will be the establishment of donor hemopoiesis in an allogeneic host. Donor hemopoiesis can be confirmed after BM transplantation by demonstrating BM donor phenotypes on blood cells and proteins. Donor and recipient can be tested before and after transplantation for the informative markers (i.e., different in donor and recipient). The karyotypic bimorphism of sex chromosomes was the first marker used for that purpose in dogs.[10] Additional markers were summarized in the joint report of the second workshop on canine immunogenetics.[126]

Red-cell markers will show a more gradual change to donor type than other blood cells because red cells have a longer survival time (±100 days) than platelets (±6 days) and granulocytes (±0.5 day). The marker for dog blood group DEA 7 is uninformative as a test for hemopoietic chimerism. The DEA 7 antigen is not an intrinsic red-cell antigen but is produced outside the hemopoietic system and thereafter is absorbed onto the red cell. The production of the antigen is radiation sensitive. Therefore that blood group disappears for a few weeks after TBI and then returns as the unchanged host phenotype.[127]

Serum proteins, such as immunoglobulins and possibly also transferrins, are also potentially useful as assays for hemopoietic chimerisms[126,128–130] but have not been used thus far for that purpose in dogs. The enzyme glutamic-oxalacetic transaminase (S-GOT) with its electrophoretic variant has been used to demonstrate that complete hemopoietic chimerism in dogs also includes alveolar macrophages.[130] Other polymorphic red- and white-cell enzymes that have been used to demonstrate chimerism are phosphoglucomutase-2, peptidase D, and superoxide dismutase.[34,126]

Complete chimerism does not guarantee a normalization of the hemopoietic or immune system. Even 2 years after allogeneic BM transplantation, the number of in vitro colony forming units (CFU-C) can be abnormally low, up to 50% lower than in normal dogs or in animals after autologous BM transplantation (H. M. Vriesendorp, unpublished observation). Peripheral blood counts and clinical status remain normal despite the low CFU-C values. The cause of that defect is still obscure. Another abnormality observed in long-surviving dogs after autologous peripheral blood HSC transplantation is that marrow spaces are reduced in volume.[131]

Immunological reconstitution after BM transplantation has been studied in dogs receiving TBI or CY conditioning.[132,133] Granulocytes formed after BM transplantation were always normal in function. Also, antigen clearance by the reticuloendothelial system was normal after TBI with or without BM transplantation.[133] IgG levels were slightly decreased in short-lived chimeras and elevated in long-lived chimeras. Primary humoral immunity was lowest in short-lived radiation chimeras (i.e., within 200 days after BM transplantation) and still significantly below normal in long-lived radiation chimeras. An appreciable IgM to IgG switch in secondary humoral immunity was only found in long-lived radiation chimeras.[133] Naturally occurring antibodies in dogs returned to preirradiation levels in radiation chimeras in about 40 days (A. Zweibaum, Hôpital Broussais, Paris, France, and H. M. Vriesendorp, unpublished observations). First and second set skin allograft survival was prolonged in short- and long-lived radiation chimeras. In contrast, in incomplete chimeras after CY conditioning, skin allograft rejection times were normal.[132] The in vitro lymphocyte response of chimeras to lectins (phytohemagglutinin and Con A) returned earlier than the response to allogeneic lymphocytes. Depression of in vitro cellular immunity assays was more prolonged in dogs conditioned with CY than in dogs conditioned with TBI. That difference might be the result of the more-sensitive assays used in the CY study as well as immunobiological differences between complete and incomplete chimeras.

An attempt to transfer immunity (i.e., a transplant of "immunological memory") in dogs to bacteriophage ϕX 174 failed.[133] However, GvH activity was increased when BM cells and leukocytes were taken from donors that had rejected a skin graft of the recipient.[57,101] The scarce data on transfer of immunological memory in the dog support data for other species that such transfer is dependent on the transplanted cell dose and type and on the strength of the antigen.[30] Lymphocytes are needed for memory transfer, and memory for strong antigens is more easily transplanted than memory for weak antigens.

Complete hemopoietic chimerism includes the complete removal and replacement of the host immune and hemopoietic system. This will effectively terminate host anti-donor activity. Lack of anti-donor activity is illustrated by the acceptance of organs of the BM donor by complete chimeras (Ref. 134 and H. M. Vriesendorp, unpublished observations). In incomplete chimeras or reversals, BM donor organs are sometimes rejected.[135] The reverse reaction of donor anti-host activity is modified in time; the chance of developing GvH disease decreases with increasing survival times after TBI. Also, GvH disease gradually becomes less severe in most surviving animals without treatment. Humoral or cellular blocking factors that might mediate that change in GvH disease either cannot be found[136,137] or cannot be reproduced.[138] Recent observations suggest that a cellular mechanism suppresses the development of GvH disease in long-lived survivors.[139] In that study, the infusion of donor lymphocytes 7 months after irradiation did not lead to lethal GvH disease unless the donor was preimmunized against the recipient by skin grafts. Incomplete chimerism can remain intact only if the donor and recipient hemopoiesis can coexist, but the mechanisms maintaining such chimerism have not been elucidated.

Another aim of BM transplantation can be tumor eradication. The anti-tumor response of the conditioning regimen and the BM transplantation procedure has been studied in dogs with spontaneously occurring lymphomas or solid tumors.[140,141] Lymphoma animals showed significant anti-tumor response after TBI and allogeneic BM transplantation, but many animals succumbed to early GI toxicity or infectious complications. Long-term survival was curtailed by GvH disease. Effective tumor eradication schedules without nonhematological toxicity have not been developed.

Another endpoint in transplantation has been the correction of genetic defects in hemopoiesis. Cyclic neutropenia and pyruvate kinase deficiencies have been corrected completely by BM transplantation from an unaffected donor dog.[142,143]

## Late Effects

Studies have been reported for TBI but not for other conditioning agents. In animals other than the dog, late effects of TBI included lens cataract formation, early epiphysial closure of bones in growing animals, skin pigmentation changes, decrease in fertility, and shortening in life span largely because of the earlier occurrence of cancer. Of those possibilities, only changes in pigmentation, fertility, and life span have been investigated in dog chimeras.

Pigmentation changes in the dog take the form of premature graying of the coat. The severity of those pigment changes is related to the dose of TBI; 350 and 500 rads of TBI produces less graying than 750 rads of TBI.

There were no gross changes in the fertility of male and female radiation chimeras as compared with the breeding characteristics in a large kennel in the Netherlands (Central Laboratory for Breeding Animals for Laboratory Research TNO), and the sex ratio was not changed, which suggests that TBI did not induce a large number of lethal mutations in the X chromosome (see Table 16). Obviously, the number of irradiated animals involved is too small for the detection of small differences. Increased neonatal mortality in the offspring of irradiated parents was caused by a still unexplained high incidence of prolonged delivery times caused by secondary uterine inertia. Such delivery problems have been reported earlier in animals treated with lower (sublethal) doses of TBI.[144]

The latency periods for tumors were decreased in dogs exposed to TBI.[145] Also, progressive intercapillary glomerulosclerosis occurred earlier in irradiated animals.[146]

TABLE 16

FERTILITY AFTER TOTAL-BODY IRRADIATION

| Treatment | Sex | Fertile | Sex ratio of offspring, males:females | Neonatal mortality | |
|---|---|---|---|---|---|
| | | | | Males | Females |
| 750 rads* | Male<br>Female | 4/4<br>5/7 (0.71) | 10:12 (0.83) | 90% | 25% |
| None | Female | 187/211 (0.89) | 595:615 (0.97) | 15% | 18% |

*300-kV X rays, 16 rads/min, half-value layer, 3.0 mm Cu. Animals were 8 to 18 months old at time of exposure.

## CONCLUSIONS

Research of BM transplantation in dogs has clearly defined their use in the experimental situation. The dog is the only animal model that regularly provides MHC-identical sibling combinations, the only standard allogeneic donor–recipient combination in human BM transplantation. Studies with dogs have defined the optimal conditioning dose (i.e., 500 rads of TBI) and the optimal BM cell dose (i.e., $2 \times 10^8$ BM cells/kg, without lymphocytes). Female BM donors have been incriminated as giving more GvH disease than male donors. Selective decontamination and gradient separation of BM are the most promising approaches for solving GvH disease in MHC-identical donor–recipient pairs. Sources of HSC's other than BM are unattractive alternatives because of their low HSC concentration or high lymphocyte contamination or both. The HvG reaction is indistinguishable from solid organ allograft rejection; therefore the term "resistance," introduced by investigators of murine hemopoiesis who suggested that BM cells were rejected in a way different from that of solid organ grafts, is invalidated. The application of donor lymphocytes in abrogating HvG reactions deserves further study because it is insufficiently defined and one of the few active principles next to TBI. Histocompatibility studies of the genetic control of HvG and GvH reactions show that in the dog multiple loci control HvG and GvH reactions and that the genes of a proportion of those loci cannot be recognized yet. Those observations indicate that the risks of BM transplantation in unrelated donor–recipient pairs remain unacceptably high, and the future for a solution of those problems by donor selection is dim. The related, one MHC haplotype mismatched, donor might be a more realistic alternative and subject for further study.

Prior sensitization in DLA-identical donor–recipient pairs can be diagnosed prospectively and also abrogated effectively by additional conditioning before BM transplantation.

The most important differences between man and dog are the fewer postthymic T lymphocytes in BM cell suspensions and the greater susceptibility to cyclophosphamide in dogs. For effective extrapolations to man, the T-cell deficiency can be corrected by artificial contamination of dog BM cells with T lymphocytes.

An intensified search for relevant disease models in dogs might lead to the establishment of models in which the applicability of BM transplantation in human diseases can be evaluated. Many recommendations for human BM transplantation originate from dog studies. Some of the recent recommendations have not even been tested in man. That thrust is an indication of the current strength and promise of the dog model.

## REFERENCES

1. V. P. Bond, T. M. Fliedner, and J. O. Archambeau, *Mammalian Radiation Lethality. A Disturbance in Cellular Kinetics*, Academic Press, New York, 1965.
2. S. Abramason, R. G. Miller, and R. G. Philips, The Identification in Adult Bone Marrow of Pluripotent and Restricted Stem Cells of the Myeloid and Lymphoid Systems, *J. Exp. Med.*, 145: 1567 (1977).
3. D. Metcalf and M. A. S. Moore, *Hemopoietic Cells*, North-Holland Publishing Co., Amsterdam, 1971.
4. J. J. Trentin and W. J. Fahlberg, An Experimental Model in Studies of Immunologic Competency in Irradiated Mice Repopulated with "Clones" of Spleen Cells, in *Conceptual Advances in Immunology and Oncology*, 16th Annual Symposium, Texas, 1962, Harper & Row Publishers, New York, 1963.
5. A. M. Wu, J. E. Till, L. Siminovitch, and E. A. McCulloch, A Cytological Study of the Capacity for Differentiation of Normal Hemopoietic Colony Forming Cells, *J. Cell. Physiol.*, 69: 177 (1967).
6. D. W. van Bekkum, M. J. van Noord, B. Maat, and K. A. Dicke, Attempts at Identification of Hemopoietic Stem Cells in Mouse, *Blood*, 38: 547 (1971).
7. D. W. van Bekkum, The Appearance of the Multipotential Hemopoietic Stem Cell, in *Experimental Hematology Today*, S. J. Baum and G. D. Ledney (Eds.), p. 3, Springer Verlag, New York, 1977.
8. K. A. Dicke, M. J. van Noord, B. Maat, U. W. Schaefer, and D. W. van Bekkum, Identification of Cells in Primate Bone Marrow Resembling the Hemopoietic Stem Cell in the Mouse, *Blood*, 42: 195 (1973).
9. W. H. Ross, M. Korbling, W. Nothdruft, W. Calvo, and T. M. Fliedner, Characterization of Bone Marrow and Lymph Node Repopulating Cells by Transplanting Mononuclear Cells into Radiated Dogs, in *Experimental Hematology Today*, S. J. Baum and G. D. Ledney (Eds.), p. 29, Springer Verlag, New York, 1977.
10. R. B. Epstein, J. Bryant, and E. D. Thomas, Cytogenetic Demonstration of Permanent Tolerance in Adult Outbred Dogs, *Transplantation*, 5: 267 (1967).
11. S. F. Goldmann et al., Hemopoietic and Lymphopoietic Split Chimerism in Severe Combined Immune Deficiency Disease (SCID), *Transplant. Proc.*, 11: 225 (1979).
12. S. Strober et al., Transplantation Tolerance After Total Lymphoid Irradiation, *Transplant. Proc.*, 11: 1032 (1979).

13. W. Welling, O. Vos, W. W. H. Weyzen, and D. W. van Bekkum, Identification and Follow-Up of Homologous and Heterologous Bone Marrow Transplants in Radiation Chimeras, *Int. J. Radiat. Biol.*, 1: 145 (1959).
14. H. M. Vriesendorp and D. W. van Bekkum, Role of Total Body Irradiation in Conditioning for Bone Marrow Transplantation, in *Immunobiology of Bone Marrow Transplantation,* S. Thierfelder (Ed.), p. 349, Springer Verlag, New York, 1980.
15. Advisory Committee of the International Bone Marrow Transplant Registry. Severe Combined Immunodeficiency Disease. Characterization of the Disease and Results of Transplantation, *J. Am. Med. Assoc.*, 238: 591 (1977).
16. R. Storb et al., Allogeneic Marrow Grafting for Treatment of Aplastic Anemia, *Blood,* 43: 157 (1974).
17. E. D. Thomas et al., One Hundred Patients with Acute Leukemia Treated by Chemotherapy, Total Body Irradiation and Allogeneic Marrow Transplants, *Blood,* 49: 511 (1977).
18. L. P. Colly and T. Hagenbeek, Experimental Chemotherapy: A Rat Model for Human Acute Myeloid Leukemia, in *Experimental Hematology Today,* S. J. Baum and G. D. Ledney (Eds.), p. 211, Springer Verlag, New York, 1977.
19. D. Hoelzer, W. Calvo, K. D. Meyer-Hamme, and E. B. Harris, Cell Distribution and Proliferation Pattern of a Transferable Acute Leukemia in Rats, *J. Natl. Cancer Inst.*, 50: 1545 (1973).
20. T. C. McQuire, M. J. Poppie, and K. L. Banks, Combined (B- and T-Lymphocyte) Immune Deficiency: A Fatal Genetic Disease in Arabian Foals, *J. Am. Med. Assoc.*, 164: 70 (1974).
21. L. M. Schiffer et al., Extracorporeal Irradiation of the Blood in Humans: Effects upon Erythrocyte Survival, *Blood,* 27: 831 (1966).
22. H. L. Lochte, S. Kasakura, M. Karetzky, J. W. Ferrebee, and E. D. Thomas, Infusion of Marrow in the Mouse and Dog After Thio-TEPA, *Blood,* 21: 424 (1963).
23. J. Abb, B. Netzel, H. V. Rodt, and S. Thierfelder, Autologous Bone Marrow Grafts in Dogs Treated with Lethal Doses of 1-(2-Chloroethyl)-3-Cyclohexy-1-Nitro-Urea, *Cancer Res.*, 38: 2157 (1978).
24. H. J. Kolb et al., Immunologic, Toxicologic and Marrow Transplantation Studies in Dogs Given Dimethyl Myleran, *Biomedicine,* 20: 341 (1974).
25. R. B. Epstein, R. Storb, R. A. Clift, and E. D. Thomas, Autologous Bone Marrow Grafts in Dogs Treated with Lethal Doses of Cyclophosphamide, *Cancer Res.*, 29: 1072 (1969).
26. H. J. Kolb et al., Marrow Grafts in LD-SD Typed Dogs Treated with Cyclophosphamide, *Transplant. Proc.*, 8: 555 (1976).
27. H. M. Vriesendorp, B. Lowenberg, T. P. Visser, S. Knaan, and D. W. van Bekkum, Influence of Genetic Resistance and Silica Particles on Survival After Bone Marrow Transplantation, *Transplant. Proc.*, 8: 483 (1976).
28. E. D. Thomas, S. Kasakura, and A. P. Cristoffanini, Isogeneic Marrow Infusion Following Nitrogen Mustard in the Dog, *Proc. Soc. Exp. Biol. Med.*, 118: 1019 (1965).
29. E. D. Thomas, J. A. Collins, E. C. Herman, and J. W. Ferrebee, Marrow Transplants in Lethally Irradiated Dogs Given Methotrexate, *Blood,* 19: 217 (1962).
30. D. W. van Bekkum and M. J. de Vries, *Radiation Chimaeras,* Academic Press, New York/London, 1967.
31. J. J. Broerse, D. W. van Bekkum, C. F. Hollander, and J. A. G. Davids, Mortality of Monkeys After Exposure to Fission Neutrons and the Effect of Autologous Bone Marrow Transplantation, *Int. J. Radiat. Biol.*, 34: 253 (1978).
32. R. Storb, P. L. Weiden, T. C. Graham, K. G. Lerner, N. Nelson, and E. D. Thomas, Hemopoietic Grafts Between DLA-Identical Canine Littermates Following Dimethyl Myleran, *Transplantation,* 24: 349 (1977).

33. I. Rieder et al., Leukocyte Transfusions for the Modification of Host-Versus-Graft Reaction in Dogs, in *Experimental Hematology Today,* S. J. Baum and G. D. Ledney (Eds.), Springer Verlag, New York, 1978.

34. H. M. Vriesendorp et al., Take and Graft vs Host Reactions of Allogeneic Bone Marrow in Tissue Typed Dogs, *Transplant. Proc.,* 7 (Suppl. 1): 849 (1975).

35. P. L. Weiden, R. Storb, T. C. Graham, G. E. Sale, and E. D. Thomas, Resistance to DLA-Nonidentical Marrow Grafts in Lethally Irradiated Dogs, *Transplant. Proc.,* 9: 285 (1977).

36. G. F. Whitmore and J. E. Till, Quantitation of Cellular Radiobiological Responses, *Annu. Rev. Nucl. Sci.,* 14: 347 (1964).

37. S. J. Sharkis and G. W. Santos, Bone Marrow Transplantation in a BN Rat Model of Acute Myelogenous Leukemia (AML), *Leukemia Research,* 1: 251 (1977).

38. E. Lorenz, D. E. Uphoff, T. R. Reid, and E. Shelton, Modification of Irradiation Injury in Mice and Guinea Pigs by Bone Marrow Injections, *J. Natl. Cancer Inst.,* 12: 197 (1951).

39. J. J. Broerse et al., The Survival of Colony-Forming Units in Mouse Bone Marrow After In Vivo Irradiation with D–T Neutrons, X- and Gamma Radiation, *Int. J. Radiat. Biol.,* 19: 101 (1971).

40. E. J. Hall, *Radiobiology for the Radiologist,* 2nd ed., Harper & Row Publishers, Hagerstown, Md., 1978.

41. W. E. Liversage, A General Formula for Equating Protracted and Acute Regimens of Radiation, *Br. J. Radiol.,* 42: 432 (1969).

42. R. J. Garner, R. D. Phemister, G. M. Angleton, A. C. Lee, and R. W. Thomassen, Effect of Age on the Acute Lethal Response of the Beagle to Cobalt-60-Gamma Radiation, *Radiat. Res.,* 58: 190 (1974).

43. H. D. Maillie, W. J. Quinlan, and H. Mermagen, On the Importance of Beagle Size on Response to Whole Body 1000 kVp X-Irradiation, *Health Phys.,* 13: 253 (1967).

44. G. M. Angleton, Dose Assignment Criteria for Beagles Exposed Bilaterally to $^{60}Co$ Gamma Rays, *Health Phys.,* 35: 625 (1978).

44a. P. Sonneveld and D. W. van Bekkum, The Effect of Whole Body Irradiation on Skeletal Growth in Rhesus Monkeys, *Radiology,* 130: 789 (1979).

44b. P. Sonneveld, E. Peperkamp, and D. W. van Bekkum, Incidence of Cataracts in Rhesus Monkeys Treated with Whole Body Irradiation, *Radiology,* 133: 227 (1979).

45. J. F. Fowler and E. L. Travis, The Radiation Pneumonitis Syndrome in Half-Body Radiation Therapy, *Int. J. Radiat. Oncol., Biol. Phys.,* 4: 1111 (1978).

46. M. I. Bull, G. P. Herzig, R. G. Graw, G. R. F. Krueger, and C. A. Bowles, Canine Allogeneic Bone Marrow Transplantation Technique and Variables Influencing Engraftment, *Transplantation,* 22: 150 (1976).

47. V. D. Courtenay, Studies on the Protective Effect of Allogeneic Marrow Grafts in the Rat Following Whole Body Irradiation at Different Dose Rates, *Br. J. Radiol.,* 36: 440 (1963).

48. H. J. Kolb et al., Bone Marrow Transplantation in DLA-Haploidentical Canine Littermates—Fractionated Total Body Irradiation and In Vitro Treatment of the Marrow Graft with Anti-T-Cell Globulin, in *Immunobiology of Bone Marrow Transplantation,* S. Thierfelder (Ed.), p. 65, Springer Verlag, New York, 1980.

49. E. Frei III, Combination Cancer Therapy. Presidential Address, *Cancer Res.,* 32: 2593 (1972).

50. L. J. Cole and E. L. Alpen, Homologous Bone Marrow Transplantation in Dogs Receiving X-Radiation Plus Urethan or 6-Mercaptopurine, *Blood,* 21: 373 (1963).

51. E. D. Thomas, S. Kasakura, J. A. Cavins, and J. W. Ferrebee, Marrow Transplants in Lethally Irradiated Dogs: The Effect of Methotrexate on Survival of the Host and the Homograft, *Transplantation,* 1: 571 (1963).

52. H. M. Vriesendorp, C. Zurcher, and D. W. van Bekkum, Engraftment of Allogeneic Dog Bone Marrow, *Transplant. Proc.*, 7: 465 (1975).
53. I. Corneci, T. Andrian, O. Costachel, and A. Feteanu, Antilymphoid Globulin, The Therapy of Secondary Disease, *Transplant. Proc.*, 3: 422 (1971).
54. G. Herzig, R. Carolla, T. Alvegard, and R. G. Graw, Allogeneic Bone Marrow Transplantation Following Total Body Irradiation and L-Asparaginase, *Clin. Res.*, 20: 490 (1972).
55. A. C. M. van Kessel and H. M. Vriesendorp, Resistance Against Allogeneic Dog Bone Marrow, *Exp. Hematol.*, 5 (Suppl. 2): 43 (1977).
56. P. J. Creaven and E. Mihich, The Clinical Toxicity of Anticancer Drugs and Its Prediction, *Semin. Oncol.*, 4: 147 (1977).
57. G. Cudkowicz and M. Bennett, Peculiar Immunobiology of Bone Marrow Allografts. I. Graft Rejection by Irradiated Responder Mice, *J. Exp. Med.*, 134: 83 (1971).
58. R. B. Livingston and J. K. Carter, *Single Agents in Cancer Chemotherapy*, IFI/Plenum, New York, 1970.
59. A. A. Gratwohl, M. I. Bull, R. G. Graw, Jr., L. Norton, and T. Knutsen, Methotrexate and Citrovorum Factor After Histoincompatible Allogeneic Bone Marrow Transplantation in Dogs, *Acta Haematol.*, 60: 233 (1978).
60. J. S. Najarian et al., Seven Years Experience with Antilymphocyte Globulin for Renal Transplantation from Cadaver Donors, *Ann. Surg.*, 184: 352 (1976).
61. H. M. Abaza, B. Nolan, J. G. Watt, and M. F. A. Woodruff, Effect of Antilymphocytic Serum on the Survival of Renal Homotransplants in Dogs, *Transplantation*, 4: 618 (1966).
62. T. R. De Meester, N. D. Anderson, and C. F. Schaffer, The Effect of Heterologous Anti-Lymphocyte Serum on Mouse Hemopoietic Stem Cells, *J. Exp. Med.*, 127: 731 (1968).
63. T. M. Feltkamp-Vroom and H. Balner, Glomerular Basement Membrane Reactive Antibodies in Antilymphocyte Sera In Vitro and In Vivo Characteristics, *Eur. J. Immunol.*, 2: 166 (1972).
64. F. T. Rapaport, T. Shimada, and K. Watanabe, Prolongation of Canine Renal Allograft Survival by L-Asparaginase, *Transplantation*, 12: 217 (1971).
65. L. J. Old, E. A. Boyse, H. A. Campbell, R. S. Brodey, J. Fidler, and J. D. Teller, Treatment of Lymphosarcoma in the Dog with L-Asparaginase, *Cancer*, 20: 1066 (1967).
66. M. Tavassoli, The Marrow—Blood Barrier, *Br. J. Haematol.*, 41: 297 (1979).
67. D. W. van Bekkum, B. Lowenberg, and H. M. Vriesendorp, in *Immunological Engineering*, D. W. Hirsch (Ed.), p. 179, MTP Press, Lancaster, England.
68. P. C. Nowell, L. J. Cole, P. L. Roan, and J. G. Habermeyer, The Distribution and In Situ Growth Pattern of Injected Rat Marrow in X-Irradiated Mice, *J. Natl. Cancer Inst.*, 18: 127 (1957).
69. D. E. Pegg, *Bone Marrow Transplantation*, Lloyd-Luke Ltd., London, 1966.
70. U. W. Schaefer, D. P. B. M. Snel, and G. W. Boorman, Intraperitoneal Administration of Allogeneic Bone Marrow Cells in Lethally Irradiated Rhesus Monkeys, *Exp. Hematol. (Oak Ridge)*, 22: 98 (1972).
71. D. W. van Bekkum, Hostile Grafts, in *Advances in Transplantation*, J. Dausset, J. Hamburger, and G. Mathe (Eds.), p. 565, Munksgaard, Copenhagen, 1968.
72. D. W. van Bekkum, The Double Barrier in Bone Marrow Transplantation, *Semin. Hematol.*, 11: 325 (1974).
73. N. C. Gorin, G. Herzig, M. I. Bull, and R. G. Graw, Jr., Long Term Preservation of Bone Marrow and Stem Cell Pool in Dogs, *Blood*, 51: 257 (1978).
74. J. W. M. Visser, H. M. Vriesendorp, and T. P. Visser, Cryopreservation of Dog Bone Marrow Cells, in *Cryoimmunology*, D. Simatos, D. N. Strong, and J. M. Ture (Eds.), p. 113, INSERM, Paris, 1977.

75. R. B. Epstein, R. Storb, R. A. Clift, and E. D. Thomas, Transplantation of Stored Allogeneic Bone Marrow in Dogs Selected by Histocompatibility Typing, *Transplantation,* 8: 496 (1969).
76. F. R. Appelbaum et al., Study of Cell Dose and Storage Time on Engraftment of Cryopreserved Autologous Bone Marrow in a Canine Model, *Transplantation,* 26: 245 (1978).
77. U. W. Schaefer, K. A. Dicke, and D. W. van Bekkum, Recovery of Haemopoiesis in Lethally Irradiated Monkeys by Frozen Allogeneic Bone Marrow Grafts, *Rev. Eur. Etud. Clin. Biol.,* 17: 483 (1972).
78. J. Sprent, Migration and Lifespan of Lymphocytes, in *B and T Cells in Immune Recognition,* F. Loor and G. E. Roelants (Eds.), p. 59, John Wiley & Sons, London.
79. D. M. Hume, B. T. Jackson, C. F. Zukoski, H. M. Lee, H. M. Kauffman, and R. H. Egdahl, The Homotransplantation of Kidneys and of Fetal Liver and Spleen After Total Body Irradiation, *Ann. Surg.,* 152: 354 (1960).
80. E. D. Thomas, J. A. Collins, S. Kasakura, and J. W. Ferrebee, Lethally Irradiated Dogs Given Infusions of Fetal and Adult Hemopoietic Tissue, *Transplantation,* 1: 514 (1963).
81. E. C. Saltzstein, M. M. Bortin, A. A. Rimm, and J. L. Husey, Long Lived Canine Allogeneic Radiation Chimera Produced with Combined Fetal Liver and Thymus Cells, *Transplantation,* 18: 463 (1974).
82. R. B. Epstein, T. C. Graham, C. D. Buckner, J. Bryant, and E. D. Thomas, Allogeneic Marrow Engraftment by Cross Circulation in Lethally Irradiated Dogs, *Blood,* 28: 692 (1966).
83. R. Storb, R. B. Epstein, H. Ragde, J. Bryant, and E. D. Thomas, Marrow Engraftment by Allogeneic Leukocytes in Lethally Irradiated Dogs, *Blood,* 30: 805 (1967).
84. T. M. Fliedner et al., Treatment of Aplastic Anemia by Blood Stem Cell Transfusion: A Canine Model, *Haematologica (Pavia),* 61: 141 (1976).
85. R. Storb, R. B. Epstein, and E. D. Thomas, Marrow Repopulating Ability of Peripheral Blood Cells Compared to Thoracic Duct Cells, *Blood,* 32: 662 (1968).
86. D. E. Uphoff, Preclusion of Secondary Phase of Irradiation Syndrome by Inoculation of Fetal Hematopoietic Tissue Following Lethal Total Body Irradiation, *J. Natl. Cancer Inst.,* 20: 625 (1958).
87. S. S. Sarpel, A. R. Zander, L. Harvath, and R. B. Epstein, The Collection Preservation and Function of Peripheral Blood Hematopoietic Cells in Dogs, *Exp. Hematol.,* 7: 113 (1979).
88. B. J. Torok-Storb, R. Storb, T. C. Graham, R. L. Prentice, P. L. Weiden, and J. W. Adamson, Erythropoiesis In Vitro: Effect of Normal Versus "Transfusion-Sensitized" Mononuclear Cells, *Blood,* 52: 706 (1978).
89. R. Storb et al., Marrow Graft Rejection in DLA Identical Canine Littermates, *Transplant. Proc.,* 11: 504 (1979).
90. P. L. Weiden et al., Preceding Transfusions and Marrow Graft Rejection in Dogs and Man, *Transplant. Proc.,* 8: 551 (1976).
91. R. B. Epstein, R. Storb, H. Ragde, and E. D. Thomas, Cytotoxic Antisera for Bone Marrow Grafting in Littermate Dogs, *Transplantation,* 6: 45 (1968).
92. H. M. Vriesendorp, A. B. Bijnen, C. Zurcher, and D. W. van Bekkum, Donor Selection and Bone Marrow Transplantation in Dogs, in *Histocompatibility Testing 1975,* F. Kissmeyer-Nielsen (Ed.), p. 963, Munskgaard, Copenhagen, 1975.
93. H. M. Vriesendorp, A. B. Bijnen, D. L. Westbroek, and D. W. van Bekkum, Genetics and Transplantation Immunology of the DLA Complex, *Transplant. Proc.,* 9: 293 (1977).

94. R. Storb et al., Effects of Prior Blood Transfusion in Marrow Grafts; Abrogation of Sensitization by Procarbazine and Antithymocyte Serum, *J. Immunol.*, 112: 1508 (1974).

95. H. M. Vriesendorp, A. B. Bijnen, A. C. M. van Kessel, H. Obertop, and D. L. Westbroek, Minor Histocompatibility Systems in Dogs, in *Experimental Hematology Today*, S. J. Baum and G. P. Ledney (Eds.), p. 109, Springer Verlag, New York.

96. R. Storb, R. B. Epstein, J. Bryant, H. Ragde, and E. D. Thomas, Marrow Grafts by Combined Marrow and Leukocyte Infusions in Unrelated Dogs Selected by Histocompatibility Typing, *Transplantation*, 6: 587 (1968).

97. H. M. Vriesendorp, W. D. H. Hendriks, B. Hogeweg, and D. W. van Bekkum, Whole Body Irradiation and Bone Marrow Transplantation in Dogs, *Int. J. Radiat. Biol.*, 29: 581 (1976).

98. M. N. Swift and S. T. Taketa, Modification of Acute Intestinal Radiation Syndrome Through Shielding, *Am. J. Physiol.*, 185: 85 (1956).

99. V. von Sebestik, H. Pfisterer, D. Huhn, and W. Stich, Die Wirkung von $^{60}$Co Gamma Strahlen auf die Lebensdauer und Feinstruktur peripherer Thrombocyten, *Blut*, 14: 261 (1967).

100. G. Meuret et al., Die Behandlung der chronischlymphatischen Leukamie durch extrakorporale Blutbestrahlung unter Verwendung konventioneller Strahlentherapiegerate, *Strahlentherapie*, 137: 529 (1969).

101. F. R. Appelbaum, C. A. Bowles, R. W. Makuch, and A. B. Deisseroth, Granulocyte Transfusion Therapy of Experimental Pseudomonas Septicemia: Study of Cell Dose and Collection Technique, *Blood*, 52: 323 (1978).

102. P. L. Weiden, R. Storb, M. S. Tsoi, T. C. Graham, K. G. Lerner, and E. D. Thomas, Infusion of Donor Lymphocytes into Stable Canine Radiation Chimeras: Implications for Mechanisms of Transplantation Tolerance, *J. Immunol.*, 116: 1212 (1976).

103. R. E. Ruthe, B. R. Andersen, B. L. Cunningham, and R. B. Epstein, The Efficacy of Granulocyte Transfusions in the Control of Systemic Candidiasis in the Leukopenic Host, *Blood*, 52: 493 (1978).

104. H. M. Vriesendorp, W. Klapwijk, T. Visser, P. J. Heidt, and D. W. van Bekkum, Graft Versus Host Disease in Canine Major Histocompatibility Complex Identical Littermates, *Exp. Hematol.*, 7(Suppl.6): 98 (1979).

105. W. L. Elkins, Cellular Immunology and the Pathogenesis of Graft Versus Host Reactions, *Prog. Allergy*, 15: 78 (1971).

106. H. Rodt et al., GvHD Suppression by Incubation of Bone Marrow Grafts with Anti T-Cell Globulin: Effect in the Canine Model and Application to Clinical Bone Marrow Transplantation, *Transplant. Proc.*, 11: 962 (1979).

107. P. R. Graze, W. Ho, and R. P. Gale, Chronic Graft Versus Host Disease in Man, *Exp. Hematol.*, 6 (Suppl. 3): 44 (1978).

108. R. Masters, A. F. Hood, and B. Cosini, Chronic Cutaneous Graft vs. Host Reaction Following Bone Marrow Transplantation, *Arch. Dermatol.*, 111: 1526 (1975).

109. A. Zweibaum, R. Oriol, N. Feingold, E. Dussaulx, M. Rousset, and G. Chevalier, Studies on Canine Secretory Alloantigens (CSA), *Tissue Antigens*, 4: 115 (1974).

110. G. E. Sale, R. Storb, and H. Kolb, Histopathology of Hepatic Acute Graft Versus Host Disease in the Dog, *Transplantation*, 26: 103 (1978).

111. R. Storb et al., Marrow Grafts Between DL-A Matched Canine Littermates, *Transplantation*, 15: 92 (1973).

112. R. Storb, R. H. Rudolph, and E. D. Thomas, Marrow Grafts Between Canine Siblings Matched by Serotyping and Mixed Lymphocyte Culture, *J. Clin. Invest.*, 50: 1273 (1971).

113. R. Storb, Decrease in the Graft Rejection Rate and Improvement in Survival After Marrow Transplantation for Severe Aplastic Anemia, *Transplant. Proc.*, 11: 196 (1979).
114. G. Klein, *Biology of the Mouse Histocompatibility-2 Complex*, Springer Verlag, New York, 1975.
115. R. Storb, P. L. Weiden, M. L. Schroeder, T. C. Graham, K. G. Lerner, and E. D. Thomas, Marrow Grafts Between Canine Littermates Homozygous or Heterozygous for Lymphocyte Defined Histocompatibility Antigens, *Transplantation*, 21: 299 (1976).
116. R. Storb et al., Marrow Grafts Between Histoincompatible Canine Family Members, *Rev. Eur. Etud. Clin. Biol.*, 17: 680 (1972).
117. D. W. van Bekkum and S. Knaan, Role of Bacterial Microflora in Development of Intestinal Lesions from Graft Versus Host Reactions, *J. Natl. Cancer Inst.*, 58: 787 (1977).
118. K. A. Dicke and D. W. van Bekkum, Allogeneic Bone Marrow Transplantation After Elimination of Immunocompetent Cells by Means of Density Gradient Centrifugation, *Transplant. Proc.*, 3: 666 (1971).
119. W. Muller-Ruchholtz, H. U. Woltge, and H. K. Muller-Hermelink, Bone Marrow Transplantation in Rats Across Strong Histocompatibility Barriers by Selective Elimination of Lymphoid Cells in Donor Marrow, *Transplant. Proc.*, 8: 537 (1976).
120. D. W. van Bekkum, H. Balner, K. A. Dicke, F. G. van den Berg, G. H. Prinsen, and C. F. Hollander, The Effect of Pretreatment of Allogeneic Bone Marrow Graft Recipients with Antilymphocytic Serum on the Acute Graft-Versus-Host Reaction in Monkeys, *Transplantation*, 13: 400 (1972).
121. H. J. Kolb, R. Storb, T. C. Graham, H. Kolb, and E. D. Thomas, Antithymocyte Serum and Methotrexate for Control of Graft Versus Host Disease in Dogs, *Transplantation*, 16: 17 (1973).
122. R. Storb, R. B. Epstein, T. C. Graham, and E. D. Thomas, Methotrexate Regimens for Control of Graft-Versus-Host Disease in Dogs with Allogeneic Marrow Grafts, *Transplantation*, 9: 240 (1970).
123. R. Storb, R. B. Epstein, T. C. Graham, R. Shiurba, and E. D. Thomas, Treatment of Canine Graft Versus Host Disease with Methotrexate and Cyclophosphamide Following Bone Marrow Transplantation from Histocompatible Donors, *Transplantation*, 10: 165 (1970).
124. R. Storb, T. C. Kolb, T. C. Graham, H. Kolb, P. L. Weiden, and E. D. Thomas, Treatment of Established Graft-Versus-Host Disease in Dogs by Antithymocyte Serum or Prednisone, *Blood*, 42: 601 (1973).
125. R. Stórb et al., Aplastic Anemia Treated by Allogeneic Bone Marrow Transplantation: A Report on 49 New Cases from Seattle, *Blood*, 48: 817 (1976).
126. Joint Report Second Workshop Canine Immunogenetics, *Transplant. Proc.*, 8: 289 (1976).
127. R. W. Bull, S. W. Swisher, A. Zweibaum, and H. M. Vriesendorp, Inapplicability of CEA7 as a Bone Marrow Transplantation Marker, *Transplant. Proc.*, 7: 575 (1975).
128. M. E. Phillips and G. J. Thorbecke, Studies on the Serum Proteins of Chimeras, *Int. Arch. Allergy Appl. Immunol.*, 29: 553 (1966).
129. R. Saison and D. Colling, An Immunoglobulin (IgM) Marker in Dogs, *Transplant. Proc.*, 7: 383 (1975).
130. P. L. Weiden, R. Storb, and M. Tsoi, Marrow Origin of Canine Alveolar Macrophages, *J. Reticuloendothel. Soc.*, 17: 342 (1975).
131. W. Calvo, T. M. Fliedner, I. Steinbach, V. Alcober, W. Nothdruft, and I. Fache, Morphologic Alterations in Canine Marrow of Long Term Survivors After 1200 Rad Whole-Body X-Irradiation and Autologous Blood Leukocyte Engraftment, *Am. J. Pathol.*, 95: 379 (1979).

132. J. Abb, H. J. Kolb, H. Grosse-Wilde, I. Rieder, and S. Thierfelder, Cell Mediated Immunity in Canine Marrow Graft Recipients Given Cyclophosphamide, *Exp. Hematol.*, 6: 58 (1978).

133. H. D. Ochs et al., Immunologic Reactivity in Canine Marrow Graft Recipients, *J. Immunol.*, 113: 1039 (1974).

134. F. T. Rapaport, F. D. Cannon, D. A. Blumenstock, K. Watanabe, and J. W. Ferrebee, Induction of Unresponsiveness to Canine Renal Allografts by Total Body Irradiation and Bone Marrow Transplantation, *Nature (London) New Biol.*, 235: 190 (1972).

135. J. von Scheel et al., Kidney Transplantation Following Treatment with Cyclophosphamide and Bone Marrow Grafting, *Transplant. Proc.*, 9: 261 (1977).

136. M. L. Schroeder, R. Storb, T. C. Graham, and P. L. Weiden, Canine Radiation Chimeras. An Attempt to Demonstrate Serum Blocking Factors by an In Vivo Approach, *J. Immunol.*, 114: 540 (1975).

137. M. S. Tsoi, R. Storb, P. L. Weiden, T. C. Graham, M. L. Schroeder, and E. D. Thomas, Canine Marrow Transplantation: Are Serum Blocking Factors Necessary to Maintain the Stable Chimeric State, *J. Immunol.*, 114: 531 (1975).

138. I. Hellstrom, K. E. Hellstrom, R. Storb, and E. A. Thomas, Colony Inhibition of Fibroblasts from Chimeric Dogs Mediated by the Dogs' Own Lymphocytes and Specifically Abrogated by Their Serum, *Proc. Natl. Acad. Sci.*, 66: 65 (1970).

139. F. T. Rapaport et al., Histocompatibility Studies in a Closely Bred Colony of Dogs. V. Mechanisms of Cellular Adaptation in Long-Term DL-A Identical Radiation Chimeras, *J. Exp. Med.*, 142: 120 (1975).

140. P. L. Weiden, R. Storb, K. G. Lerner, G. F. Kao, T. C. Graham, and E. D. Thomas, Treatment of Canine Malignancies by 1200 r Total Body Irradiation and Autologous Marrow Rats, *Exp. Hematol.*, 3: 123 (1975).

141. P. L. Weiden, R. Storb, G. E. Sale, T. C. Graham, and E. D. Thomas, Allogeneic Hematopoietic Grafts After Total Body Irradiation in Dogs with Spontaneous Tumors, *J. Natl. Cancer Inst.*, 61: 353 (1978).

142. J. B. Jones, T. J. Yang, J. B. Dale, and R. D. Lange, Canine Cyclic Hematopoiesis: Marrow Transplantation Between Littermates, *Br. J. Haematol.*, 30: 215 (1975).

143. P. L. Weiden, R. Storb, T. C. Graham, and M. C. Schroeder, Severe Hereditary Haemolytic Anaemia in Dogs Treated by Marrow Transplantation, *Br. J. Haematol.*, 33: 357 (1976).

144. A. C. Andersen and L. S. Rosenblatt, Reproductive Ability of Female Dogs (Beagles) Surviving a Single Midlethal, Whole Body X-Ray Exposure, *Radiat. Res.*, 68: 361 (1976).

145. A. C. Andersen and L. S. Rosenblatt, The Effect of Whole Body X-Irradiation on the Median Lifespan of Female Dogs (Beagles), *Radiat. Res.*, 39: 177 (1969).

146. P. H. Guttman and A. C. Andersen, Progressive Intercapillary Glomerulosclerosis in Aging and Irradiated Beagles, *Radiat. Res.*, 35: 45 (1968).

# Kidney Transplantation

10

A. B. BIJNEN
*and*
H. OBERTOP
*Laboratory for Experimental Surgery,*
*Erasmus University, Rotterdam, The Netherlands*

The transplantation of cadaver kidneys in man would not have been possible without two major developments in the field of transplantation—the advent of pharmacological immunosuppression and the development of techniques for short-term kidney preservation.

The dog has long been used for transplantation studies because of its size and temperament, and experiments with this animal formed the basis of the two above developments. Drugs that have been found to have an immunosuppressive effect in small laboratory animals were tested extensively in dogs before their clinical use in humans. The dog has been almost the sole experimental animal for research on kidney preservation.

The role of the dog as a preclinical model for organ transplantation created the need to obtain a knowledge of its immunogenetics equivalent to the knowledge gathered for man. Presently the different histoincompatibility situations encountered in man can be imitated in the dog model.[1] Thus the effect of donor selection on organ allograft survival can be tested in all different histoincompatibility situations in a controlled manner in the dog, and these results theoretically can be translated to histocompatibility matching for improving transplantation results in man. Immunological enhancement of allografts, which has been successful in rodents, has not been reproducible in the dog.[2] Any attempt to induce enhancement in man should be preceded by the development of reproducible protocols in the dog or other preclinical models.

The dog can also be used as a "postclinical model"; phenomena observed in retrospective studies in humans can be investigated prospectively in a controlled situation. For example, retrospective studies in humans indicated that blood transfusions have a beneficial effect on subsequent renal allografts,[3] but it was uncertain whether blood transfusions alone were responsible for graft protection. In subsequent prospective studies with a controlled dog model[4] as well as in experiments with rhesus monkeys,[5] random blood transfusions indeed prolonged kidney allograft survival. New protocols are presently being studied in experimental animals to establish the best transfusion policy.

The rhesus monkey is an excellent experimental animal for preclinical transplantation research[5-8] since its major histocompatibility complex (MHC) is almost as well known as that of man and inbreeding is supposed to be less than in the dog. However, the monkey's pedigree is usually not known, and small family size hampers extensive intrafamilial testing of allografts in monkeys. In dogs large litters allow transplantation experiments between sibs. Furthermore, since sibs that are identical for the antigens of the MHC of the dog, the so-called DLA complex, are easily available, better controlled transplantation experiments can be performed in unrelated animals by using the triplet model (Fig. 1): the kidney of a third-party donor can be transplanted to one of a pair of identical sibs; this sib receives special treatment, and the other serves as an untreated control. The coexistence of experimental models in different species, such as dogs and rhesus monkeys, is beneficial since such models may supplement each other, and findings in one model may stimulate similar research with other animals before conclusions are drawn for humans.

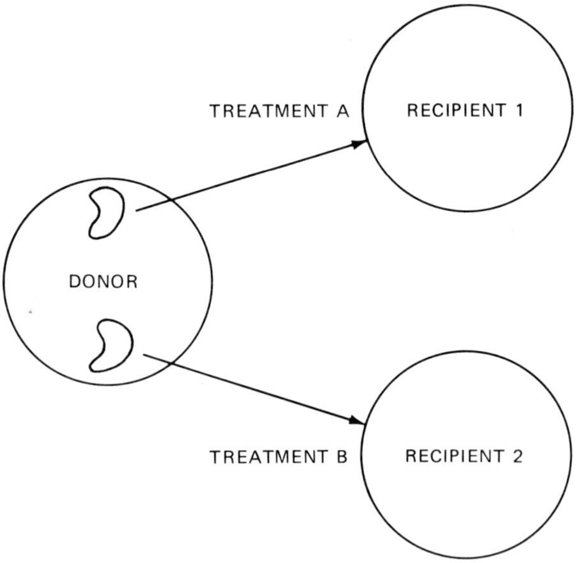


**Fig. 1 Triplet model. Recipients 1 and 2 are MHC-identical sibs. Differences in graft survival in recipient 1 and recipient 2 cannot be caused by differences in match grade for MHC antigens and are therefore more likely to be caused by differences between treatment A and treatment B.**

In addition to kidney transplantation, many vascularized organs (small intestine, pancreas, lung, heart, and liver) and nonvascularized tissues (skin, bone marrow, and frozen nerve) have been transplanted in dogs.[9] Skin transplantation has been used in many species to detect immunogenetic principles and as such has served as a model for the allografting of other organs, but skin grafts from well-matched donors are generally rejected more rapidly than similarly matched vascularized allografts.[10] Consequently conclusions drawn from skin-grafting experiments may not be valid for allografting of vascularized organs, such as the kidney.

## SURGICAL TECHNIQUE OF KIDNEY TRANSPLANTATION

Our standard transplantation procedure begins with a median laparotomy. The renal artery is anastomosed end to end to the external iliac artery, the renal vein is anastomosed end to side to the common iliac vein, and an endovesical ureteroneocystostomy with a submucosal tunnel is performed. When the donor kidney has two arteries, the tail artery can be used. We do not use an aortic patch since we generally exchange kidneys between recipients, and the donor has to survive. A bilateral nephrectomy of the recipient's own kidneys at the time of transplantation provides simple and objective criteria for graft rejection; rejection is indicated by animal death from renal insufficiency or serum creatinine above 1000 $\mu M$/liter. The clinical diagnosis of rejection is confirmed histologically.

In another technique of renal allografting,[11,12] an accessory kidney is grafted into the neck of the recipient. Renal function of the transplanted kidney can be studied while the recipient's own kidneys are kept in place, or a second kidney can be transplanted to the groin. In yet another experimental model,[13] the renal vein is anastomosed to the portal vein, providing a different pathway of the venous outflow. In that model the rejection rate of kidneys was not different from kidneys anastomosed to the systemic circulation.

## KIDNEY PRESERVATION

The dog has figured most prominently in the development of kidney preservation techniques although a few other experimental animals (e.g., the rabbit[14]) have also been used. Two different methods of kidney preservation are now widely used in clinical applications. Belzer, Ashby, and Dunphy[15] developed a machine for the continuous pulsatile perfusion of isolated donor organs in 1967; they were able to preserve canine kidneys for 72 hr and human kidneys for more than 48 hr. Subsequent refinement of the method of Belzer, Ashby, and Dunphy, through use of other and smaller machines and other perfusion fluids, has extended the preservation of canine kidneys up to 96 hr with immediate restoration of normal function.[16] A few years after the introduction of the perfusion machine by Belzer, Ashby, and Dunphy, Collins[17] developed another method to preserve the kidney without continuous perfusion. The donor organ is removed, flushed with cooled preservation fluid, and stored in melting ice. With such cold storage in preservation fluid, a canine kidney can be preserved up to 30 hr without loss of function.

The original composition of the preservation fluid, which resembled the intracellular milieu, subsequently has been changed.[18,19] Additives, such as heparin, procaine, and phenoxybenzamine, were shown to be unnecessary or maybe even toxic to the kidney. The currently recommended solution is a slightly hyperosmolar fluid without additives.[20] Cold storage of the kidney without machine perfusion seems as good as machine preservation for periods up to 36 hr.

## IMMUNOSUPPRESSION

The prime cause of failure of a transplanted kidney is immunological rejection of the graft. The graft will not be rejected if the donor is genetically identical to the recipient (isogeneic transplant), as in the case of monozygotic twins. Graft rejection can always occur when the donor and recipient are genetically different (allogeneic transplant) since such individuals will carry different histocompatibility antigens. The immune reaction evoked by the antigens of an allograft must be prevented or at least suppressed if rejection of allogeneic transplants is to be prevented.

Many forms of immunosuppressive treatment, especially pharmacological immunosuppression, have been first investigated in skin allograft models in rodents and subsequently tested in kidney transplantation in dogs.[21-26] Other methods, such as thymectomy, splenectomy, lymphadenectomy, total-body irradiation, extracorporeal irradiation of blood, and thoracic duct drainage, proved either unsuccessful or too hazardous to apply to man. The present mainstays of immunosuppression are pharmaceuticals, such as corticosteroids and antimetabolites.[27] Some immunosuppressive agents may have severe side effects when used at a therapeutic level in single drug protocols. In clinical practice, combinations of drugs are used to achieve a therapeutic effect with doses below the toxic dose.

The immunosuppressive effect of corticosteroids has been investigated in dogs since 1951. Initially, no effect of cortisone alone was seen,[28] but some effect of this drug was noticed when it was combined with nitrogen mustard.[29] The dosage was probably low according to present standards (50 mg intramuscularly daily; the weight of the dogs was not mentioned). In a later experiment[30] canine kidney allograft survival was successfully prolonged by prednisolone alone at a comparatively high dose (30 mg per dog per day). In that experiment the antimetabolites azathioprine, 6-methyl-mercaptopurine, and 6-mercaptopurine each had a beneficial effect comparable to that of prednisolone alone, but azathioprine (5 mg/kg-day) and 6-methyl-mercaptopurine (5 mg/kg-day) did not enhance the effect of prednisolone alone in the few dogs tested. Pierce, Varco, and Good[21] also reported that 6-mercaptopurine prolonged kidney allograft survival in dogs.

Many different immunosuppressive regimes have been tested by Calne, Alexandre, and Murray[31]; azathioprine repeatedly produced the most satisfactory results in different drug combinations, especially in combination with actinomycin D. Calne, Alexandre, and Murray used relatively high doses of azathioprine (10 mg/kg-day in the first two postoperative days, followed by 4 mg/kg-day). A marked drug toxicity was noticed, and several animals died of bone-marrow depression. Cyclophosphamide,[22] procarbazine hydrochloride,[25] and L-asparaginase,[26] all of which have been effective in the treatment of lymphoproliferative diseases, have been tested for their effect on canine renal allograft survival. None of these proved to be more effective than azathioprine, and none of them is widely used for human patients. However, the favorable results of L-asparaginase in dogs and the limited toxicity of that drug in man may warrant further consideration of its use as an immunosuppressant in man.[26]

Antilymphocytic serum (ALS) prolonged skin homograft survival in rats[32] and was subsequently tested in dogs.[23] Horse anti-dog antilymphocyte serum given intravenously daily (3 ml per kilogram of initial body weight) from the day before transplantation for 60 days significantly prolonged allograft survival. Of eight dogs, only two died of uremia, but the other six died from gastrointestinal hemorrhage or infectious disease within 80 days of transplantation. Monaco et al.[24] administered ALS at a lower dosage [1 ml/kg-day subcutaneously (sc)] for 1 to 3 weeks before transplantation. Graft survival was significantly prolonged; two of seven dogs survived more than 350 days and had normal renal function. A short course of three doses of horse anti-dog ALS (4 ml/kg at 2 days before and 2 ml/kg at day 0 and at 2 days after transplantation) also prolonged graft survival.[33–35]

Other successful protocols have also been described,[36,37] and ALS is now being subjected to clinical trial.

A fungicidal drug, Cyclosporin A, with a potent immunosuppressive action at a dose of 25 mg/kg, prevents kidney allograft rejection in dogs[38] and rabbits,[39] with almost no serious irreversible side effects reported so far. At higher doses it may be hepatotoxic and nephrotoxic. Bone-marrow toxicity has not been reported. The mode of action of this drug is unknown, although it seems to be acting through a specific effect on T cells.[40] Preliminary data show promising results in transplantation in man,[41] but more evidence must be accumulated before the drug can be recommended for general use in organ transplantation in man.

Another mode of immunosuppressive therapy which warrants prospective testing in the dog model is irradiation of the lymph nodes. Such "total nodal irradiation" is used in the treatment of patients with Hodgkin's disease and seems to have a potent immunosuppressive effect.[42]

With present immunosuppressive therapy, about 50% of the cadaveric renal allografts are still rejected within a year.[43] Therefore other approaches to improving renal allograft survival must be pursued.

## DONOR PRETREATMENT

Kidney allografts carry donor leukocytes in the interstitium at the time of transplantation. These so-called passenger leukocytes may play a major role in the sensitization of the recipient against the donor antigens.[44,45] Data from experiments with rats,[46] humans,[47,48] and dogs[49,50] indicate that the elimination of such passenger leukocytes prior to transplantation by means of high doses of antimetabolites may considerably improve renal allograft survival. In vitro perfusion of the canine kidney with antilymphocyte serum[51] or mitogens[52] also has a beneficial effect.

Donor pretreatment in dogs does not always favorably influence graft survival. For instance, donor pretreatment with procarbazine and methylprednisolone abolishes the beneficial effect obtained by host preimmunization by means of third-party spleen cells in conjunction with low doses of prednisone and azathioprine.[53] Suppressor T cells, which might be induced by the administration of third-party spleen cells, possibly are stimulated by passenger leukocytes. Such a dual effect of passenger leukocytes, the induction of rejection or specific unresponsiveness, may explain why no effect of donor pretreatment was found in a prospective randomized trial in humans.[54] In another study (H. L. F. Brom et al., Leyden University, The Netherlands, unpublished data), the favorable effect of donor pretreatment was abolished by flushing the kidney with mannitol before transplantation. Pretreated kidneys initially fared well but were rejected at the same rate as kidneys from untreated donors. The favorable effect of donor pretreatment in that experiment may have resulted from an accumulation of the antimetabolite in the kidney rather than from elimination of passenger leukocytes.

The complex processes resulting from donor pretreatment must be studied further before optimal protocols can be defined and universal donor pretreatment can be advised for renal transplantation in man.

## HISTOCOMPATIBILITY MATCHING

Because immunosuppressive treatment is not always successful or may have serious side effects, donor–recipient matching for the most important histocompatibility antigens has been tested as another experimental tack for preventing graft rejection. Genetic matching may reduce doses of immunosuppressive drugs necessary to prevent rejection.

Histocompatibility antigens are coded by many genes on many chromosomes. For instance, skin-graft survival in mice is controlled by at least 29 different genes.[55] One chromosomal region that codes for important histocompatibility antigens has been identified in many species. This region is called the major histocompatibility complex (MHC) (see Chap. 8). It is unknown just how many histocompatibility systems influence renal allograft survival. In most species, two kinds of genetic systems have been defined within the MHC, the serologically defined (SD) antigens and the lymphocyte defined (LD) antigens.[56] In man, as well as in experimental animals, matching for antigens of the MHC has proved important for survival of many different organ allografts in related donor–recipient combinations. For example, in humans, matching for SD antigens drastically improves kidney allograft survival in related combinations.[57–61] Matching for the SD antigens in unrelated donor–recipient combinations improves renal allograft survival to a much lesser extent.[61–67] The defined antigens of the MHC, the SD and LD antigens, may in themselves be relevant for allograft survival, or other genetic systems linked to the SD and LD systems may be more important.

Our experiments on kidney allograft survival in different histocompatibility situations in unimmunized dogs have included both a series without postoperative immunosuppression to avoid the influence of the extra variables[68–70] and a series with immunosuppressive drugs.

## Without Postoperative Immunosuppressive Treatment

**In Littermate Nonrecombinant Beagles.** In 1970 Rapaport et al.[71] published the results of 70 kidney transplants in littermate and nonlittermate DLA-compatible and -incompatible beagles and outbred mongrel dogs. Mean transplant survival times were 28.6 days for leukocyte group compatible littermates (n = 13), 14.8 days for leukocyte group incompatible littermates (n = 8), 28.3 days for leukocyte group compatible nonlittermates (n = 9), 12.4 days for leukocyte group incompatible nonlittermates (n = 19), and 9.5 days for outbred mongrels (n = 21). In a continuation of those experiments, leukocyte group haplo-identical littermates were transplanted; these survived for 24.1 days (Ref. 72). Our comparable groups of dogs had similar transplant survival times except that we found longer graft survival times in DLA-matched littermates (Fig. 2; Table 1). In both studies grafts between DLA-identical littermate donor–recipient combinations survived significantly longer than littermate donor–recipient combinations that were different by one or two haplotypes. Thus the MHC antigens of dogs are very important for renal allograft survival.

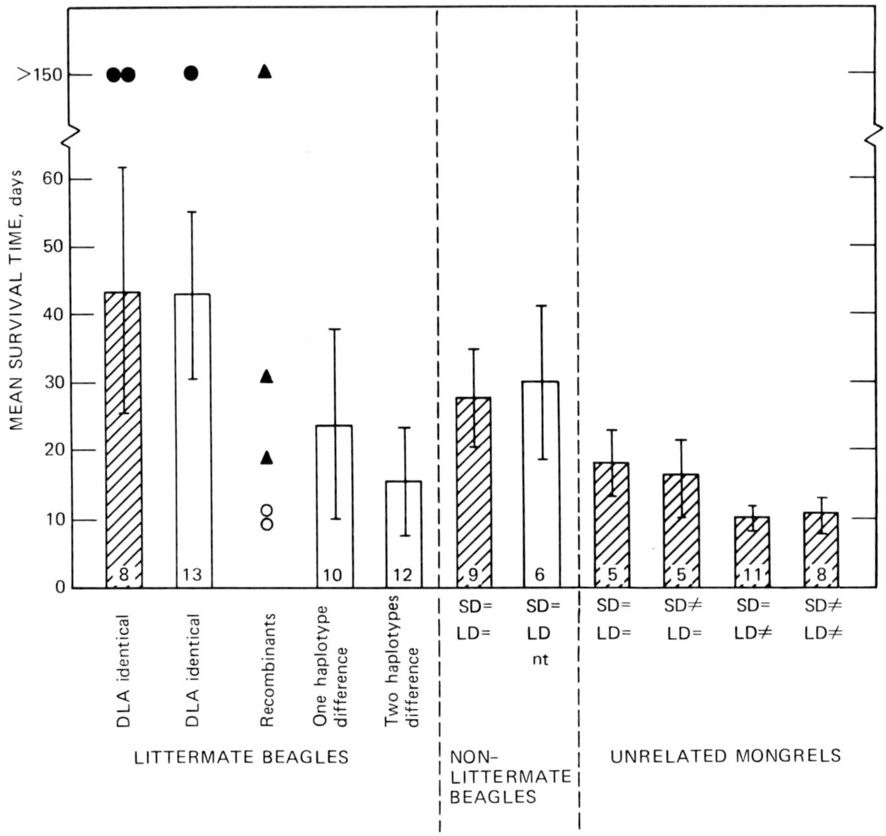


**Fig. 2** Mean survival times of recipients of kidneys from tissue-typed littermate beagles, nonlittermate beagles, and unrelated mongrels. The number of experimental animals in each group is indicated inside the bars, and standard deviations are indicated by vertical lines. Unhatched bars indicate dogs that were typed by serological methods only (SD, serologically defined). Hatched bars indicate dogs that were typed by serological methods and tested in mixed lymphocyte cultures (LD, lymphocyte defined). ●, long-term survivors (not included in the calculation of mean survival times). ○, DLA-A/B recombinant. △, SD/LD recombinant. ▲, GT recombinant (see Fig. 3).

Even in DLA-matched pairs, most renal allografts were eventually rejected; so other non-DLA genes must also be relevant for kidney allograft survival. The graft survival difference between groups with one and two haplotypes difference is not significant but suggests a gene dosage effect.

In littermate donor–recipient combinations, the graft survival times are the same whether matching is performed by SD typing or in combination with mixed lymphocyte reactions (MLR), since the recombination frequency between SD and LD genes is low (0.7%) (Ref. 73).

TABLE 1

KIDNEY ALLOGRAFT SURVIVAL TIMES AND SOME RELEVANT p VALUES FOR DIFFERENCES IN SURVIVAL TIMES AS OBTAINED BY WILCOXON RANK SUM TESTS

[SD represents serologically defined antigens in donor and recipient, equal (=) or not equal (≠); and LD represents antigens defined by mixed lymphocyte culture in donor and recipient, equal (=), not equal (≠), or not tested (nt); •, not calculated]

| | Littermate beagles | | | | Nonlittermate beagles | | Unrelated mongrels | | | |
|---|---|---|---|---|---|---|---|---|---|---|
| | 0 haplotype difference | | | | | | | | | |
| | SD= LDnt | SD= LD= | 1 haplotype difference | 2 haplotypes difference | SD= LDnt | SD= LD= | SD= LD= | SD≠ LD= | SD= LD≠ | SD≠ LD≠ |
| Unrelated mongrels | | | | | | | | | | |
| SD≠ LD≠ | • | 0.0005 | • | 0.06 | 0.0004 | 0.0005 | 0.003 | 0.008 | 0.39 | |
| SD= LD≠ | • | • | • | • | • | • | 0.001 | 0.016 | | |
| SD≠ LD= | • | • | • | • | • | • | 0.07 | | | |
| SD= LD= | • | • | 0.95* | • | • | 0.02 | | | | |
| Nonlittermate beagles | | | | | | | | | | |
| SD= LD= | • | 0.0005 | 0.4* | 0.005* | 0.66 | | | | | |
| 2 haplotypes difference | <0.001 | • | 0.10 | | | | | | | |
| 1 haplotype difference | <0.005 | • | | | | | | | | |
| 0 haplotype difference (LDnt) | 0.45 | | | | | | | | | |
| Number of transplants | 13 | 8 | 10 | 12 | 6 | 9 | 5 | 5 | 11 | 8 |
| Median survival time, days | 41 | 49.5 | 18.5 | 12 | 29.5 | 29 | 17 | 14 | 10 | 10.5 |
| Mean survival time, days | 43.8† | 43.7† | 23.8 | 15.6 | 30.8 | 27.8 | 18.4 | 16.3 | 10.1 | 10.4 |
| ± standard deviation | 12.5 | 18.0 | 13.8 | 7.8 | 11.4 | 7.1 | 4.6 | 5.2 | 1.9 | 2.7 |

*Double-sided Wilcoxon rank sum test; all other p values are obtained by use of a one-sided Wilcoxon rank sum test.

†Long-term survivors excluded.

**In DLA Recombinants.** Kidneys were exchanged among three dogs with a recombination in the MHC and one or two of their respective littermates to determine the relative importance of the subregions of the DLA complex for renal survival. Donors and recipients were selected to be identical for the nonrecombinant haplotype. The pairs showed a difference only for part of the recombinant haplotype. The site of the recombination and the graft survival times (Fig. 3) show that donor–recipient pairs with identical antigens in the DLA-D region yield a much longer graft survival than those with a difference in the DLA-D region. These data suggest an important role of antigens dependent on the DLA-D region but, of course, do not exclude a possible influence of antigens dependent on other regions of the chromosome.

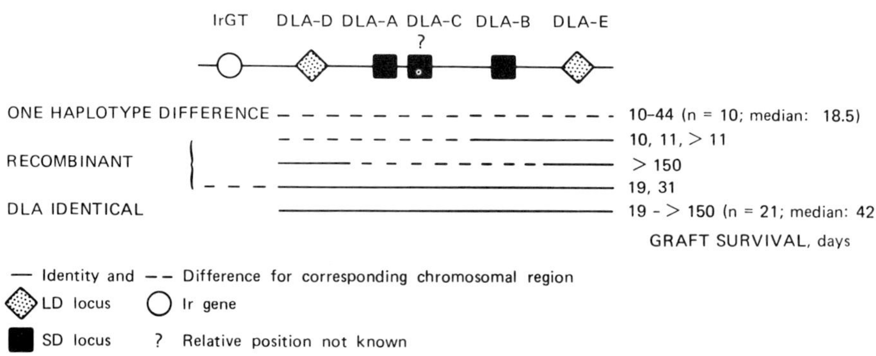


**Fig. 3 Model of the genetic structure of the major histocompatibility complex (MHC) in the dog and the relation of MHC to renal allograft survival in littermate donor–recipient pairs. A solid line under an antigen of the MHC indicates identity between donor and recipient for that** chromosomal region.

**In Unrelated Mongrels.** Matching for LD antigens significantly prolongs renal allograft survival in unrelated mongrels (Fig. 2; Table 1). Serologically defined matching without lymphocyte defined matching apparently does not influence kidney graft survival but does cause a slight improvement in LD-matched combinations. The effect of SD and LD matching in unrelated mongrels is disappointing when compared with the effect of DLA matching in beagle littermate donor–recipient pairs. Several factors may account for the shorter survival times in unrelated MLR-negative mongrels. First, kidney allograft survival has not been studied in DLA-identical littermate mongrels. If mongrels are more heterogeneous than beagles, then minor (non-MHC) histocompatibility systems might be stronger, and shorter survival times would be expected in littermate mongrels. Thus mongrels and beagles may not represent comparable populations. Second, the sensitivity of the MLR technique may have been insufficient, and LD-matched mongrels, in fact, might not be LD identical.[74] Third, the cumulative effect of non-MHC histocompatibility systems may be stronger in unrelated, as compared with related, donor–recipient pairs. Finally, other genetic systems of the chromosome that carries the MHC (located within or outside the MHC) may exert their own influence on renal allograft survival. The relative importance of the third and fourth possibilities has been further studied in nonlittermate beagles.

**In Nonlittermate Beagles.** Eighty-three percent (48 of 58) of kidneys transplanted to DLA-identical unimmunosuppressed littermate beagles are rejected within 100 days (Ref. 70); this emphasizes the relevance of minor histocompatibility systems in MHC-identical pairs. Calculations with the formula of Simonsen[75] show that a minimum of two minor histocompatibility systems must be assumed to explain that rate of rejection, but a larger number of minor histocompatibility systems is a more likely assumption.[70] Minor histocompatibility systems may also play a role in MHC-mismatched combinations since related beagles mismatched by two haplotypes survive longer than mismatched unrelated mongrels. None of these minor histocompatibility systems has been identified, although sex-linked differences may play a role in immunized recipients.[76] No such evidence for sex-linked histocompatibility antigens is available in unimmunized dogs. Grafts from DLA- SD-identical nonlittermate beagles survive for significantly shorter times than those from DLA-identical littermate beagles and about as long as those from littermate donor–recipient pairs that are different by one haplotype. Testing with MLR does not improve the results. Those results imply that the cumulative strength of the undefined histocompatibility antigens (MHC or non-MHC) is of the same magnitude as that of the defined MHC antigens. A strong linkage disequilibrium exists between the defined antigens of the MHC (Refs. 77 to 80). Therefore a similarly strong linkage disequilibrium may exist between defined and undefined antigens of the MHC. In that case SD- and LD-identical littermates would, in fact, be MHC identical, which would mean that non-MHC antigens would be very important. The presence of many important minor histocompatibility systems would consequently be a serious impediment for donor selection procedures. However, these conclusions are derived from a model without immunosuppressants and are not a priori true for situations or models that include the use of immunosuppressive drugs.

## With Postoperative Immunosuppressive Treatment

In a second series of experiments, the influence of immunosuppressive postoperative therapy was tested in kidney transplants between DLA- SD-, and LD-identical littermates, DLA- SD-, and LD-compatible nonlittermates, and littermates different by one haplotype.

All animals received 1 mg of prednisolone per kilogram of body weight and 2 mg of azathioprine per kilogram of body weight intravenously (iv) daily for 100 days. In contrast to the findings in unimmunosuppressed dogs, none of the seven DLA-identical siblings rejected their allografts during that period, but one animal with a normally functioning kidney died 73 days after transplantation. None of the seven DLA-identical nonlittermates rejected their allografts. During the 100-day period, all recipients in those two groups had normal kidney function. Of the seven recipients that were different by one haplotype, four rejected their kidneys within 16 days, and three other recipients survived for more than 100 days but had impaired kidney function. Thus, with immunosuppressive therapy, SD and LD matching in nonlittermates yields excellent transplant results and non-MHC differences play no significant role.

## SPECIFIC UNRESPONSIVENESS FOR RENAL ALLOGRAFTS

Specific unresponsiveness, the lack of immunologic rejection of a graft by a normai immune system, can be divided into immunological enhancement and tolerance. Enhancement is defined by Kaliss[81] as an antibody-dependent phenomenon leading to prolonged allograft survival. Graft protection by cellular mechanisms can also be included in the definition of enhancement. Humoral and cellular mechanisms can be labeled enhancement whenever antidonor immunity is present and allograft protection is specific (i.e., third-party allografts are regularly rejected whereas grafts from a specific donor survive indefinitely, even without the help of additional immunosuppressive treatment). Tolerance is defined as the absence of an immune response to donor antigens while other immune responses function normally.[82] Separation of specific unresponsiveness into tolerance and enhancement sometimes seems artificial.[83]

Specific tolerance in graft recipients can be induced by transplanting the immune system of the graft donor into the recipient. In rodents tolerance can also be induced by injecting donor antigen during the neonatal period.[82,83] In the dog permanent acceptance of skin and vascularized organ grafts can be obtained only when transplantation follows 3 to 6 months after supralethal total-body irradiation and bone-marrow transplantation.[84] Host-vs.-graft and graft-vs.-host reactions can be overcome by matching the donor and the recipient for the major histocompatibility complex.

Enhancement can be considered as the opposite of graft rejection; both mechanisms are induced by immunity against alloantigens governed by major and minor histocompatibility differences between the donor and the recipient. Both hyperacute or accelerated graft rejection and graft enhancement can follow pretreatment of the recipient with donor antigen. Recipient response is possibly dependent on the activation of T helper cells.[85] Separation of the protective and the destructive effect of donor antigen immunization is essential before enhancement can be exploited in the human.

In the mouse, antigens controlled by or associated with the immune region (Ia-antigens) can induce enhancing antibodies.[86–88] In humans[89] and rhesus monkeys,[90] B-cell antibodies directed against the so-called DR (D-related) antigens, which are associated in family and population studies with the antigens governing the outcome of the MLR, are the analogues of Ia antibodies in mice. These B-cell antibodies do not induce hyperacute graft rejection, and their enhancing activity in vivo has yet to be proved. Ia-like antigens have not been studied in the dog.

Immunological enhancement of vascularized organs has been studied most extensively in rodents.[91–97] So-called "active" enhancement can be achieved by actively immunizing the recipients with antigens of the prospective donor. Passive enhancement can result from injection of recipient serum or serum fractions that have been immunized with donor antigens. Antibodies or antigen–antibody complexes can be responsible for the enhancement of allografts.[85] Whereas graft prolongation by immunological enhancement can be easily obtained in rodents, well-documented reports of enhancement in the dog are rare. The successful results of rat enhancement protocols cannot be extrapolated to the dog. Whereas inbred strains of rats offer a reproducible immunogenetic situation, in the dog a different histoincompatibility characterizes each donor–recipient combination. Even the best results in the dog show only a small and unpredictable effect. In several cases successful results by one researcher could not be reproduced by others. In the few cases of successful significant prolongation of renal allograft survival in dogs by active enhancement, postoperative immunosuppression was added.[98–101]

## Subcutaneous Administration of Donor Blood

Successful enhancement of kidney allografts in mongrel dogs was reported by Halasz, Orloff, and Hirose[102]; recipients that were immunized by subcutaneous injections of small amounts of donor blood survived significantly longer than a control group did. Unfortunately, those results could not be reproduced in other laboratories. For instance, pretreatment of the recipients in our laboratory, precisely following the protocol of Halasz, Orloff, and Hirose, did not prolong renal allograft survival in beagle littermates different by two DLA haplotypes (Fig. 4, group f), and, in fact, dogs pretreated in that manner showed a shorter mean graft survival time than did nonpretreated recipients and recipients that received one transfusion of donor blood intravenously.[103] Sutherland et al.,[104] using both subcutaneous and intravenous injections of donor blood, have also repeated the experiments of Halasz, Orloff, and Hirose in tissue-typed beagles. With untreated dogs as controls, median graft survival time was shorter in the pretreated dogs regardless of the route of blood injection. Some pretreated dogs, however, survived longer than controls. The unique observation of enhancement of renal allografts by Halasz, Orloff, and Hirose

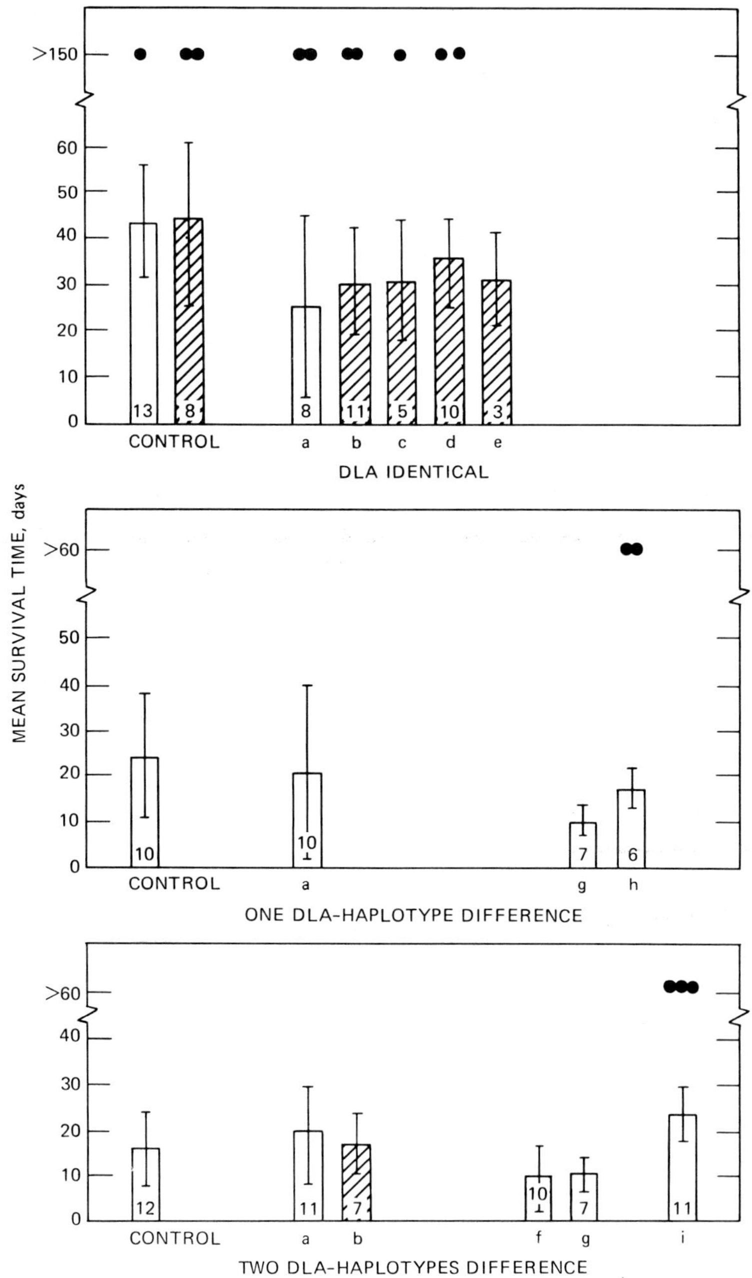
MEAN SURVIVAL TIME, days
>150
60
50
40
30
20
10
0
13
8
8
11
5
10
3
CONTROL
a
b
c
d
e
DLA IDENTICAL
>60
50
40
30
20
10
0
10
10
7
6
CONTROL
a
g
h
ONE DLA-HAPLOTYPE DIFFERENCE
>60
40
30
20
10
0
12
11
7
10
7
11
CONTROL
a
b
f
g
i
TWO DLA-HAPLOTYPES DIFFERENCE

could have resulted from chance differences in histocompatibility match grade between experimental and control groups.

## Intravenous Administration of Donor Blood

Several investigators have studied the effect of intravenous injection of donor blood on renal allograft survival in the dog. Multiple blood transfusions from the prospective kidney allograft donor led to antibody production in five out of ten mongrel recipients of unknown tissue type and accelerated graft rejection in four. Immunosuppressive treatment after transplantation did not effect any long-term graft survival.[105] Sutherland et al.[104] reported that two small transfusions of donor blood led to a shorter median graft survival time in tissue-typed beagles. Harder et al.[106] obtained a small, but statistically significant, prolongation of renal allograft survival in unrelated beagles when recipients had received multiple transfusions of donor blood or plasma in combination with preoperative administration of procarbazine and rabbit anti-dog anti-thymocyte serum. No postoperative immunosuppression treatment was given.

We[107] tested the effect of one single infusion of donor blood on the survival of renal allografts in tissue-typed beagle littermates with three different donor–recipient DLA combinations (DLA identical, different by one DLA haplotype, and different by two DLA haplotypes) (Fig. 4, group a). The effect of tissue typing was evident in both transfused and nontransfused dogs, but donor blood transfusion did not affect mean survival time. Of the 29 transfused dogs, 5 produced lymphocytotoxic antibodies against their donor with positive cross-match tests, which led to accelerated graft rejection. However, exclusion of

**Fig. 4 Mean survival times of recipient of kidneys from tissue-typed littermates. Standard deviations are indicated by vertical lines. Unhatched bars indicate dogs that were typed by serological methods only. Hatched bars indicate dogs that were typed by serological methods and mixed lymphocyte cultures. •, long-term survivors. Control, nontreated recipients.**

- a 200 ml of donor blood iv 14 days before transplantation
- b 100 ml of pooled third-party blood iv 28, 21, and 14 days before transplantation
- c 200 ml of parental blood iv 14 days before transplantation
- d 100 ml of third-party blood iv 7, 14, 21, and 28 days after transplantation
- e 1 to 3 ml of placenta eluate iv 1 day before and 1 and 2 days after transplantation
- f 2 ml of donor blood sc 10 and 5 days before transplantation
- g 6 weekly doses of $1 \times 10^8$ donor lymphocytes sc before transplantation
- h 20 ml of antidonor serum iv daily beginning 1 day before transplantation
- i Horse anti-dog ALS, 4 ml per kilogram of body weight at days −2, 0, and 2; donor bone marrow, $2.4 \times 10^8$ cells per kilogram of body weight at days 10 and 20

these cross-match positive animals did not make the transfusion effect statistically significant.

## Intravenous Administration of Parental Blood

The effect of one transfusion of blood from one of the parents of the kidney donor and the recipient was studied in DLA-identical beagle littermate donor–recipient combinations without immunosuppressive treatment.[108] Such a protocol can be used to determine whether a recipient is able to produce DLA antibodies without the risk of a positive cross match between kidney donor and recipient. Parental blood could theoretically induce immunization not only against DLA but also against non-DLA antigens. The effect of immunization against non-DLA antigens on graft survival can thus be studied. Kidney graft survival was not prolonged by transfusion of parental blood (Fig. 4, group c). Although antibodies directed against the kidney donors were not found, antibodies against a random panel of dogs were detected in four of six transfused dogs. The dog with the highest immune reactivity had the shortest graft survival.

## Administration of Other Donor Antigens

Donor antigen sources other than blood have been used by several investigators in attempts to obtain enhancement of renal allografts in the dog. Pretreatment of mongrel dogs with repeated intravenous injections of spleen cells or subcellular fragments of spleen cells of the prospective kidney donor slightly prolonged renal allograft survival.[109] Using similar protocols, Wilson et al.[98,99] and Holl-Allen et al.[100,101] were able to prolong renal allograft survival significantly when low doses of immunosuppressive drugs were administered postoperatively. Their immunosuppressive regime alone was ineffective. Calne et al.[110] could not induce renal graft enhancement in the dog by pretreatment with spleen cells, lymph node cells, bone marrow, or a semisoluble donor antigen, even when postoperative immunosuppressive drugs were used. Harder et al.[106] have reported a small beneficial effect of donor liver extract in combination with procarbazine and ATS on transplanted canine kidneys.

A combination of donor bone marrow and ALS, which has been shown effective in enhancing renal allografts in rodents, has also proved valuable in mongrel dogs.[111] In our laboratory the beneficial effect of a short preoperative course of horse anti-dog lymphocyte serum on kidney graft survival in DLA-matched beagle littermates could not be improved further by pretreatment of the prospective kidney recipient with donor bone marrow (Fig. 4, group i) (Ref. 35). Passive vs. active enhancement of renal allografts was tested by Jeekel et al.[2,112] with tissue-typed littermate donor–recipient combinations in a triplet model (see Fig. 1). One recipient was actively immunized with donor lymphocytes, whereas a DLA-identical sibling was treated with serum of the immunized dog. Both actively and passively immunized dogs were grafted with a kidney from a nonidentical littermate donor. No postoperative immunosuppressive drugs were given. The actively treated dogs rejected kidneys faster than normal (Fig. 4, group g). Two of six passively treated beagles were long-term survivors (Fig. 4, group h); no such long-term survivors were found in the control group. In an attempt to enhance kidney grafts in beagles, J. Jeekel (Laboratory for Experimental Surgery, Erasmus University, The Netherlands, unpublished data) pretreated recipients with an eluate of the placenta of their mothers before grafting a kidney from a DLA-identical sibling donor. That treatment did not affect graft survival in these three dogs (Fig. 4, group e).

The effect of a pooled kidney antigen preparation, which would include all possible antigens, on renal allograft survival in dogs was studied by Stephens et al.[113] as a model for possible application to human clinical situations. That pretreatment did not improve graft survival. Cytotoxic antibodies to donor kidney cells were detected in all dogs, and graft rejection was accelerated.

## THIRD-PARTY BLOOD TRANSFUSIONS AND KIDNEY TRANSPLANT SURVIVAL

Blood transfusions given to human patients with end-stage renal failure and to experimental animals can affect a subsequent kidney graft.[114] If no cytotoxic antibodies are detected after blood transfusions in the recipients, allograft survival in those patients is better than in patients without preformed antibodies who did not receive blood transfusions.[115–118] If antibodies against a potential kidney donor are produced, and thus the cross-match test between the recipient's serum and the donor's lymphocytes is positive, transplantation cannot be performed, because of a high risk of hyperacute graft rejection.[119,120] If antibodies not directed against the kidney donor are formed, the average survival time of the kidney graft may be shorter than that for patients without antibodies,[41,121–124] but other investigators[125–129] have not supported that relationship.

Cytotoxic antibody formation in patients on chronic dialysis reduces their chance of being accepted as a kidney graft recipient.[130] Because of a possible adverse effect, many physicians and hospitals have a policy of withholding blood transfusions from prospective kidney recipients. Retrospective analyses of transplantation data have suggested that the observed downward trend for graft survival rates since about 1972 might be attributed in part to this transfusion policy.[116,131] A prospective randomized trial is needed to test the validity of this statement.

Because the effect of third-party blood transfusions on the transplanted kidney cannot be easily determined in humans, animal models must be used to mimic the human situation.

## Third-Party Blood Transfusions Without Postoperative Immunosuppressive Treatment

In one experiment,[132] transfusions were administered after kidney transplantation to pairs of DLA-identical littermates to avoid a possible effect of preimmunization on kidney graft survival. Antibody response to third-party blood was not a DLA-linked genetic trait, nor was it related with a shortened kidney graft survival. Kidney graft survival was not significantly influenced by the postoperative administration of blood (Fig. 4, group d).

In a second experiment,[133] three pooled third-party blood transfusions were given before transplantation. Four beagle families were transfused, and the immune response in all dogs against blood donors and a panel of 20 unrelated dogs was studied. An immune reactivity score was computed. All dogs produced lymphocytotoxic antibodies after three transfusions. Immune reactivity was not genetically controlled or linked to DLA, sex, red cell markers (DEA), or canine secretory alloantigens (CSA) (Table 2). Graft survival in the DLA-identical and -nonidentical donor–recipient pairs was not improved by the blood transfusions (Fig. 4, group b).

TABLE 2

RELATION OF IMMUNE REACTIVITY AFTER BLOOD TRANSFUSIONS TO SEX AND DIFFERENT GENETIC MARKERS IN ONE TYPICAL DOG FAMILY

(I and II Stage Lymphocytotoxicity Test)

| | Immune reactivity* | | | | | |
|---|---|---|---|---|---|---|
| Dogs | I stage | II stage | Sex | DLA haplotypes | DEA† | CSA‡ |
| Parents | | | | | | |
| ♂ | Not tested | Not tested | ♂ | AB | Not tested | |
| ♀ | High | High | ♀ | CD | Not tested | |
| Offspring | | | | | | |
| 1 | Low | Low | ♂ | AC | 4.5.6.7 | AY |
| 2 | High | High | ♂ | BC | 1.1. 4.5.6.7 | AY |
| 3 | High | High | ♀ | AC | 1.1. 4. 6.7 | AY |
| 4 | Low | High | ♀ | BC | 1.1. 4.5.6.7 | AY |
| 5 | Low | High | ♀ | AC | 1.1. 4. 6.7 | AY |

*Against blood donors and a panel of 20 unrelated dogs.[133,134]
†Erythrocyte marker.[135]
‡Canine secretory alloantigens.[136]

## Third-Party Blood Transfusions with Postoperative Immunosuppressive Treatment

A beneficial effect of third-party blood transfusion with postoperative immunosuppression on renal allograft survival has been reported in dogs[4,137,138] and rhesus monkeys.[139,140]

Fabre et al.[138] reported that, for dogs, a combination of immunosuppressive treatment administered after transplantation and pooled third-party blood transfusions resulted in a high immune response and a broad spectrum of leukocytotoxins against potential kidney donors. Kidneys transplanted in the presence of a positive cross match were rejected in an accelerated, though not hyperacute, manner. Only one dog pretreated with pooled blood transfusion did not produce cytotoxins, and that dog survived as long as 63 days. When blood from an individual third-party donor was transfused into prospective kidney recipients, fewer cytotoxins were formed. Recipients having a negative cross-match test did not reject the graft, although all died from causes not related to graft rejection.

The effect of unpooled third-party blood transfusions on renal allograft survival was studied in a controlled experimental protocol[4] (Fig. 1).

One mongrel dog was used as a kidney donor for one transfused and one nontransfused beagle littermate. Before transplantation, three 100-ml blood transfusions from three unrelated blood donors were given to the prospective recipients at weekly intervals. Kidneys were grafted 2 weeks after the last transfusion. Low doses of immunosuppressive drugs (azathioprine, 2 mg/kg iv; and prednisolone, 1 mg/kg iv) were given daily after transplantation. This immunosuppressive regimen by itself gives only a slight protection of renal allografts. Cytotoxic antibodies against blood donors and a random panel of dogs were detected in all transfused dogs, but no positive cross-match tests were found between the kidney donor and the recipient. Graft survival was significantly longer in transfused dogs than in the nontransfused controls (Fig. 5). Survival times of renal transplants in immunosuppressed mongrel dogs are also given (Fig. 5).

Preliminary results of pretransplant transfusions of erytnrocyte and leukocyte suspensions, given in a quantity comparable to 100 ml of third-party whole blood, showed positive cross-match tests in five of ten erythrocyte-treated and three of ten leukocyte-treated recipients. Four of those eight dogs showed accelerated graft rejection. Renal allograft survival was prolonged in half the dogs that received third-party leukocyte infusions and in 30% of the erythrocyte-treated dogs. Surprisingly, two of the long-term survivors pretreated with erythrocytes and leukocytes had positive cross matches.

The degree of immune reactivity immediately before transplantation was clearly correlated with graft rejection in recipients that were pretreated with transfusions of whole blood but not in recipients that were pretreated with erythrocytes or leukocytes.

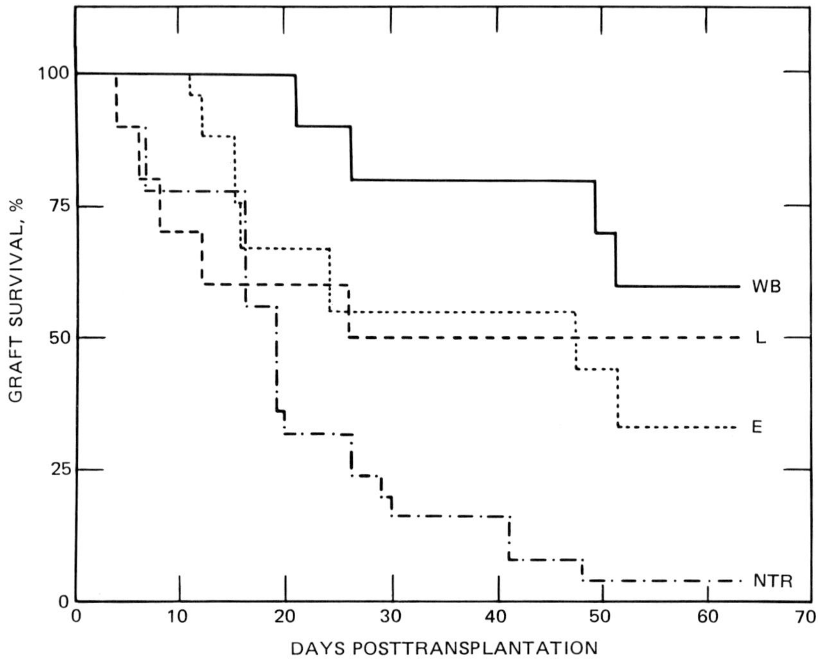


**Fig. 5** Renal allograft survival in transfused and nontransfused beagle littermates with postoperative immunosuppression (azathioprine, 2 mg per kilogram of body weight iv; prednisolone, 1 mg per kilogram of body weight iv). WB, beagles transfused with third-party whole blood and given postoperative immunosuppression (IS). L, beagles transfused with third-party leukocytes and IS. E, beagles transfused with third-party erythrocytes and IS. NTR, nontransfused beagles and mongrels with IS.

The mechanism of the protective effect of blood transfusions on renal graft survival in the dog is still unknown. Our data do not support a genetically defined unresponsiveness, nor do they support unresponsiveness by blood transfusions as being a specific (i.e., induced by donor antigen) mechanism.

## CONCLUSIONS

The dog is an excellent experimental animal for transplantation studies. The recently developed simple and efficient kidney preservation system of storing the organ in melting ice after perfusion with an easily available solution is the result of research in the dog. The dog has proved valuable in the testing of immunosuppressive drugs before their use in transplantation in man. Similarly, other new drugs can be evaluated for their effect on kidney transplantation in the dog. Since human cadaver grafts are frequently rejected, methods must be developed to improve graft survival results. The effect of donor pretreatment on graft survival is still uncertain in the dog, and more studies must be undertaken to develop a clinically applicable protocol. In both dogs and humans, the value of tissue typing by serological methods and by mixed lymphocyte cultures has been recognized in related and, to a lesser extent, unrelated donor–recipient pairs. Identity for lymphocyte-defined antigens apparently is more relevant to graft survival than identity for serologically defined antigens. The importance of minor histocompatibility systems has been suggested in models without immunosuppressive treatment, but these minor histocompatibility differences can probably be overcome by immunosuppressive agents. The development of tissue typing in the dog parallels the development of tissue typing in man. Consequently the clinical situation in man is reflected in an outbred animal model, and the effect of drugs, preimmunization, and several other measures can be tested in this preclinical model. Because successful enhancement of graft survival in rodents could not be repeated in the dog, attempts to induce specific unresponsiveness in man should await the outcome of further enhancement studies in outbred animals, such as dogs. The beneficial effect of blood transfusions on renal allograft survival in man could be repeated in the dog but only if immunosuppression were given after transplantation. The dog model offers an opportunity to develop an optimal blood transfusion policy in man.

After so many years of clinical kidney transplantation, only little improvement of graft survival has been achieved, and, despite optimal care for the condition of the organ and the recipient before, during, and after transplantation, the graft survival figures still are depressingly low. The dog, especially the tissue-typed well-controlled model, probably will continue to aid the search for methods to improve human renal graft survival.

## ACKNOWLEDGMENT

The experiments described in this chapter were performed in collaboration with H. M. Vriesendrop and D. L. Westbroek and financed in part by the Nierstichting Nederland (Dutch Kidney Foundation).

## REFERENCES

1. H. M. Vriesendorp, H. Grosse-Wilde and M. E. Dorff, The Major Histocompatibility Complex of the Dog, in *The Major Histocompatibility System in Man and Animals*, D. Götze (Ed.), pp. 129-163, Springer-Verlag, New York, 1977.
2. J. Jeekel, H. Obertop, H. M. Vriesendorp, I. McDicken, and D. L. Westbroek, Effect of Anti-Donor Serum (ADS) and Donor Cells on Renal Allograft Survival in DL-A Tissue-Typed Littermate Beagles, *Transplant. Proc.*, 7: 435-438 (1975).

3. G. Opelz, D. P. S. Sengar, M. R. Mickey, and P. I. Terasaki, Effect of Blood Transfusions on Subsequent Kidney Transplants, *Transplant. Proc.*, 5: 253-259 (1973).
4. H. Obertop, A. B. Bijnen, H. M. Vriesendorp, and D. L. Westbroek, Prolongation of Renal Allograft Survival in DLA Tissue Typed Beagles After Third-Party Blood Transfusions and Immunosuppressive Treatment, *Transplantation*, 26: 255-259 (1978).
5. A. A. van Es, R. L. Marquet, J. J. van Rood, M. W. Kalff, and H. Balner, Blood Transfusions Induce Prolonged Kidney Allograft Survival in Rhesus Monkeys, *Lancet*, 1: 506-509 (1977).
6. H. Balner, Current Knowledge of the Histocompatibility Complex of Rhesus Monkeys, *Transplant. Rev.*, 15: 50-61 (1973).
7. H. Balner, W. van Vreeswijk, and J. H. Roger, Ia-Like Antigens of Rhesus Monkeys: Current State of Serology and Genetics, *Transplant. Rev.*, 30: 3-17 (1976).
8. A. A. van Es, R. L. Marquet, W. van Vreeswijk, B. Tank, and H. Balner, Influence of Matching for RhLA (SD) Antigens and of Mixed Lymphocyte Reactivity on Allograft Survival in Unrelated Rhesus Monkeys, *Transplant. Proc.*, 9: 257-260 (1977).
9. H. M. Vriesendorp, A. B. Bijnen, D. L. Westbroek, and D. W. van Bekkum, Genetics and Transplantation Immunology of the DLA Complex, *Transplant. Proc.*, 9: 293-296 (1977).
10. H. M. Vriesendorp, Applications of Transplantation Immunology in the Dog, in *Basic and Clinical Aspects of Veterinary Immunology*, C. A. Brandley and C. E. Cornelius (Eds.), Advances in Veterinary Sciences and Comparative Medicine, Vol. 23, pp. 229-265, Academic Press, New York, 1979.
11. J. Miller, B. Hattler, M. Davis, and M. C. Johnson, Cellular and Humoral Factors Governing Canine Mixed Lymphocyte Cultures After Renal Transplantation, *Transplantation*, 12: 65-76 (1971).
12. D. R. Kahn, F. H. Bach, J. H. Dufek, P. M. Buikholder, and G. C. Dacumos, Predictive Value of Mixed Leukocyte Culture Testing on Allograft Survival in Unrelated Dogs, *Surgery*, 78: 1-5 (1975).
13. G. Mazzoni, J. Benichou, K. A. Porter, and T. E. Starzl, Renal Homotransplantation with Venous Outflow or Infusion of Antigen into the Portal Vein of Dogs or Pigs, Transplantation at Portal Site, *Transplantation*, 24: 268-273 (1977).
14. E. H. Jensen, Preservation of Rabbit Kidneys Without Perfusion. The Significance of the $Na^+/K^+$ Ratio, the Phosphate Concentration and the Dextrose Concentration in the Washout Fluid, in *Organ Preservation*, D. E. Pegg (Ed.), Churchill Livingstone, Edinburgh, 1973.
15. F. O. Belzer, B. S. Ashby, and J. E. Dunphy, 24-Hour and 72-Hour Preservation of Canine Kidneys, *Lancet*, 2: 536-539 (1967).
16. A. E. Osaki et al., Successful 96-Hour Preservation of Canine Kidneys Using a New Machine, *Transplant. Proc.*, 9: 247-249 (1977).
17. G. M. Collins, M. B. Bravo-Shugarman, and P. I. Terasaki, Kidney Preservation for Transportation, *Lancet*, 2: 1219-1222 (1969).
18. S. A. Sacks, P. H. Petrisch, and J. J. Kaufman, Canine Kidney Preservation Using a New Perfusate, *Lancet*, 1: 1024-1028 (1973).
19. G. M. Collins and N. A. Halasz, Current Aspects of Renal Preservation, *Urology (Suppl.)*, 10: 22-32 (1977).
20. J. F. W. B. Rijkse, *Preservation of Canine Kidneys*, Thesis, Leiden University, 1978.
21. J. C. Pierce, R. L. Varco, and R. A. Good, Prolonged Survival of a Renal Allograft in a Dog Treated with 6-Mercaptopurine, *Surgery*, 50: 186-195 (1961).
22. G. B. Reams, Use of Cyclophosphamide in an Attempt to Modify the Canine Renal Homograft Response, *Nature*, 197: 713-714 (1963).

23. H. M. Abaza, B. Nolan, J. G. Watt, and M. F. A. Woodruff, Effect of Antilymphocytic Serum on the Survival of Renal Homotransplants in Dogs, *Transplantation,* 4: 618-632 (1966).
24. A. P. Monaco et al., Antiserum to Lymphocytes: Prolonged Survival of Canine Renal Allografts, *Science,* 153: 1264-1267 (1966).
25. A. S. MacDonald, C. C. Chan, and C. F. Faldy, Use of Procarbazine Hydrochloride (Natulan) and Procarbazine Hydrochloride and Azathioprine in Canine Renal Allografts, *Transplantation,* 11: 103-104 (1971).
26. F. T. Rapaport, T. Shimada, and K. Watanate, Prolongation of Canine Renal Allograft Survival by L-Asparaginase, *Transplantation,* 12: 217-222 (1971).
27. R. Weil III, G. P. J. Schröter, J. C. Wert, and T. E. Starzl, A 14-Year Experience with Kidney Transplantation, *World J. Surg.,* 1: 145-156 (1977).
28. L. Persky and S. Jacob, Effect of ACTH and Cortisone on Homogeneous Kidney Transplants, *Proc. Soc. Biol. Med.,* 77: 66-68 (1951).
29. R. Baker, R. Gordon, J. Huffer, and G. H. Miller, Experimental Renal Transplantation. I. Effect of Nitrogen Mustard, Cortisone, and Splenectomy, *Arch. Surg.,* 65: 702-705 (1951).
30. C. F. Zukoski, J. M. Callaway, and W. G. Rhea, Prolonged Acceptance of a Canine Renal Allograft Achieved with Prednisolone, *Transplantation,* 3: 380-386 (1965).
31. R. Y. Calne, G. P. J. Alexandre, and J. E. Murray, A Study on the Effects of Drugs in Prolonging Survival of Homologous Renal Transplants in Dogs, *Ann. N.Y. Acad. Sci.,* 99: 743-761 (1962).
32. M. F. A. Woodruff and N. A. Anderson, Effect of Lymphocyte Depletion by Thoracic Duct Fistula and Administration of Antilymphocytic Serum on the Survival of Skin Homografts in Rats, *Nature,* 200: 702 (1963).
33. J. Jeekel, H. Obertop, F. Bauer, H. M. Vriesendorp, and D. L. Westbroek, The Effect of Antibodies and of Active Immunization on the Survival of Renal Allografts in DL-A Tissue Typed Littermate Beagles, *Eur. Surg. Res.,* 5: 28 (1973).
34. D. L. Westbroek, H. M. Vriesendorp, J. G. van den Tweel, J. de Gruyl, and H. van Urk, Influence of SD and LD Matching on Kidney Allograft Survival in Unrelated Mongrel Dogs, *Transplant. Proc.,* 7: 427-429 (1975).
35. J. Jeekel, H. Obertop, D. L. Westbroek, and H. M. Vriesendorp, Prolonged Survival of Renal Allografts in Tissue Typed Beagles with ALS and Donor Bone Marrow, *Transplantation,* 22: 534-537 (1976).
36. G. J. A. Clunie, B. Nolan, K. James, J. G. Watt, and M. F. A. Woodruff, Prolongation of Canine Renal Allograft Survival with Antilymphocytic Serum, *Transplantation,* 6: 459-467 (1968).
37. R. K. Lawson, L. R. Ellis, D. Kirchheim, and C. V. Hodges, The Prolongation of Canine Renal Homograft Function Using Antilymphocyte Serum as an Immunosuppressive Agent, *Transplantation,* 5: 169-183 (1967).
38. R. Y. Calne and D. J. G. White, Cyclosporin A. A Powerful Immunosuppressant in Dogs with Renal Allografts, *IRCS Med. Sci.,* 5: 595 (1977).
39. C. J. Green and A. C. Allison, Extensive Prolongation of Rabbit Kidney Allograft Survival After Short Term Cyclosporin-A Treatment, *Lancet,* 1: 1182-1183 (1978).
40. D. J. G. White, A. M. Plumb, G. Pawelec, and G. Brons, Cyclosporin-A. An Immunosuppressive Agent Preferentially Active Against Proliferating T Cells, *Transplantation,* 27: 55-58 (1979).
41. R. Y. Calne et al., Cyclosporin-A in Patients Receiving Renal Allografts from Cadaver Donors, *Lancet,* 2: 1323-1327 (1978).
42. S. Strober et al., Transplantation Tolerance After Total Lymphoid Irradiation, *Transplant. Proc.,* 11: 1032-1038 (1979).

43. Advisory Committee to the Renal Transplant Registry, The 13th Report of the Human Transplant Registry, *Transplant. Proc.*, 9: 6-29 (1977).
44. R. D. Guttmann, R. R. Lindquist, and S. A. Ockner, Renal Transplantation in the Inbred Rat. IX. Hematopoietic Origin of Immunogenic Stimulus of Rejection, *Transplantation*, 8: 472-484 (1969).
45. D. Steinmuller and E. A. Hart, Passenger Leukocytes and Induction of Allograft Immunity, *Transplant. Proc.*, 3: 673-674 (1971).
46. R. D. Guttman and R. R. Lindquist, Renal Transplantation in the Inbred Rat. XI. Reduction of Allograft Immunogenicity by Cytotoxic Drug Pretreatment of Donors, *Transplantation*, 8: 490-495 (1969).
47. R. D. Guttmann, J. G. Beaudoin, and D. D. Morehouse, Reduction of Immunogenicity of Human Cadaver Renal Allografts by Donor Pretreatment, *Transplant. Proc.*, 5: 663-665 (1973).
48. R. D. Guttman et al., Donor Pretreatment as an Adjunct to Cadaver Renal Allotransplantation, *Transplant. Proc.*, 7: 117-121 (1975).
49. F. P. Stuart, T. Garrick, A. Holter, A. Lynch, and E. Bastien, Delayed Rejection of Renal Allografts in the Rat and Dog by Reduction of Passenger Leucocytes, *Surgery*, 70: 128-134 (1971).
50. H. Zincke and J. E. Woods, Attempted Immunological Alteration of Canine Renal Allograft Donors, *Transplantation*, 18: 480-486 (1974).
51. C. O. Callender, R. L. Simmons, L. H. Toledo-Pereyra, E. A. Santiago-Delphin, and J. S. Najarian, Prolongation of Kidney Allografts Perfused by Antilymphocyte Globulin In Vitro, *Transplantation*, 16: 377-378 (1975).
52. R. L. Simmons, L. H. Toledo-Pereyra, R. Nowygrad, and J. S. Najarian, Marking of Kidney-Graft Antigens with Concanavalin A., *Transplant. Proc.*, 7 (Suppl. 1): 677-679 (1975).
53. C. J. van der Linden, W. A. Buurman, A. S. Fabricant, and J. Jeekel, Renal Allograft Survival in Dogs After Host Preimmunization and Donor Pretreatment, *Brit. J. Surg.*, 65: 822 (1978).
54. J. R. Jeffery, A. Downs, J. W. Graham, C. Lye, E. Ramsey, and A. E. Thomson, A Randomized Prospective Study of Cadaver Donor Pretreatment in Renal Transplantation, *Transplantation*, 25: 287-289 (1978).
55. R. J. Graff and D. W. Baily, The Non-H2 Histocompatibility Loci and Their Antigens, *Transplant. Rev.*, 15: 26-49 (1973).
56. D. Götze (Ed.), *The Major Histocompatibility System in Man and Animals*, Springer-Verlag, New York, 1977.
57. J. J. van Rood, A. van Leeuwen, and J. W. Bruining, The Relevance of Leucocyte Antigens for Allogenic Renal Transplantation, *J. Clin. Pathol.*, 20: 504-512 (1967).
58. D. P. Singal, M. R. Mickey, and P. I. Terasaki, Serotyping for Homotransplantation. XXIII. Analysis of Kidney Transplants from Parental Versus Sibling Donors, *Transplantation*, 7: 246-258 (1969).
59. H. Brynger et al., Effect of HLA Compatibility on Primary Kidney Grafts from Living Related and Cadaveric Donors, *Transplant. Proc.*, 9: 479-482 (1977).
60. G. Opelz and P. I. Terasaki, Studies on the Strength of HLA Antigens in Related Donor Kidney Transplants, *Transplantation*, 24: 105-111 (1977).
61. B. G. Solheim, A. Flatmark, E. Enger, J. Jewell, and E. Thorsby, Influence of HLA-A, -B, -C and -D Matching on the Outcome of Clinical Kidney Transplantation, *Transplant. Proc.*, 9: 475-478 (1977).
62. J. Dausset et al., Serologically Defined HL-A Antigens and Long-Term Survival of Cadaver Kidney Transplants, A Joint Analysis Performed by France Transplant and the London Transplant Group, *New Engl. J. Med.*, 290: 979-984 (1974).

63. H. Festenstein, J. A. Sachs, A. M. I. Paris, G. D. Pegrum, and J. F. Moorehead, Influence of HLA Matching and Blood Transfusion on Outcome of 502 London Transplant Group Renal Graft Recipients, *Lancet,* 1: 157-161 (1976).

64. G. Opelz, M. R. Mickey, and P. I. Terasaki, Calculations on Long Term Graft and Patient Survival in Human Kidney Transplantation, *Transplant. Proc.,* 9: 27-30 (1977).

65. J. A. Sachs, H. Festenstein, V. A. Tuffnel, and A. M. I. Paris, Collaborative Scheme for Tissue Typing and Matching in Renal Transplantation. IX. Effect of HL-A, -B, and -D Locus Matching, Pretransplant Transfusion and Other Factors on 612 Cadaver Renal Grafts, *Transplant. Proc.,* 9: 483-486 (1977).

66. Scandiatransplant Report, HLA Matching and Kidney Graft Survival, *Lancet,* 1: 240-242 (1975).

67. J. P. van Hooff, G. J. van der Steen, H. M. A. Schippers, and J. J. van Rood, Efficacy of HL-A Matching in Eurotransplant, *Lancet,* 2: 1385-1388 (1972).

68. D. L. Westbroek et al., The Influence of DL-A Histocompatibility on the Function and Pathohistological Changes in Unmodified Canine Renal Allografts, *Transplantation,* 14: 582-589 (1972).

69. A. B. Bijnen, D. L. Westbroek, H. Obertop, and H. M. Vriesendorp, Genetics of Kidney Allograft Survival in Dogs. I. Relevance of Subregions of the Major Histocompatibility Complex in Recipients Without Immunosuppressive Therapy, *Transplantation,* 28: 186-190 (1979).

70. A. B. Bijnen, D. L. Westbroek, H. Obertop, H. van Urk, and H. M. Vriesendorp, Genetics of Kidney Allograft Survival in Dogs. II. Relevance of Minor Histocompatibility Systems in Recipients Without Immunosuppressive Therapy, *Transplantation,* 28: 191-195 (1979).

71. F. T. Rapaport, T. Hanaoka, T. Shimada, F. D. Cannon, and J. W. Ferrebee, Histocompatibility Studies in a Closely Bred Colony of Dogs. I. Influence of Leukocyte Group Antigens upon Renal Allograft Survival in the Unmodified Host, *J. Exp. Med.,* 131: 881-893 (1970).

72. J. Dausset, F. T. Rapaport, F. D. Cannon, and J. W. Ferrebee, Histocompatibility Studies in a Closely Bred Colony of Dogs. III. Genetic Definition of the DL-A System of Canine Histocompatibility, with Particular Reference to the Comparative Immunogenicity of the Major Transplantable Organs, *J. Exp. Med.,* 34: 1222-1237 (1971).

73. A. B. Bijnen, H. M. Vriesendorp, H. Grosse-Wilde, and D. L. Westbroek, Polygenic Control of Mixed Lymphocyte Reactions in Dogs, *Tissue Antigens,* 9: 187-194 (1977).

74. A. B. Bijnen, A. M. Dekkers-Bijma, H. M. Vriesendorp, and D. L. Westbroek, Value of the Mixed Lymphocyte Reaction in Dogs as a Genetic Assay, *Immunogenetics (N. Y.),* 8: 287-297 (1979).

75. M. Simonsen, Immunogenetical Speculations on Human Kidney Transplants, *Transplantation,* 4: 354-364 (1966).

76. H. M. Vriesendorp, A. B. Bijnen, A. C. M. Van Kessel, H. Obertop, and D. L. Westbroek, Minor Histocompatibility Systems in Dogs, *Transplant. Immunol.,* 12: 109-117 (1978).

77. R. Bachvaroff et al., Phenotypic Expression of the Main Histocompatibility Complex (DL-A) in Randomly Selected Mongrel Dogs. I. Serologically Detectable (SD) DL-A Antigen and Mixed Leukocyte (MLC) Reactivity, *Transplant. Proc.,* 5: 1561-1565 (1973).

78. H. Grosse-Wilde et al., Immunogenetics of Seven LD Alleles of the DL-A Complex in Mongrels, Beagles and Labradors, *Transplant. Proc.,* 7: 159-164 (1975).

79. Joint Report of Second International Workshop on Canine Immunogenetics, *Transplant. Proc.,* 8: 289-311 (1976).

80. J. G. van den Tweel, H. M. Vriesendorp, A. Termijtelen, D. L. Westbroek, M. I. Bach, and J. J. van Rood, Genetic Aspects of Canine Mixed Lymphocyte Cultures, *J. Exp. Med.*, 140: 825-835 (1974).
81. N. Kaliss, Immunological Enhancement of Tumor Homografts in Mice, A Review, *Cancer Res.*, 18: 992-1003 (1958).
82. P. W. Medawar, Tolerance Reconsidered. A Critical Survey, *Transplant. Proc.*, 5: 7-9 (1973).
83. R. H. Levey, Immunological Tolerance and Enhancement. A Common Mechanism, *Transplant. Proc.*, 3: 41-48 (1971).
84. H. M. Vriesendorp and D. W. van Bekkum, Susceptibility to Total Body Irradiation, *Exp. Hematol.*, 6: 19 (1978).
85. J. R. Batchelor, The Riddle of Kidney Graft Enhancement, *Transplantation*, 26: 139-141 (1978).
86. D. A. L. Davies and B. Alkins, What Abrogates Heart Transplant Rejection in Immunological Enhancement, *Nature*, 247: 294-296 (1974).
87. D. A. L. Davies and N. A. Staines, Ia Antibodies in Enhancement, *Transplant. Rev.*, 30: 18-39 (1976).
88. J. L. J. Jansen, R. A. P. Koene, G. J. van Kamp, F. H. M. Hagemann, and P. G. A. B. Wijdeveld, Hyperacute Rejection and Enhancement of Mouse Skin Grafts by Antibodies with a Distinct Specificity, *J. Immunol.*, 115: 392-394 (1975).
89. R. B. Ettinger et al., Successful Renal Allografts Across a Positive Cross-Match for Donor B-Lymphocyte Allo-Antigens, *Lancet*, 2: 56-58 (1976).
90. R. L. Marquet, A. A. van Es, G. A. Heystek, R. H. van Leersum, and H. Balner, Attempts to Induce Immunological Enhancement for Kidney Allografts in Rhesus Monkeys. I. Effect of Various Allo-Antisera, Including B-Cells, *Transplantation*, 25: 188-192 (1978).
91. J. W. Fabre and P. J. Morris, The Effect of Donor Strain Blood Pretreatment on Renal Allograft Rejection in Rats, *Transplantation*, 14: 608-617 (1972).
92. J. W. Fabre and P. J. Morris, Experience with Passive Enhancement of Renal Allografts in a (DA x Lewis) $F_1$ to Lewis Strain Combination, *Transplantation*, 13: 604-608 (1972).
93. J. W. Fabre and P. J. Morris, The Mechanism of Specific Immunosuppression of Renal Allograft Rejection by Donor Strain Blood, *Transplantation*, 14: 634-640 (1972).
94. R. L. Marquet, G. A. Heystek, and W. J. Tinbergen, Specific Inhibition of Organ Allograft Rejection by Donor Blood, *Transplant. Proc.*, 3: 708-710 (1971).
95. R. L. Marquet and G. A. Heystek, The Induction of Abolition of Specific Immunosuppression of Heart Allografts in Rats by Use of Donor Blood and Cyclophosphamide, *J. Immunol.*, 115: 405-408 (1975).
96. R. L. Marquet, G. A. Heystek, B. Tank, and A. A. van Es, Prolongation of Heterotopic Heart Allograft Survival in Rats by Use of Antigen—Antibody Complexes, *Transplantation*, 21: 454-459 (1976).
97. R. L. Marquet, G. A. Heystek, and R. H. van Leersum, The Effect of Antidonor Serum and Antigen—Antibody Complexes on the Survival of Heart Allografts in Rats Presensitized with Donor Blood, *Transplantation*, 23: 119-123 (1977).
98. R. E. Wilson et al., Active Enhancement of Renal Allograft in Dog and Rat by Subcellular Antigen, *Transplant. Proc.*, 3: 705-707 (1971).
99. R. E. Wilson, A. Rippin, R. K. Dagher, P. Kinnaert, and G. J. Busch, Prolonged Canine Renal Allograft Survival After Pretreatment with Solubilized Antigen, *Transplantation*, 7: 360-371 (1971).
100. R. T. J. Holl-Allen, A. F. Scharli, P. R. Maggs, G. J. Busch, S. J. Simonian, and R. E. Wilson, Active Enhancement of Canine Renal Allografts Using Solubilized Lymphocyte Antigen, *Transplantation*, 10: 472-483 (1970).

101. R. T. J. Holl-Allen, The Effect of Antigen Pretreatment on Canine Renal Allograft Survival, *Ann. R. Coll. Engl.*, 49: 1-17 (1971).
102. N. A. Halasz, M. J. Orloff, and F. Hirose, Increased Survival of Renal Homografts in Dogs After Injection of Graft Donor Blood, *Transplantation,* 2: 453-458 (1964).
103. H. Obertop, A. B. Bijnen, H. M. Vriesendorp, and D. L. Westbroek, Effect of Subcutaneous Injections of Kidney Donor Blood on Renal Allograft Survival in DLA-Typed Dogs, *Transplantation,* 26: 201-202 (1978).
104. D. E. R. Sutherland, C. T. Lum, W. P. Payne, and J. S. Najarian, Effect of Pretreatment with Donor Blood on Renal Allograft Survival in Dogs Stratified by DLA Match and MLC Reactivity, *Transplant. Proc.,* 11: 160-166 (1979).
105. C. B. Currier, Jr., and J. C. Pierce, Effects of Blood Transfusions on Antibody Formation and Canine Renal Allograft Survival, *Surg. Gynecol. Obstet.,* 139: 241-244 (1974).
106. F. Harder, A. L. Floersheim, J. Jeekel, M. Mihatson, Th. Ruedi, and P. Tondelli, Prolongation of Canine Renal Allografts with Preoperative Immunosuppression and Donor Specific Immunization, *Transplant. Proc.,* 11: 978-981 (1979).
107. H. Obertop, J. Jeekel, H. M. Vriesendorp, I. MacDicken, and D. L. Westbroek, The Effect of Donor Blood on Renal Allograft Survival in DL-A Tissue Typed Beagle Littermates, *Transplantation,* 20: 49-52 (1975).
108. H. Obertop, A. B. Bijnen, H. M. Vriesendorp, and D. L. Westbroek, The Effect of Prior Parental Blood Transfusions on the Survival of Renal Allografts from a DLA Identical Sibling, *Transplantation,* 26: 388-390 (1978).
109. B. S. Linn, Renal Allograft and Donor Spleen Cell, *Ann. Surg.,* 164: 223-226 (1966).
110. R. Y. Calne, D. R. Davis, P. Medawar, and J. R. Wheeler, Effect of Donor Antigen on Dogs with Renal Allograft Survival, *Transplantation,* 4: 742-746 (1966).
111. D. T. Caridis, A. Liegeois, I. Barrett, and A. P. Monaco, Enhanced Survival of Canine Renal Allograft of ALS-Treated Dogs Given Bone Marrow, *Transplant. Proc.,* 5: 671-674 (1973).
112. J. Jeekel, H. Obertop, F. Bauer, H. M. Vriesendorp, and D. L. Westbroek, Attempts to Induce Enhancement of Canine Kidney Allografts, *Tissue Antigens,* 3: 171 (1973).
113. F. O. Stephens, K. C. Cochrum, D. F. Scott, D. C. Price, and F. O. Belzer, Attempted Renal Allograft Enhancement by Pooled Kidney Antigen Pretreatment, *Ann. Surg.,* 175: 41-47 (1972).
114. J. J. van Rood, H. Balner, and P. J. Morris, Blood Transfusion and Transplantation, *Transplantation,* 26: 275-277 (1978).
115. R. M. R. Barnes, G. B. Williams, G. D. Pegrum, E. M. Gordon, and M. E. de Wardener, Assessment of Human Renal Transplantation, *Lancet,* 2: 1040-1044 (1974).
116. J. P. van Hooff, M. W. Kalff, and A. E. van Poelgeest, Blood Transfusions and Kidney Transplantation, *Transplantation,* 22: 306-308 (1976).
117. G. Opelz and P. I. Terasaki, Prolongation Effect of Blood Transfusions on Kidney Graft Survival, *Transplantation,* 22: 380-383 (1976).
118. G. Opelz and P. I. Terasaki, Enhancement of Kidney Graft Survival by Blood Transfusions, *Transplant. Proc.,* 9: 121-122 (1977).
119. F. Kissmeyer-Nielsen, S. Olsen, V. P. Petersen, and O. Fjeldborg, Hyperacute Rejection of Kidney Allografts Associated with Pre-Existing Humoral Antibodies Against Donor Cells, *Lancet,* 1: 662-665 (1966).
120. R. Patel and P. I. Terasaki, Significance of the Positive Cross Match Test in Kidney Transplantation, *New Engl. J. Med.,* 280: 735-739 (1969).
121. M. Heideman and G. Claes, Influence of Clinically Determined Immunologic Reactivity on Renal Transplantation, *Transplant. Proc.,* 9: 77-79 (1977).

122. R. T. D. Oliver, J. A. Sachs, H. Festenstein, G. D. Pelgrum, and J. F. Moorehead, Influence of HL-A Matching, Antigenic Strength and Immunoresponsiveness on Outcome of 349 Cadaver Renal Grafts, *Lancet,* 2: 1381-1384 (1972).
123. R. T. D. Oliver, J. A. Sachs, and H. Festenstein, A Collaborative Scheme for Tissue Typing and Matching in Renal Transplantation. VI. Clinical Relevance of HL-A Matching in 349 Cadaver Renal Transplants, *Transplant. Proc.,* 5: 245-251 (1973).
124. J. J. van Rood et al., Graft Survival in Unrelated Donor-Recipient Pairs Matched for MLC and HL-A, *Transplant. Proc.,* 5: 409-414 (1973).
125. F. O. Belzer et al., Is HL-A Typing of Clinical Significance in Cadaver Renal Transplantations? *Lancet,* 1: 774-777 (1974).
126. F. O. Belzer, O. Salvatierra, Jr., K. C. Cochrum, and H. A. Perkins, Good Kidney Graft Survival in Hyperimmunized Patients, *Transplant. Proc.,* 7: 71-74 (1975).
127. J. D. Briggs et al., Influence of HLA Matching and Blood Transfusion on Renal Allograft Survival, *Transplantation,* 25: 80-85 (1978).
128. C. O. Callender, R. L. Simmons, E. J. Yunis, L. H. Toledo-Pereyera, M. T. M. Deshazo, and J. Najarian, Anti HL-A Antibodies: Failure to Correlate with Renal Allograft Rejection, *Surgery,* 76: 573-580 (1974).
129. R. M. Ferguson, H. Noreen, E. J. Yunis, R. L. Simmons, and J. S. Najarian, Does "Responder"/"Non Responder" Status Influence Renal Allograft Success? *Transplant. Proc.,* 9: 69-72 (1977).
130. T. C. Fuller, F. L. Delmonico, A. B. Cosimi, C. E. Huggings, M. King, and P. S. Russell, Impact of Blood Transfusion on Renal Transplantation, *Ann. Surg.,* 187: 211-218 (1978).
131. P. I. Terasaki, G. Opelz, and M. R. Mickey, Analysis of Yearly Kidney Transplant Survival Rates, *Transplant. Proc.,* 8: 139-144 (1976).
132. A. B. Bijnen, H. M. Vriesendorp, J. de Gruyl, H. van Urk, I. McDicken, and D. L. Westbroek, Host Responses to Blood Transfusions After Transplantation of a DL-A Identical Kidney, *Transplant. Proc.,* 7: 431-433 (1975).
133. R. W. Bull et al., The Effect of Prior Third-Party Blood Transfusions on Canine Renal Allograft Survival, *Transplantation,* 26: 249-254 (1978).
134. S. Ahrons and F. Kissmeyer-Nielsen, *Symp. Ser. Immunobiol. Stand.,* 18: 95-98 (1973).
135. S. N. Swisher and L. E. Young, The Blood Grouping Systems of Dogs, *Physiol. Rev.,* 41: 495-520 (1961).
136. A. Zweibaum et al., Studies on Canine Secretory Alloantigens (C.S.A.), *Tissue Antigens,* 4: 115-129 (1974).
137. G. M. Abouna et al., Effect of Pretreatment with Multiple Blood Transfusions and with Skin Grafts on the Survival of Renal Allografts in Dogs, *Transplant. Proc.,* 9: 265-268 (1977).
138. J. W. Fabre et al., A Study of Three Protocols of Blood Transfusions Before Renal Transplantation in the Dog, *Transplantation,* 26: 94-98 (1978).
139. R. L. Marquet, A. A. van Es, G. A. Heystek, R. H. van Leersum, and H. Balner, Prolongation of Baboon to Rhesus Kidney Xenograft Survival by Pretransplant Rhesus Blood Transfusion, *Transplantation,* 25: 165-166 (1978).
140. A. A. van Es, R. L. Marquet, J. J. van Rood, and H. Balner, Influence of a Single Blood Transfusion on Kidney Allograft Survival in Unrelated Rhesus Monkeys, *Transplantation,* 26: 325-330 (1978).

# Autoimmune Diseases of Canines

# 11

THOMAS E. FRITZ

*Division of Biological and Medical Research, Argonne National Laboratory, Argonne, Illinois*

*and*

NOEL R. ROSE

*Department of Immunology and Microbiology, Wayne State University, School of Medicine, Detroit, Michigan*

In a small proportion of the dog population, spontaneous disorders of the immune system have been recognized as autoimmune diseases. The name implies a failure of the recognition of "self" and suggests a lack of, or circumvention of, the normal tolerance of host antigens.

Autoimmunity can be defined as an abnormal state characterized by a self-directed immunologic response that may lead to disease.[1] Cell or tissue injury or destruction is a prominent feature in autoimmune disease. Sometimes autoimmunity is self-limited; in other cases the condition is self-perpetuating and, once begun, may continue indefinitely, commonly with remissions and exacerbations.

Immunologists differ in their definition of the requirements for designating a disease or lesion as an autoimmune response. One of us (Rose) has pointed out that the crucial concept in establishing the autoimmune basis for a human disease is similar to establishing the microbial etiology of a human disease by fulfilling Koch's postulates.[2] In the case of autoimmune disease, the investigator must (1) demonstrate antibodies or evidence of cell-mediated immunity in patients, (2) identify the antigen, (3) produce the antibodies or cell-mediated activity in experimental animals, and (4) show that the pathologic changes are the same in the human spontaneous disease and in the experimental animals. Thus animal models, through their role in identifying the putative antigen and demonstrating its autoantigenic activity, are essential in establishing the autoimmune nature of human diseases. Several recent reviews and discussions have examined the problems, definitions, explanations, and hypotheses concerned with the development and production of spontaneous and experimentally induced autoimmune diseases.[2–7]

A discussion of autoimmune diseases would be difficult without beginning with the term "horror autotoxicus," coined by Ehrlich and Morgenrath[8] in 1900 to describe the fundamental rule that animals do not produce antibodies to antigens of their own tissues. According to their postulate, violation of that rule would precipitate tissue damage and the animal's own death. Just how prophetic they were has been realized dramatically in the intervening years as more information on a variety of autoimmune diseases has accumulated.

During the past 10 years, the great progress in identifying autoimmune diseases in dogs has been the subject of several comprehensive reviews.[9–12] Because the autoimmune diseases in dogs are counterparts of human diseases, they have value as biomedical research models. This chapter surveys the several spontaneous disease problems in dogs that are counterparts of similar problems in man and refers to more comprehensive and specific references dealing with both diagnosis and therapy.

## AUTOIMMUNE HEMOLYTIC ANEMIA

Autoimmune hemolytic anemia (AHA) in dogs as well as in its counterpart in man is among the oldest recognized autoimmune diseases. It was identified as a clinical disease in dogs long before its pathogenesis was understood.[13] Lewis, Schwartz, and Henry[14] first recognized that this hematologic disease commonly was accompanied by other immunologic problems, a finding common to other autoimmune diseases.

The disease is characterized by acute hemolytic crises with signs of systemic anemia due to an increased rate of destruction of erythrocytes.[15–16] Accompanying clinical signs include splenomegaly, lymphadenopathy, fever, anorexia, polydipsia, and, occasionally, vomiting and diarrhea.

The classic work of Coombs, Mourent, and Race,[17] who developed the antiglobulin test (Coombs' test), has made possible the study of both human and canine cases and demonstration of the true autoimmune nature of the problem. Boorman, Loutit, and Steabban[18] were among the first to demonstrate that the surfaces of the erythrocytes of affected animals and man contained immunoglobulins that were directed against those erythrocytes. Although in the dog AHA may take several forms or degrees of severity,[12] all cases show immunoglobulins on the erythrocyte cell surfaces.

The disease is seen in many breeds and all ages, but, as shown in published reports, the problem predominates in females.[12,14] The immediate etiology is not known.

## AUTOIMMUNE THROMBOCYTOPENIA

Episodes of depressed circulating platelets and related hemorrhage described as idiopathic thrombocytopenic purpura (ITP) have been reported in the dog by Magrane, Magrane, and Ross[19] and by Waye.[20] Halliwell[12] discussed the same subject but referred to the condition as immune-mediated thrombocytopenia. The clinical signs, which include thrombocytopenia, petechiae, ecchymoses, melena, and epistaxis, may occur alone or in combination with AHA (Ref. 21). Halliwell[12] and Wilkins et al.[22] have detailed the clinical features, which may vary from minor episodes of petechiation to life-threatening episodes resulting from acute internal or external hemorrhage or progressive anemia from repeated minor to moderate hemorrhages.

In 1951, on the basis of the discovery of a humoral factor in plasma of human patients with thrombocytopenia that was capable of significantly lowering the platelet counts of volunteers, Jarrington et al.[23] initially suggested that certain forms of thrombocytopenia represented an immunologic disorder. The immunologic basis for the destruction of platelets has been described by Wilkins, Hurvitz, and Dodds-Laffin[22] and by Joshi and Jain,[24] using the platelet factor 3 (PF-3) releasing assay of "immunoinjury" developed by Karpatkin and Siskind[25] in 1969. In the presence of antibody directed against the platelet membrane, PF-3 is released from the platelets. The PF-3 test measures the amount of procoagulant activity in normal platelet suspensions after incubation with either patient or control screen globulin fractions. Halliwell[12] reported less confidence in the PF-3 test because of wide ranges in normal clotting times (60 to 120 sec) in dogs. Halliwell[12] also questioned the role of the complement system, which is not considered in the PF-2 test. The PF-3 test seems indicative of serum globulins that cause platelet damage, but both false-positive and false-negative reactions cannot be eliminated if other causes are eliminated in cases of thrombocytopenia and hemorrhage.

## SYSTEMIC LUPUS ERYTHEMATOSIS

Although this disease is discussed comprehensively in Chapter 13, a general overview of systemic lupus erythematosis (SLE) in this chapter is necessary to relate it to the other autoimmune diseases. Halliwell[12] has recently reviewed the present knowledge of canine lupus, and Quimby and Schwartz[26] have made specific comparisons of the human and canine versions of the syndrome.

SLE, a disease of multiple organs and tissues, has been recognized by physicians for more than a century.[27] The name of the disease was derived from the facial cutaneous manifestations in man that result in a wolf-like appearance. The many and varied clinical features in man and their uncoordinated sequential or random development sometimes make a precise early diagnosis difficult. Cohen and Caruso[28] have proposed a set of 14 criteria as diagnostic aids, which taken separately could define other individual autoimmune responses but collectively are diagnostic of SLE. These lesions include arthritis, kidney disease, hemolytic anemia, thrombocytopenia, and skin reactions, including erythema, alopecia, and photosensitivity. The musculoarticular symptoms and cutaneous lesions are the most common clinical signs and problems that stimulate human patients or dog owners to seek evaluations, whereas renal disease is the most common cause of death in animals and man.

In man, a search for the lupus erythematosis (LE) cell was a first step in diagnosis and in understanding the pathogenesis of the disease.[29] The LE cell is a neutrophil that has phagocytized nuclear fragments from degenerating cells opsonified by antinuclear factors. The factor, IgG antibody to DNA–protein, has been demonstrated in animals, including mice and dogs, in addition to man.[14]

The simultaneous or sequential development of multiple autoimmune lesions or signs, including AHA, ITP, polyarthritis, and nephritis, first led to the recognition of the syndrome in the dog.[14]

Lewis and Schwartz[30] described the genetic implications of the syndrome in a colony established for the perpetuation of the disease. The study of that colony is being continued and extended by Quimby and Schwartz.[26] Although multiple serologic or immunologic abnormalities, including LE cells, antinuclear antibodies, direct antiglobulins (Coombs' positive), and rheumatoid factor (RF), have been demonstrated in that colony, clinical signs of disease have not developed after several generations of breeding and analysis.

The genetic predisposition to SLE in man, dogs, and mice is not questioned, but the mechanisms and conditioning factors are not clear.[4,11,26] Quimby and Schwartz[26] postulate the existence of two classes of genes: (1) those which permit a general predisposition to autoimmunity and (2) those which determine the phenotype of the disease. Thus in SLE the two classes of genes are interacting and complementary; a "normal" individual could have either, but not both, yet remain free of symptoms. Reports by Lewis[11] and Quimby and Schwartz[26] on the transmission of a cell-free "factor" that results in serologic abnormalities in the recipients suggest that a replicating agent, a virus, is associated with horizontal and vertical transmission of SLE. The required combination of viral, genetic, and other factors which will allow experimental induction of the clinical disease has not yet been determined.

## RHEUMATOID ARTHRITIS

Rheumatoid arthritis (RA) is a common chronic crippling disease of man whose etiology is still unknown and whose pathogenesis is obscure even after many years of investigation. It is commonly associated with a serum factor, rheumatoid factor (RF), which is, in reality, a diverse group of antibodies (mostly measured as IgM) against IgG. RF is not specific for rheumatoid arthritis in man or dogs and can occur in any chronic long-standing process when there is constant antigen–antibody interaction.

Until the last 10 years, arthritic conditions of the dog were classified as degenerative in nature or infectious in origin, and no disease process comparable to RA was recognized. The first well-documented case of rheumatoid-like arthritis of the dog was described by Lewis and Borel[31] in 1971. More recently, Pedersen et al.[32] and Newton et al.[33] have reported on several additional well-documented cases of RA in dogs which fit the criteria for RA established by the American Arthritis Association.[34,35] Those reports and others[36,37] point out similarities between RA in man and the condition in dogs and consequently document a suitable animal model for the disease.

As in man, RF in dogs is associated with RA as well as with SLE (Refs. 38 and 39); so the differential diagnosis between the two depends on clinical signs, pathologic lesions, and application of other criteria.

There appears to be no sex, breed, or age predilection to RA in the dog. The disease process is chronic and insidious, and affected dogs are usually depressed, anoretic, and pyretic in addition to being lame. The symptoms tend to recur as intermittent episodes that become progressively severe.

As the disease progresses, the joints become enlarged, warm, and sensitive on manipulation. The lesions are mostly concentrated in the peripheral joints with the axial skeleton relatively spared. The histologic lesions are similar to those in RA in man and include degenerative and inflammatory changes in the bone, cartilage, synovial membranes, and supporting tissues. Hematologic changes in dogs, as in man, include leukocytosis, eosinophilia, hypergammaglobulinemia, and generalized lymphadenopathy.

Halliwell[12] has proposed a scheme for pathogenesis of the lesions and production of RF in dogs, but he admits that the stimulus to initiate the sequence is missing.

## AUTOIMMUNE SKIN DISEASES

The bullous skin diseases of man had no recognized counterpart until recently, when Stannard, Gribble, and Baker[40] and Hurvitz and Feldman[41] published papers on canine pemphigus vulgaris. In the bullous skin diseases, as the name implies, fluid-filled bullae or blisters form as a result of localized separation of layers of the epidermis. In man, several forms of the disease are recognized, but the two basic clinical forms are pemphigus vulgaris and pemphigus foliaceous, which differ basically in the location of the lesions. The same terminology has been applied to lesions in the dog, but, because of the thinness of epidermis in areas of predilection in the dog, the two lesions are often difficult to distinguish. Vulgaris is characterized by the formation of soft bullae that tend to rupture and leave raw denuded areas that grow by peripheral extension. The lesions result from acantholysis, which is the loss of coherence of epidermal cells leading to a fissure or cleft. The fissures are suprabasal, with the basal cells remaining in contact with the dermis. In contrast, the foliaceous form has bullae that are located in the stratum corneum or superficial layers of the epidermis. Because of the superficial location, acantholytic bullae are less commonly observed, and the ulcerations tend to be more shallow.

Another related condition has been termed bullous pemphigoid; the clinical features are similar, but the bullae are subepidermal in origin, and the entire epidermis is elevated. Acantholysis is not a prominent histological feature.

The vulgaris form shows a predilection for the oral mucosa and the mucocutaneous junctions. In some cases the base of the nails is involved, which leads to loss of the nails. The condition is painful, and the affected animals can be febrile and appear generally depressed. The foliaceous variant, described by Halliwell and Goldschmidt,[42] does not have the predilection for the mucosae but more commonly affects the face and ears or may be generalized. Rather than weeping lesions, these are crusting, scaling lesions that may become confluent or extensive.

Numerous immunofluorescence studies in man have confirmed the presence of autoantibodies specific for the intercellular cement substance of skin and the mucosa. Both in the dog and in man, these IgG antibodies react specifically at the site of the primary histologic lesion. Halliwell[12] has reviewed the lesions and diagnostic features in man and dogs.

## THYROIDITIS

Thyroiditis (Hashimoto's disease) is a commonly encountered and significant disease in man and has its counterpart in several species of animals, including the dog.[43] Tucker,[44] in 1962, first identified thyroiditis in the dog during routine examinations among experimental beagles. Mawdesley-Thomas,[45] Musser and Graham,[46] and Fritz, Zeman, and Zelle[47] also described similar lesions in beagles from experimental colonies. A familial occurrence of the disease suggests a genetic or hereditary predisposition.[46,47] The several descriptions of the histologic changes in the thyroid gland include focal, nodular, and diffuse infiltration of the thyroid with lymphocytes, plasma cells, macrophages, and, to a lesser degree, neutrophils and giant cells. The degree of involvement varied widely from single isolated nodules to overwhelming destruction of the gland. Lymphoid nodules with germinal centers have been described as an infrequent characteristic in most reports. The varied degree of damage to the parenchyma, follicles, and colloid was generally related to the cellular reaction; cases showing extensive cellular reactions had distorted or obliterated follicles with so few remaining in some cases that the glandular nature was unrecognizable. Many of the more severe cases were also characterized by marked hyperplasia and hypertrophy of the parafollicular cells.[47,48]

More cases of spontaneous thyroiditis have been described in beagles than in other breeds,[44–47,49] but this apparent predominance may be misleading. Because of the selection of the beagle for breeding and use in long-term experiments, particularly in toxicity studies, the beagle has received closer examination, which has led to recognition of otherwise undetected asymptomatic cases. In contrast, cases of thyroid disease in other breeds are recognized only when symptoms indicate a clinically apparent hypofunction of the gland.[50] A survey of any closed breeding group might disclose an incidence greater than that suggested by symptomatic illness alone.

Fritz, Zeman, and Zelle[47] reviewed the pedigrees of all dogs in their colony with thyroiditis and discovered that the disease was associated with one progenitor line of the colony, referred to as "A" line. The incidence of thyroiditis increased with the percent of genes derived from A-line progenitors. The observation that the disease is more frequent among inbred dogs added more evidence to the hypothesis that the disease is under some measure of genetic control.

Several other features characterize thyroiditis in the dog, particularly the beagle. In most cases the thyroids are not grossly affected. The dog does not share with man the distinctly higher incidence in the female. The disease is not progressive nor does it lead to fibrosis.

Thyroid function is only occasionally subnormal in dogs with thyroiditis, and even then it occurs only in association with the most severe lesions. In one study measurements of protein-bound iodine (PBI) and $^{131}I$ thyroid-function tests did not detect significant differences between beagles with and without thyroiditis, but $T_3$ levels were decreased in those with lesions.[46] Another study of the same colony related lowered PBI and $^{131}I$ uptake values with severe lesions.[49] In a third study of the same colony, Mizejewski, Baron, and Poissant[51] identified autoantibodies to canine thyroid by immunodiffusion, passive hemagglutination, complement fixation, immunofluorescence, and skin tests. The skin test for delayed hypersensitivity was positive in a few cases, but immunofluorescence was the most consistently positive test. A combination of all the known thyroid antigens, including thyroglobulin, CA-2 (the nonthyroglobulin colloidal antigen), and microsomal antigen in the follicular cells, was involved in canine thyroiditis.

Fritz, Zeman, and Zelle[47] showed a high correlation between lower uptake of $^{131}I$ and shorter biological half-times of $^{131}I$ in the thyroid of asymptomatic dogs with severe lesions. Sera from the affected animals described by Fritz,

Zeman, and Zelle were tested by Rose and Witebsky[52] and found to contain antibodies for thyroglobulin.

## ORCHITIS

Testicular lesions and sterility have been induced by experimental immunizations of several species of animals with antigens derived from spermatozoa or testicular tissues[53–55] and, more recently, by in vitro conditioned lymphocytes.[56] These studies have attempted to explain the origin of spontaneous testicular degeneration, inflammation, and related sterility in man and animals; explore the usefulness of immunologic techniques for fertility control; and investigate the long-term effects of vasectomy on the testes.[57–60]

Spontaneous lymphocytic orchitis was described in a closed breeding colony of beagles by Fritz et al.[61] The incidence of orchitis in that colony was 32% of 69 males over 1 year of age. Orchitis was genetically correlated with thyroiditis; both diseases had a familial incidence that increased with increasing degrees of relatedness to three sibling progenitors or a partially inbred line in the colony. Among dogs that derived all their genes from the three progenitors, 85% had thyroiditis and 65% had orchitis.

Histologic changes in the testes of affected dogs varied considerably, and the term lymphocytic orchitis was used to characterize the lesions, which included lymphoid cell infiltrates as well as tubular degeneration. Nodular and diffuse forms of lymphoid infiltration and aggregations of lymphoid cells were observed in histologic preparations of the affected testes. A few specimens showed plasma cells, histiocytes, and a few multinucleated giant cells, and neutrophils were rare. The lymphocytic infiltrations were located in the epididymis; all portions of the testes, including the interstitium; and within the seminiferous epithelium, including the tubular lumina. In addition, some dogs had significant tubular atrophy without a cellular reaction.

The gross changes were limited to testicular atrophy, which resulted in reduction in size and weight, and were related to a clinical history that showed declining reproductive performance and reduced sperm counts.

Both the resemblance of the lesions in dogs to those of other well-defined autoimmune processes, particularly thyroiditis, and correlation of orchitis with the familial incidence of thyroiditis suggest that the thyroiditis–orchitis disease complex in this beagle colony is another example of the simultaneous occurrence of multiple lesions in an autoimmune disorder.

Compared with the literature on other spontaneous diseases with autoimmune characteristics, few reports document the incidence or pathogenesis of orchitis in animals and man.[4,57,62] The comparison of the lesions in dogs with those in man would be difficult because of the lack of definition of the incidence and lesions in the spontaneous disease in both species.

## DISCUSSION

The specific diseases described in this chapter represent only those few which have been recognized or reasonably identified in the dog on the basis of general criteria used to define an autoimmune disease in man. Because of the economics of veterinary medicine, the identification and description of additional canine diseases of an autoimmune nature have been slow. With the recognition of the usefulness of animal models for studying some of the more clinically important but poorly defined diseases in man, however, more attention is being given to identifying comparable spontaneous or experimentally induced diseases in dogs. Rose[2] lists more than 20 human diseases, in addition to those already described here in the dog, as having autoimmune manifestations. Diseases such as Addison's,[63] which has been shown to have circulating antibodies to adrenocortical components, have yet to be described in the dog.

A common disease of the dog, glomerulonephritis, known to be the result of an aberrant immune mechanism,[64] needs intensive examination before it can be exploited as a model for medical research for extrapolation to the problem in man. Sjögren's syndrome (a form of keratoconjunctivitis and xerostomia), recently described in dogs,[65] in man is associated with a wide variety of autoimmune syndromes, including rheumatoid arthritis and SLE as well as lymphoid reactions varying from benign focal proliferation to widespread lymphoreticular malignancies.[66,67] Further development of such models and characterization of such syndromes in animals would undoubtedly lead to increased knowledge of the clinically important human counterparts with obscure etiology.

Demyelinating encephalopathies are probably the most extensively studied of the experimental autoimmune diseases in animals,[2] but those animal models have yielded little information on the basic cause of even the most common spontaneous demyelinating disease of man, multiple sclerosis. Lampert,[68] in a recent review, discussed how viral infections may be related to autoimmune reactions in demyelinating diseases of man and animals, including canine distemper and old dog encephalitis.[68,69] Examples of infectious animal diseases exhibiting autoimmune responses are Aleutian disease in mink and infectious equine anemia, both known to be of viral origin.[11,70]

Although the autoimmune diseases represent diverse problems of altered cellular physiologic mechanisms, a unifying concept of their basic cause may be the reward of continued search. Several diseases in the dog, including thyroiditis, orchitis, and SLE, have a predisposing genetic component, and spontaneous autoimmune diseases in other species, including man, also have a familial pattern of inheritance. Whether this hereditary or genetic disposition might be a vertical transmission of an infectious agent (virus) or a more classical form of genetic control needs further definition. The genetic basis for classical autoimmune diseases has been greatly strengthened by the evolving information on experimentally induced autoimmune diseases.[71]

Further studies of the autoimmune diseases of dogs, including the pathogenesis, alterations in immune mechanisms, and analyses of retrospective data, as well as experimental breedings and transmission studies, should lead to a better understanding of the corresponding human disorders.

## ACKNOWLEDGMENT

This work was supported in part by the U. S. Department of Energy under contract No. W-31-109-ENG-38(TEF).

## REFERENCES

1. W. Dameshek, Autoimmunity: Theoretical Aspects. Autoimmunity—Experimental and Clinical Aspects, *Ann. N. Y. Acad. Sci.,* 124: 6-28 (1965).
2. N. R. Rose, Autoimmune Diseases, in *Principles of Immunology,* N. R. Rose, F. Milgrom, and C. J. VanOss (Eds.), pp. 436-475, Macmillan Publishing Co., Inc., New York, 1979.
3. W. Dameshek, E. Witebsky, and F. Milgrom (Eds.), Autoimmunity—Experimental and Clinical Aspects, *Ann. N. Y. Acad. Sci.,* 124(I and II): 1-890 (1965).
4. N. R. Rose, in *The Inflammatory Process,* Vol. III, B. J. Zweifach, L. Grant, and R. T. McClusky (Eds.), Academic Press, Inc., New York, 1974.
5. M. Burnet, *Auto-Immunity and Auto-Immune Disease,* F. A. Davis Co., Philadelphia, 1972.
6. N. Talal (Ed.), *Autoimmunity,* Academic Press, Inc., New York, 1977.
7. R. A. Good and E. Yunis, Association of Autoimmunity Immunodeficiency and Aging in Man, Rabbits and Mice, *Fed. Proc.,* 33: 2040-2050 (1974).
8. P. Ehrlich and J. Morgenrath, Ueber Haemolysine, *Verl. Klin. Worchenschr.,* 37: 453-538 (1900).
9. R. O. Jacoby and R. A. Griesemer, Immunological Injury to Dogs, *Vet. Sci. Comp. Med.,* 14: 149-207 (1970).
10. F. W. Quimby, C. A. Jensen, D. Nawrocki, and P. Scollin, Selected Autoimmune Diseases in the Dog, *Vet. Clin. N. Am.,* 8: 665-682 (1978).
11. R. M. Lewis, Spontaneous Autoimmune Diseases of Domestic Animals, *Rev. Exp. Pathol.,* 13: 55-82 (1974).
12. R. W. W. Halliwell, Autoimmune Diseases in the Dog, *Adv. Vet. Sci. Comp. Med.,* 22: 221-263 (1978).
13. G. Miller, G. F. Firth, S. N. Swisher, and L. E. Young, Studies on the Destruction of Red Blood Cells by Canine Autoantibodies in Normal Dogs and in a Dog with Naturally Occurring Autoimmune Hemolytic Disease, *Am. J. Dis. Child.,* 93: 35-36 (1957).
14. R. M. Lewis, R. S. Schwartz, and W. B. Henry, Canine Systemic Lupus Erythematosis, *Blood,* 25: 143-160 (1965).
15. R. W. Bull, A. J. Bowdler, and R. Schirmer, Autoimmune Hemolytic Disease in the Dog, *J. Am. Vet. Med. Assoc.,* 159: 880-883 (1971).

16. M. D. Avolt, J. E. Lund, and J. C. Pickett, Autoimmune Hemolytic Anemia in a Dog, *J. Am. Vet. Med. Assoc.,* 162: 45-46 (1973).
17. R. R. A. Coombs, A. E. Mourent, and R. R. Race, Detection of Weak and "Incomplete" Rh Agglutinins: New Test, *Lancet,* 2: 15-16 (1945).
18. K. E. Boorman, J. F. Loutit, and D. B. Steabban, Poly-Agglutinable Red Cells, *Nature,* 158: 446-447 (1946).
19. H. J. Magrane, G. W. Magrane, and J. R. Ross, Idiopathic Thrombocytopenia in a Dog, *J. Am. Vet. Med. Assoc.,* 135: 520-522 (1959).
20. J. W. Waye, Idiopathic Thrombocytopenic Purpura in a Dog, *Can. Vet. J.,* 1: 569-571 (1969).
21. R. M. Lewis, W. B. Henly, G. W. Thorton, and C. E. Gilmore, Syndrome of Autoimmune Hemolytic Anemia and Thrombocytopenia in Dogs, in *Proceedings of the 100th Annual Meeting of the American Veterinary Medical Association,* New York, July 28–Aug. 1, 1963, pp. 140-16 3, American Veterinary Medical Association, Chicago, Ill., 1963.
22. R. J. Wilkins, A. I. Hurvitz, and W. J. Dodds-Laffin, Immunologically Mediated Thrombocytopenia in the Dog, *J. Am. Vet. Med. Assoc.,* 163: 277-288 (1973).
23. W. J. Jarrington, V. Minich, J. W. Hollingsworth, and C. V. Moore, Demonstration of a Thrombocytopenia Factor in the Blood of Patients with Thrombocytopenia Purpura, *J. Lab. Clin. Med.,* 38: 1-16 (1951).
24. B. C. Joshi and N. C. Jain, Detection of Antiplatelet Antibody in Serum and on Megakaryocytes of Dogs with Autoimmune Thrombocytopenia, *Am. J. Vet. Res.,* 37: 681-685 (1976).
25. S. Karpatkin and G. W. Siskind, In Vitro Detection of Platelet Antibody in Patients with Idiopathic Thrombocytopenia Purpura, *Blood,* 33: 795-817 (1969).
26. F. W. Quimby and R. S. Schwartz, The Etiopathogenesis of Systemic Lupus Erythematosis, *Pathobiol. Ann.,* 18: 35-59 (1978).
27. W. Osler, On the Visceral Complications of Erythema Exudativum Multiforme, *Am. J. Med. Sci.,* 110: 629-646 (1895).
28. A. S. Cohen and J. J. Caruso, Criteria for the Classification of Systemic Lupus Erythematosis—Status, *Arthritis Rheum.,* 15: 540-543 (1972).
29. M. M. Hargraves, H. Richmond, and R. Morton, Presentation of Two Bone Marrow Elements: "Tart" Cell and "L. E." Cell, *Proc. Staff Meet. Mayo Clin.,* 23: 25-28 (Jan. 21, 1948).
30. R. M. Lewis and R. S. Schwartz, Canine Systemic Lupus Erythematosis, *J. Exp. Med.,* 134: 417-438 (1971).
31. R. M. Lewis and Y. Borel, Canine Rheumatoid Arthritis, *Arthritis Rheum.,* 14: 67-74 (1971).
32. N. C. Pedersen, R. R. Pool, J. J. Castles, and K. Weisner, Noninfectious Canine Arthritis: Rheumatoid Arthritis, *J. Am. Vet. Med. Assoc.,* 169: 295-303 (1976).
33. C. D. Newton, A. J. Lipowitz, R. E. W. Halliwell, H. L. Allen, D. N. Biery, and H. R. Schumacker, Rheumatoid Arthritis in Dogs, *J. Am. Vet. Med. Assoc.,* 168: 113-131 (1976).
34. G. P. Rodnan (Ed.), *Primer on Rheumatic Diseases,* 7th ed., American Rheumatism Association Foundation, New York, 1973.
35. M. W. Ropes, G. A. Bennett, S. Cobbs, R. Jacox, and R. A. Jesser, 1958 Revision of Diagnostic Criteria for Rheumatoid Arthritis, *Ann. Rheum. Dis.,* 33: 357-360 (1974).
36. S. Liu, P. F. Sutter, C. A. Fischer, and H. D. Dorfman, Rheumatoid Arthritis in a Dog, *J. Am. Vet. Med. Assoc.,* 154: 495-502 (1969).
37. R. E. W. Halliwell, R. B. Lavelle, and K. M. Bott, Canine Rheumatoid Arthritis: A Review and Case Report, *J. Small Anim. Pract.,* 13: 239-243 (1972).
38. R. M. Lewis and J. E. Hathaway, Canine Systemic Lupus Erythematosis Presenting with Symmetrical Polyarthritis, *J. Small Anim. Pract.,* 8: 273-283 (1967).

39. R. Winchester, Tests for Detection of the Rheumatoid Factors, in *Manual of Clinical Immunology*, N. R. Rose and H. Friedman (Eds.), pp. 665-668, American Society for Microbiology, Washington, D.C., 1976.
40. A. A. Stannard, D. H. Gribble, and B. B. Baker, A Mucocutaneous Disease in the Dog, Resembling Pemphigus Vulgaris in Man, *J. Am. Vet. Med. Assoc.*, 166: 575-582 (1975).
41. A. I. Hurvitz and E. Feldman, A Disease in Dogs Resembling Human Pemphigus Vulgaris: Case Reports, *J. Am. Vet. Med. Assoc.*, 166: 585-590 (1975).
42. R. E. W. Halliwell and M. H. Goldschmidt, Pemphigus Foliaceous in the Canine: A Case Report and Discussion, *J. Am. Anim. Hosp. Assoc.*, 13: 431-436 (1977).
43. P. E. Bigazzi and N. R. Rose, Spontaneous Autoimmune Thyroiditis in Animals as a Model of Human Disease, in *Progress in Allergy*, P. Kallos, B. H. Waksman, and A. deWeck (Eds.), pp. 245-274, S. Karger, Basel, 1975.
44. W. E. Tucker, Thyroiditis in a Group of Laboratory Dogs, *Am. J. Clin. Pathol.*, 38: 70 (1962).
45. L. E. Mawdesley-Thomas, Lymphocytic Thyroiditis in the Dog, *J. Small Anim. Pract.*, 9: 539 (1968).
46. E. Musser and W. R. Graham, Familial Occurrence of Thyroiditis in Purebred Beagles, *Lab. Anim. Care*, 18: 58 (1968).
47. T. E. Fritz, R. C. Zeman, and M. R. Zelle, Pathology and Familial Incidence of Thyroiditis in a Closed Beagle Colony, *Exp. Mol. Pathol.*, 12: 14-30 (1967).
48. L. E. Mawdesley-Thomas and D. W. Jolly, Autoimmune Disease in the Beagle, *Vet. Rec.*, 80: 535 (1967).
49. W. H. Beierwaltes and R. H. Nishiyama, Dog Thyroiditis: Occurrence and Similarity to Hashimoto's Struma, *Endocrinology*, 83: 501 (1968).
50. B. M. Bush, Thyroid Disease in the Dog. II. A Review, *J. Small Anim. Pract.*, 10: 185 (1969).
51. G. J. Mizejewski, J. Baron, and G. Poissant, Immunologic Investigations of Naturally Occurring Canine Thyroiditis, *J. Immunol.*, 107: 1152 (1971).
52. N. R. Rose and E. Witebsky, Experimental Thyroiditis, in *Immunological Diseases*, M. Samter (Ed.), p. 1179, Little, Brown & Co., Boston, 1971.
53. D. W. Bishop, R. Narbaitz, and M. Tessof, Induced Aspermatogenesis in Adult Guinea Pigs Injected with Testicular Antigen and Adjuvant in Neonatal Stages, *Dev. Biol.*, 3: 444-485 (1961).
54. P. G. Brown, L. E. Glynn, and E. J. Holbora, The Pathogenesis of Experimental Allergic Orchitis in Guinea Pigs, *J. Pathol. Bacteriol.*, 86: 505-520 (1963).
55. G. J. Losos, A. J. Winter, and K. McEntee, Induction of Testicular Degeneration in Bulls by Autoimmunization, *Am. J. Vet. Res.*, 29: 2295-2306 (1968).
56. H. Wekerle and M. Begeman, Experimental Autoimmune Orchitis: In Vitro Induction of an Autoimmune Disease, *J. Immunol.*, 116: 159-161 (1976).
57. K. S. K. Tung, Human Sperm Antigens and Antisperm Antibodies. I. Studies on Vasectomy Patients, *Clin. Exp. Immunol.*, 20: 93-104 (1975).
58. K. S. K. Tung and A. J. Woodroffe, Immunopathology of Experimental Allergic Orchitis in the Rabbit, *J. Immunol.*, 120: 320-328 (1978).
59. M. J. Alexander and K. S. K. Tung, Immunological and Morphological Effects of Vasectomy in the Rabbit, *Anat. Rec.*, 188: 339-350 (1977).
60. R. E. Hackett and K. Waterhouse, Vasectomy-Reviewed, *Am. J. Obstet. Gynecol.*, 116: 435-455 (1973).
61. T. E. Fritz, L. S. Lombard, S. A. Tyler, and W. P. Norris, Pathology and Familial Incidence of Orchitis and Its Relation to Thyroiditis in a Closed Beagle Colony, *Exp. Mol. Pathol.*, 24: 142-158 (1976).
62. B. H. Wakeman, The Distribution of Experimental Auto-Allergic Lesions, *Am. J. Pathol.*, 37: 673-693 (1960).

63. J. Nerup, V. Andersen, and G. Bendixen, Anti-Adrenal, Cellular Hypersensitivity in Addison's Disease, *Clin. Exp. Immunol.*, 4: 355-363 (1969).
64. Glomerular Disease in the Dog, *Vet. Rec.*, 95: 61-62 (1974).
65. F. W. Quimby, R. S. Schwartz, T. Poskitt, and R. M. Lewis, A Disorder of Dogs Resembling Sjogren's Syndrome, *Clin. Immunol. Immunopathol.*, 12: 471-476 (1979).
66. L. G. Anderson and N. Talal, The Spectrum of Benign to Malignant Lymphoproliferation in Sjogren's Syndrome, *Clin. Exp. Immunol.*, 9: 199-221 (1971).
67. S. S. Kassan and M. Gardy, Sjogren's Syndrome: An Update and Overview, *Am. J. Med.*, 64: 1037-1046 (1978).
68. P. W. Lampert, Autoimmune and Virus-Induced Demyelinating Diseases, *Am. J. Pathol.*, 91: 175-208 (1978).
69. D. D. Adams, A Theory of the Pathogenesis of Rheumatic Fever, Glomerulonephritis and Other Autoimmune Diseases Triggered by Infection, *Clin. Exp. Immunol.*, 5: 105-115 (1969).
70. R. W. Leader, Viruses and Autoimmune Diseases in Animals Other than Man, *Am. J. Clin. Pathol.*, 56: 315-319 (1971).
71. N. R. Rose, Autoimmunity Revisited, *Nature*, 245: 88-90 (1978).

# 12 Systemic Lupus Erythematosus

ROBERT M. LEWIS
*Department of Pathology,*
*New York State College of Veterinary Medicine,*
*Cornell University, Ithaca, New York*

Systemic lupus erythematosus (SLE) is a complex, multisystem immunologic disease that has been recognized in three species: man, mice, and dogs. Although the symptoms and lesions have been well characterized, the cause of SLE remains unknown. Efforts to define the etiology and pathogenesis of human SLE have been substantially aided by investigations of the naturally occurring disease in New Zealand mice, random outbred dogs, and the inbred progeny of canine lupus patients. The usefulness of the murine model to students of systemic lupus has been summarized in recent reviews by Talal[1] and Quimby and Schwartz.[2] This chapter reviews the canine form of SLE and examines how studies of this animal model have influenced a more thorough understanding of the disease in man.

## CLINICAL FEATURES

Systemic lupus erythematosus, originally described as a spontaneous canine disease by Lewis, Schwartz, and Henry[3] in 1965, is now a well-recognized entity in canine medicine.[4–11] Clinically the disease is characterized by the simultaneous or sequential development of hemolytic anemia, thrombocytopenic purpura, polyarthritis,* proteinuria, and dermatitis. As such, the disease in random outbred house pets closely parallels the clinical findings in man.[12,13]

*Immune mediated arthropathies of the dog are detailed by Niels C. Pedersen and Roy Pool in Chapter 13.

Historically the combined occurrence of hemolytic anemia and thrombocytopenic purpura in a single patient was an important clinical indication that often led to the diagnosis of SLE (Ref. 14). However, the growing awareness of the disease by practicing veterinarians has increased the frequency of its diagnosis, and currently symmetrical polyarthritis is the most frequent presenting sign of this canine disorder.[15,16] No breed predisposition has been determined. Females are affected more frequently than males, and most affected animals are in the young adult age group of 2 to 5 years.

## SEROLOGIC FEATURES

Hyperglobulinemia is a consistent finding in canine SLE, and autoantibodies to a variety of intrinsic antigens play an important role in both the diagnosis and pathogenesis of the disease.[4,11,17,18] Thus, in patients with hemolytic anemia, the direct antiglobulin test is positive, which indicates sensitization of host erythrocytes by host immunoglobulin. Eluates prepared from affected erythrocytes sensitize normal dog red blood cells to an indirect antiglobulin test, which confirms that the antierythrocyte antibody is an autoantibody. Similarly, patients with thrombocytopenic purpura contain in their plasma immunoglobulin directed against platelets, as demonstrated by the Platelet Factor III Assay.[19] Although platelet elution techniques have not been performed to confirm the autoantibody nature of this immunoglobulin, the immunopathogenesis of the thrombocytopenic purpura in canine SLE is considered similar to that of the hemolytic anemia.

Antibodies to nuclear antigens are a hallmark of SLE in all three species that suffer from the disease (man, mice, and dogs). The LE cell test is highly specific for canine SLE and serves as the most useful diagnostic test in a clinical setting.[20] In this assay a polymorphonuclear neutrophil (PMN) ingests nucleoprotein that has reacted with antinuclear antibody (Fig. 1). Fluorescent tests for antinuclear antibody (ANA) are frequently positive in patients with canine SLE, although this test is less specific for the disease than the LE cell test.[18,21] Diffuse, rim, and speckled nuclear staining patterns can be observed in positive tests (Fig. 2), but the predominant pattern is that of diffuse nuclear fluorescence. Low titers of circulating ANA have been observed in canine patients with other autoimmune diseases, such as rheumatoid arthritis, autoimmune hemolytic anemia, immunopathic thrombocytopenic purpura, and Sjögren's syndrome. A recent study by Quimby et al.[18] indicates that the mean titers of ANA in canine SLE are approximately 1 : 200, whereas in other autoimmune diseases the mean titers are less than 1 : 50. Consequently, although ANA is not specific for SLE, a high titer of ANA strongly supports that diagnosis.

Antibodies to native double-stranded DNA (N-DNA) and heat-denatured single-stranded DNA (D-DNA) are also present in canine SLE (Ref. 22). The

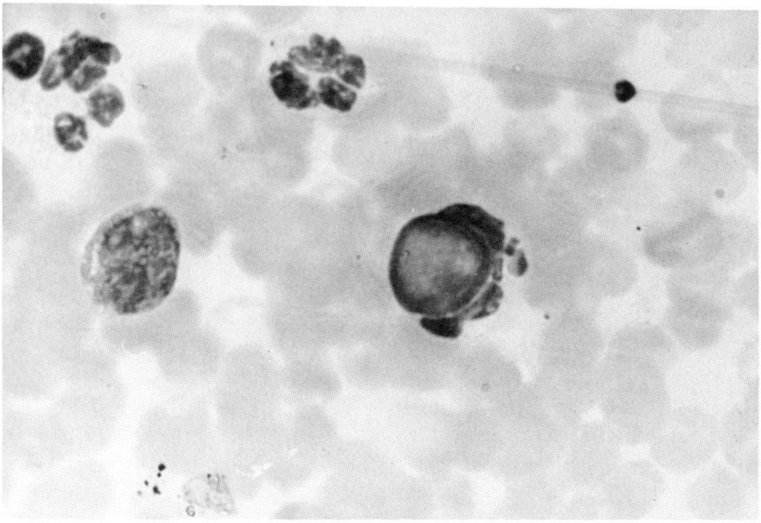

Fig. 1 An LE cell from a dog with SLE. A neutrophil has ingested a large homogeneous mass of nucleoprotein. The tinctorial and morphologic changes of the nuclear material follow the cell's reaction with antibodies to nuclear antigens in the patient's serum.

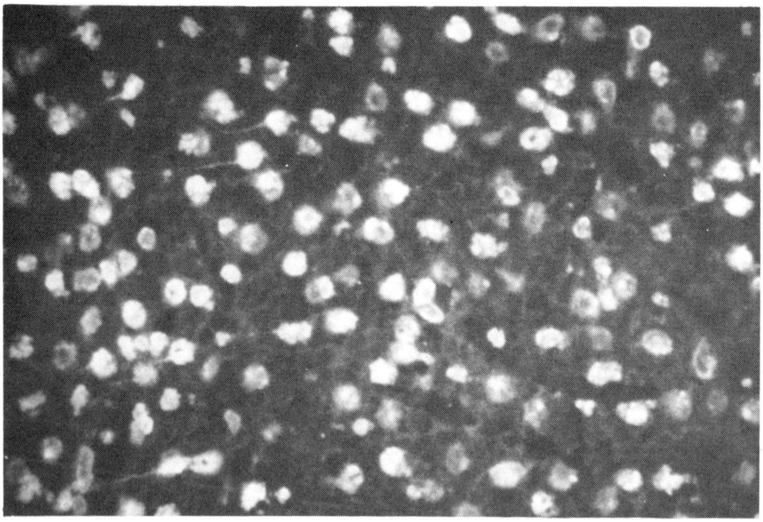

Fig. 2 Antinuclear antibody detected in the serum of a dog with SLE. Cryostat sections of rat liver have been incubated with patient serum and stained with a fluoresceinated polyvalent antiserum to canine immunoglobulin. Diffuse, peripheral, and speckled nuclear staining patterns are visible in this preparation.

presence of anti-N-DNA antibody is unique to patients with SLE regardless of species and is highly specific for the diagnosis of SLE. In addition to their diagnostic importance, anti-DNA antibodies are functionally significant in mediating glomerulonephritis induced by immune complexes in canine SLE. Both anti-D-DNA and anti-N-DNA antibodies have been recovered from eluates prepared from the renal cortices of canine lupus patients with chronic glomerulonephritis.[23]

Rheumatoid factor (Rf) may also be present in the sera of dogs with SLE, particularly those with clinical, radiographic, or pathologic evidence of symmetrical polyarthritis.[16] In detailed studies of one case by Lewis and Borel,[24] rheumatoid factor in serum and joint fluid of an affected dog had greater affinity for autologous IgG than for homologous IgG. This heightened reactivity to intrinsic antigen was interpreted as evidence supporting the autoimmune basis of the disease. Autoantibodies to other tissue antigens, such as thyroglobulin or thyroid microsomal antigen, may also be present in selected cases.[25]

The status of lymphocytotoxic antibodies in canine SLE remains unknown and represents an area urgently needing investigation. Information from current studies in both human and murine SLE indicates that the absolute or functional loss of specific lymphocyte populations may be pivotal to the loss of normal immunoregulation. One result of selective lymphocyte depletion may be the production of autoantibodies, which subsequently mediate the immunologic damage to the host. Thus, if antibodies to host lymphocytes should be a component of canine SLE, they would amplify and maintain a state of autoreactivity.

## PATHOLOGIC FEATURES

The lesions encountered in canine SLE vary and, in general, are related to the presenting signs.[26] For example, animals with hemolytic anemia will express extramedullary hematopoiesis, erythrophagocytosis, and other tissue changes consistent with hemolytic disease. Similarly, if thrombocytopenic purpura is a component of the disease, megakaryocytic hyperplasia and multifocal hemorrhages will be evident. The joint lesion in symmetrical polyarthritis primarily affects the soft tissues of the joint capsule and is characterized by villous hypertrophy of the synovium, focal infiltrations of lymphocytes and plasma cells, proliferation of synovial lining cells, and periarticular fibrosis (Figs. 3 and 4). The skin lesion responsible for the focal exfoliative dermatitis observed clinically is characterized by hyperkeratosis, parakeratosis, vacuolar degeneration of basal epithelia, dermal congestion, focal hemorrhage, and infiltration by lymphocytes and plasma cells (Fig. 5).

Central to the pathology of canine SLE are changes in lymphoid tissue and blood vessels. Enlarged lymph nodes exhibit prominent primary and secondary

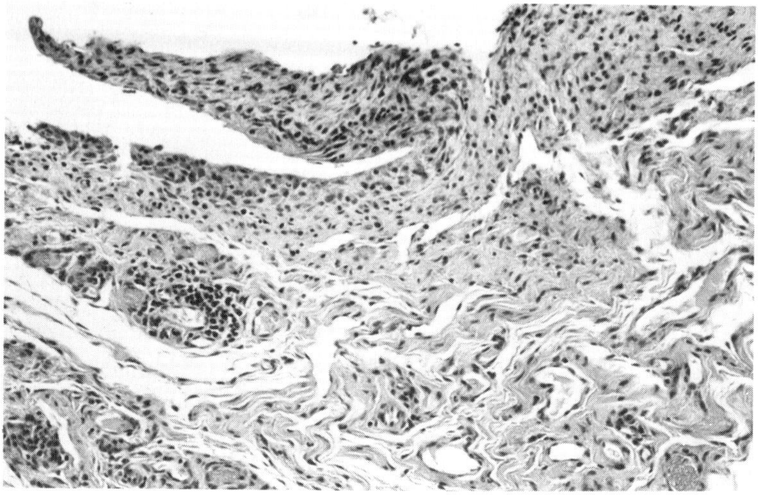

Fig. 3 Joint capsule from a dog with SLE and symmetrical polyarthritis. Villous proliferation of the synovial membrane, replication of synovial lining cells, diffuse lymphocytic synovitis, congestion, and perivascular accumulations of lymphocytes and plasma cells characterize the lesion.

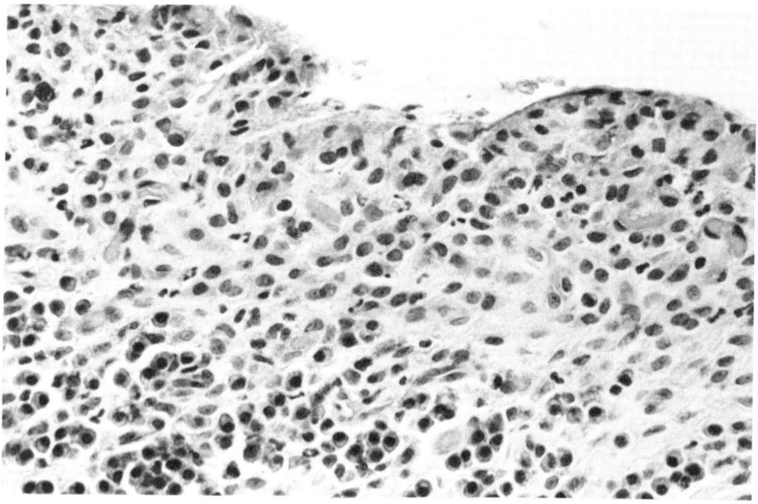

Fig. 4 Synovial membrane from a dog with SLE. The thickened synovia contains proliferating synovial lining cells, occasional neutrophils, and focal accumulations of plasma cells. At the joint surface fine strands of fibrin adhere to the lining cells.

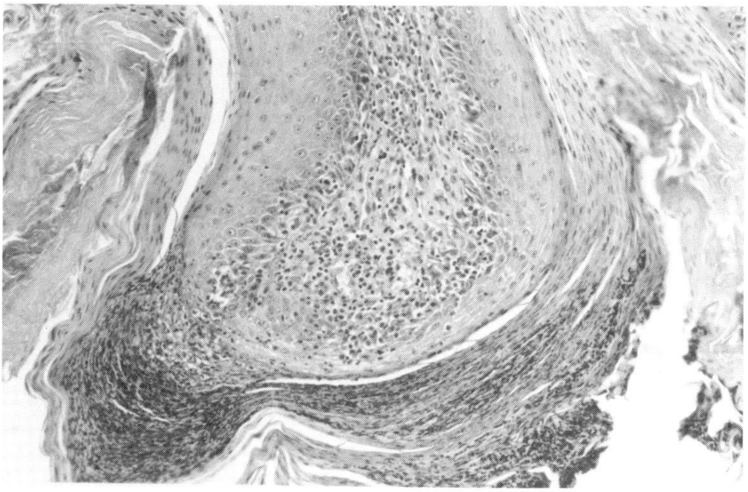

Fig. 5 Biopsy of malar rash from a dog with SLE. The lesion is characterized by hyperkeratosis, parakeratosis, vacuolar degeneration of the basal epithelial cells, congestion, hemorrhage, and, predominantly, a lymphocytic infiltrate in the dermis.

lymphoid follicles with active germinal centers, densely cellular paracortical zones, and medullary cords packed with lymphocytes and plasma cells. These morphologic changes typify reactive hyperplasia of both T and B lymphocytes. In addition, the thymus often contains single or multiple lymphoid follicles with active germinal centers (Fig. 6). Similar thymic lesions have been reported in human and murine SLE as well as in some cases of human myasthenia gravis.[2 7]

Vascular lesions can affect any organ. Typically, small muscular arterioles are affected with lesions ranging from acute necrotizing vasculitis to chronic segmental fibrinoid change (Figs. 7 and 8). Of critical importance is the renal lesion, chronic diffuse membranoproliferative glomerulonephritis (Fig. 9). By light microscopy the glomerular lesion is characterized by cellular proliferation, thickening of the capillary walls, increased mesangial matrix, and intraglomerular adhesions. During active stages of the lesion, focal necrosis and neutrophilic infiltrates occur in glomerular tufts. Immunofluorescent techniques reveal the focal deposition of host immunoglobulin (IgG, IgA, and IgM) and complement (beta 1c globulin) along glomerular capillary basement membranes. These deposits range from discrete fine aggregates to more coarse accumulations of immunoglobulin and complement (Fig. 10). Often the mesangial zones will also contain dense aggregates of these immune reactants. Ultrastructurally the glomerular lesion is characterized by fusion of epithelial cell foot processes, focal electron densities in the capillary basement membrane, proliferation of mesangial cells, and increased amounts of mesangial matrix (Fig. 11). The

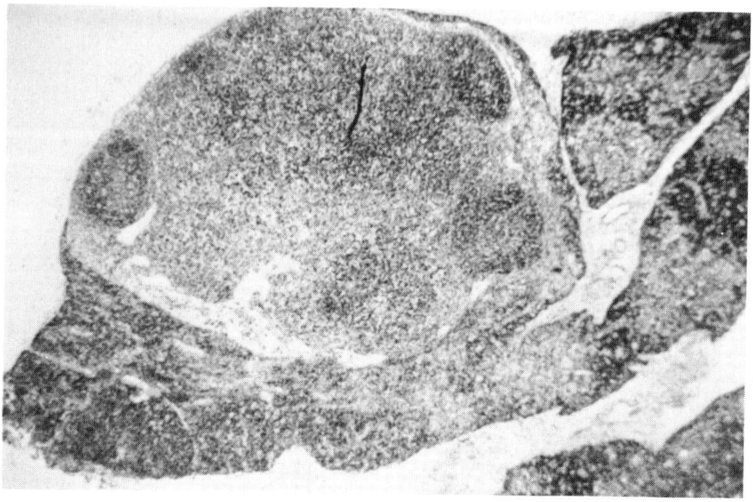

Fig. 6 Thymus from a dog with SLE. Most of one thymic lobule is distorted by multiple lymphoid follicles, two of which contain active germinal cancers.

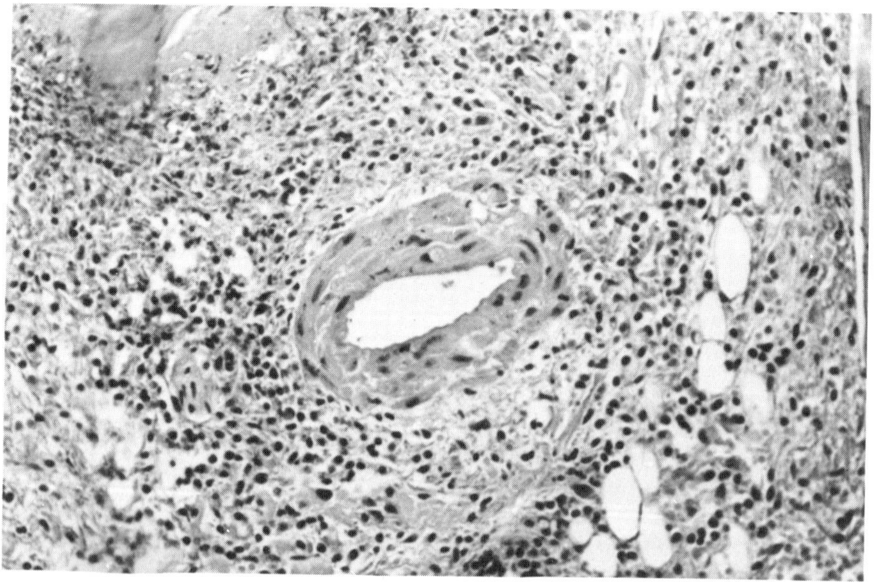

Fig. 7 Acute necrotizing arteritis of a small muscular arteriole in canine SLE. In addition to necrosis of the tunica media, a dense mass of mixed cells has infiltrated the tunica adventitia and adjacent periarteriolar tissue.

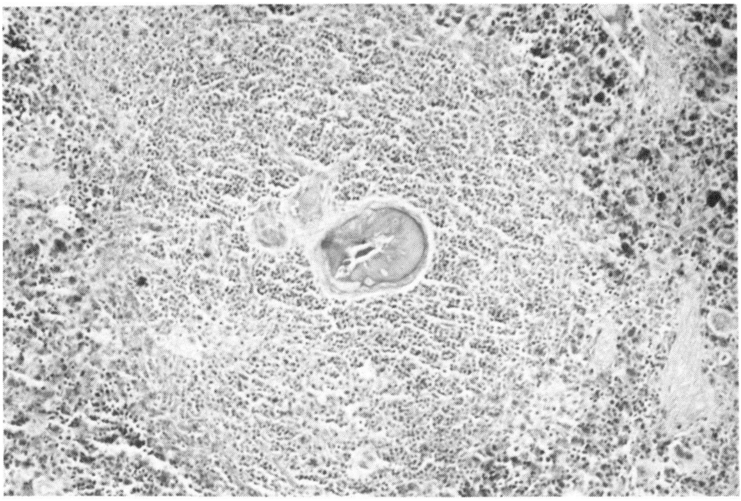

Fig. 8 Segmental fibrinoid change in the tunica media of a central arteriole from the spleen of a dog with SLE. Accumulation of hemosiderin by phagocytes at the periphery of the splenic corpuscle indicates the hemolytic component of the disease in this patient.

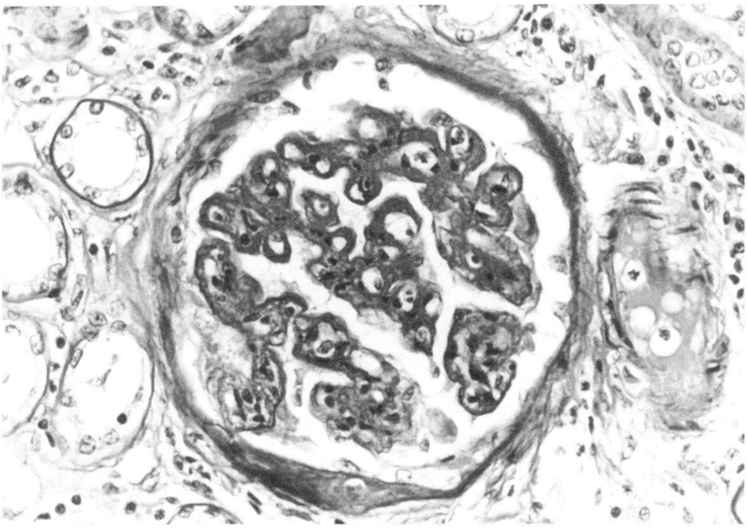

Fig. 9 Chronic membranoproliferative glomerulonephritis in canine SLE (periodic acid–Schiff reagent). The lesion is characterized by marked thickening of capillary walls (wire loop lesions), increased mesangial matrix, intraglomerular adhesions, and periglomerular fibrosis of Bowman's capsule. In addition, focal segmental necrosis of glomerular tufts with neutrophilic inflitrates indicates the active nature of the lesion.

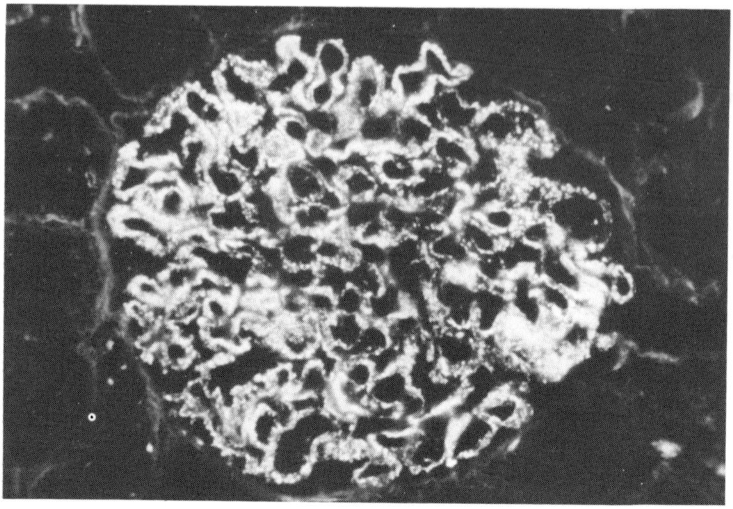

Fig. 10 Severe glomerulonephritis mediated by the immune complex in canine SLE. The cryostat section of a glomerulus was stained with a fluoresceinated polyvalent antiserum to canine immunoglobulins. Discrete focal deposits of immunoglobulin are visible within the capillary walls and along the basement membranes of all capillary loops. Focal accumulations of larger deposits are also visible in the mesangium. A similar staining pattern is achieved with fluoresceinated reagents to canine beta 1C globulin (complement).

fluorescent staining pattern and the ultrastructural features of the renal lesion are characteristic of glomerulonephritis mediated by immune complexes. Acid eluates prepared from selected nephritic patients have contained antibody with specificity to both native DNA and denatured DNA (Ref. 23). Thus nucleic acids represent at least one antigenic component of the glomerular lesion in canine SLE.

## ETIOLOGY

### Genetic Experiments

In one study of the possible role of genetic factors in canine SLE under controlled laboratory conditions, a large colony of affected dogs was assembled to test the hypothesis that SLE was inherited as an autosomal genetic defect.[21] In six different breeding experiments, inbred, outbred, and backcrossed progeny were studied for clinical, serological, and pathological evidence of SLE. A positive LE cell test was chosen as the phenotypic marker for the hypothetical gene. The results of serial examinations of 480 inbred offspring indicated that

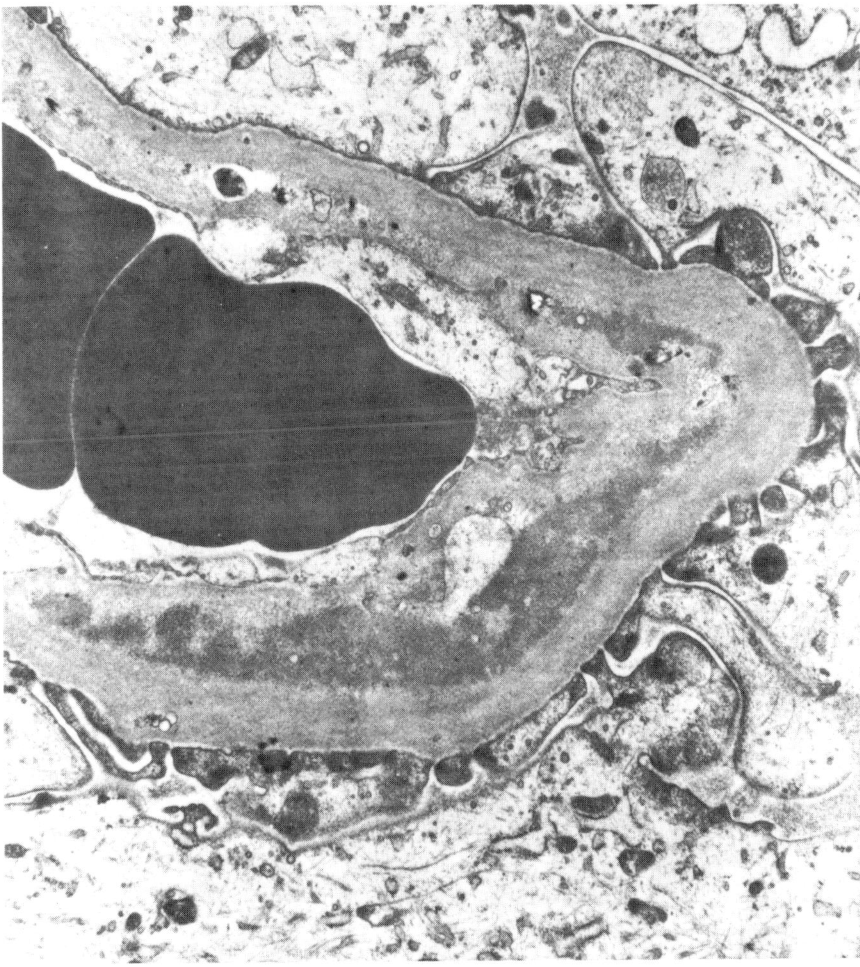

Fig. 11 Ultrastructural changes associated with nephritis mediated by the immune complex typify the renal lesion of canine SLE. Irregular subendothelial electron dense deposits are present within the glomerular capillary basement membrane. Fusion of overlying epithelial cell foot processes accompanies the changes within the basement membrane.

95% of these animals had the SLE marker by 18 months of age. Additional studies on backcross and outcross matings also revealed a high incidence of positive tests. When subjected to extensive genetic analysis, the data from only two of the six breeding experiments supported the original hypothesis. Moreover, the results of two backcross experiments, which should be more reliable in predicting the segregation of genes than inbreeding, failed to support a single-gene theory.

An alternative hypothesis of SLE transmission through a series of interacting genes also seems unlikely. Multiple genes located on a single chromosome would have been inherited in exactly the same fashion as a single gene. If the genes had been located on different chromosomes, the frequency of the inherited trait would have been less than that of the single gene since the chance for random combination of different genes is inversely proportional to the number of genes involved. In the face of a very high incidence of the SLE marker, the results cannot be explained by any conventional genetic mechanism.

## Viral Experiments

To test the hypothesis that a viral agent associated with SLE might have been transmitted from one generation to the next, Lewis et al.[27] prepared cell-free filtrates from the spleens of SLE colony dogs that had positive LE cell tests and antinuclear antibody. These filtrates were passed through 220-$\mu$m filters and injected into newborn $CAF_1$ mice, W/Fu rats, and normal beagle puppies. A high percentage of normal mice and puppies, but not rats, receiving this material developed characteristic serologic abnormalities: antinuclear antibody, positive LE cell tests, and antibody to native DNA. Although the actual disease was not reproduced in these experiments, the development of antibody to N-DNA is a highly specific marker for SLE.

Further experiments on the transmissibility of antibodies to nuclear antigens yielded the following results. Newborn $CAF_1$ mice injected with cell-free filtrates prepared from lymphoid tissue of dogs with ANA develop ANA at about 9 months. Once detected, ANA persists and rises in titer. The latent period before ANA appears can be reduced to as little as 1 month by serial passage of spleen cells from ANA-positive mice into newborn syngeneic mice. Serial passage of viable lymphoid cells or cell-free filtrates from ANA-positive primary recipient mice induces the formation of ANA and antibodies to N-DNA in secondary and tertiary recipients. Cell-free filtrates prepared from the spleens of normal dogs, cats, rats, rabbits, guinea pigs, and mice have none of these effects when injected into newborn $CAF_1$ mice.

Apparently a replicating agent is associated with SLE which can be detected by the appearance of antibodies to nuclear antigens, particularly N-DNA. The agent was present in the original members of the SLE dog colony, and it has been transmitted vertically and perhaps horizontally to their progeny, which accounts for the high incidence of serologic abnormalities in the colony dogs. The host range of this agent includes mice, but not rats, and its presence in a susceptible host induces the formation of ANA.

Native DNA is nonimmunogenic, and numerous efforts to induce anti-DNA antibody in laboratory animals by experimental immunization procedures have not been successful. Yet antibody to N-DNA occurs regularly in one instance, SLE, regardless of the species affected (man, mouse, or dog). Further,

DNA/anti-DNA immune complexes mediate some of the lesions that characterize the disease, glomerulonephritis, vasculitis, and dermatitis. Thus anti-N-DNA antibody assumes a prominent role in the pathogenesis of SLE by forming circulating immune complexes. Therefore, if a transmissible agent (virus) is associated with SLE, its function must differ from that of more-conventional infectious agents, and its effects must in some way be associated with the development of anti-N-DNA antibody.[28–30]

## Activation of Murine Leukemia Virus

The development of antinuclear antibody in mice injected with canine cell-free filtrates is followed in some mice by the appearance of malignant lymphomas (lymphoblastic lymphoma, reticulum cell tumor, or myeloma) which contain murine leukemia virus (MuLV), as determined by electron microscopy, the presence of diagnostic cell surface antigens, an in vitro plaque assay, and mixed cell cytopathogenicity tests.[27] The association between the presence of ANA and lymphoma is specific for the dog-induced tumors and is not found in other MuLV-containing murine lymphoid tumors. Cell-free filtrates prepared from the dog-induced murine lymphomas will induce the formation of ANA and tumors in syngeneic mice and ANA alone in allogeneic mice or normal newborn puppies.

One of the tumors that developed in mice injected with cell-free filtrates from SLE dogs was a plasmacytoma that produced a monoclonal protein with immunologic specificity for native DNA. This tumor was established as a long-term tissue-culture cell line (SP-104). As had been observed in vivo, the cultured tumor cells (plasmablasts) produced not only large numbers of C-type virus particles but also a monoclonal IgA (k) immunoglobulin that had antibody specificity for native DNA (Ref. 31).

The C-type virus associated with the SP-104 plasmacytoma is a B-tropic virus with an AKR envelope coat. The virus cannot infect cells from heterologous species and thus has the characteristics of an ecotropic virus. Zenotropic virus production was not detected in these tumor cells. Purified virions from the SP-104 culture supernatant induced the development of antinuclear antibodies, including anti-DNA antibody, in 75% of recipient $CAF_1$ mice. In addition, these virions are associated with low oncogenicity, with only 5% of the $CAF_1$ recipients developing malignant lymphoma after being inoculated with the purified virus.

The SP-104 virions have also been analyzed by hybridization techniques and found to lack significant nucleotide sequences in common with RNA or DNA from dogs or human beings. However, SP-104 does contain nucleotides that hybridize with other well-characterized murine viruses.[32]

Evidently the original canine cell-free filtrates contained an agent, possibly a virus, capable of infecting murine cells in vivo. The presence of this agent conceivably activated latent endogenous murine leukemia virus, and this resulted in the production of antibodies to nuclear antigens and the development of lymphoid malignancies in the murine host.[33] The fact that one virus can activate another has been clearly shown with polyoma virus and lymphocytic choriomeningitis virus, both of which are capable of activating latent C-type RNA viruses in mice.[34,35]

## Viral Immunology of Canine SLE

Serologic evidence for the presence of C-type RNA virus in lymphocytes from dogs with SLE was studied in first-generation offspring from the SLE dog colony.[23] Purified lymphocytes from a 10-year-old female German shepherd with positive LE cell tests, ANA, and persistent proteinuria were used in an indirect immunofluorescent assay with three antisera: rabbit anti-SP-104 (murine) virion antibody, rabbit anti-P-30 (feline) antibody, and goat anti-P-30 (feline) antibody. Before use in the assay, the antisera were absorbed three times with purified lymphocyte preparations from normal dogs. Lymphocytes from six normal dogs failed to react with any of the reagents used, but lymphocytes from the dog with SLE stained with both of the antisera specific for viral antigen P-30 but not with the anti-SP-104 antiserum. Thus some circulating lymphocytes from this symptomatic animal possessed on the cell membrane an antigen that reacted with antisera prepared to the interspecies viral antigen P-30, an antigen common to all known mammalian C-type virions.

Shortly after these immunofluorescent assays, the SLE-affected dog developed clinical signs of the nephrotic syndrome. Pathologic analysis and immunofluorescent study of one kidney revealed morphologic and immunologic changes characteristic of membranous glomerulonephritis induced by immune complexes.

Although initial attempts to demonstrate the presence of viral antigen in the renal lesion with standard techniques were inconclusive, specific staining reactions were obtained with the rabbit and goat antisera to viral antigen P-30 after elution of frozen kidney sections with acid glycine buffer (pH 3.0). In addition, the glomerular lesion also reacted with the rabbit antisera to SP-104 virions. (The use of these reagents on frozen sections of normal dog kidneys, both with and without acid-glycine elution, produced no visible stain.) Presumably, viral antigen in the kidney resulted from the deposition of circulating immune complexes, whereas viral antigen on the membrane of circulating lymphocytes may well indicate intracellular presence of the virus or its nucleic acid.

An acid-glycine eluate was prepared from renal cortex and tested for antibody specificity to C-type RNA viral envelope antigen, N-DNA, and D-DNA by radioimmunoassay to demonstrate the presence of antibodies to viral antigen in the renal lesion. When tested against radiolabeled native and/or denatured DNA, the eluate consistently bound greater than 90% of the labeled nucleic acid in each assay. When tested against radiolabeled murine leukemia virus virions, the eluate bound 34% of the available label, an indication of a moderate amount of anti-envelope antibody activity. Prior incubation of the eluate with antisera to canine immunoglobulins reduced the binding values in all tests to background levels, and prior incubation of the eluate with purified native DNA, denatured DNA, or SP-104 virions specifically inhibited the binding in each radioimmunoassay. Thus antibodies with specificity for nucleic acid and viral antigen were detected in the eluate prepared from the kidney of a dog with SLE.

## Viral Immunology of Human SLE

The detection of C-type viral antigen in two different tissues of a symptomatic canine patient with SLE led to a similar study with human SLE (Ref. 34). Serum from a 12-year-old girl with active, untreated SLE was precipitated with ammonium sulfate, and the immunoglobulin-rich fraction was conjugated with fluorescein isothiocyanate. The resulting conjugate was then used to stain purified, living lymphocytes from other SLE patients. Live lymphocytes were incubated with the fluorescein conjugate for 30 min at 4°C, washed in buffered saline, fixed in 95% ethanol, and examined by dark-field fluorescent microscopy. Although lymphocytes obtained from six normal people and from six patients with diseases other than SLE failed to react with the conjugate, that SLE reagent did bind to the cytoplasmic membranes of lymphocytes from eight of ten patients with SLE. The pattern of fluorescence was typical of membrane staining [Fig. 12(a)]. If the stained-cell preparation was warmed to 37°C before fixation, the stained membrane determinants migrated to one pole of the cell [Fig. 12(b)]. The number of lymphocytes stained by the conjugate ranged from only 6 to 650 cells per $10^5$ lymphocytes. The reactivity of the reagent was abolished after incubation with the rabbit antihuman immunoglobulin serum. Immunoglobulins fractionated from normal human serum by ammonium sulfate precipitation and conjugated to fluorescein isothiocyanate did not bind to any lymphocytes from five patients with SLE. These experiments demonstrated that an antibody in the serum of an SLE patient reacts with a membrane determinant on a few lymphocytes from other patients with SLE. This antibody is not the anti-lymphocyte antibody described in SLE since it does not react with normal lymphocytes. It is not an anti-IgG that stains the antigen receptors of B lymphocytes for the same reason. Further, it does not stain phytohemagglutinin-transformed normal human lymphocytes, nor is it an antibody against the Epstein–Barr virus. Moreover, the determinant on the test lymphocyte is not an anti-IgG that might bind to the SLE reagent, because fluorescein isothiocyanate-conjugated normal IgG did not react with the test cells.

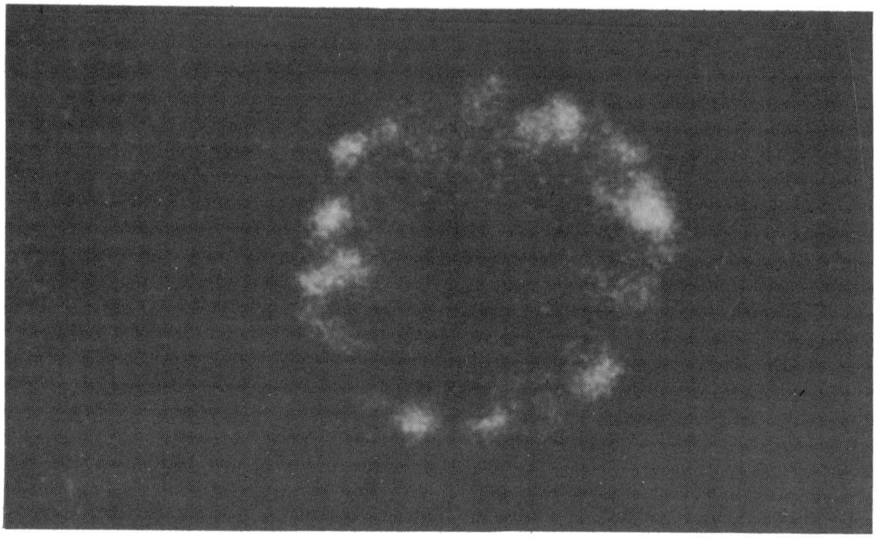

Fig. 12(a) Peripheral blood lymphocyte from a canine SLE patient. The cell was stained live at 4°C with a fluorescein conjugated reagent to C-type viral antigen (anti-SP-104). Discrete focal aggregates of viral antigen are present on the cytoplasmic membrane.

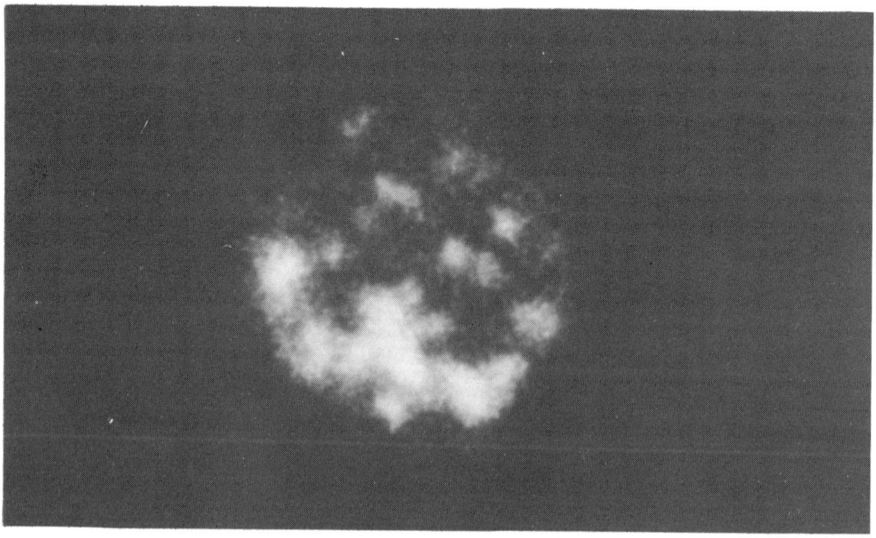

Fig. 12(b) Another lymphocyte from the patient shown in Fig. 12(a). This cell was stained as described in Fig. 12(a) and then warmed to 37°C. The stained membrane determinants have migrated to one pole of the cell, a process designated as polar capping.

The immunoglobulin fraction of serum from the patient with SLE was absorbed with double- and single-stranded DNA until DNA-binding antibodies were removed, as determined by radioimmunoassay, to further define the nature of this SLE antibody. The absorbed fraction, after conjugation with fluorescein isothiocyanate, still bound to the lymphocytes of other patients with SLE. In sharp contrast, absorption of the SLE reagent with a purified virus preparation from the supernatant of the SP-104 culture completely inhibited the staining reaction between the reagent and the lymphocytes of five of seven other SLE patients. The results of this inhibition study suggested that an antigen on the cytoplasmic membrane of lymphocytes from SLE patients shared antigenic determinants with the C-type virus particles produced by the SP-104 culture. Further, an antibody to this determinant was also present in serum of the untreated SLE patient. When the rabbit antibody to SP-104 virus was similarly applied to purified peripheral blood lymphocytes obtained from human SLE patients, typical membrane fluorescence was observed in five of the six SLE patients, and the proportion of lymphocytes stained by the rabbit antiserum was similar to that stained by the human SLE reagent. Lymphocytes from six patients with diseases other than SLE and twelve normal blood donors failed to react with the anti-SP-104.

Thus the reactivity in the SLE conjugate is abolished by absorption with C-type particles obtained from the unique SLE dog-induced murine lymphoma cell line that also produces antibody to native DNA, the antibody characteristic of SLE. In addition, the rabbit antiserum to the virions produced by that same murine cell line reacts with lymphocytes of patients with SLE in a manner identical to that of the SLE reagent. Consequently the lymphocytes of patients with SLE likely express a membrane antigen that was shared by the C-type RNA virus of the murine plasmacytoma. The presence of such a determinant might indicate the host cell "reservoir of infection" by this type of virus.

The relationship of C-type viral antigen to human SLE has been studied further by several laboratories. Antigen P-30 has been found in spleen extracts from two lupus patients[36] and in the glomerular lesion of a patient with SLE (Ref. 37). Antigens related to C-type virus have been identified in renal lesions from 11 lupus patients.[38] Further studies by Panem et al.[39] and Quimby and Schwartz[2] have verified the presence of a C-type viral antigen in the renal lesions of 32 of 33 human cases of SLE. Thus the initial observations in canine SLE have now been extended to and confirmed in human patient material, and the evidence for a viral role in the etiology of SLE continues to mount.

The demonstration of interspecies C-type viral antigen in the tissues of SLE-affected members of two cohabitant species (man and dog) has prompted another line of investigation related to viruses and SLE. Beaucher, Garman, and Condemi[40] studied two families with multiple cases of SLE in which their household dogs had clinical and serologic findings suggestive of SLE. Most important, the dogs had circulating antibodies to N-DNA. In one family the dog's illness antedated the diagnosis of human SLE, whereas in the other family the dog was acquired several years after the family members were affected. Observations such as this have led to more formal sero-epidemiologic studies, which may add to our understanding of the role played by C-type viruses in SLE.

## Multiple Factors of SLE Etiology

The combined evidence from investigations of all three affected species (man, mice, and dogs) points to a requirement for the interaction of three important etiologic factors in the development of SLE: (1) genetic susceptibility, (2) defective immunoregulation, and (3) activation or expression of C-type virus and/or its antigens. On the basis of continuing studies of the SLE dog colony, Quimby and Schwartz[2] have recently postulated that the biological events during the development of SLE may be explained by the existence of two classes of genes: (1) those which allow a general disposition to autoimmunity (Class I genes) and (2) those which determine the phenotype of the disease (Class II genes). Class I genes are related to immunoregulation, and they control the activities of T and B lymphocytes. Permissive alleles of these genes lead to the production of autoantibodies. Class II genes control the manifestation of the disease, e.g., rash, nephritis, and hemolytic anemia. Alleles of Class II genes would determine such things as the size of immune complexes, the functions of the complement system, and the nature of cellular reaction to inflammatory stimuli. Both classes of genes are complex and consist of multiple unlinked dominant and recessive components. Quimby and Schwartz further speculate that individuals with SLE have both classes of genes. Normal individuals may have either class but not both. Genes of Class II cannot be expressed in the absence of Class I genes. Type C virus could interact with either class or both classes of genes and thus modify their expression and subsequent biologic effects.

The dual system of interacting genes could explain how a canine dam with autoimmune hemolytic anemia, immunopathic thrombocytopenia purpura, and polyarthritis, when mated to a normal sire, could produce descendants with Sjögren's syndrome and lymphocytic thyroiditis;[25] why autoantibodies are found in the healthy relatives of canine and human patients with SLE; why family studies of human and canine SLE yield "erratic" results; and why multiple, apparently unrelated autoimmune diseases cluster in a single patient.

The genetic hypothesis presented by Quimby and Schwartz attempts in broad terms to provide an explanation for many of the phenomena observed in long-term studies of the lupus dog colony. Alternative explanations for some of these observations are possible, and arguments against the dual-gene theory are presented in their review.[2] Nevertheless, that working hypothesis should provide a conceptual base for future experiments. Hopefully, "man's best friend" will continue to serve as useful substrate in the search for the elusive cause(s) of SLE.

## REFERENCES


1. N. Talal, Disordered Immunologic Regulation and Autoimmunity, *Transplant. Rev.*, 31: 240-263 (1976).
2. F. W. Quimby and R. S. Schwartz, The Etiopathogenesis of Systemic Lupus Erythematosus, in *Pathobiology Annual*, Vol. 8, H. L. Ioachim (Ed.), pp. 31-65, Raven Press, New York, 1978.
3. R. M. Lewis, R. Schwartz, and W. B. Henry, Canine Systemic Lupus Erythematosus, *Blood*, 25: 143-160 (1965).
4. G. Trautwein, Autoantibodies and Autoimmune Diseases in Animals, *Dtsch. Tieraerztl. Wochenschr.*, 79: 280-284 (1972).
5. R. J. Slapendel, A. W. Kersjes, and A. Rijnberk, Canine Systemic Lupus Erythematosus Treated with Prednisone, *Zentralbl. Veterinaermed., Reihe A*, 19: 23-34 (1972).
6. R. D. Schultz, Immunologic Disorders in the Dog and Cat, *Vet. Clin. North Am.*, 4: 153-174 (1974).
7. J. W. Alexander, J. W. George, and J. V. Moffa, A Case of Systemic Lupus Erythematosus in a Dog, *Vet. Med. Small Anim. Clin.*, 70: 147-154 (1975).
8. D. J. Moore, Canine Systemic Lupus Erythematosus, the Disease, Clinical Manifestations and Treatment, *J. S. Afr. Vet. Assoc.*, 47: 267-275 (1976).
9. B. Utroska, Autoimmune Hemolytic Anemia Associated with Lupus Erythematosus, *Vet. Med. Small Anim. Clin.*, 71: 1247-1249 (1976).
10. S. H. Krum, G. H. Cardinet III, B. C. Anderson, and T. A. Holliday, Polymyositis and Polyarthritis Associated with Systemic Lupus Erythematosus in a Dog, *J. Am. Vet. Med. Assoc.*, 170: 61-64 (1977).
11. J. C. Monier, D. Schmitt, M. Perraud, C. Fleury, M. Gioud, and M. Lapras, Antibody to Soluble Nuclear Antigens in Dogs (German Shepherd) with a Lupus Like Syndrome, *J. Dev. Comp. Immunol.*, 2: 161-174 (1978).
12. R. M. Lewis, R. S. Schwartz, and C. E. Gilmore, Autoimmune Diseases in Domestic Animals, *Ann. N. Y. Acad. Sci.*, 124: 178-200 (1965).
13. R. M. Lewis, Spontaneous Autoimmune Diseases of Domestic Animals, *Int. Rev. Exp. Pathol.*, 13: 55-82 (1974).
14. R. W. Bull, Animal Models of Autoimmune Hemolytic Disease, *Semin. Hematol.*, 13: 349-353 (1976).
15. N. C. Pedersen, K. Weisner, J. J. Castles, G. V. Ling, and G. Weiser, Noninfectious Canine Arthritis: The Inflammatory, Nonerosive Arthritides, *J. Am. Vet. Med. Assoc.*, 169: 304-310 (1976).
16. R. M. Lewis and J. E. Hathaway, Canine Systemic Lupus Erythematosus; Presenting with Symmetrical Polyarthritis, *J. Small Anim. Pract.*, 8: 273-284 (1967).
17. J. S. Johnson, J. H. Vaughan, and S. N. Swisher, Canine Immunoglobulins. II. Antibody Activities in Six Immunoglobulin Classes, *J. Immunol.*, 98: 935-940 (1967).
18. F. W. Quimby, C. A. Smith, M. Brushwein, and R. M. Lewis, The Efficacy of Serodiagnostic Tests in the Recognition of Canine Immune Disorders, *Am. J. Vet. Res.*, (in press).
19. R. J. Wilkins, A. I. Hurvitz, and W. J. Dodds-Laffin, Immunologically Mediated Thrombocytopenia in the Dog, *J. Am. Vet. Med. Assoc.*, 163: 277-282 (1973).
20. R. M. Lewis, An Evaluation of the Clinical Usefulness of the LE Cell Phenomenon in Dogs, *J. Am. Vet. Med. Assoc.*, 147: 939-941 (1965).

21. R. M. Lewis and R. S. Schwartz, Canine Systemic Lupus Erythematosus. Genetic Analysis of an Established Breeding Colony, *J. Exp. Med.*, 134: 417-438 (1971).
22. R. M. Lewis, B. D. Stollar, and E. B. Goldberg, A Rapid, Sensitive Test for the Detection of Antibodies to DNA, *J. Immunol. Methods*, 3: 365-374 (1973).
23. R. M. Lewis, Evidence for a Virus in Canine Systemic Lupus Erythematosus, in *Experimental Models of Chronic Inflammatory Diseases*, L. E. Glynn and H. D. Schlumberger (Eds.), pp. 71-76, Springer-Verlag, Berlin, 1977.
24. R. M. Lewis and Y. Borel, Canine Rheumatoid Arthritis: A Case Report, *Arthritis Rheum.*, 14: 67-72 (1971).
25. F. W. Quimby, C. Jensen, D. Nawrocki, and P. Scollin, Selected Autoimmune Diseases in the Dog, *Vet. Clin. North Am.*, 8: 665-682 (1978).
26. R. M. Lewis, Specific Immunologic Lesions, in *Pathology of Laboratory Animals*, Vol. II, K. Benirschke, F. M. Garner, and T. C. Jones (Eds.), pp. 1953-1978, Springer-Verlag, New York, 1978.
27. R. M. Lewis, J. Andre-Schwartz, G. S. Harris, M. S. Hirsch, P. H. Black, and R. S. Schwartz, Canine Systemic Lupus Erythematosus. Transmission of Serologic Abnormalities by Cell-Free Filtrates, *J. Clin. Invest.*, 52: 1893-1907 (1973).
28. R. S. Schwartz, Viruses and Systemic Lupus Erythematosus, *New Engl. J. Med.*, 293: 132-136 (1975).
29. R. M. Lewis, Lessons Learned from Man's Best Friend: Canine Systemic Lupus Erythematosus, *Int. J. Dermatol.*, 16: 117-119 (1977).
30. R. M. Lewis and R. S. Schwartz, Canine Systemic Lupus Erythematosus: A Communicable Disease? in *Infection and Immunology in the Rheumatic Diseases*, D. O. Dumonde (Ed.), pp. 259-263, Blackwell Scientific Publications, Oxford, 1976.
31. J. A. Dixon, S. Sugai, and N. Talal, An Unusual Mouse Myeloma Protein Binding Native DNA, *Clin. Exp. Immunol.*, 19: 347-354 (1975).
32. F. W. Quimby et al., Characterization of a Retrovirus that Cross-Reacts Serologically with Canine and Human Systemic Lupus Erythematosus (SLE), *Clin. Immunol. Immunopathol.*, 9: 194-210 (1978).
33. R. S. Schwartz, *Autoimmunity, Malignancy, and Viruses*, paper presented at Baylor Medical School, Houston, Tex., Mar. 8, 1973.
34. R. M. Lewis, W. Tannenberg, C. Smith, and R. Schwartz, C-Type Viruses in Systemic Lupus Erythematosus, *Nature*, 252: 78-79 (1974).
35. M. B. A. Oldstone, T. Aoki, and F. J. Dixon, Activation of Spontaneous Murine Leukemia Virus-Related Antigen by Lymphocyte Choriomeningitis Virus, *Science*, 174: 843-845 (1971).
36. M. Strand and J. T. August, Type C RNA Virus Gene Expression in Human Tissue, *J. Virol.*, 14: 1584-1596 (1974).
37. R. C. Mellors and J. W. Mellors, Antigen Related to Mammalian Type C RNA Viral P-30 Proteins Is Located in Renal Glomeruli in Human Systemic Lupus Erythematosus, *New Engl. J. Med.*, 295: 470-475 (1976).
38. S. Panem, N. Ordonez, and W. Kirstein, C-Type Virus Expression in Systemic Lupus Erythematosus, *New Engl. J. Med.*, 295: 470-475 (1976).
39. S. Panem, N. G. Ordonez, A. I. Katz, B. H. Spargo, and W. H. Kirsten, Viral Immune Complexes in Systemic Lupus Erythematosus, Specificity of C Type Viral Complexes, *Lab. Invest.*, 39: 413-420 (1978).
40. W. M. Beaucher, R. H. Garman, and J. J. Condemi, Familial Lupus Erythematosus. Antibodies to DNA in Household Dogs, *New Engl. J. Med.*, 296: 982-984 (1977).

# Immune-Mediated Arthropathies of the Canine

# 13

NIELS C. PEDERSEN
*and*
ROY POOL
*Department of Medicine and Department of Pathology, School of Veterinary Medicine, University of California, Davis*

Over the last decade there has been a renewed interest in immunologically mediated joint disease of animals. These entities are common in humans but were previously thought to be nonexistent in animals.[1,2] Among the commonly studied veterinary species, the dog appears to be most like man in the diversity of its joint disorders.[3] Canine immunologic joint diseases include a rheumatoid-like arthritis,[4-10] ankylosing spondylitis,[11] arthritis of systemic lupus erythematosus,[12-16] and an idiopathic arthritis.[15]

Although there are some differences among veterinary researchers on how to classify canine arthropathies of apparent immune origin, we have chosen to classify them as follows:

- I. Inflammatory joint disease of apparent immune origin
  - A. Erosive (deforming)
    1. Canine rheumatoid arthritis
  - B. Nonerosive
    1. Idiopathic arthritis
    2. Arthritis associated with systemic lupus erythematosus
    3. Arthritis as a secondary manifestation of chronic infectious disease
    4. Enteropathic arthritis

## CANINE RHEUMATOID ARTHRITIS

Canine rheumatoid arthritis predominantly afflicts small or toy breeds of dogs, such as the Shetland sheepdog, toy poodle, Pekinese, Welsh corgi, miniature Doberman pinscher, and various small crossbreeds. The disorder

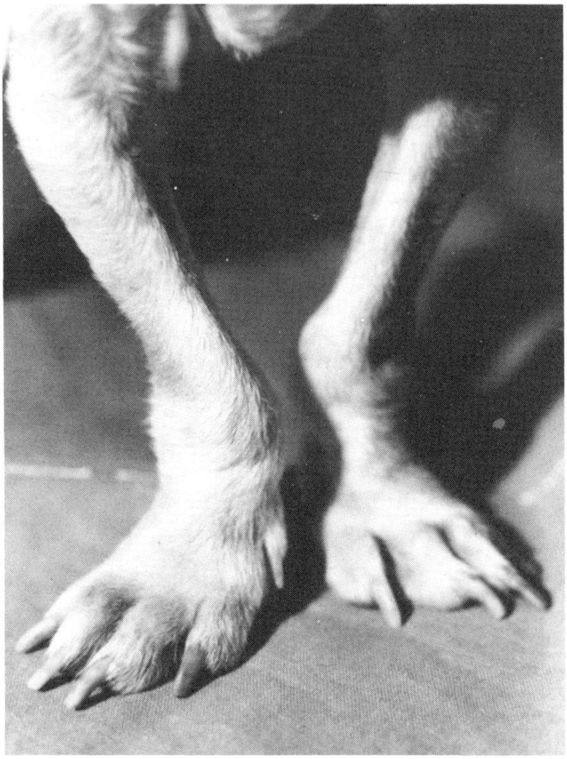

Fig. 1 Dog with rheumatoid arthritis. Notice joint swelling and angular deformities of carpal joints.

occurs more frequently in females and has an age at onset of from 1 to 8 years, with an average at around 4½ years. The disease is relatively uncommon, with only about two to four cases per 10,000 canine visits.

Canine rheumatoid arthritis is manifested initially as a shifting lameness that is often associated with a low-grade fever and a slight generalized lymphadenopathy. After several months the migratory nature of the joint disease becomes less pronounced, and a chronic, unremitting, generally symmetrical arthritis remains. Within several weeks to several months or more, deformities occur, usually in distal joints of the limbs, especially in the carpal and tarsal joints. These are manifested initially as angular deformities (Fig. 1) that eventually progress to more pronounced deformities or complete luxation of the joint (Fig. 2).

The earliest radiographic changes consist of periarticular soft-tissue swelling and periarticular osteoporosis (Fig. 3). Lucent cyst-like areas are frequently found in the subchondral bone (Fig. 4). After several weeks to months from the

Fig. 2 Dog with rheumatoid arthritis. Disease is quite advanced, and there are severe deformities of carpal joints.

onset of clinical signs, the prominent radiographic lesion consists of erosions at the margins of cartilage and subchondral bone and in the area of synovial attachments (Fig. 5). Associated with this process is a loss of articular cartilage manifested by a narrowing of the joint spaces (Figs. 4 and 5). In advanced cases (Fig. 4), this loss of articular cartilage and collapse of the underlying subchondral bone causes a characteristic mushrooming of the proximal ends of the metacarpal and metatarsal bones, and fibrous ankylosis occurs, particularly in the carpal and tarsal joints. Deformities, resulting from the destruction of the joint surfaces and increased laxity of the joint capsule and supporting ligament,

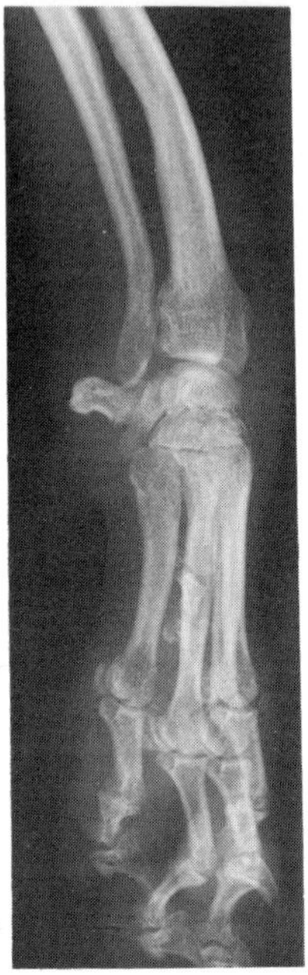

Fig. 3 Lateral radiograph of the distal forelimb of a dog with early rheumatoid arthritis. There is soft-tissue swelling around carpal joint, and moderate periarticular osteoporosis in the distal ulna and radius, carpal bones, and in the distal and proximal metacarpal and phalangeal bones.

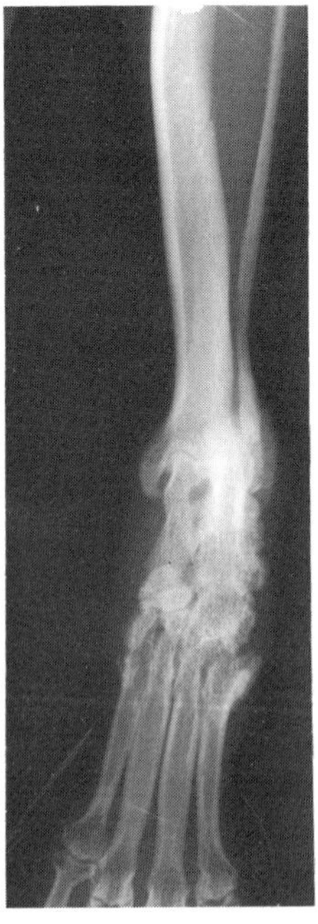

Fig. 4 Anterior–posterior radiograph of the left tarsal joint of a dog with rheumatoid arthritis. Notice severe periarticular osteoporosis, complete collapse of the tarsal and tarsometatarsal joint spaces, and mushrooming of the proximal ends of the metatarsal bones. Lucent cyst-like areas are evident in the subchondral bone.

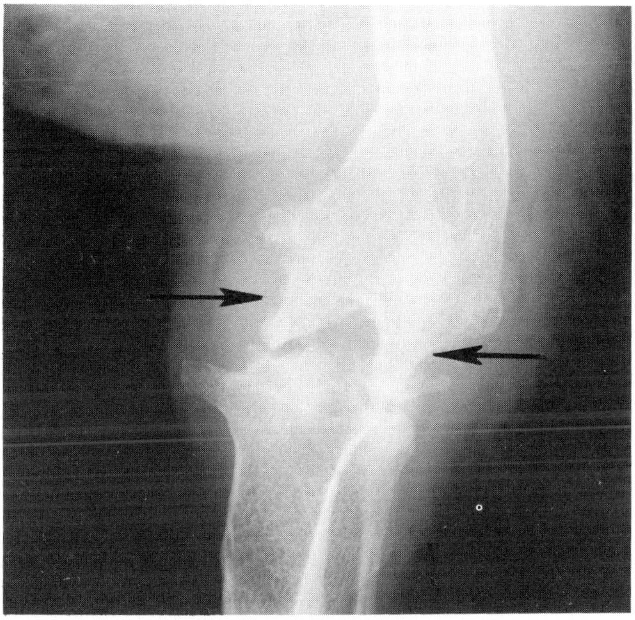

(a)

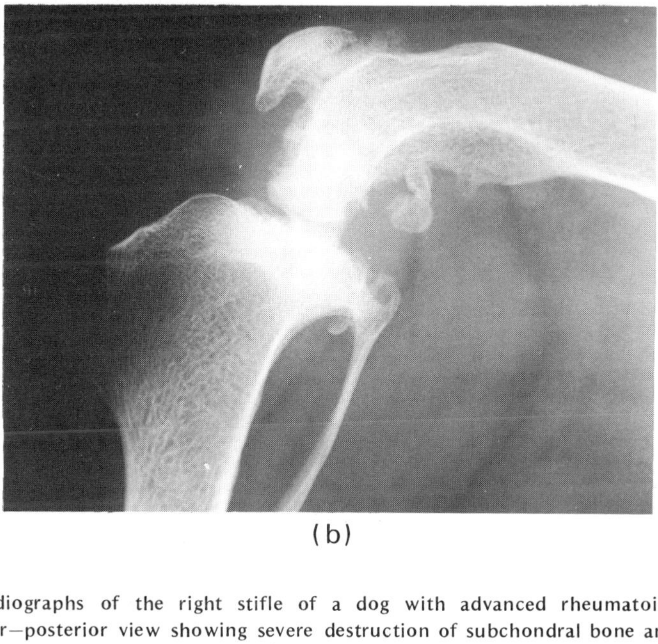

(b)

Fig. 5 Radiographs of the right stifle of a dog with advanced rheumatoid arthritis. (a) Anterior–posterior view showing severe destruction of subchondral bone and articular cartilage at the margins of the joints (arrows) and angular deformity of the articulation. (b) Lateral view of same joint, showing severe marginal erosion.

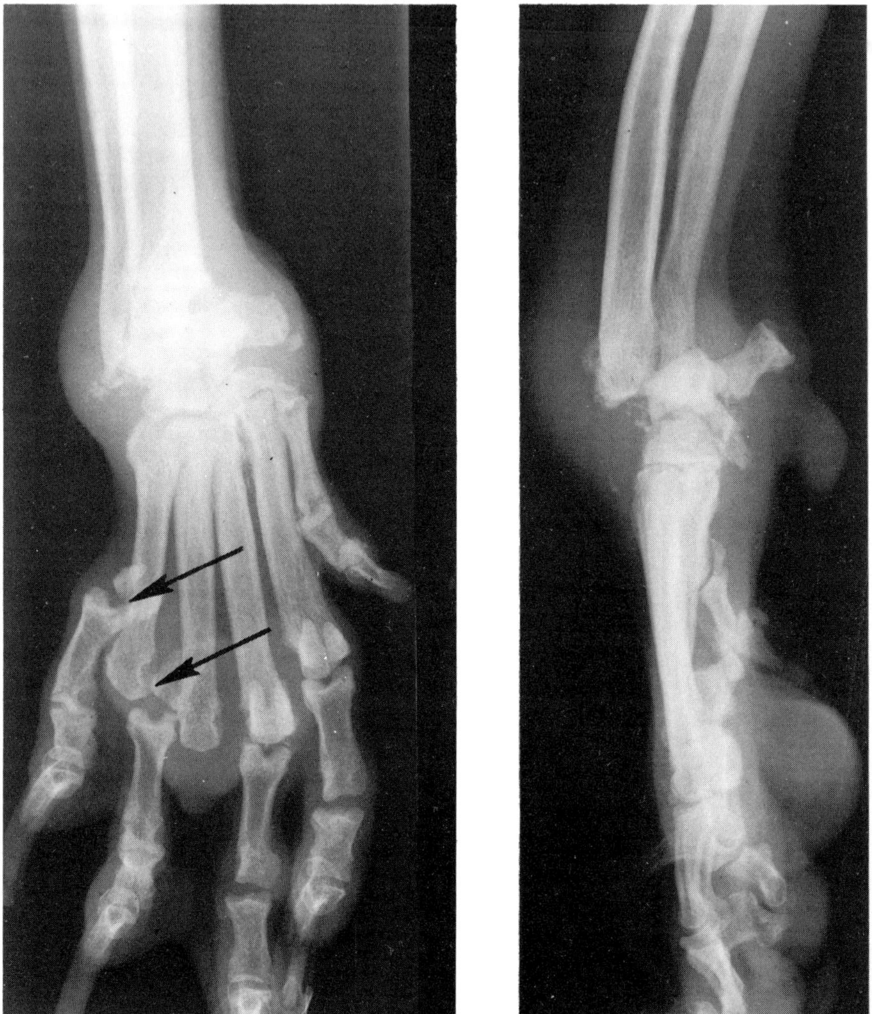

(a) (b)

Fig. 6 Radiographs of the carpal joint of a dog with rheumatoid arthritis. (a) Anterior–posterior view. The carpal joint spaces have collapsed, and there is luxation of the fourth and fifth metacarpophalangeal joints (arrows). The joint spaces of some of the interphalangeal joints are wider than normal. (b) Lateral view showing anterior luxation of the radius and ulna caused by destruction of the radiocarpal joint surfaces.

are usually most pronounced in the interphalangeal, carpal, and tarsal joints (Fig. 6). Radiographic changes are usually most pronounced in the carpus and tarsus and in the smaller joints of the digits, followed by the elbows and stifles. Radiographic changes are usually not discernible in the shoulder and hip joints until very late in the course of the disease. Involvement of the apophyseal joints of the cervical spine is relatively infrequent.

Clinicopathologic abnormalities associated with rheumatoid arthritis are diverse. The hematocrit is either normal or slightly below normal. The anemia, when present, is of the normocytic–normochromic type. Affected animals tend to have an elevated white blood cell count with an absolute neutrophilia. Plasma fibrinogen levels are elevated in about half of the dogs, and the corrected erythrocyte sedimentation rate is frequently accelerated. Concentrations of serum gamma globulin are either normal or slightly elevated. Alpha-2 globulin levels that are frequently elevated reflect the inflammatory nature of the disease. Serum albumin concentrations are depressed in some animals.

Serologic tests for the conventional IgM rheumatoid factor, best detected with sheep red blood cells or latex particles coated with dog or rabbit IgG, are positive in one-third to one-half of the dogs. Even when positive, the titers are rarely over 1 : 128 and are usually in the range of 1 : 16 to 1 : 32. Fluorescent antinuclear antibody tests are positive in a small percentage of dogs. Direct Coomb's antibody tests are negative. Serum complement levels are either normal or elevated. In contrast, synovial fluid complement levels are often depressed when adjustments are made for the relative serum protein content.

The number of white blood cells in the synovial fluid is significantly increased, with cell counts ranging from 4000 to 75,000 cells/mm$^3$. The increase in cells is attributed mainly to neutrophils, although absolute numbers of large and small mononuclear cells are also increased. The quality of the mucin clot formed after treatment of the synovial fluid with acetic acid is often diminished, as would be expected with an inflammatory reaction. When stained by the fluorescent antibody technique, from 1 to 20% of the synovial fluid leukocytes contain IgG inclusions (Fig. 7) which resemble those in "ragocytes" in human rheumatoid arthritis; in dogs the IgG inclusions are often in mononuclear cells. Unlike inhuman rheumatoid arthritis, C3 inclusions are rare in synovial fluid cells of dogs with rheumatoid arthritis.

Cultures of synovial fluid and synovial tissue are negative for bacteria, mycoplasma, chlamydia, and viruses. Virus-like particles have been found in the synovial tissue of a dog with a disorder resembling human juvenile rheumatoid arthritis.[17]

Early gross changes within the joint include increased amounts of a straw-colored turbid synovial fluid, congestion of the synovial membrane, and pitting and dullness of the cartilage along the margins of articular cartilage. In later stages the joint capsules of more severely affected diarthroses are thickened and fibrotic. Villous hypertrophy of the synovium is often grossly apparent (Fig. 8).

Histopathologic abnormalities of various degrees are found in all diarthroidal joints. Apophyseal and costovertebral joints of the spine show mild alterations; the most severe changes are in joints of the limbs, especially in the carpal and tarsal joints. In most instances the synovial lining cells are hypertrophic, forming

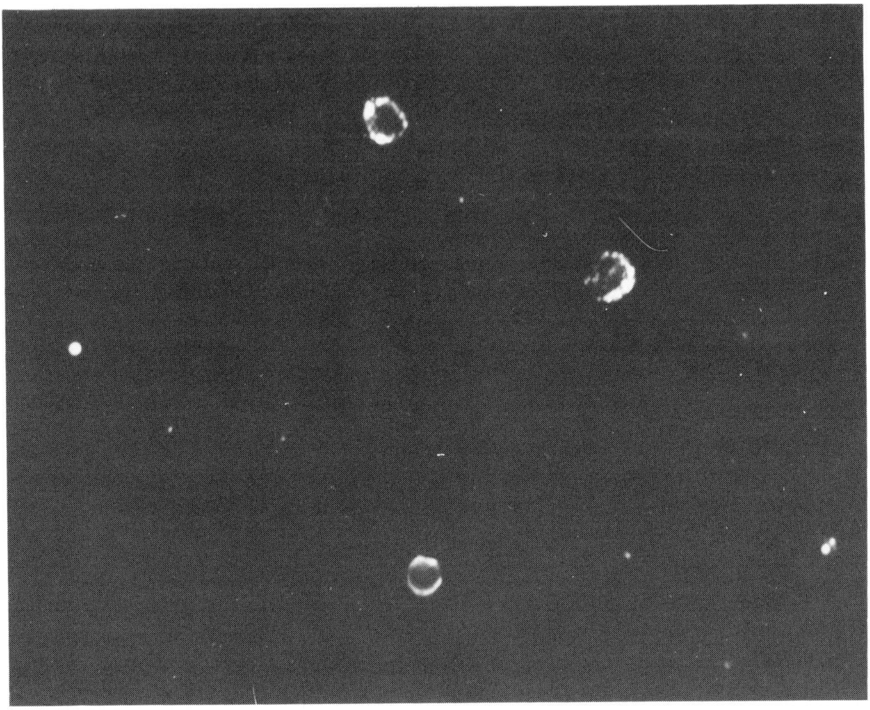

Fig. 7 Synovial fluid leukocytes from a dog with rheumatoid arthritis. The leukocytes have been fixed in acetone and stained with rabbit anti-dog IgG globulin–fluorescein isothiocyanate. Some mononuclear cells and, to a lesser extent, polymorphonuclear cells contain IgG inclusions. (Magnification, 450 x.)

a layer several cells thick (Fig. 9). Lining cells are large, ovoid, and hyperchromatic and have granular material in the cytoplasm. Clumps of fibrin are adherent to the synovial surface. The inflammatory response in the subsynovial connective tissue consists mainly of plasma cells and lymphocytes dispersed through the stroma and aggregated around small blood vessels (Fig. 9). Distinctive lymphoid follicles and foci of collagen necrosis are not found. Synovitis of similar microscopic appearance also affects many tendon sheaths in the distal portions of the limbs.

In advanced arthritic lesions the marked erosion of articular cartilage and loss of subchondral bone at the margins of joint surfaces are caused by granulation tissue arising from an inflamed synovium, extending across the borders of the articular surface as a pannus, and simultaneously undermining the bony support of the articular cartilage by eroding the subchondral bone (Fig. 10). Destruction of articular cartilage leads to fibrous ankylosis in the carpal and tarsal joints of some dogs. Cartilage is sometimes lost from central regions of the joint (Fig. 11). This loss is caused by granulation tissue arising in

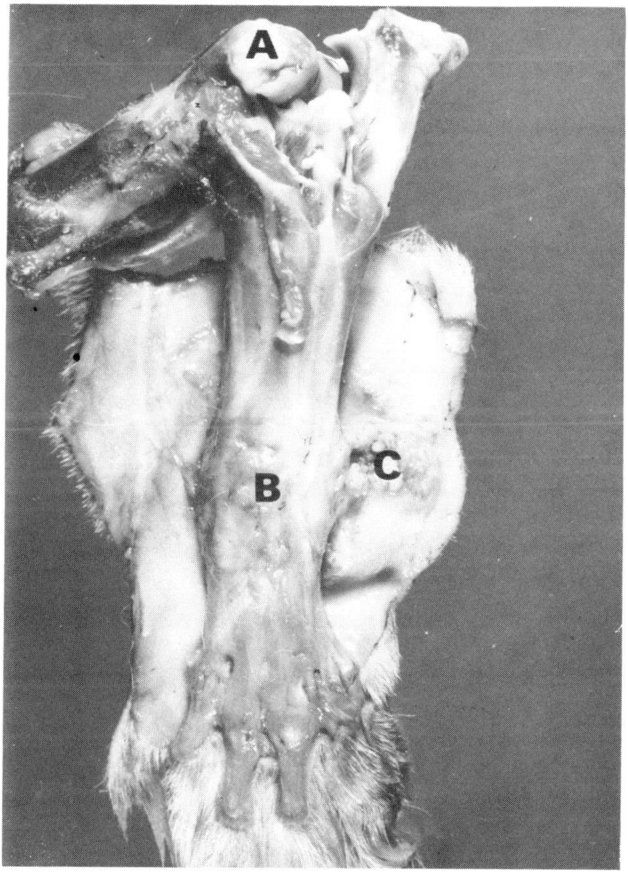


Fig. 8 Right forelimb of a dog with rheumatoid arthritis. Specimen has been partially skinned, and the elbow joint has been opened up. Note deep erosion of cartilage and bone of lateral condyle of humerus (A), ankylosed radiocarpal joint (B), and bursa-like structure with villous synovial lining arising from joint capsule (C).

the marrow cavity of the epiphysis and eroding through the subchondral bone. Blood vessels and granulation tissue then penetrate through the articular cartilage, reach the joint cavity, and spread across the articular surface as a pannus. Granulation tissue arising from the marrow cavity is responsible for the lucent cyst-like areas on radiographs of the subchondral bone.

Some clinicians advocate the use of salicylates to treat dogs with rheumatoid arthritis whereas others prefer glucocorticoids. In our experience the disease cannot be appreciably altered by salicylates. Glucocorticoids will bring about dramatic initial improvement, but with dogs, as with humans, the joint disease progresses in spite of therapy, albeit at a slower pace. In addition, progressively higher doses of glucocorticoids are needed to sustain the improvement; if the

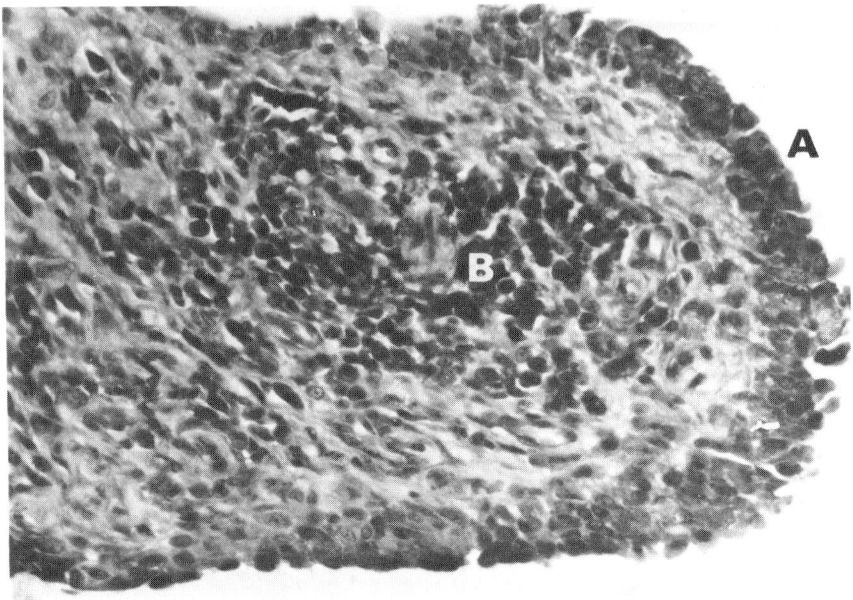


Fig. 9 Photomicrograph of a dog with rheumatoid arthritis, showing hyperplasia of synovial lining cells (A) and plasma cells and lymphocytes dispersed in stroma or aggregated around small blood vessels (B). (Hematoxylin and eosin stain; magnification, 370 x.)

drug is discontinued, the joint disease recurs more severely than if it had not been treated. A more effective treatment for canine rheumatoid arthritis is prednisolone, azathioprine, and cyclophosphamide used in combination to induce a remission; prednisolone alone is used to maintain remission.[9] A complete remission of clinical signs and synovial fluid abnormalities can usually be achieved in 2 to 9 months with this combination of drugs. Once in remission, the disease can usually be controlled by giving the dog prednisolone on alternating days. All drugs can be withdrawn if the disease has remained in complete remission for 3 to 6 months. About half the dogs will sustain their remission after the drugs have been withdrawn. Radiographic improvement of joint pathology occurs if treatment is begun when angular deformities are absent or minimal. Badly damaged joints often have to be treated with surgical fusions once the joint disease is brought under control. Secondary degenerative joint disease is a frequent sequela in dogs that have been successfully treated for the rheumatoid arthritis.

## IDIOPATHIC ARTHRITIS

Idiopathic arthritis is the most common of noninfectious, nonerosive, inflammatory joint diseases of dogs.[15] For example, in our clinic idiopathic

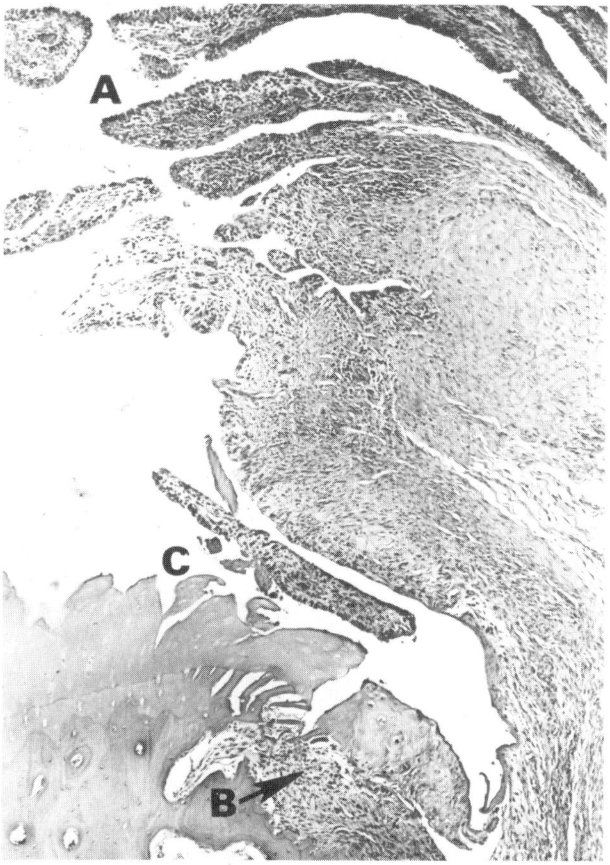


Fig. 10 Photomicrograph of specimen from stifle joint of a dog with rheumatoid arthritis. There is villous hypertrophy (A), destruction of subchondral bone by granulation tissue arising from inflamed synovium (B), and fibrillation of the articular cartilage (C). (Hematoxylin and eosin stain; magnification, 60 x.)

arthritis is about 25 times as prevalent as rheumatoid arthritis in canines. Idiopathic arthritis affects dogs as young as 6 months of age, but the average age of onset is about 4 years. German shepherds, Brittany spaniels, Great Danes, Doberman pinschers, and other setters and retrievers appear to be particularly predisposed. Unlike rheumatoid arthritis, idiopathic arthritis is less common in small breeds, but, when it occurs in small dogs, it is most often seen in toy poodles, Lhasa apsos, Yorkshire terriers, Welsh corgis, and other small crossbreeds.

The disease is usually polyarticular in nature, although pauciarticular involvement is also occasionally seen; the disease is rarely monoarticular. Signs of joint disease include generalized stiffness in dogs with pauciarticular or

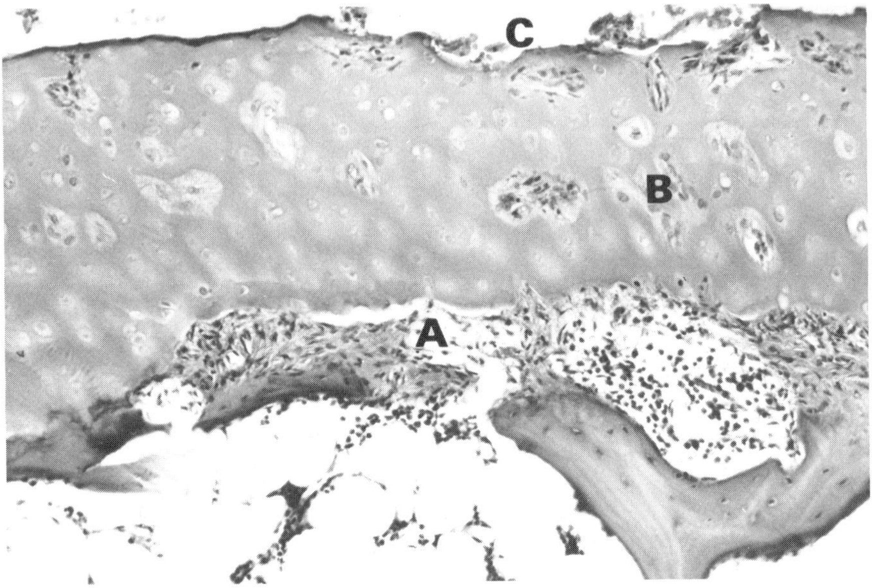


Fig. 11 Histologic appearance of cartilaginous defect in the trochlear groove of the femur of a dog with rheumatoid arthritis. Note destruction of subchondral bone by granulation tissue arising in the marrow cavity (A), extension of vessels through the articular cartilage (B), and destruction of articular cartilage by pannus originating in the marrow cavity (C). (Hematoxylin and eosin stain; magnification, 115 x.)

monoarticular involvement. Signs are often cyclic and occur concurrently with fever in most cases. Periarticular swelling, redness, and pain are sometimes apparent. As with rheumatoid arthritis, distal limb joints are most severely involved.

The only radiographic change is periarticular soft-tissue swelling. Some new periosteal bone is formed after many months of disease in some dogs.

Clinicopathologic abnormalities include leukocytosis, neutrophilia, hyperfibrinogenemia, and elevated erythrocyte sedimentation rate. Serologic tests for antinuclear antibodies are negative. Serum and synovial fluid complement ($CH_{50}$) levels are normal.

Synovial fluid from dogs with idiopathic arthritis is indistinguishable from the fluid from dogs with other noninfectious and infectious inflammatory arthropathies. Synovial fluid leukocytes usually do not contain inclusions with IgG and C3. Cultures of synovial fluid and synovial membrane are negative for bacteria, mycoplasma, chlamydia, or viruses.

Histopathologic abnormalities in the synovial membrane of dogs with idiopathic arthritis differ from those in dogs with rheumatoid arthritis. The synovial membrane in dogs with idiopathic arthritis is mildly hyperplastic, and the primary infiltrating cells in the superficial layers of the synovium are neutrophils. The synovium may be covered with a thin layer of fibrin exudate containing entrapped neutrophils and red blood cells. There is hyperemia of superficial blood vessels. The few plasma cells and lymphocytes tend to form small aggregates around blood vessels deep in the synovium.

The condition is treated successfully with glucocorticoid alone in more than half the dogs. High levels (2 mg per kilogram of body weight) of prednisolone are given daily for the first several weeks. Gradually the dose is decreased, and the drug is given on alternate days once remission has occurred. Drug therapy is discontinued after the disease has been in remission for 3 to 6 months. About half the dogs will not have a recurrence of disease. Dogs that fail to respond well to prednisolone or require high chronic levels of the drug (>1 mg/kg daily on alternating days) can be treated with lower dosages of prednisolone in combination with cyclophosphamide and/or azathioprine. Once in remission, the disease usually can be contained with prednisolone alone.

## CANINE SYSTEMIC LUPUS ERYTHEMATOSUS

The clinical manifestations of systemic lupus erythematosus (SLE) in canines are very similar to those in humans.[15] One of the most frequent clinical manifestations of canine SLE is a nonerosive arthritis, which is usually polyarticular or pauciarticular. The joint disease may be the sole manifestation of disease or may be part of a systemic disease that includes autoimmune hemolytic anemia, thrombocytopenia, skin disease, glomerulonephritis, polyneuritis, polymyositis, or myelopathy.[13,15]

When arthritis is the sole or predominant manifestation of SLE, the overall disease resembles idiopathic arthritis except that antinuclear antibody can be found in the serum of dogs with SLE. In fact, both disorders tend to occur in the same breeds of dogs, and idiopathic arthritis possibly is a seronegative variant of SLE. Radiographic changes and histologic changes in the synovium are identical in dogs with SLE and idiopathic arthritis. Serum and synovial fluid complement levels usually are normal except in cases with neuromuscular involvement in which serum complement levels often are depressed.

Theories on the etiology and pathogenesis of canine SLE, its similarities to human SLE, and descriptions of its other systemic features are covered in detail in Chapter 13.

## ARTHRITIS SECONDARY TO CHRONIC INFECTIOUS DISEASE

A nonerosive type of arthritis is sometimes seen in dogs with a number of different types of chronic infectious diseases,[15] including subacute bacterial endocarditis, chronic otitis externa and media, follicular vaginitis, pyometra, chronic actinomyces granulomas (foxtail grass awn migration or other plant material), and chronic fungal infections, such as coccidioidomycosis. Nonerosive arthritis also has been found in a dog with an X-linked bactericidal dysfunction of peripheral leukocytes. The disease is usually monoarticular or pauciarticular, the carpal and tarsal joints are most frequently involved, and synovial fluid changes and synovial membrane histopathology resemble those found in the arthritis of systemic lupus erythematosus or idiopathic arthritis.

The role of the overlying infectious process in the joint disease is unknown, but it is likely an immune complex manifestation. The possibility that organisms are present in the synovial tissue cannot be excluded completely despite negative cultures from the synovial tissue and fluid.

## ENTEROPATHIC ARTHRITIS

A nonerosive arthritis has been found in some dogs with ulcerative colitis[15] but has not been well documented. Synovial fluid changes are identical to those of other inflammatory arthropathies of the dog.

## CONCLUSIONS

Experimental models that closely resemble human immune-mediated joint diseases are not available, but naturally occurring joint diseases of dogs may prove valuable. The most valuable model would be canine rheumatoid arthritis, which closely resembles the human disease. Unfortunately the rarity of naturally occurring canine rheumatoid arthritis limits the number of dogs available for research. Canine SLE is already a model for human SLE, and the interest in naturally occurring canine SLE as a model for SLE of man is growing. Idiopathic polyarthritis of dogs is the most common and therefore potentially the most valuable model to use in studies of human joint disease provided that the human analogue can be identified.

The greatest current value of naturally occurring cases of canine rheumatoid arthritis, SLE, and idiopathic arthritis is probably in the area of pharmacology and therapeutics. Drugs and combinations of drugs that would be difficult to use in man can be tested in dogs. Indeed, the clinical use of potent immunosuppressive drugs to control rheumatoid arthritis is probably more advanced in dogs than in man.

Selective breeding studies initiated with dogs affected with joint disease not only would be beneficial in determining the heritability of these disorders but also could suggest means for breeding large numbers of dogs with immune-mediated joint disease and thus open the way for more detailed laboratory studies.

## REFERENCES


1. C. M. Pearson, Arthritis in Animals, in *Arthritis and Allied Conditions*, J. L. Hollander and D. J. McCarty (Eds.), pp. 195-207, Lea and Febiger, Philadelphia, Pa., 1972.
2. J. Sokoloff, Comparative Pathology of Arthritis, *Adv. Vet. Sci.*, 6: 194-250 (1960).
3. N. C. Pedersen and R. Pool, Canine Arthritis, *Vet. Clin. North Am.*, 8: 465-493 (1978).
4. R. E. W. Halliwell, R. B. LaVelle, and K. M. Butt, Canine Rheumatoid Arthritis—A Review and Case Report, *J. Small Anim. Pract.*, 13: 239-243 (1972).
5. R. M. Lewis and Y. Borel, Canine Rheumatoid Arthritis—A Case Report, *Arthritis Rheum.*, 14: 67-74 (1971).
6. S. Liu, P. F. Suter, C. A. Fisher, and H. D. Dorfman, Rheumatoid Arthritis in a Dog, *J. Am. Vet. Med. Assoc.*, 154: 495-502 (1969).
7. C. D. Newton, A. J. Lipowitz, R. E. Halliwell, D. N. Biery, and H. L. Allen, Canine Rheumatoid Arthritis, in *25th Annual Gaines Veterinary Symposium*, Fort Collins, Colo., Oct. 23, 1975, pp. 17-25.
8. C. D. Newton, A. J. Lipowitz, R. E. Halliwell, H. L. Allen, D. N. Biery, and H. R. Schumacher, Rheumatoid Arthritis in Dogs, *J. Am. Vet. Med. Assoc.*, 166: 113-121 (1976).
9. N. D. Pedersen, R. R. Pool, J. J. Castles, and K. Weisner, Noninfectious Canine Arthritis, Rheumatoid Arthritis, *J. Am. Vet. Med. Assoc.*, 169: 295-303 (1976).
10. R. L. Rudy, R. B. Hohn, and J. B. Harrison, Rheumatoid and Osteoarthritis in the Dog, in *22nd Annual Gaines Veterinary Symposium*, Tuskegee, Ala., Sept. 17, 1972, pp. 27-32.
11. D. Sikes, F. A. Hayes, A. K. Priestwood, and J. F. Smith, Ankylosing Spondylitis and Polyarthritis of the Dog: Physiopathologic Changes of Tissues, *Am. J. Vet. Res.*, 31: 703-712 (1970).
12. R. M. Lewis and J. E. Hathaway, Canine Systemic Lupus Erythematosus Presenting with Symmetrical Polyarthritis, *J. Small Anim. Pract.*, 8: 273-284 (1967).
13. R. M. Lewis, R. S. Schwartz, and W. B. Henry, Canine Systemic Lupus Erythematosus, *Blood*, 25: 143-160 (1965).
14. G. V. Ling and O. W. Schalm, The L. E. Cell Phenomena in the Dog, *Calif. Vet.*, 24: 20-25 (December 1970).
15. N. C. Pedersen, K. Weisner, J. J. Castles, G. V. Ling, and G. Weiser, Noninfectious Canine Arthritis: The Inflammatory Nonerosive Arthridities, *J. Am. Vet. Med. Assoc.*, 169: 304-310 (1976).
16. R. J. Slappendel, A. W. Kersjes, and A. Rijnberk, Canine Systemic Lupus Erythematosus Treated with Prednisolone, *Zbl. Vet. Med. A.*, 19: 23-24 (1972).
17. C. D. Newton, H. L. Allen, R. E. Halliwell, and H. R. Schumbacher, Jr., Clinicopathologic Conference: Juvenile Rheumatoid-Like Arthritis, *J. Am. Vet. Med. Assoc.*, 165: 459-464 (1974).

# 14

# Canine Cyclic Hematopoiesis

R. D. LANGE
*and*
J. B. JONES
*Department of Medical Biology, University of Tennessee Memorial Research Center and Hospital, Center for the Health Sciences/Knoxville Units, Knoxville, Tennessee*

In 1958 a short clinical note in the veterinary literature described the early death of a silver-gray English collie.[1] Ten years later, Lund, Padgett, and Ott[2] described canine cyclic neutropenia as an inherited, semilethal condition of collies. The coat color of the affected dogs was gray with white markings. It is now well established that these hematologic aberrations are inherited in a classic autosomal recessive pattern and are accompanied by dilution of all pigmented hair or skin so that black areas appear charcoal and sable areas appear silver-gray or light tan.[3,4]

A colony of dogs genetically heterozygous for this condition was established[4] at the University of Tennessee Memorial Research Center in 1971. Dr. John Lund supplied a heterozygous male collie–beagle hybrid as part of the foundation stock. This male was produced from a successful crossbreeding of a homozygous gray collie male to a beagle bitch. The collie–beagle male was mated to collie bitches heterozygous for the gray gene, and a rigid Mendelian inheritance pattern was observed (Fig. 1). Puppies with the diluted coat color invariably exhibited cyclic neutropenia, an indication either of multiple effects from a single gene or of closely linked alleles.

The laboratory-maintained colony provided data on birth weights, neonatal histopathology, and disease incidence as well as the opportunity to study hematopoiesis throughout the life of the dogs. Individual birth weights among affected puppies (Fig. 2) were highly variable. Usually pups weighing half, or less than half, the weight of their smallest normal littermate died within a few days of birth.[4] These lightweight puppies may exemplify a more extensive expression of the genetic defect. Histologically, their tissues exhibited physiologic immaturity. Whenever the birth weights of the puppies approached those of the normal

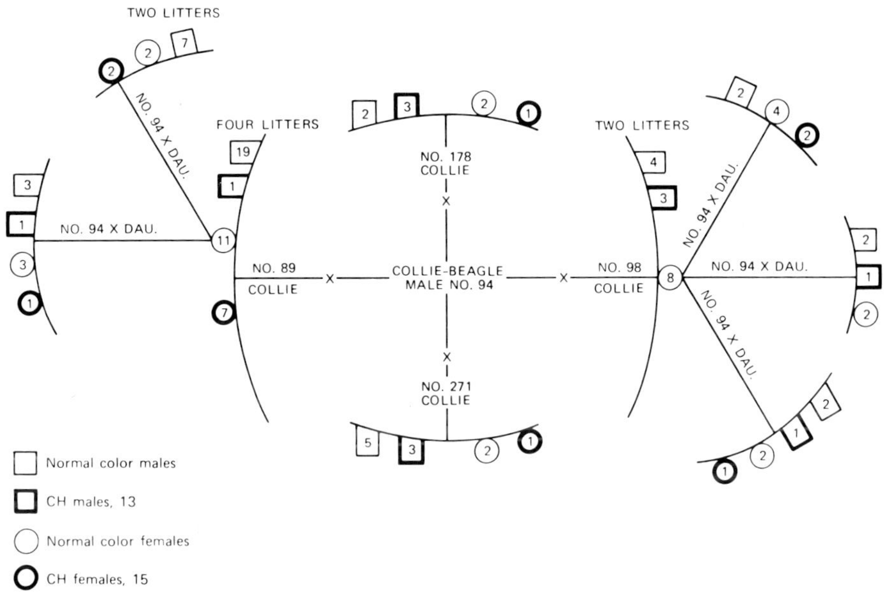


**Fig. 1 Family relationships and members affected with cyclic hematopoiesis (CH) among 110 offspring of male No. 94. All parents were heterozygous for CH. [From J. Jones, R. Lange, and E. Jones, Cyclic Hematopoiesis in a Colony of Dogs,** ***J. Am. Vet. Med. Assoc.,*** **166: 365 (1975).]**

littermates, if optimal care was provided and litter size was reduced to decrease nursing competition, survival rates were high. Leukocyte counts were started on the 8th day of life, and periodic neutropenia in affected pups was noted within the following 12 to 14 days. The first neutropenic episode almost certainly occurs within the first 12 days of life. Infections of the umbilicus, lungs, and even septicemia often occur in gray newborns. Also, these puppies often have gingivitis, conjunctivitis, and arthritic manifestations during neutropenic episodes. However, many newborn gray puppies grow vigorously with no signs of disease, and the dam's colostral antibodies and a clean environment may account for early life vigor. As the puppy approaches weaning and its passive immunity wanes, enteritis, pneumonia, and pyogenic abscesses become increasingly common.

Maintaining and rearing cyclic neutropenic puppies requires careful nursing and constant vigilance against infectious diseases. Broad-spectrum antibiotics are used in cases of obvious infections. Immunizations against canine distemper–hepatitis are administered as for normal puppies, although their antibody responses to these antigens have not been evaluated. Parasite control must be rigid because of the predisposition to enteritis during neutropenic episodes, and

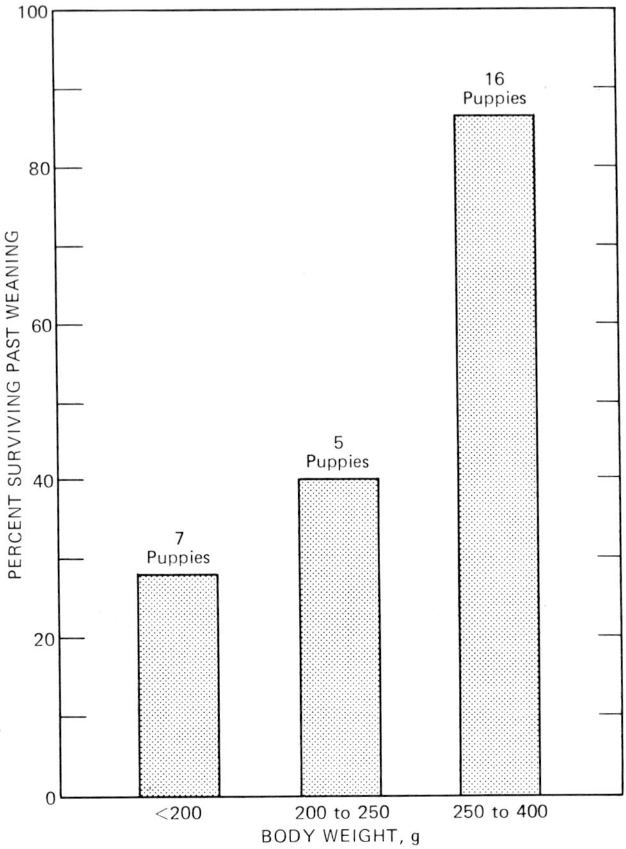


**Fig. 2 Survival percentages of 28 CH pups of three different birth weight categories. [From J. Jones, R. Lange, and E. Jones, Cyclic Hematopoiesis in a Colony of Dogs,** *J. Am. Vet. Med. Assoc.,* **166: 366 (1975).]**

rigorous sanitation must be practiced or most of the puppies will die during the first 14 weeks of life.

Peripheral blood cell counts in early and adult life have been published in several reports.[2,5-8] A neutrophilia following the neutropenia occurs in many, but by no means all, of the cyclic dogs, and this inconsistency is probably related to intercurrent infections. Furthermore, the reticulocytosis reported in affected puppies becomes less marked in the mature dog, which no longer has an expanding erythron mass. Such inconsistencies must be considered whenever the dogs are used as models of hematologic diseases and cellular proliferation. Peripheral hemograms show (1) a profound recurrent neutropenia with 11- to 14-day intervals, (2) a monocytosis peaking just as the neutropenia abates, (3) a variable thrombocytosis occurring 1 to 2 days before the neutropenia, and (4) a moderate reticulocytosis (1.5 to 10%) occurring during the neutropenia. The first day the peripheral neutrophil count falls below 1600 neutrophils/$mm^3$ is the most consistent change in the hemogram and therefore is referred to as day 1 of the cycle. With this reference point, successive cycles and cycles among several dogs can be compared. Since all three major cell types cycle, the disease is probably more correctly labeled canine cyclic hematopoiesis (CH).

Yang, Dale, and Jones[9] have shown that the peripheral blood of CH dogs spontaneously incorporated tritiated thymidine ($^3$H–TdR) in a cyclic pattern that varied inversely with daily total leukocyte counts. This cyclic change was thought to represent an influx of immature leukocytes from the bone marrow at the nadir of neutropenia. Recently Angus, Wyand, and Yang[10] found variable responses of lymphocytes of CH dogs to phytohemagglutinin—a T-lymphocyte mitogen. In general, the lymphocyte responses of CH dogs were reduced when compared with those of normal dogs, but one had a decreased response when the polymorphonuclear neutrophil (PMN) count was normal but a markedly increased response during neutropenic periods. This finding points to a possible heterogeneity in the ability of the lymphoid system to compensate for the deficit in peripheral neutrophils.

In CH dogs platelet sizes varied inversely with platelet counts.[11] Platelet adhesiveness and clot retraction values were below normal levels throughout the cycle and were not influenced by neutropenia.[12] Results of clot retraction studies, in which platelets and plasma of CH dogs, normal dogs, and CH dogs transplanted with normal dog marrow were studied in all possible combinations, indicated that a plasma factor was involved in the decreased platelet function of CH dogs.

Fluctuations in peripheral blood cell values sometime suggest that cellular life spans may be shortened. A single report[8] describing PMN life-span studies indicates that neutrophil life span is normal or very near normal.

Although the affected puppies generally are mildly anemic,[13] we have had several mature affected dogs whose erythrocyte values were within normal ranges. Erythrocyte production probably can fulfill demands of a mature dog whose total erythrocyte mass is not expanding, whereas demand probably exceeds production in the rapidly growing puppy. This theory is compatible with findings of a normal erythrocyte life span and iron use but a recurrent, transient failure to produce erythrocytes.[14] As expected, the serum iron of affected dogs peaked when the marrow was in its myeloid phase and reached its nadir during the erythroid phase.

Amyloidosis occurs in CH puppies that have not reached puberty.[15,16] The amyloid is histologically like that of secondary amyloidosis of humans. Many major organs may be affected, and dogs reaching maturity may have renal amyloidosis sufficient to create azotemia and a concurrent anemia along with a lack of erythropoietic stimulation. This complication is important to many hematologic studies. The cause of the amyloidosis is speculative but may involve some peculiar sensitivity of the lymphoid system. Machado et al.[16] believe these dogs may be a useful model for the study of human amyloidosis.

Sixty percent of dogs with CH have bone and joint reactions that seem to be unrelated to infection. The lesions appear radiographically as dystrophic bone formation and histologically as an agenesis with inflammatory infiltrates.

The dilute pigmentation in gray collies results from diminished formation of melanin from tyrosin in melanocytes.[17] A relationship between canine neutropenia and the merle gene, however, seems unlikely.[3] Yang[18] has recently observed that gray collies, which had been transplanted with normal bone marrow 2 years previously, showed normal coloration of the hair coat on the tails and on certain other body areas. Yang speculated that a "super stem cell" in the donor bone marrow, the mutagenic effect of radiation, or the occurrence of metabolic cooperation between donor cells and recipient hair-follicle cells might be involved in this unexpected bone-marrow transplantation effect.

Bone-marrow morphology in CH dogs has been studied in a number of laboratories. Lund[19] found (1) a decrease in the relative number of immature granulopoietic cells prior to the decrease in PMN's in the marrow; (2) an increase in the relative number of myelocytes, promyelocytes, and myeloblasts when the number of PMN's was low; and (3) a decrease in the relative number of erythropoietic cells. In 1972 Dale et al.[8] found a single wave of myeloid proliferation in their dogs during each cycle. They noted a proliferation of PMN precursors before the return of PMN's in the bone marrow. When the peripheral count of PMN's was normal, the percentage of myeloid precursors declined. Bone-marrow reserves of neutrophils were found to rise and fall with each cycle. Marrow erythroid precursors, particularly normoblasts, fluctuated with the cycle.

Scott et al.[20] studied the ultrastructure of bone marrow of six gray collies and confirmed that neutrophilic granulocyte differentiation was a synchronized event. They divided the bone-marrow changes of neutrophil differentiation and maturation into five phases as follows:

1. Days 1 to 3, stem cells, erythrocyte precursors, and immature neutrophils predominate.

2. Days 4 to 5, primary granule production and granule condensation; stem cells and immature neutrophils continued to develop.

3. Days 6 to 7, initiation of specific granule production associated with development of less-mature neutrophil granulocytes.

4. Days 7 to 9, cessation of stem-cell differentiation; evidence of cellular injury with continuation of secondary granule production and beginning of granule condensation.

5. Days 10 to 12, no further neutrophil differentiation but continued maturation of the remaining neutrophil granules.

Morphologic studies on daily bone-marrow aspirates from six CH dogs in our colony have shown that myeloid cells compose the majority of the bone marrow during days 3 through 10 (Figs. 3 and 4). The wave of myelopoiesis starts with myeloblasts; by day 8 to day 9, the majority of cells become mature bands and segmented forms. From day 10 through day 14 and on days 1 and 2, the majority of the cells are erythroid with the myeloid series represented by immature, undifferentiated cells.

Marrow transplantation studies have been performed in three laboratories. Dale and Graw[21] transplanted bone-marrow cells from a normal dog into a gray collie with resultant normal granulopoiesis in the CH recipient. Weiden et al.[22] exposed two normal collie dogs to 1200 R total body irradiation by $^{60}Co$ sources and then infused bone marrow from CH littermates. The recipients subsequently showed regular fluctuations of granulocyte and reticulocyte counts. In our laboratory, four pairs of dogs were used for transplantation studies.[23,24] After being irradiated with 1250 R by $^{60}Co$ sources, two normal dogs were engrafted with compatible CH bone marrow and two CH dogs were engrafted with normal bone marrow. No further cycles of neutropenia were found in the CH dogs following transplantation with normal bone marrow; platelets and reticulocytes ceased to fluctuate cyclically. The normal dogs that received CH marrow showed cyclic rises and falls in the number of platelets, reticulocytes, and neutrophils. The normal dogs transplanted with CH marrow became anemic, whereas the hematocrits of CH dogs transplanted with normal bone marrow approached normal levels[25] (Fig. 5).

## HUMORAL STIMULATORS

One of the possible pathogenic mechanisms for CH is through aberrant hormonal regulatory function of the long-range humoral regulators. In 1971 Dale et al.[26] reported that a factor for the stimulation of granulocytic colony growth appeared cyclically in the urine in gray collies. The peak colony stimulating activity (CSA), as measured by the effect of test materials on mouse bone marrow in solid-phase methyl cellulose cultures, developed during the neutropenic phase and was not detectable when the blood neutrophil counts reached normal levels. Yang et al.[27] found that bone-marrow cells from normal dogs, treated with serum from CH or normal dogs, produced large numbers of colonies in soft agar. The largest number of colonies was formed when the test serum was obtained from CH dogs at the nadir of their neutropenic episodes. Serum CSA was inversely related to peripheral neutrophil counts, which suggests a negative feedback control.

Studies have also related serum erythropoietin (Epo) activity to cyclic hematopoiesis. Adamson, Dale, and Elin[28] demonstrated definite cycling of Epo after inducing a profound anemia by phlebotomy in gray collies. In other

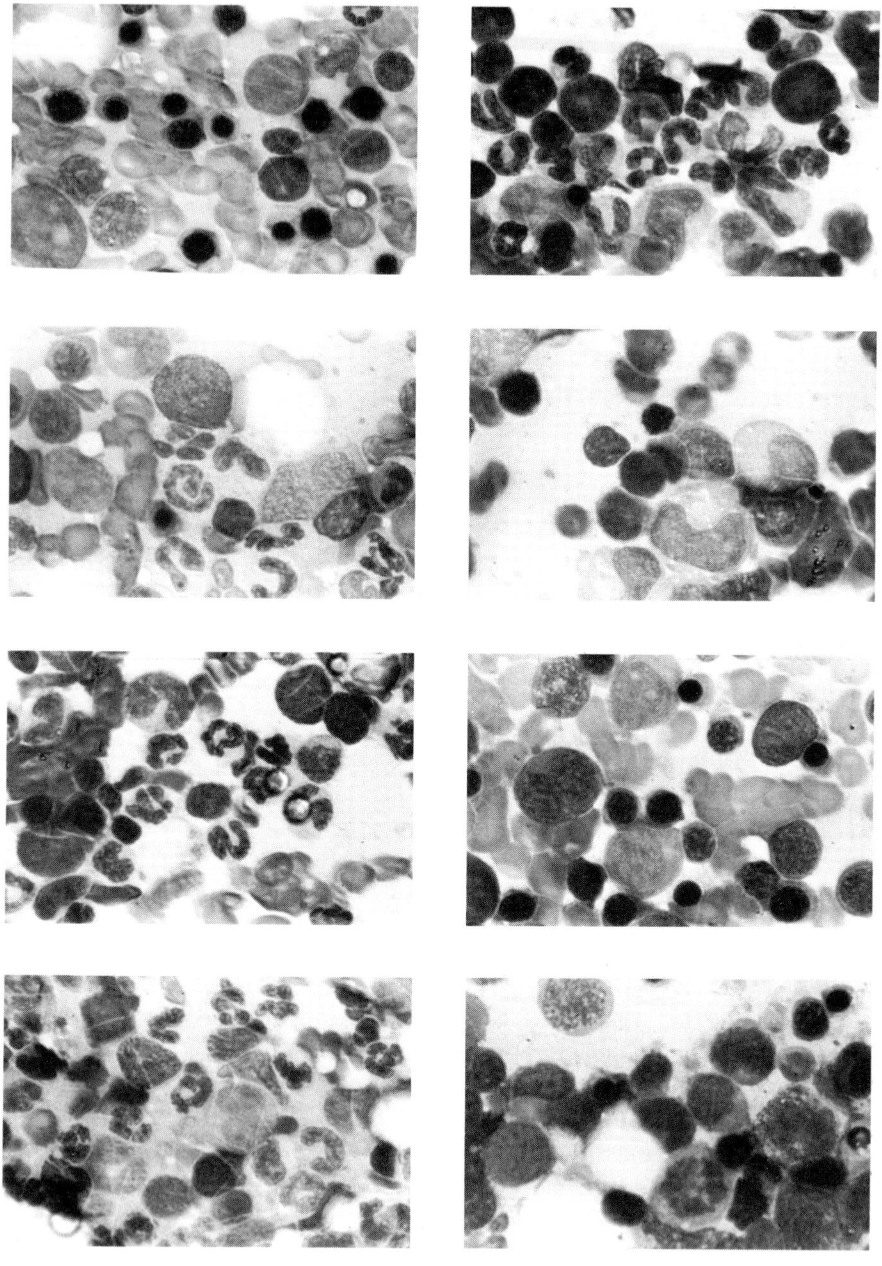

Fig. 3 Bone-marrow morphology from CH dogs during different cycle days. Left-hand figures (top to bottom), days 1, 2, 4, and 6; right-hand figures (top to bottom), days 8, 9, 10, and 12.

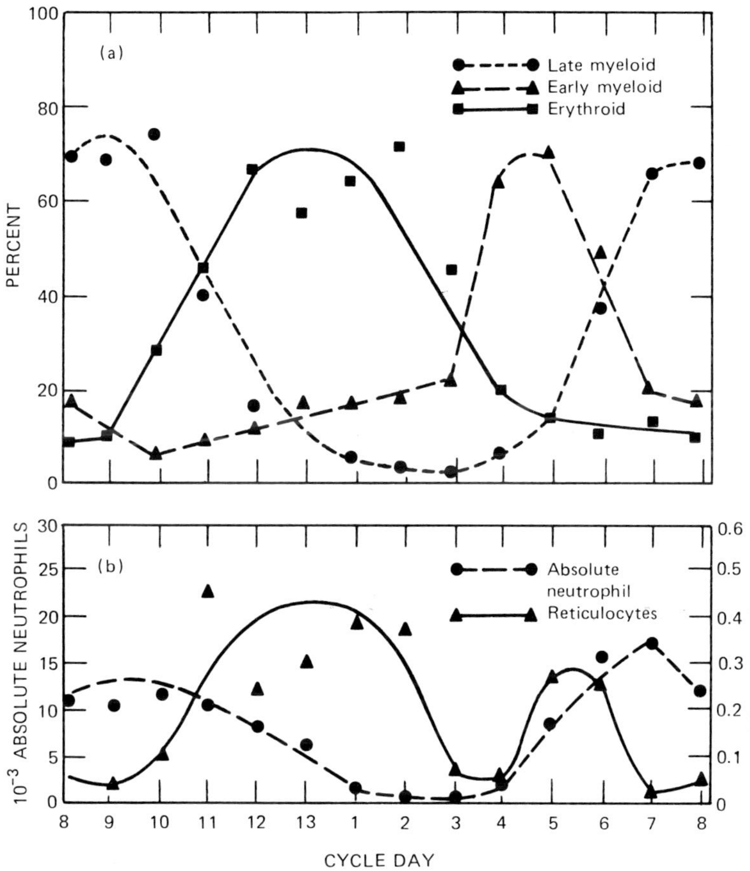


**Fig. 4 Average bone-marrow (a) and peripheral blood (b) counts for two CH dogs throughout a 13-day cycle.**

experiments in our laboratory,[13,23] dogs were exposed to a constant stimulus of hypoxia, and their plasma Epo was found to cycle at 11- to 12-day intervals (Fig. 6). In some dogs minor mid-cycle peaks were observed, and the amount of Epo produced varied between different animals. The peaks of Epo characteristically appeared approximately 5 days after the onset of neutropenia. Normal puppies maintained in the hypobaric chamber experienced only a single, initial rise in Epo. When marrow from CH dogs was grafted into normal dogs, Epo cycled in response to hypoxia.[13] Also, in two CH dogs transplanted with normal bone marrow, Epo responses to hypoxia resembled those of an untransplanted normal dog. These experiments show that the CH serum Epo cycles observed under constant hypoxia are mediated by the CH marrow itself.

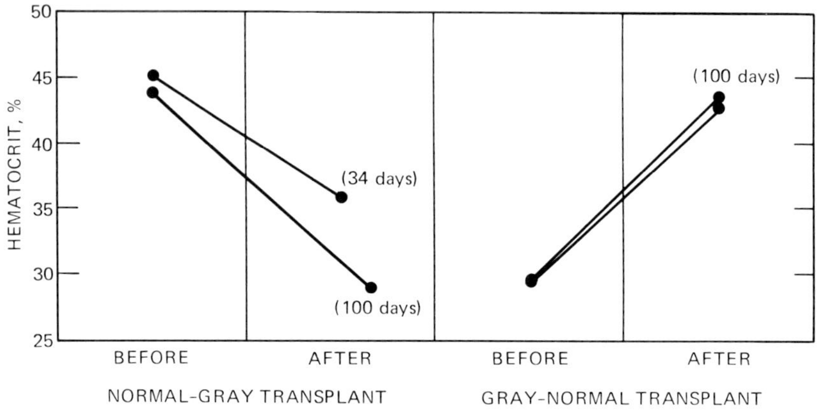


**Fig. 5 Hematocrit values of bone-marrow chimeras before and after transplant.** Normal–gray transplant refers to a normal dog transplanted with bone marrow from a CH gray littermate. Gray–normal refers to a CH gray dog transplanted with bone marrow from a normal-colored littermate.

Adamson, Dale, and Elin[28] studied the effect of serum from CH dogs on an in vitro system of hemoglobin synthesis. In this system the heme synthesis of dog marrow cells exposed to Epo or serum was measured. The serum of an unstressed CH dog demonstrated definite cyclic activity of Epo that varied from 0.02 to 0.04 unit/ml. These values were obtained by simultaneously determined Epo dose–response curves.

Dunn et al.[29] have measured the serum erythroid stimulating activity (ESA) in unstressed dogs by the fetal mouse liver cell assay system. Six cycles from four CH dogs were observed. The ESA titers were below the normal range except for days 11 and 12 when values approximated those of normal dogs. In two CH dogs engrafted with normal marrow, ESA activity was not cyclic.

In studies to test for other humoral factors, significantly elevated levels of thrombopoietin have been found in CH serum fractions at the beginning of active thrombopoiesis.[11] T. Yang et al. (unpublished data) have shown a periodic variation of lymphokine (migration inhibition factor) output in the serum of CH dogs. This evidence suggests a relationship between lymphocyte–macrophage–granulocyte interactions indicative of a lymphoid participation in the regulation of granulopoiesis.

The serum of CH dogs was examined by White et al.[30] for its ability to stimulate hemoglobin synthesis by normal canine bone marrow. Hemoglobin synthesis varied during the cycle, which indicated the presence of a cycling agent affecting hemoglobin synthesis in CH dogs. The agent responsible for the changes is not known.

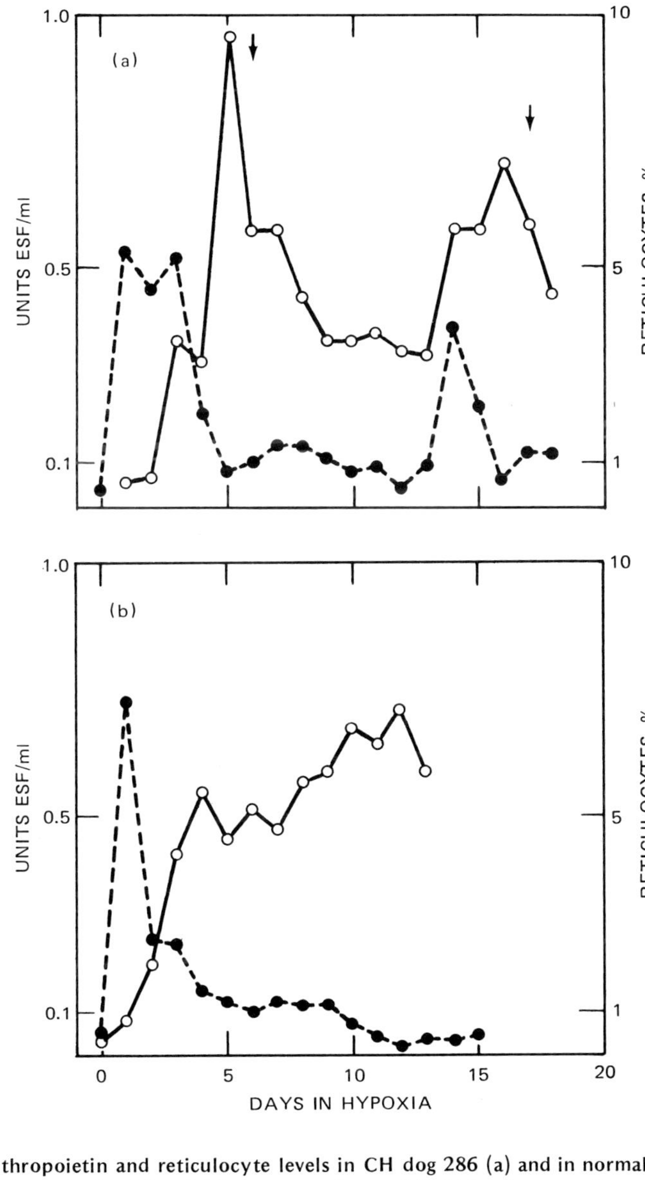


Fig. 6 Erythropoietin and reticulocyte levels in CH dog 286 (a) and in normal dog 360 (b) exposed to hypoxia for 15 and 18 days. Starting and finishing hematocrits on dog CH 286 were 35% and 41%, respectively. For the normal dog the values were 46% and 50%, respectively. Arrows indicate start of neutropenic periods. ○——○, reticulocytes. ●– – –●, Epo units/ml. In this one study the Epo peak did not appear 5 days after onset of neutropenia as was true in the majority of dogs investigated.[13] It has been postulated that the Epo peaks may be affected by the day the dog is placed in the hypoxic environment. [From R. Lange and J. Jones, Hormonal Control of Erythropoiesis in Canine Cyclic Haematopoiesis, *Scand. J. Haematol.*, 16: 59 (1976).]

## PROGENITOR CELLS

The results of bone-marrow allografts showed that canine CH was of marrow origin,[21-24] and the Epo studies showed that the CH marrow influenced Epo cycles. A reliable marrow pluripotential stem-cell assay is not available for dogs. Committed stem cells, both erythrocytic and granulocytic, have been studied to determine indirectly whether or not the CH multipotential stem cells are fluctuating cyclically. For a review of cloning assays in dogs, see Chapter 1.

The in vitro colony forming unit (CFU-C) of granulocytic cells was studied in several CH dogs on each day of the cycle.[29] These studies involved the aspiration of marrow cells from the long bones, iliac crests, and ribs, separation of nucleated cells on Ficoll–Hypaque, and the subsequent culture of nucleated marrow cells. The number of CFU-C's, relative to a constant number of nucleated marrow cells cultured, fluctuated from above that of concurrent normal dog marrow cell cultures to near zero. The greatest number of CFU-C's was in marrow taken just prior to or during the neutropenic episodes. When tested halfway between neutropenic episodes (days 6 to 8 in the cycle), cultures of CH marrow treated with the serum of normal dogs formed more colonies per plate than did cultures treated with CH serum high in CSA. This pattern was the reverse of that found on cycle days 10 to 14 and 1 to 2, when the CH serum stimulated many more colonies than did the normal serum. The cause for this reversal has not been found, but CH marrow cells possibly are involved in a positive response to CSA at days 10 to 14 and 1 to 2 and then fail to grow in vitro in the presence of high stimulation on days 6 to 8. Kurland and Moore[31] have recently shown that bone-marrow adherent cells are capable of producing negative CFU-C influences following excess production of positive growth influences on CFU-C. They propose a CSF-prostaglandin $E_2$ interaction.

The CFU-E, like the CFU-C, is also highest at days 10 to 14 and 1 to 2 of the CH cycle and lowest during days 6 to 8 (Ref. 32). Surprisingly, the CFU-E peak occurs during the period of peripheral reticulocytosis. Differences between the erythrocyte–granulocyte cell maturation, storage pool size, and cell releasing times could then account for the coincidental reticulocytosis–neutropenia.

The diffusion-chamber culture system, as first described by Bøyum and Borgstrøm,[33] will support dog marrow cell growth at about the same level as it will human marrow cell growth. This system is thought to support the growth of the granulocytic committed cell and perhaps multipotential stem cells.[34] The CH dog's marrow cultured in diffusion chambers implanted into lethally irradiated mice produced an 8-day cellularity that was greatest with marrow obtained on days 10 to 14 and 1 to 2 and least with marrow obtained on days 6 to 8 (Ref. 35).

In summary, all three progenitor cells, CFU-E, CFU-C, and diffusion-chamber progenitors, are most abundant on days 10 to 14 and 1 to 2 of the CH cycle (Fig. 7). These facts, together with the marrow cytology, transplantation

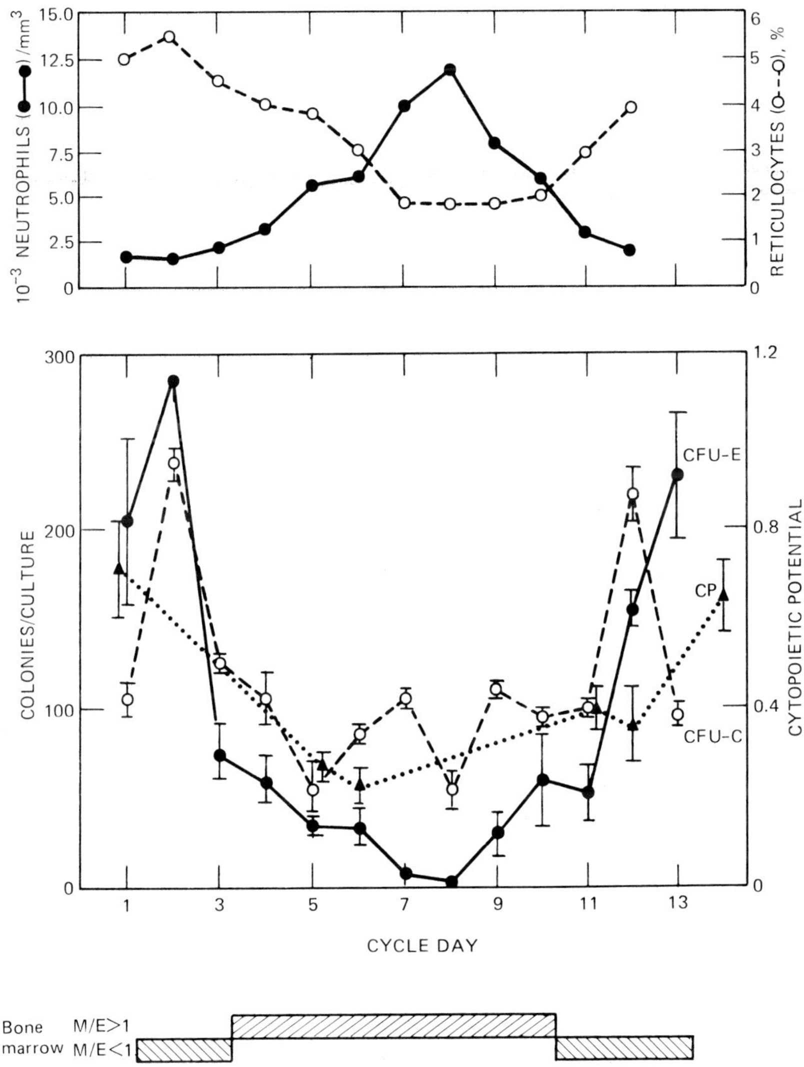


Fig. 7 The interrelationships of marrow progenitor cells with peripheral blood elements in canine CH. CFU-E data represent combined results from five dogs. CFU-C and reticulocyte data are from Dunn et al.[29]; neutrophil counts, marrow myeloid/erythroid (M/E) ratios, and cytopoietic potentials (CP) of the marrow to populate diffusion chambers are from Dunn et al.[35] Values in the marrow of hematologically normal dogs were: CFU-E, $97 \pm 5/2.4 \times 10^5$ cells; CFU-C, $90 \pm 8/1.2 \times 10^6$ cells; and the cytopoietic potential, $0.77 \pm 0.05$. (The marrow CP's are the slopes of the regression lines relating the number of cells harvested from the diffusion chambers to the number of cells in the inoculum.) Vertical bars indicate ± standard error of the mean. [From C. Dunn, J. Jolly, J. Jones, and R. Lange, Erythroid Colony Formation In Vitro from the Marrow of Dogs with Cyclic Hematopoiesis: Interrelationship of Progenitor Cells, *Exp. Hematol.*, 6: 706 (1978).]

studies, and humoral regulator studies, show that CH marrow is intrinsically cyclic. Further, the intrinsic cycle appears to affect, simultaneously, all measurable progenitor cells. Therefore it is believed that a cell (or cells) exists in the CH marrow which is capable of elaborating an inhibitor or a stimulator of multipotential stem-cell replication.

## PATHOGENESIS

Various theories have been proposed to explain the pathogenesis of CH. Among the theories are (1) competition at the stem-cell level, (2) defect in the limiting stem-cell pool with an abnormality of short-range cell–cell interaction, (3) periodic interruption of hematopoietic elements, (4) regulatory feedback abnormality, (5) cycling of stem cells with production of a "poietin-controlling factor," and (6) cycling of stem cells.

Any theoretical explanation of CH must consider and support the following facts:

1. It is possible to produce or abrogate the CH syndrome by bone-marrow transplantation.
2. The apogees of monocyte, platelet, and reticulocyte counts occur near the nadir of neutrophil counts.
3. The highest levels of CSF occur at the nadir of the granulocyte counts.
4. The highest levels of erythropoietin (in erythroid stress experiments) occur starting 5 to 6 days after the onset of neutropenia.
5. Each of the cycles of neutrophils, platelets, and erythroid elements is independent of the cycles of the other two elements in relation to peripheral stimulation by phlebotomy, hypoxia, and endotoxin.[13]

The original suggestion that the cyclic behavior of CH resulted from competition at the stem-cell level was made by Patt, Lund, and Maloney.[36] They proposed that the predisposing condition in the gray collie was a functionally restricted stem-cell pool with an oscillatory efflux governed primarily by a neutrophil feedback loop operating against a background of stem-cell competition. However, work in both Adamson's laboratory[28] and our own[37] has shown that the erythroid compartment can be markedly stimulated without changing the cycles of neutrophils. In one study a small dog was exposed to the stimulus of constant hypoxia at 0.4 atm (Ref. 13). Daily for 32 days, 50 ml of blood was removed for erythropoietin assays. In spite of these combined erythroid stimuli of hypoxia and blood loss, the cycles of neutropenia persisted, and neutrophilia developed even though the hematocrit was maintained at a stable level. Neither Adamson's study nor our own study would support the hypothesis of competition at the stem-cell level in CH.

Maloney, Lund, and Patt[38] have reported that the neutropenia can be overcome with endotoxin without altering the reticulocyte cycle and have suggested that in CH the defect is basically in the limiting stem-cell pool with an abnormality of short-range cell–cell interaction leading to wide fluctuations of neutrophil-committed stem cells. They thought that recurring stem-cell competition coinciding with a periodic "turning on" of a marked outflow to neutrophil commitment as the pool of committed cells reaches its nadir could account for the oscillations of reticulocytes and platelets.

Cell–cell interaction brings about communications in populations of cells. McCulloch et al.[39] stated in 1974: "It seems likely that proteins mediating intracellular communication are specific gene products; the activation of such genes is necessary for the development of regulatory mechanisms acting at the level of committed stem cells." Indeed it has been established that mononuclear phagocytes (Mϕ) have the capacity to elaborate biological regulators of granulopoiesis.[40] Thus an optimal number of Mϕ is associated with an increase in the number of CFU-C, whereas higher numbers of Mϕ are inhibitory, probably because of production of PGE. Kurland[40] stated: "Central surveillance and effector role of Mϕ represents an exquisite self-regulating unit that functions to prevent the excessive myelopoiesis in response to an elevation of CSF." However, more than one form of CSF has been found, and Metcalf[41] has called attention to the fact that there is a heterogeneity of the granulocyte–macrophage progenitor cell compartment, which suggests both a capacity for subtle modulation in these cells and the existence of heterogeneous subsets of end-cell polymorphs and monocyte macrophages.

Lajtha and Wright[42] also have found evidence for the production of stimulatory and inhibiting factors by living bone-marrow cells; stem-cell proliferation can be controlled by short-range inhibitory action. In tissue culture a 50,000 to 100,000 molecular-weight fraction of a crude extract of a cell-free supernatant fluid of noncycling normal bone marrow caused inhibition of stem-cell cycling without inhibiting committed precursor-cell cycling. A 30,000 to 50,000 molecular-weight fraction from regenerating marrow (high cycling activity) induced stem-cell cycling.

Dale, Alling, and Wolff[7] thought that the pathogenesis of CH resided in recurring failure of hematopoietic elements, perhaps a cessation of stem-cell differentiation. On the basis of ultrastructural studies, Scott et al.[20] felt that abnormal neutrophil maturation was another factor in CH and was perhaps linked to cessation of stem-cell differentiation through an abnormal feedback regulatory mechanism for controlling neutrophil production. Dale, Alling, and Wolff[7] considered that a difference in maturation time accounts for the diametric phases of reticulocyte and platelet apogees on the one hand and neutrophil counts on the other. However, indirect evidence suggests that the different marrow cellular elements have approximately the same transit time. The mean granulocyte transit time of 82.1 hr found in normal dogs[43] agrees with the time of peak appearance of $^3$H–TdR-labeled neutrophils in gray collies.[36] After a severe hemorrhage in dogs, neutrophils peaked on day 3 to day 5, reticulocytes responded on day 4 with a peak on day 6, and platelets responded on day 4 but peaked on day 9 (Ref. 44). In irradiated dogs platelet depletion was followed by a return to normal counts on day 6 or day 7 (Ref. 45). Therefore it seems unlikely that marrow transit time is a cause for the 180° phase difference between reticulocytes and granulocytes. Unequal storage pools could account for an abnormality of progenitor-cell production but not for the diametric phase difference in the peripheral mature cells.

In another hypothesis a defect of hematopoietic regulation at the stem-cell level purportedly leads to the periodicity of the cellular elements, and CH is a genetic disorder of hematopoietic control mechanisms.[26] Adamson, Dale, and Elin,[28] for instance, have suggested that erythropoietic cycling is caused by a regulatory feedback from proliferating cells on humoral erythropoietic production. Cycling of marrow cellular elements could also account for cycling erythropoietin and colony stimulating activity.

We have suggested an hypothesis[13] that a humoral factor is produced in the bone marrow of cyclic hematopoietic dogs. This factor leads to the production of specific long-range stimulators, which, in turn, act on specific response cells to trigger their entrance into their normal developmental pathway. The possibility has also been raised of a decrease in the proliferation or in the sensitivity of the stem cells to humoral and/or cellular stimuli. Since the syndrome can be produced by bone-marrow transplantation, an abnormality of the stem cell is suggested, and, further, the stem-cell defect may cause the stem cell to cycle. Stem-cell cycling would be in agreement with the studies that have shown that both the myeloid and erythroid cells cycle in the same phase.[29] The diffusion-chamber work confirmed that the basic fault probably lies in the multipotential stem cells and that the numbers of these cells can be reduced.[35] The long-term culture system recently reported by Dexter and others[46–48] may be useful in studying the CH stem cell's replication and its microenvironment.

## CONCLUSIONS

Although the basic defect in CH has not been defined, CH is a good paradigm of hematopoiesis; if the hematopoietic controls in CH dogs can be ascertained, the mechanisms that control and adjust normal hematopoiesis may be better understood.

## ACKNOWLEDGMENTS

These studies were supported in part by the National Heart, Lung, and Blood Institute grants numbers HL 10567-12 and HL 15647-04. The assistance of present and former colleagues, which include C.D.R. Dunn, J. E. Fuhr, E. A.

Machado, T. P. McDonald, and T. J. Yang, is gratefully acknowledged. D. Long and L. Simpson provided excellent stenographic aid.

## REFERENCES


1. L. Ford, Possible Pleiotrophic Effects of the "Gray" Gene in Collie Dogs, *Proc. 10th Int. Congr. Hum. Genet.*, 83: 1-2 (1958).
2. J. Lund, G. Padgett, and R. Ott, Cyclic Neutropenia in Grey Collie Dogs, *Blood*, 29: 452-461 (1967).
3. J. Lund, G. Padgett, and J. Gorham, Additional Evidence on the Inheritance of Cyclic Neutropenia in the Dog, *J. Hered.*, 61: 47-49 (1970).
4. J. Jones, R. Lange, and E. Jones, Cyclic Hematopoiesis in a Colony of Dogs, *J. Am. Vet. Med. Assoc.*, 166: 365-367 (1975).
5. J. Jones, E. Jones, and R. Lange, Early-Life Hematologic Values of Dogs Affected with Cyclic Neutropenia, *Am. J. Vet. Res.*, 35: 849-852 (1974).
6. N. Cheville, The Gray Collie Syndrome, *J. Am. Vet. Med. Assoc.*, 152: 620-630 (1968).
7. D. Dale, D. Alling, and S. Wolff, Cyclic Hematopoiesis: The Mechanism of Cyclic Neutropenia in Grey Collie Dogs, *J. Clin. Invest.*, 51: 2197-2204 (1972).
8. D. Dale, S. Ward, H. Kimball, and S. Wolff, Studies of Neutrophil Production and Turnover in Grey Collie Dogs with Cyclic Neutropenia, *J. Clin. Invest.*, 51: 2190-2196 (1972).
9. T. Yang, J. Dale, and J. Jones, Cyclic Activity of DNA Synthesis of Peripheral Blood Leukocytes in Canine Cyclic Neutropenia, *Exp. Mol. Pathol.*, 25: 121-130 (1976).
10. K. Angus, D. Wyand, and T. Yang, Impaired Lymphocyte Response to Phytohemagglutinin in Dogs Affected with Cyclic Neutropenia, *Clin. Immunol. Immunopathol.*, 11: 39-51 (1978).
11. T. McDonald, R. Clift, and J. Jones, Canine Cyclic Hematopoiesis: Platelet Size and Thrombopoietin Level in Relation to Platelet Count, *Proc. Soc. Exp. Biol. Med.*, 153: 424-428 (1976).
12. M. Reese, T. McDonald, and J. Jones, Platelet Function Studies in Dogs with Cyclic Hematopoiesis, *Proc. Soc. Exp. Biol. Med.*, 153: 324-329 (1976).
13. R. Lange and J. Jones, Hormonal Control of Erythropoiesis in Canine Cyclic Haematopoiesis, *Scand. J. Haematol.*, 16: 56-65 (1976).
14. R. Lange, J. Jones, C. Chambers, Y. Quirin, and J. Sparks, Erythropoiesis and Erythrocytic Survival in Dogs with Cyclic Hematopoiesis, *Am. J. Vet. Res.*, 37: 331-334 (1976).
15. R. Gregory, E. Machado, and J. Jones, Animal Model: Amyloidosis Associated with Canine Cyclic Hematopoiesis in the Gray Collie Dog, *Am. J. Pathol.*, 87: 721-724 (1977).
16. E. Machado, R. Gregory, J. Jones, and R. Lange, The Cyclic Hematopoietic Dog: A Model for Spontaneous Secondary Amyloidosis, *Am. J. Pathol.*, 92: 23-34 (1978).
17. J. Lund and D. Barkman, Color Dilution in the Gray Collie, *Am. J. Vet. Res.*, 35: 265-268 (1974).
18. T. Yang, Recovery of Hair Coat Color in Gray Collie (Cyclic Neutropenia)—Normal Bone Marrow Transplant Chimeras, *Am. J. Pathol.*, 91: 149-153 (1978).
19. J. Lund, *Canine Cyclic Neutropenia*, Ph.D. Thesis, Washington State University, Pullman, Wash., 1969.
20. R. Scott, D. Dale, A. Rosenthal, and S. Wolff, Cyclic Neutropenia in Grey Collie Dogs, Ultrastructural Evidence for Abnormal Neutrophil Granulopoiesis, *Lab. Invest.*, 28: 514-525 (1973).

21. D. Dale and R. Graw, Transplantation of Allogeneic Bone Marrow in Canine Cyclic Neutropenia, *Science,* 183: 83-84 (1974).
22. P. Weiden, B. Robinett, T. Graham, J. Adamson, and R. Storb, Canine Cyclic Neutropenia: A Stem Cell Defect, *J. Clin. Invest.,* 53: 950-953 (1974).
23. J. Jones, R. Lange, T. Yang, H. Vodopick, and E. Jones, Canine Cyclic Neutropenia: Erythropoietin and Platelet Cycles After Bone Marrow Transplantation, *Blood,* 45: 213-219 (1975).
24. J. Jones, T. Yang, J. Dale, and R. Lange, Canine Cyclic Haematopoiesis: Marrow Transplantation Between Littermates, *Br. J. Haematol.,* 30: 215-233 (1975).
25. R. Lange and J. Jones, Erythropoiesis in Dogs with Cyclic Hematopoiesis, in *Topics in Haematology,* Proceedings of the 16th International Congress of Hematology, Kyoto, Japan, Sept. 5–11, 1976, pp. 788-791, International Society of Hematology, Asian-Pacific Division, Mizuho, Nagoya, Japan, 1976.
26. D. Dale, C. Brown, P. Carbone, and S. Wolff, Cyclic Urinary Leukopoietic Activity in Gray Collie Dogs, *Science,* 173: 152-153 (1971).
27. T. Yang, J. Jones, E. Jones, and R. Lange, Serum Colony-Stimulating Activity of Dogs with Cyclic Neutropenia, *Blood,* 44: 41-48 (1974).
28. J. Adamson, D. Dale, and R. Elin, Hematopoiesis in the Grey Collie Dog. Studies of the Regulation of Erythropoiesis, *J. Clin. Invest.,* 54: 965-973 (1974).
29. C. Dunn, J. Jones, J. Jolly, and R. Lange, Progenitor Cells in Canine Cyclic Hematopoiesis, *Blood,* 50: 1111-1120 (1977).
30. J. White, J. Jones, R. Lange, and J. Fuhr, Evidence for the Existence of an Agent in the Serum of the Cyclic Hematopoietic Dog Which Influences Hemoglobin Synthesis, *Experientia,* 34: 1367-1368 (1978).
31. J. Kurland and M. Moore, Modulation of Hemopoiesis by Prostaglandins, *Exp. Hematol.,* 5: 357-373 (1977).
32. C. Dunn, J. Jolly, J. Jones, and R. Lange, Erythroid Colony Formation In Vitro from the Marrow of Dogs with Cyclic Hematopoiesis: Interrelationship of Progenitor Cells, *Exp. Hematol.,* 6: 701-708 (1978).
33. A. Bøyum and R. Borgstrøm, The Concentration of Granulocytic Stem Cells in Mouse Bone Marrow, Determined with Diffusion Chamber Technique, *Scand. J. Haematol.,* 7: 294-303 (1970).
34. D. Squires and L. Lamerton, The Effect of Various Cytotoxic Agents on Bone Marrow. Progenitor Cells as Measured by Diffusion Chamber Assays, *Br. J. Haematol.,* 29: 31-42 (1975).
35. C. Dunn, J. Jones, J. Jolly, and R. Lange, Cell Proliferation of Canine Cyclic Hematopoietic Marrow in Diffusion Chambers, *Proc. Soc. Exp. Biol. Med.,* 158: 50-53 (1978).
36. H. Patt, J. Lund, and M. Maloney, Cyclic Hematopoiesis in Grey Collie Dogs: A Stem-Cell Problem, *Blood,* 42: 873-884 (1973).
37. R. Lange, Stimulated Erythropoiesis in Dogs with Cyclic Hematopoiesis, *Blood Cells,* 1: 599-603 (1975).
38. M. Maloney, J. Lund, and H. Patt, Modification of Canine Cyclic Hematopoiesis by Endotoxin, *Proc. Soc. Exp. Biol. Med.,* 147: 205-210 (1974).
39. E. McCulloch, T. Mak, G. Price, and J. Till, Organization and Communication in Populations of Normal and Leukemic Hemopoietic Cells, *Biochim. Biophys. Acta,* 355: 260-299 (1974).
40. J. Kurland, The Mononuclear Phagocyte and Its Regulatory Interactions in Hemopoiesis, in *Experimental Hematology Today,* S. Baum and G. Ledney (Eds.), pp. 47-60, Springer-Verlag, New York, 1978.

41. D. Metcalf, The Control of Neutrophil and Macrocyte Production at the Progenitor Cell Level, in *Experimental Hematology Today,* S. Baum and G. Ledney (Eds.), pp. 35-46, Springer-Verlag, New York, 1978.
42. L. Lajtha and E. Wright, Regulation of Stem Cell Proliferation, in *Experimental Hematology Today,* S. Baum and G. Ledney (Eds.), pp. 3-8, Springer-Verlag, New York, 1978.
43. K. Deubelbeiss, J. Dancey, L. Harker, and C. Finch, Neutrophil Kinetics in the Dog, *J. Clin. Invest.,* 55: 833-839 (1975).
44. V. Perman, D. Sorensen, E. Usenik, V. Bond, and E. Cronkite, Hemopoietic Regeneration in Control and Recovered Heavily Irradiated Dogs Following Severe Hemorrhage, *Blood,* 19: 738-742 (1962).
45. C. Craddock, W. Adams, S. Perry, and J. Lawrence, Dynamics of Platelet Production as Studied by Depletion Technique in Normal and Irradiated Dogs, *J. Lab. Clin. Med.,* 45: 906-919 (1955).
46. T. Dexter and L. Lajtha, Proliferation of Haemopoietic Stem Cells In Vitro, *Br. J. Haematol.,* 28: 525-530 (1974).
47. T. Dexter and N. Testa, Differentiation and Proliferation of Haemopoietic Cells in Culture, in *Methods in Cell Biology,* Vol. 14, M. Prescott (Ed.), pp. 387-405, Academic Press, Inc., New York, 1976.
48. M. Moore and T. Dexter, Stem Cell Regulation in Continuous Hematopoietic Cell Culture, *Transplant. Proc.,* 10: 83-90 (1978).

# 15 Lymphoid Neoplasia in the Canine

CHARLES A. HOLMBERG
*Department of Veterinary Pathology, College of Veterinary Medicine, Texas A&M University, College Station, Texas*
*and*
FLOYD D. WILSON
*Laboratory for Energy-Related Health Research, School of Veterinary Medicine, University of California, Davis*

Lymphoid neoplasms are one of the more-frequent naturally occurring malignant neoplasms in the canine. The incidence of lymphoid neoplasms of approximately 24 per 100,000 in one population of dogs is similar to that in man.[1-3] Lymphoid neoplasms have been diagnosed in 0.13 to 0.36% of all dogs presented for medical care at veterinary hospitals,[4-6] which reflects the higher occurrence among "sick" animals. These figures indicate the availability of this spontaneous neoplasm for biomedical research.

The specific aspects of lymphoid neoplasms emphasized in this chapter are the pathological variations observed in the dog and the immunological aspects, including the functional nature of the neoplastic cells composing the neoplasms.

The general clinical and epidemiology features of lymphoid neoplasia in the canine have been reported.[1,7-16] Evidence of breed predisposition has been clearly documented for the boxer breed only.[17] No distinct differences in sex distribution have been recognized. Although lymphoid neoplasia has been recognized in all age groups, the neoplasms are most frequent in dogs 7 to 10 years old.[8,18]

The etiology of spontaneous lymphoid neoplasms in the dog has not been defined. The methods that successfully demonstrated a viral etiology of lymphoid neoplasms in other animals have been applied to canine lymphoid neoplasms without success in several laboratories. Infrequent cases containing either possible C-type virus particles or herpesvirus particles apparently reflect etiologically unassociated passenger virus infections.[8,9,19,20] The absence of recognizable oncogenic viruses implies that canine lymphoid neoplasia may be more analogous to human lymphoid neoplasia than other animal models used for immunological studies.

Serial transplantation of cells from canine lymphoid neoplasms to fetal and neonatal recipients has not induced neoplasia of recipient cellular origin.[21–23] The transplantation studies provide, however, a model system for analyzing host response in tumor engraftment–rejection phenomena. The histocompatibility antigens of the transplanted canine tumor cells provide a definable tumor specific antigen for experimental studies in an animal large enough to be monitored individually with time.

## PATHOLOGICAL CLASSIFICATION OF LYMPHOID NEOPLASIA IN THE CANINE

Rapidly changing concepts of the functional nature of lymphoid neoplasia obviate the value of a detailed discussion and comparison of all the past anatomical classification schemes. The variety of classification schemes and variations in nomenclature of canine lymphoid neoplasia, as with lymphoid neoplasia in man and other species, is the result of both clinical–pathological variations and different conceptual viewpoints of investigators. A binomial classification system, which has been proposed with minor variations by several research groups, provides a usable framework of anatomical classification necessary for comparison with and evaluation of the immunological characterization of the neoplastic cell populations.[7,9,24,25] In that system each lymphoid neoplasm is designated by a two-part name; the first is based on the gross anatomical distribution of lesions in the body, and the second is based on the histological evaluation of the cell type and cellular organization in lesions (Table 1). Further studies simultaneously comparing the anatomical and immunological features of lymphoid neoplasms should lead to combined classifications such as those currently being evaluated for the human.[26–31] Further, the observations that come from combined anatomical and immunological studies will add to the usefulness of naturally occurring canine lymphoid neoplasia as an animal model for understanding the biology of lymphoid neoplasia in man.

### Anatomical Distribution

**Thymic Form of Malignant Lymphoma.** The major lesion is located within the anterior mediastinal tissue, which reflects the apparent thymic origin of the neoplasm. Thymic epithelial remnants can usually be demonstrated within the lymphoid neoplasm, but histological examination will differentiate that lymphoma from the true thymoma, which consists entirely or predominantly of epithelial cells.

The thymic lymphoma is a relatively uncommon form of malignant lymphoma in the dog; for example, thymic lymphomas composed 2% of the

TABLE 1

BINOMIAL CLASSIFICATION SYSTEM FOR CANINE LYMPHOID NEOPLASMS

**Anatomical Classification**

1. Thymic form of malignant lymphoma.
2. Alimentary form of malignant lymphoma.
3. Multicentric form of malignant lymphoma.
4. Lymphocytic leukemia.
5. Miscellaneous forms of malignant lymphoma (solitary and skin).

**Histological Classification**

1. Histological pattern of lesion
   a. Diffuse.
   b. Nodular (follicular).
2. Predominant cell type
   a. Undifferentiated (stem cell).
   b. Histiocyte (reticulum cell sarcoma).
   c. Poorly differentiated lymphocytic (lymphoblastic).
   d. Well-differentiated lymphocytic (lymphocytic).
   e. Mixed cell (histiocytic—lymphocytic).
   f. Plasmacytoid differentiation.

canine lymphoid neoplasms of one group of dogs.[8] Canine thymic lymphomas are apparently composed of T lymphocytes,[32] as has been shown for similar cases in man[27–33] and cats.[34] Progression to lymph nodes and other organs, though variable, typically is more similar to the human forms, with early leukemic dissemination,[35] than to the feline form, with minimal extra-thymic dissemination, until the terminal stages of the disease.[34]

**Alimentary Form of Malignant Lymphoma.** With this form of lymphoma, the predominant tumor mass is located within the wall of the alimentary canal and the immediately associated mesenteric lymph nodes. The neoplastic involvement may progress to lymph nodes and other organs throughout the body, but the alimentary tract lesions remain prominent, which reflects the apparent anatomical origin of the neoplasm.

The alimentary form, which comprises 7 to 45% of the cases, is the second most frequent anatomical type of lymphoma in the dog, depending on geographic location.[8,9]

**Multicentric Form of Malignant Lymphoma.** The multicentric form of malignant lymphoma is characterized by bilaterally symmetrical enlargement of most, if not all, superficial lymph nodes, visceral lymph nodes, spleen, and liver. These tissues are usually uniformly affected at the onset of clinically recognized disease. Progression of the disease can lead to lesions in any organ, but the most common organs for lesions are the tonsils, lungs, kidneys, alimentary canal, and bone marrow. Neoplastic cells in the blood generally associated with bone-marrow invasion[9] are detectable in 7 to 13% of animals at the time the clinical disease is recognized. In contrast to animals with the true leukemia form, animals with the multicentric form have either none or relatively few abnormal lymphocytes in the blood until the terminal stage of the multicentric disease forms.

The multicentric malignant lymphoma, which comprises 45 to 84% of the cases, is the most common anatomical form in the dog, depending on the geographic location.[7–9]

The multicentric, alimentary, and thymic forms of malignant lymphoma have similar histological patterns of neoplastic cell localization in organs. Early lymph node localization occurs predominantly through the cortex with involvement of the medulla, which results in retention of recognizable medullary cords and sinusoids. Splenic involvement is initially within the white pulp compartment (follicles and/or periarteriolar lymphatic sheaths), although specific localization cannot be determined in advanced lesions. The early splenic lesions may be recognized as either focally distinct follicular involvement or linear accumulations within the periarteriolar lymphatic sheaths. Neoplastic cell accumulations in the liver are usually more prominent within the portal canals, with less severe involvement along hepatic cords. Occasional cases show diffuse hepatic cord (sinusoidal) locations throughout the acinar units without the portal canal accumulations. Lung lesions are typically most prominent in periarteriolar and peribronchiolar locations.

**Leukemic Form (Lymphocytic Leukemia).** Lymphocytic leukemia is restricted in this anatomical classification to those lymphoid neoplasms which have from inception marked involvement of the blood, bone marrow, liver, spleen, and lymph nodes. Lymphocyte counts in the blood of affected animals range from 30,000 to 150,000. Early tissue localization of the leukemic form, unlike other malignant lymphoma forms, is in the splenic cords (Billroth's cords) in red pulp, in the medullary cords of lymph nodes, and in a diffuse sinusoidal pattern in the liver. The bone marrow is diffusely involved from the onset of disease.

Lymphocytic leukemia is the least common form in the dog (apparently less than 1% of lymphoid neoplasms). Two biologically different forms, acute (or blastic) and chronic (or small lymphocytic), have not been well characterized.

**Other Miscellaneous Forms.** Less than 10% of canine lymphoid neoplasms do not fit in one of the four preceding categories. These miscellaneous forms include solitary or regionally restricted lesions and perhaps skin forms. The spectrum of cases with significant skin involvement has not been well characterized in the dog, although that form comprises up to 7% of the lymphoma cases.[8] Cases with multifocal skin lesions may be variants of

TABLE 2

HISTOLOGICAL DIAGNOSIS AND AGE OF 94 CASES OF CANINE MALIGNANT LYMPHOMA

| Diagnosis | Number of cases | Average age ± standard deviation |
|---|---|---|
| Diffuse undifferentiated | 1 | 8 |
| Diffuse histiocytic | 37 | 7.4 ± 3.2 |
| Follicular histiocytic | 11 | 4.8 ± 3.3 |
| Diffuse | | |
| Poorly differentiated lymphocytic | 36 | 6.8 ± 2.7 |
| Well-differentiated lymphocytic | 3 | 7 ±1 |
| Follicular | | |
| Poorly differentiated lymphocytic | 4 | 3.2 ± 0.6 |
| Well-differentiated lymphocytic | 1 | 11 |
| Diffuse mixed cell | 1 | 6 |

advanced multicentric malignant lymphomas,[9] although similarity to the mycosis fungoides complex of man should also be evaluated.[36,37]

## Histological Classification

Histological evaluation of lymphoid neoplasms includes two components, the pattern of neoplastic cell organization and the predominant cell type. The pattern of neoplastic cell organization is either follicular or diffuse.[38,39] Follicular or nodular malignant lymphomas are recognized less frequently in dogs than in man, which reflects both absence of this distinction from previous classifications in the veterinary literature[7] and the gradual disappearance of the distinction with either natural progression of the lesions or following chemically induced remission (C. A. Holmberg, Texas A&M University, unpublished). Follicular patterns characterized 1 to 16% of canine malignant lymphomas[10,39] (Table 2), as compared with approximately 40% of human cases.[40] A follicular pattern in peripheral lymph nodes from multicentric and alimentary forms of malignant lymphoma has been associated with follicular localization of the splenic lesions.[39] All canine malignant lymphomas with a nodular pattern have been of B-cell nature, analogous to the disease in the human.[30,39] In dogs the diffuse pattern in lymph node lesions has been associated with a diffuse periarteriolar lymphatic sheath localization in the spleen, even when the neoplastic cells were of B lymphocyte nature.[39]

Characterization and nomenclature of the predominant cell type in lymphoid neoplasms is the most variable aspect of the anatomical classification of lymphoid neoplasia in the dog. The terms listed in Table 1 are derived from Rappaport's classification of non-Hodgkin's lymphomas;[38] a plasmacytoid cell category and synonyms are in parentheses. The conceptual inaccuracy of the terms used in this part of the binomial classification (especially histiocytic and stem cell) differs from current immunological and hematological understanding. However, when used as adjectives to indicate morphological features of different groups, this classification still has value because of simplicity and widespread acceptance. Replacement of this classification should be based on studies characterizing the functional or biological nature of the neoplastic cell populations in a manner analogous to the newer classifications proposed for human neoplasms.[26–31] The usefulness of those different classifications for human neoplasms has not been demonstrated for the dog, and also it is not clear if they can be applied on a purely morphological basis without simultaneous use of the immunological tests.

The morphologic criteria for the cytological categories of malignant lymphoma and their use in the dog have been described[7,10,39] and are summarized briefly as follows. The undifferentiated or stem-cell type (Fig. 1) is characterized by cells that contain very uniform, round, large nuclei lacking chromatin clumping and having relatively small nucleoli. The small amount of pale cytoplasm frequently has indistinct cell borders in histological sections. Canine lymphomas of this undifferentiated cell type are uncommon among our cases.

The histiocytic cell types (Fig. 2) include a range of cell types having in common a large vesicular nuclei with large prominent nucleoli and abundant but variably staining cytoplasm. This cell type has the largest nuclei of lymphoid neoplasms, which may be round, oval, or have variable degrees of nuclear indentations. Immunologic techniques have shown that all canine lymphomas of this cell type have a lymphocytic nature, and the term histiocytic is currently used in only a descriptive sense, a remnant of the previous belief that the morphology reflected a tissue histiocyte of monocytic origin. Figure 3 is a transmission electron micrograph of a typical histiocytic malignant lymphoma. The cells lack lysosomal granules and phagocytic activity characteristic of monocytes and macrophages, which is in agreement with immunological tests demonstrating their lymphocytic nature.

The poorly differentiated lymphocytic cell type (Fig. 4) is characterized by nuclear size intermediate between the histiocytic and well-differentiated lymphocytic cell types and by more coarse chromatin staining features and pleomorphic nuclei than the undifferentiated cell type. The well-differentiated lymphocytic cell type (Fig. 5) is characterized by smaller nuclei with condensed chromatin and usually well-defined borders in histological sections.

The mixed-cell type (Fig. 6) is characterized by intermixed populations of cells compatible with the histiocytic and lymphocytic cell types but without a clear predominance of either cell type.

The plasmacytoid differentiation category in Table 2 includes the clinical–pathological entities, such as multiple myeloma, solitary plasmacytomas,

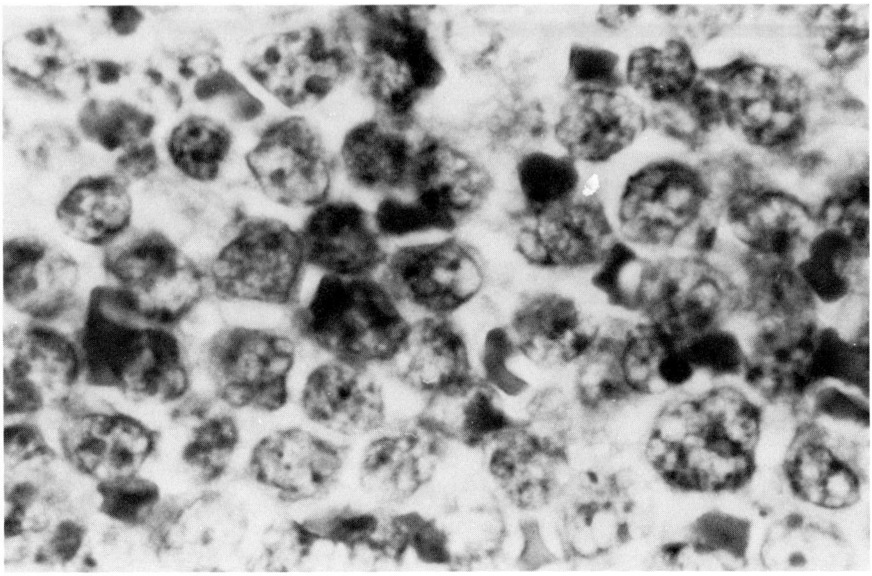

Fig. 1 Undifferentiated malignant lymphoma. (Magnification, 1000 x.)

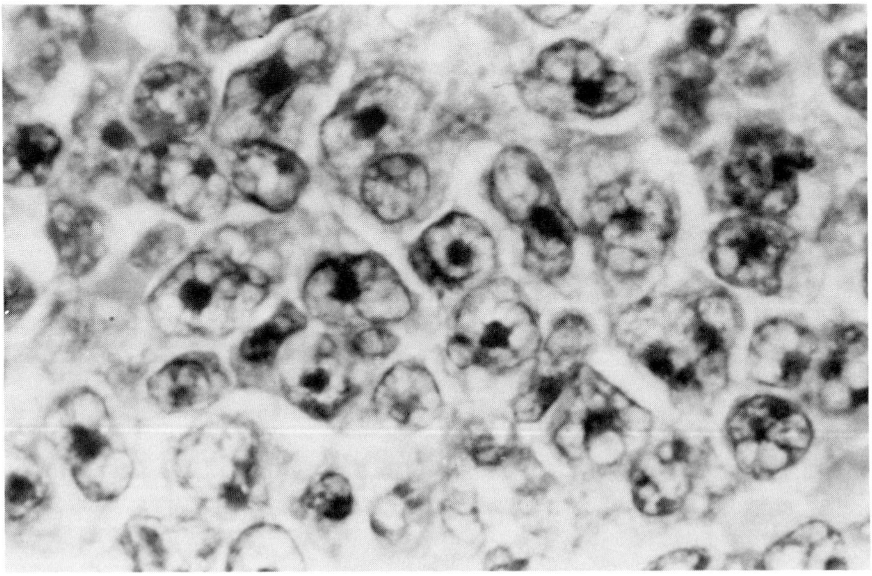

Fig. 2 Histiocytic malignant lymphoma. (Magnification, 1000 x.)

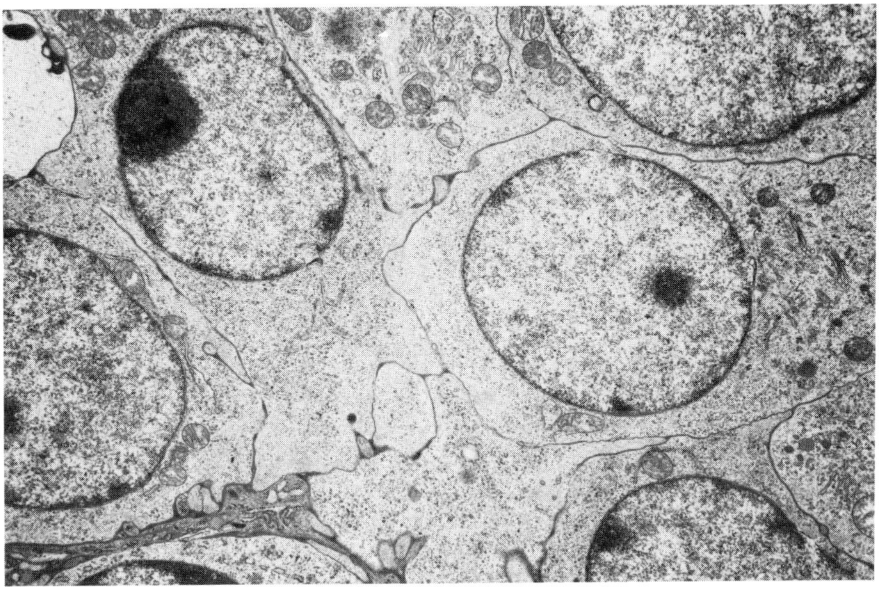

Fig. 3 Transmission electron micrograph of histiocytic malignant lymphoma.

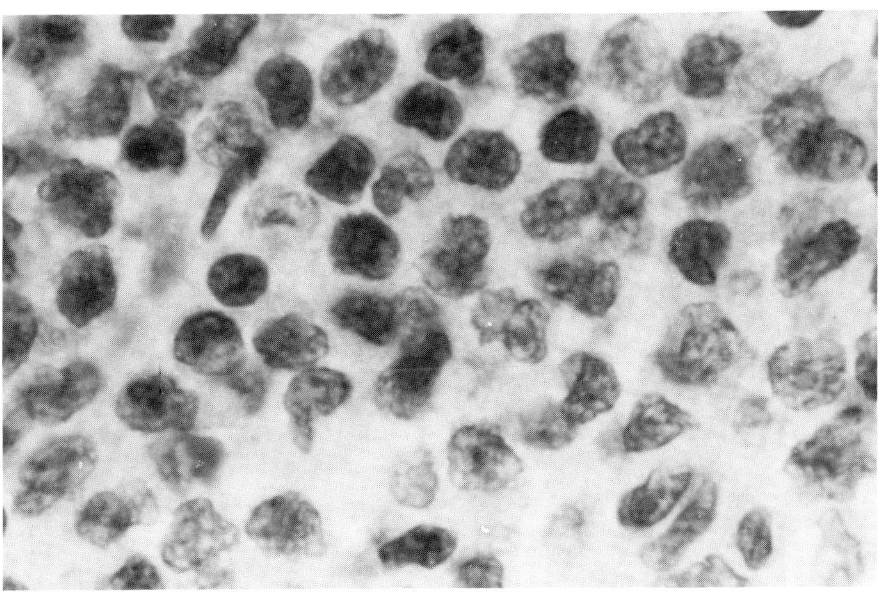

Fig. 4 Poorly differentiated lymphocytic malignant lymphoma. (Magnification, 1000 x.)

Fig. 5 Well-differentiated lymphocytic malignant lymphoma. (Magnification, 1000 ×.)

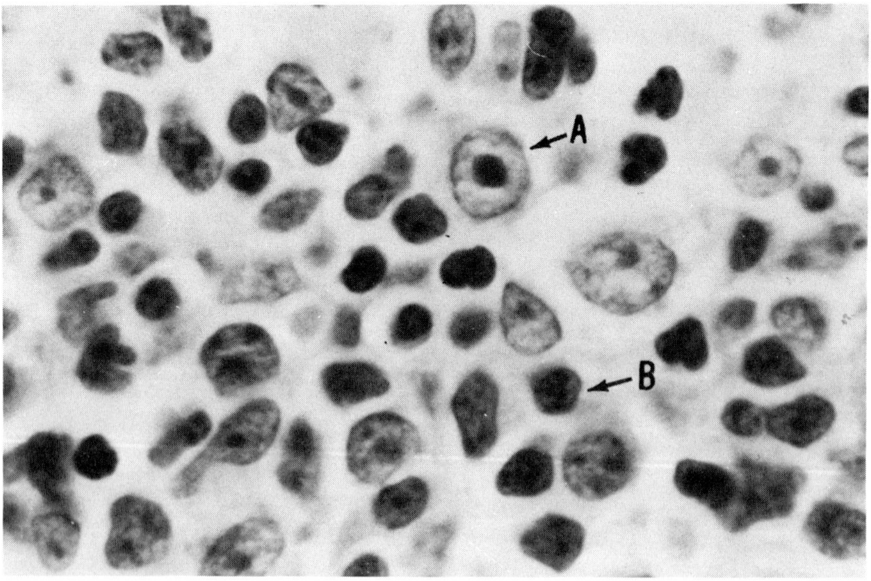


Fig. 6 Mixed-cell malignant lymphoma consisting of large "histiocytic"-type cells (A) and well-differentiated lymphocytic cells (B). (Magnification, 1000 ×.)

Waldenström-type macroglobulinemia, and other neoplasma producing specific immunoglobulin components. Lymphoid neoplasms having the various immunological, lesion distribution, and histological criteria of the plasmacytoid criteria occur in the dog. Peripheral lymph node lesion biopsy alone does not allow for differentiation of the various forms of B lymphocyte neoplasia. Also, some canine lymphoid neoplasms lack morphologically evident plasma cell differentiation, but they secrete monoclonal immunoglobulin components detectable by serum electrophoresis.

The relative frequency of the different histological cell types is indicated by the summary of 94 cases presented to the Veterinary Medicine Teaching Hospital at the University of California, Davis (Table 2). As in other reports, animal age is not specifically correlated with tumor cell type, and cell type is not associated with anatomical distribution pattern of the lesions.[9,10]

## IMMUNOLOGICAL CHARACTERIZATION OF LYMPHOID NEOPLASMS

Lymphocytes consist of two basic populations, B lymphocytes (bursal-equivalent derived) and T lymphocytes (thymus derived) and their precursors. Each of those two major populations has subpopulations with different roles in the immune response (Chapter 4). The current status of knowledge of the normal lymphocyte subpopulations and their identification in the canine are reviewed elsewhere in this book (Chapter 4). Cell suspensions from neoplastic lesions have been characterized for cell type and other immunological attributes. The procedure for T-cell identification is spontaneous rosette formation with guinea pig or human erythrocytes,[41-44] but that test does not identify all the T-cell subpopulations. B lymphocytes can be identified with a fluorescent antibody technique that detects cell-surface immunoglobulin or with the EAC (erythrocyte–antibody–complement) rosettes that identify receptors for activated complement.[41-44] Both B and T lymphocytes have Fc receptors, but the specificity for different subpopulations of each has not been defined in the dog. Mitogen responsiveness and scanning and transmission electron microscopy have also been studied as possible techniques for characterization of the biological nature of different lymphoid neoplasms.[39,45] Originally, T cells were thought to have smooth cell surfaces, which is in contrast to the numerous prominent villi on B cells, but subsequent scanning electron microscopic studies have not confirmed that difference. Both smooth and villous cell surfaces have been observed in B lymphocyte neoplasms of the dog (Figs. 7a and 7b). The possible association of these variations in cell-surface morphology with other cell-membrane-controlled functions has not been determined.

The results of immunological characterization of cells from neoplastic lymphoid lesions have been compared with the gross anatomical distribution and

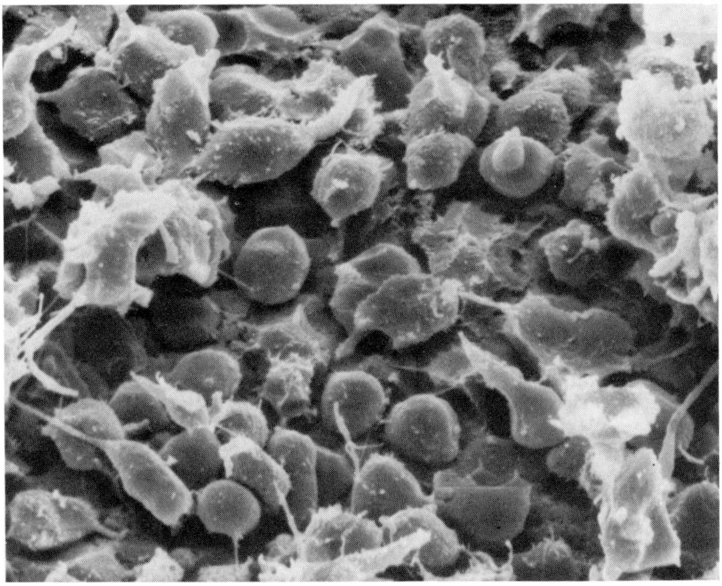

(a)

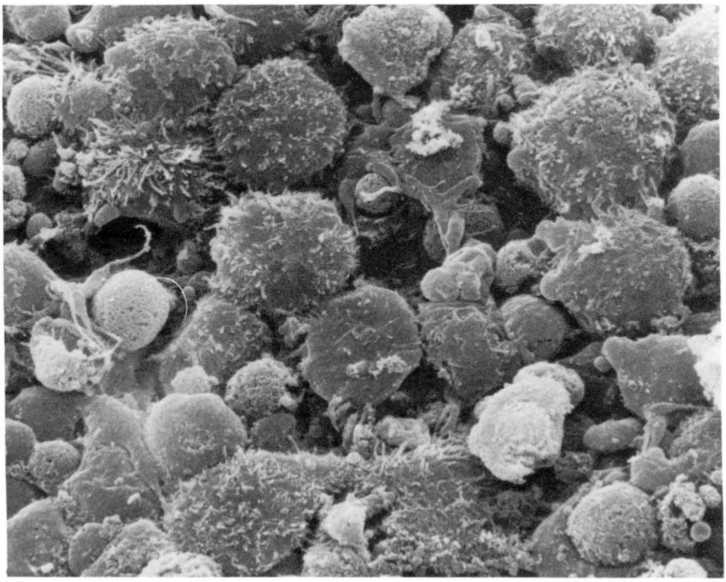

(b)

Fig. 7 Scanning electron micrographs of cells from a canine malignant lymphoma. Cells show both a smooth cell-surface morphology (a) and a villous cell-surface morphology (b), even though both groups functionally are B lymphocytes.

histological classification of the lesions. Most of these cellular studies have been conducted on multicentric malignant lymphomas. On the basis of the presence of surface immunoglobulin and/or complement receptors as reported in the literature, the majority of cases (72% of 64 cases) were of B-cell nature.[32,39,45-48] Typically, 85 to 95% of the cells were positive for B-cell markers, and less than 15% of cells were positive for T lymphocyte markers. Although most cells from these B-lymphocyte-type neoplasms were positive for both surface immunoglobulin and complement receptors, biological heterogeneity was indicated by cases that were both immunoglobulin positive and complement-receptor negative and cases that were immunoglobulin negative and complement-receptor positive[32,42,46] as well as by variable expression of Fc receptors.

Twenty-eight percent of the 64 cases of canine malignant lymphoma were not defined as distinctly B or T type. Not all the currently available cell identification tests were conducted on all the animals in those initial reports. Within that 28% were cases with cells lacking B- or T-cell markers (after correction for the expected less than 15% remaining nonneoplastic cells). Also, cases with moderate numbers (20 to 55%) of B or T cells could not be confidently classified as to neoplastic cell type. Only two cases were reported to have both B- and T-cell markers on the same cells.[32]

The few thymic lymphoma cases examined had either formed T-cell rosettes or lacked any cell-surface markers.[32] Some rosette negative thymic lymphoma cells can develop the ability to form spontaneous rosettes after in vitro culture for 3 to 5 days (C. A. Holmberg, Texas A&M University, unpublished data).

The alimentary lymphomas examined have all been of B-cell nature. The multicentric lymphomas with skin lesions, as with mycosis fungoides of man, have a T-cell nature[32] (also C. A. Holmberg, Texas A&M University, unpublished data).

The comparison of histological classification of malignant lymphomas with immunological results indicated that all neoplasms with a nodular pattern were composed of immunoglobulin positive B lymphocytes, which is analogous to findings in the human.[39,49] The only morphological differentiation for lymphoid neoplasms was cell differentiation by B lymphocytes.[39] The cytological heterogeneity among the different lymphoid neoplasms may be analogous to the "maturation blocks" and morphological variation seen during blast transformation in response to mitogen or antigen stimulation of small lymphocytes.[39]

The mitogen responsiveness of neoplastic lymphoid cell populations is quite variable, although it is generally less than that of normal lymphocyte populations from lymph nodes.[39,45] Comparisons of mitogen responsiveness with the different histological types combined with studies for surface immunoglobulin suggested that nodular histiocytic lymphomas were biologically homogeneous, whereas diffuse, poorly differentiated lymphocytic lymphomas were biologically heterogenous.[39] Also, cell-surface morphology viewed by scanning electron microscopy was not correlated with cell-surface immunoglobulin.[39]

Monoclonal immunoglobulin peaks (M components) have been detected in the sera of dogs with both neoplastic and nonneoplastic lymphoid proliferations.[50–56] The frequency of M components in the various forms of lymphoid neoplasia has not been established, although examples of IgM, IgG, and IgA M components have been demonstrated.

Two other immunological aspects of canine malignant lymphomas that have been investigated are the immunological reactivity or capability[57–59] and antigens on neoplastic cell populations.[60–65] Both the humeral and cellular immunological reactivity were lower in animals with lymphomas than in normal animals or animals with neoplasms other than lymphoma.[57,58] The test procedures included antibody production after specific immunization, clearance of foreign particles from the blood, and skin-graft rejection. Tumor specific antigens were demonstrated by antigen-induced lymphocyte blastogenesis, and an antigen common to all lymphoid tumors was demonstrated by precipitating antibody.[60–62]

The recent development of techniques that allow both the characterization of surface markers on canine lymphocytes and the growth of colonies of lymphohematopoietic elements in semisolid culture systems has provided the means for a new approach to the study of malignant lymphomas in dogs.

We recently had the opportunity to study six cases of malignant lymphoma in beagles from the Davis Laboratory for Energy-Related Health Research colony with a variety of techniques for the characterization of cell-surface receptors (Chapter 2) and for their response to stimulation of lymphohematopoietic colonies in semisolid culture systems (Chapter 2). Two of the six animals presented with malignant lymphoma in regionally restricted lymph nodes, one animal presented with simultaneous metastatic mammary carcinoma and malignant lymphoma, one animal presented with lymphoma restricted to the spleen, one presented with multicentric lymphoma with bone marrow and blood involvement, and one beagle presented with hematogenous lymphoid neoplasia compatible with chronic lymphocytic leukemia of well-differentiated lymphocytic cell type. Four of the animals received combination chemotherapy consisting of cytoxan, vincristione, and prednisolone.

No uniform profile was found for these canine malignant lymphomas. Although some animals exhibited profound and uniform suppression of all parameters, others demonstrated a selective increase in B-cell profiles.

In the four animals receiving chemotherapy, response to the cytotoxic agents was reflected by changes in both lymphocyte progenitors and bone-marrow monocyte–granulocyte progenitors.

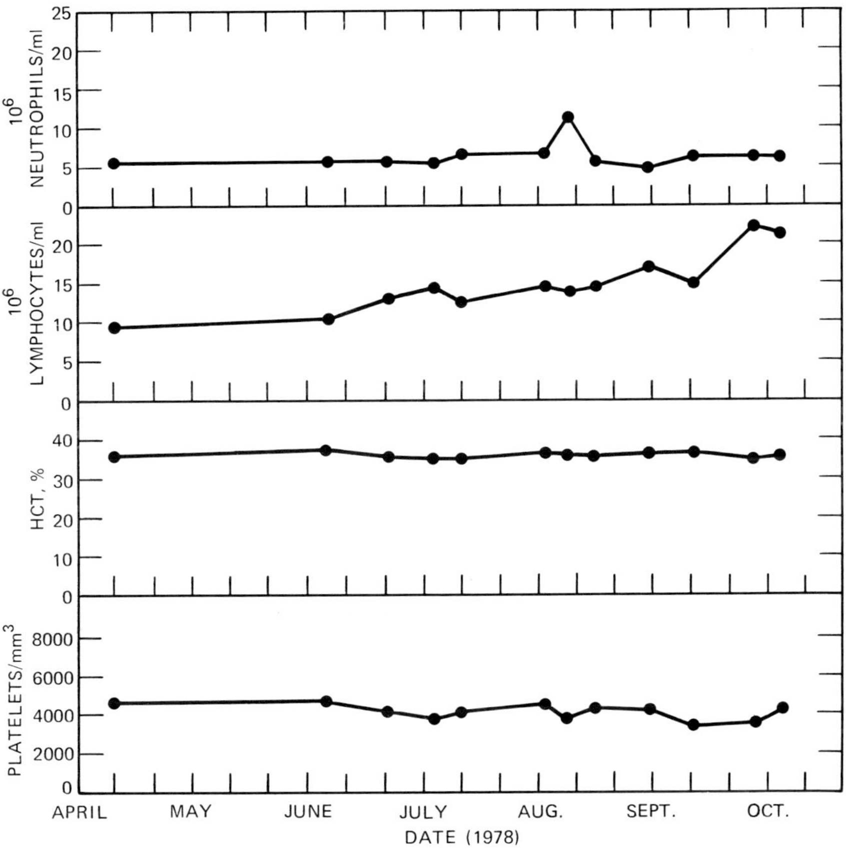


**Fig. 8 Hematological values for a dog with chronic lymphocyte leukemia of B lymphocyte nature.**

The animal with chronic lymphocytic leukemia (Fig. 8) exhibited a persistently high percentage of B cells, identified by EAC rosettes, in blood and an abnormal increase in presumptive B-cell progenitors (anti-IgG stimulated colonies) (Table 3). T-cell indicators were generally reduced. Thus this dog had a B-cell form of chronic lymphocytic leukemia, a form of lymphoid neoplasia not common in the dog but similar to the disease in man.

Simultaneous use of cell-marker studies and semisolid agar cloning techniques will synergistically provide quantitative and functional information for both malignant cells and normal host cells for characterization of malignant lymphomas. The multiparametric approach to the study of canine malignant lymphomas appears to potentially offer a new method for the investigation of mechanisms and functional properties of these hematopoietic disorders and also allows for the monitoring of response to chemotherapy.

TABLE 3

CHARACTERIZATION STUDIES ON SURFACE RECEPTORS AND T AND B LYMPHOCYTE PROGENITORS IN PERIPHERAL BLOOD

| | Surface markers (percent of cells)* | | | Lymphocyte colonies*† | | |
|---|---|---|---|---|---|---|
| Date | Percent E rosettes | Percent EAC rosettes | Percent immunoglobulin positive | Con-A | PHA-P | Anti-IgG |
| 6-29-78 | 13 | 69↑↑ | 39 | 64 | 2 | 29 |
| 7-18-78 | 12 | 48↑ | 28 | 247 | 69 | |
| 8-4-78 | 4↓ | 46↑ | 84↑↑↑ | 10↓ | | 13↓ |
| 9-28-78 | 9 | 58↑ | | >1↓ | 8↓ | 57 |
| 10-10-78 | 10 | 76↑↑ | | 13↓ | 2↓ | 1379↑↑↑ |

*Arrows following the values indicate a parameter that is markedly different from control values.

†Colonies formed per $5 \times 10^5$ cells plated with the T-cell stimulating Con-A and PHA-P or B-cell stimulating antiserum to IgG.

## TREATMENT OF LYMPHOID NEOPLASIA

Treatment of lymphoid neoplasia in the canine has included chemotherapy, immunotherapy, and radiotherapy.[8] Although the mean survival without therapy is approximately 30 days,[10,63] mean survival was increased to approximately 200 days with chemotherapy.[64,65] Longer mean survival has been obtained when chemotherapeutically induced remission is followed by vaccination with chemically altered autologous neoplastic cells. As in humans, various drug combinations yield longer survival durations than do individual drugs.[8] One important aspect of chemotherapy that can be investigated with the dog lymphoma model is the response of various neoplastic cell subpopulations to individual drugs. The relative effect of different new drugs could also be monitored for both the neoplastic cell population and the immunological function of nonneoplastic lymphoid subpopulations in animals with tumors of characterized immunological nature. Chemotherapeutic agents have a differential effect on normal lymphocyte subpopulations, but the response of individual lymphoma cases has not yet been related to the morphological lymphoma classifications.

Another treatment for canine malignant lymphomas is radiation followed by bone-marrow reconstitution. Weiden et al.[66] investigated the effect of autologous bone-marrow reconstitution in 25 dogs with malignant lymphomas or nonhematopoietic solid tumors treated with 1200-R total-body irradiation. Many of the animals died early as a result of sepsis from radiation-induced leukopenia. Only 4 of the 14 animals surviving treatment long enough for clinical evaluation showed significant response of their tumor to total-body irradiation. Seven of the eight dogs surviving more than 14 days ultimately died of recurrent lymphoma; the longest survival was 144 days. These investigators concluded that lethal doses of total-body irradiation alone are not a dependable cure for spontaneous malignant lymphomas in the dog.

## CONCLUSIONS

Naturally occurring lymphoid neoplasia in the canine provides an excellent animal model for investigating various aspects of the basic biology of this disease process. The range of clinical–pathological variation of the disease in dogs is similar to that in man, especially the non-Hodgkin's malignant lymphoma. The predominant form of lymphoid neoplasia in the canine is multicentric malignant lymphomas in middle and older age dogs. That form has a spectrum of histological cell types like the non-Hodgkin's malignant lymphomas of man. Other similarities with the human disease are their B-cell nature and lack of evidence for a viral etiology. The other less-frequent forms of spontaneous canine lymphoid neoplasia with analogies to lymphoid neoplasms in man include thymic lymphomas of T-cell nature and marked blood vascular (leukemic) involvement, various forms of plasma cell differentiating lymphomas with gammopathies, cases with skin involvement that may be comparable to mycosis fungoides, alimentary tract neoplasms, and the lymphocytic leukemias. The availability of large neoplastic and nonneoplastic tissue samples and serum from the individual animal; the ability to closely monitor natural progression and experimental therapy; and the continually growing background of knowledge of the genetics, tissue typing, immunology, and general biology of the dog provide advantages not available in other models for this disease process.

## REFERENCES


1. C. R. Dorn, D. O. N. Taylor, and H. H. Hibbard, Epizootiological Characteristics of Canine and Feline Leukemia and Lymphoma, *Am. J. Vet. Res.*, 28: 993-1001 (1967).
2. C. R. Dorn, Survey of Animal Neoplasms in Alameda and Contra Costa Counties, California. II. Cancer Morbidity in Dogs and Cats from Alameda County, *J. Natl. Cancer Inst.*, 40: 307-318 (1968).
3. W. A. Priester, Tumors in Domestic Animals, *J. Natl. Cancer Inst.*, 47: 1333-1334 (1971).
4. F. Bloom and L. M. Meyer, Malignant Lymphoma (So-Called Leukemia) in Dogs, *Am. J. Pathol.*, 21: 683-715 (1945).
5. D. Wirth, Die Leukamie Beim/Hund, *Monatsh. praktische Tierkeilkunde,* 31: 97-130 (1920).
6. A. Parodi, M. Wyers, and J. Paris, Incidence of Canine Lymphoid Leukemia: Age, Sex and Breed Distribution; Results of a Necropsic Survey, *Bild Haematol.*, 30: 263-266 (1968).

7. J. E. Moulton and D. L. Dungworth, Tumors of the Lymphoid and Hemopoietic Tissues, in *Tumors in Domestic Animals,* 2nd ed., J. E. Moulton (Ed.), pp. 150-204, University of California Press, Berkeley, 1978.
8. G. H. Theilen and B. R. Maidwell, *Veterinary Cancer Medicine,* Chap. 13, Part 2, pp. 241-252, Lea & Febiger, Philadelphia, 1979.
9. J. Anderson, W. F. H. Jarrett, and G. W. Crighton, A Classification of Lymphoid Neoplasms of Domestic Mammals, *Natl. Cancer Inst. Monogr.,* 32: 343-353 (1969).
10. R. A. Squire, M. Bush, E. C. Melby, L. M. Neeley, and B. Yarbrough, Clinical and Pathological Study of Canine Lymphoma: Clinical Staging, Cell Classification, and Therapy, *J. Natl. Cancer Inst.,* 51: 565-574 (1973).
11. T. A. Nobel, F. Nauman, and U. Klopfer, Survey of Animal Neoplasms in Israel, 1954–1959, *Refuah Vet.,* 27: 115-117 (1970).
12. A. W. Backgren, Lymphatic Leukosis in Dogs. An Epizoological, Clinical, and Haematological Study, *Acta Vet. Scand.,* Suppl. 1 (6): 1-80 (1965).
13. R. A. Squire, Spontaneous Hematopoietic Tumors of Dogs, *Natl. Cancer Inst. Monogr.,* 32: 97-116 (1969).
14. N. H. Altman and R. A. Squire, Diagnosis of Canine Lymphoma, *J. Am. Vet. Med. Assoc.,* 157: 1676-1680 (1970).
15. T. C. Jones and J. A. Berg, Comparative Pathology of Lymphocytic Malignancies, in *The Lymphocyte and Lymphocytic Tissues,* pp. 170-173, Paul B. Hoeber, Inc., New York, 1960.
16. H. A. Smith, Malignant Lymphomas in Animals, *Am. J. Clin. Pathol.,* 38: 75-87 (1966).
17. W. A. Priester, Canine Lymphoma: Relative Risk in Boxer Breed, *J. Natl. Cancer Inst.,* 38: 833-845 (1967).
18. W. A. Priester and N. Mantel, Occurrence of Tumors in Domestic Animals. Data from 12 United States and Canadian Colleges of Veterinary Medicine, *J. Natl. Cancer Inst.,* 47: 1333-1344 (1971).
19. A. L. Chapman et al., Preliminary Report on Virus-Like Particles in Canine Leukemia and Derived Cell Cultures, *Cancer Res.,* 27: 18-25 (1967).
20. T. J. Kakuk, Isolation of a Canine Herpesvirus from a Dog with Malignant Lymphoma, *Am. J. Vet. Res.,* 30(11): 1951-1960 (1969).
21. T. J. Kakuk, R. W. Hinz, R. F. Longham, and G. H. Conner, Experimental Transmission of Canine Malignant Lymphoma to the Beagle Neonate, *Cancer Res.,* 28: 716-723 (1968).
22. L. N. Owen, Transplantation of Canine Lymphocytic Leukemia, *Bibl. Haematol.,* 39: 139-148 (1973).
23. H. Cohen, A. L. Chapman, W. J. Bopp, C. E. Schmidt, and C. E. Przyblski, Pathogenesis of a Transplanted Canine Lymphocytic Leukemia, *Cancer,* 33: 1313-1324 (1974).
24. Lindsay J. Mackey and W. F. H. Jarrett, Pathogenesis of Lymphoid Neoplasia in Cats and Its Relationship to Immunologic Cell Pathways. I. Morphologic Aspects, *J. Natl. Cancer Inst.,* 49: 853-865 (1972).
25. W. F. H. Jarrett and L. J. Mackey, Neoplastic Diseases of the Haematopoietic and Lymphoid Tissues, *Bull. World Health Organ.,* 50: 21-34 (1974).
26. R. J. Lukes and R. D. Collins, Immunological Characterization of Human Malignant Lymphomas, *Cancer,* 34: 1488-1503 (1974).
27. R. J. Lukes et al., A Morphologic and Immunologic Surface Marker Study of 299 Cases of Non-Hodgkin's Lymphomas and Related Leukemias, *Am. J. Pathol.,* 90: 461-486 (1978).
28. R. C. Braylan, E. S. Jaffee, and C. W. Berard, Malignant Lymphomas: Current Classification and New Observations, in *Haematologic and Lymphoid Pathology Decennial: 1966–1976,* S. D. Sommers (Ed.), pp. 213-270, Appleton Century Crofts, New York, 1975.

29. R. F. Dorfman, Classification of Non-Hodgkin's Lymphomas, *Lancet,* 1: 1295-1296 (1974).
30. R. B. Mann, E. S. Jaffee, and C. W. Berard, Malignant Lymphomas—A Conceptual Understanding of Morphological Diversity, *Am. J. Pathol.,* 94: 105-191 (1979).
31. H. Rappaport and R. C. Brylan, Changing Concepts in the Classification of Malignant Neoplasms of the Hematopoietic System, in *The Reticuloendothelial System,* J. W. Rebuck, C. W. Berard, and M. R. Abell (Eds.), pp. 1-19, Williams & Wilkins Co., Baltimore, 1975.
32. D. E. Onions, B and T Cells in Canine Lymphosarcoma Cells, *J. Natl. Cancer Inst.,* 59: 1001-1006 (1977).
33. J. L. Smith, C. R. Barker, G. P. Clein, and R. D. Collins, Characterization of Malignant Mediastinal Lymphoid Neoplasm (Sternberg Sarcoma) as Thymic in Origin, *Lancet,* 1: 74-77 (1973).
34. C. A. Holmberg, J. S. Manning, and B. I. Osburn, Feline Malignant Lymphomas: Comparison of Morphologic and Immunologic Characteristics, *Am. J. Vet. Res.,* 37: 1455-1460 (1976).
35. A. Potena and S. Damiano, Leukemic–Lymphoid Leukemia in a Dog with Mediastinal Syndrome. The Anatomo-Pathological Findings, *Acta Med. Vet.,* 14: 223-240 (1968).
36. E. B. Howard, S. W. Nielsen, and A. J. Kenyon, Cutaneous Lymphomatosis in a Boxer Dog. Pathologic and Biochemical Observations, *Pathol. Vet.,* 6: 76-85 (1969).
37. D. F. Kelley, R. E. W. Haliwell, and R. M. Schwartzman, Generalized Cutaneous Eruptions in a Dog with a Histological Similarity to Human Mycosis Fungoides, *Brit. J. Dermatol.,* 86: 164-171 (1972).
38. H. Rappaport, *Atlas of Tumor Pathology,* Sec. III—Fasicle 8 Hematopoietic System, Armed Forces Institute of Pathology, Washington, D. C., 1966.
39. C. A. Holmberg, J. S. Manning, and B. I. Osburn, Canine Malignant Lymphomas: Comparison of Morphological and Immunological Parameters, *J. Natl. Cancer Inst.,* 56: 125-135 (1976).
40. S. E. Jones et al., Non-Hodgkin's Lymphomas. IV. Clinicopathological Correlation in 405 Cases, *Cancer,* 31: 806-823 (1973).
41. C. A. Bowles, G. S. White, and D. Lucas, Rosette Formation by Canine Peripheral Blood Lymphocytes, *J. Immunol.,* 114: 399-402 (1975).
42. C. C. Whitacre and R. W. Lang, A Technique for Separation of Canine Lymphocytes and Their Use in the Lymphocyte Toxic, Blastogenic, and Rosette Assays, *Transfusion,* 15: 346-350 (1975).
43. C. A. Miller, A. R. Carbonell, R. Peng, M. R. MacKenzie, and M. Shifrine, Cell Surface Markers on Canine Lymphocytes, *Am. J. Vet. Res.,* 39: 1191-1194 (1978).
44. A. R. Zander, N. Boopalm, and R. B. Epstein, Surface Markers on Canine Lymphocytes, *Transplant. Proc.,* 7: 369-373 (1975).
45. S. K. Dutta, M. N. Novilla, M. K. Bumgardner, and A. Ingling, Lymphocyte Responsiveness to Mitogens and Quantitation of T and B Lymphocytes in Canine Malignant Lymphoma, *Am. J. Vet. Res.,* 39: 455-458 (1978).
46. C. A. Holmberg, J. S. Manning, and B. I. Osburn, Canine Malignant Lymphomas with B Lymphocyte Characteristics, *Am. J. Vet. Res.,* 38: 1877-1879 (1977).
47. D. E. Onions, B- and T-Cells in Canine Lymphosarcoma, *Vet. Rec.,* 97: 108 (1975).
48. D. E. Onions, B and T Cells in Canine Lymphosarcoma, *Brit. J. Cancer,* 32: 241 (1975) (Abstract).
49. E. S. Jaffee, E. M. Shevach, M. M. Frank, C. W. Berard, and I. Green, Nodular Lymphoma—Evidence for Origin from Follicular B Lymphocytes, *New Engl. J. Med.,* 290: 813-818 (1974).
50. O. W. Schalm, N. C. Jain, and E. J. Carroll, *Veterinary Hematology,* pp. 550-558; 617-628, Lea & Febiger, Philadelphia, 1975.

51. E. G. MacEwen, A. I. Hurvitz, and A. Hayes, Hyperviscosity Syndrome Associated with Lymphocytic Leukemia in Three Dogs, *J. Am. Vet. Med. Assoc.*, 170: 1309-1312 (1977).
52. G. Virella, R. J. Slappendel, and J. Goudswaard, Multiple Myeloma, IgA Cryoglobulinemia and Serum Hyperviscosity in a Dog, *Int. Arch. Allergy Appl. Immunol.*, 55: 537-541 (1977).
53. K. G. Braund, R. M. Everett, and R. A. Albert, Neurologic Manifestations of Monoclonal IgM Gammopathy Associated with Lymphocytic Leukemia in a Dog, *J. Am. Vet. Med. Assoc.*, 172: 1407-1410 (1978).
54. W. C. Pires, R. B. Rojas, H. Sechin, C. D. E. Oliveira, and R. D. E. Mello, Multiple Myeloma in the Dog, *Haematol. Lat. (Milan)*, 13: 207-210 (1971).
55. A. I. Hurvitz, J. M. Kehoe, J. D. Capra, and R. Prata, Bence Jones Proteinemia and Proteinuria in a Dog, *J. Am. Vet. Med. Assoc.*, 159: 1112-1116 (1971).
56. V. J. Shepard, W. J. Dodds-Laffin, and R. J. Laffin, Gamma A Myeloma in a Dog with Defective Hemostasis, *J. Am. Vet. Med. Assoc.*, 160: 1121-1127 (1972).
57. P. L. Weiden et al., Immune Reactivity in Dogs with Spontaneous Malignancy, *J. Natl. Cancer Inst.*, 53: 1049-1056 (1974).
58. L. N. Owen, D. E. Bostock, and R. E. Halliwell, Cell Mediated and Humeral Immunity in Dogs with Spontaneous Lymphosarcoma, *Eur. J. Cancer*, 11: 187-191 (1975).
59. C. A. Bowles, D. Lucas, L. Norton, and R. G. Graw, Immunological Studies of Canine Lymphosarcoma: Mixed Leukocyte Reactivity Following Chemotherapy, *Clin. Immunol. Immunopathol.*, 9: 211-217 (1978).
60. W. D. Winters and H. D. Snow, Immunodiffusion Studies of Canine Lymphosarcoma Antigens, *J. Natl. Cancer Inst.*, 53: 1027-1032 (1974).
61. W. D. Winters and H. D. Snow, Precipitating Antibodies to Canine Lymphosarcoma-Associated Antigens, *Am. J. Vet. Res.*, 35: 1333-1335 (1974).
62. M. S. Tsoi, P. L. Weiden, and R. Storb, Immunological Reactivity to Soluble Autochthonous Tumor Extracts in Dogs with Spontaneous Malignancies, *J. Immunol.*, 116: 1134-1139 (1976).
63. B. R. Modewell, Chemotherapy for Canine Lymphosarcoma, *Am. J. Vet. Res.*, 36: 1525-1528 (1975).
64. S. E. Crow, G. H. Theilen, E. Benjamini, M. Torten, A. M. Henness, and W. C. Buhles, Chemoimmunotherapy for Canine Lymphosarcoma, *Cancer*, 40: 2102-2108 (1977).
65. G. H. Theilen, M. Worley, and E. Benjamini, Chemoimmunotherapy for Canine Lymphosarcoma, *J. Am. Vet. Med. Assoc.*, 170: 607-610 (1977).
66. P. L. Weiden, R. Storb, K. G. Lerner, G. F. Kao, T. C. Graham, and E. P. Thomas, Treatment of Canine Malignancies by 1200 R Total Body Irradiation and Autologous Marrow Grafts, *Exp. Hematol.*, 3: 124-134 (1975).

# The Canine as a Model in Radiobiologic Research

# 16

STEVEN A. BOOK
*Laboratory for Energy-Related Health Research, University of California, Davis*

Concern about the potential health effects of people from developing nuclear technologies has required that biologic effects of radiation be documented in appropriate animal models. Emphasis, however, has not been placed merely on short-term effects of very high radiation doses but rather on the long-term (late) effects of low exposures, primarily carcinogenesis. A variety of animal species has been used in experimental studies to answer the numerous questions about radiation effects and about relationships between dose, effect, and time. Ultimately, the information obtained from animal species is to be projected to humans to predict the response of people exposed to radiation.

First, the proper evaluation of the long-term effects of radiation required experimental animals with sufficient longevity to allow manifestation of the late effects. Second, those animals had to possess physiologic and anatomic characteristics that resemble human characteristics. Third, for use in laboratory situations, appropriate animal models had to be easy to obtain and handle.

The dog met those requirements and, in addition, brought with it a considerable background in anatomy, physiology, and behavior from its prior use as a laboratory animal (on a small, individual scale) and as a patient in veterinary medicine. The beagle, because of its substantial pedigree history, availability, suitability for living in a colony situation, and desirable physical characteristics and disposition, became the experimental dog of choice for radiobiologic research.[1,2]

For the past three decades, the beagle dog has been used as the primary large animal model for studies of radiation-induced carcinogenesis. For studies begun by the U. S. Atomic Energy Commission, colonies of beagles were established at the University of California at Davis, the University of Utah, the University of

Rochester, the Lovelace Foundation, Battelle Pacific Northwest Laboratories, and Argonne National Laboratory. Another beagle colony, sponsored by the U. S. Food and Drug Administration's Bureau of Radiological Health, was established at Colorado State University. These laboratories have used beagles in various radiation studies involving exposures externally to radiation from X-ray or gamma sources or internally to deposited alpha or beta emitters administered by injection, ingestion, or inhalation. Canine exposures, furthermore, have been acute or chronic, single or fractionated. Animals have been exposed to radiation as adults or as pups, including neonates and fetuses. A wide array of radionuclides in a variety of chemical forms has been administered.

Because the results of those studies would fill many volumes, this chapter represents a general overview of a single, long-term experiment on radiation toxicology currently being performed at the University of California at Davis. The purpose of this chapter is, not to review all beagle radiobiology, but instead to demonstrate the value of the canine model in radiobiologic research. It is, after all, in the field of radiation biology that the canine model—the beagle model—reaches its *succès fou.*

## COMPARATIVE TOXICITY OF STRONTIUM-90 AND RADIUM-226

Beagle dogs, administered graded dosages of $^{90}$Sr or $^{226}$Ra in early life, have been under study for nearly 20 years at the Laboratory for Energy-Related Health Research (formerly the Radiobiology Laboratory) of the University of California at Davis. The primary goal of this study is to provide information on long-term consequences from chronic exposure to bone-seeking radionuclides for application to human experiences.

### Historical Aspects

In the late 1950's the experimental design of the study was established.* The fundamental concept of the experiment, as delineated in 1960, was to evaluate

*Those who participated in the development of the experimental design included, as the Board of Consultants, Drs. R. D. Evans of the Massachusetts Institute of Technology, A. M. Brues of Argonne National Laboratory, W. H. Langham of Los Alamos Scientific Laboratory, and H. A. Blair of the University of Rochester Atomic Energy Project. Others of the "Founding Fathers" included Dr. H. D. Bruner and Mr. D. M. Gardiner of the U. S. Atomic Energy Commission; Captain J. Barton, U.S.A.F., of the Defense Atomic Support Agency; Dr. W. S. S. Jee of the University of Utah; and Drs. A. C. Andersen, M. Goldman, and F. Shultz of the University of California at Davis. Some twenty other scientists also participated in meetings prior to the establishment of the study design.

the relative toxicity of $^{90}$Sr and $^{226}$Ra in beagles. The project at Davis was to provide estimates of the effects in beagles from continuous exposure to $^{90}$Sr and from brief exposure to $^{226}$Ra. Estimates of the human effects from brief exposures to $^{226}$Ra were to be provided by other investigators from studies on people who had been exposed to $^{226}$Ra occupationally or medically. The relative toxicity of $^{90}$Sr ingestion by humans could then be estimated by the following approximate relationship:

$$\frac{\text{Effects in people from } ^{90}\text{Sr (continuous exposure)}}{\text{Effects in people from } ^{226}\text{Ra (brief exposure)}} \approx \frac{\text{Effects in beagles from } ^{90}\text{Sr (continuous exposure)}}{\text{Effects in beagles from } ^{226}\text{Ra (brief exposure)}}$$

In establishing the experimental design, the advisory group iterated the reasons for using the beagle as an experimental animal, citing the practical experience in the care and management of a large beagle colony over previous years and the physical characteristics of the beagle that provide a reasonable basis for estimating effects of radiation—body size, bone structure, metabolism, physiology, pathology, and life span—to permit a more accurate extrapolation to humans than would be possible from smaller experimental animals.

## Experimental Design

For the scaling of radiation effects to humans, appropriate models were used for the comparison of effects produced by multiple injections of $^{226}$Ra into young adult beagles with effects of $^{90}$Sr ingestion. Radium-226 was injected intravenously into dogs beginning at 435 days of age and every 2 weeks thereafter until 540 days for a total of eight injections. Six treatment levels yielded dosages with an approximately 500-fold range (R dogs, Table 1). The activity per injection ranged from 0.003 to 1.25 $\mu$Ci of $^{226}$Ra per kilogram of body weight, and the nominal quantity injected, assuming a 10-kg dog, ranged from 0.24 to 100 $\mu$Ci of $^{226}$Ra.

If 25% of the injected radium is assumed to be retained in the skeleton, the dosage at R10 represents the canine equivalent of 10 times the maximum permissible skeletal burden for man (0.1 $\mu$Ci of $^{226}$Ra).[3] The pattern of injection—from 14 to 18 months of age—was chosen to simulate the pattern of $^{226}$Ra assimilation by the radium dial painters, a population of young women who painted luminous timepiece dials in the early part of this century and received significant exposures of alpha-particle-emitting radionuclides.[4]

Strontium-90 was administered in the food, with the dietary concentration maintained at a constant level with respect to well-controlled dietary calcium levels. Dosages were administered to seven treatment groups (plus controls) over a 1500-fold range (D dogs, Table 2). Exposures of the experimental animals

TABLE 1

EXPERIMENTAL DESIGN OF $^{226}$Ra TOXICITY STUDY

$^{226}$Ra Injection Series
(eight biweekly injections starting at 435 days of age)

| Treatment code | Multiple of 10 level | Dose, μCi/kg of body weight | Total injected, μCi/kg of body weight | Nominal quantity injected, μCi |
|---|---|---|---|---|
| R00 | 0 | 0 | 0 | 0 |
| R05 | 0.3 | 0.003 | 0.024 | 0.24 |
| R10 | 1 | 0.008 | 0.064 | 0.64 |
| R20 | 6 | 0.047 | 0.376 | 3.76 |
| R30 | 18 | 0.14 | 1.12 | 11.2 |
| R40 | 54 | 0.42 | 3.36 | 33.6 |
| R50 | 162 | 1.25 | 10.0 | 100.0 |

TABLE 2

EXPERIMENTAL DESIGN OF $^{90}$Sr TOXICITY STUDY

$^{90}$Sr Ingestion Series
(in utero to 540 days of age)

| Treatment code | Multiple of 10 level | Dose, μCi $^{90}$Sr/ g dietary calcium | Total ingested, μCi/day | Nominal quantity ingested, μCi |
|---|---|---|---|---|
| D00 | 0 | 0 | 0 | 0 |
| D05 | 0.3 | 0.007 | 0.02 | 10 |
| D10 | 1 | 0.021 | 0.07 | 40 |
| D20 | 6 | 0.123 | 0.44 | 240 |
| D30 | 18 | 0.37 | 1.3 | 700 |
| D40 | 54 | 1.11 | 4 | 2,260 |
| D50 | 162 | 3.33 | 12 | 6,560 |
| D60* | 486 | 10.0 | 36 | 19,400 |

$^{90}$Sr Injection Series
(single iv injection at 540 days of age)

| Treatment code | Multiple of 10 level | Dose, μCi/kg of body weight | Total injected, μCi/kg of body weight | Nominal quantity injected, μCi |
|---|---|---|---|---|
| S20 | 6 | 3.7 | 3.7 | 37 |
| S40 | 54 | 33 | 33 | 330 |

*D60 dogs were not in the original experimental design but were added in 1967.

began in utero (21 days after mating) when their dams were given radioactive feed. Pups were nursed while their mothers continued on the dietary regimen, and pups were weaned onto the same $^{90}$Sr-containing diet. This ration was fed until the dogs were 540 days of age; thereafter the animals received the standard (nonradioactive) laboratory ration. The nominal quantity ingested was 10 to 19,400 $\mu$Ci of $^{90}$Sr.

The time of administration—from prior to birth until 18 months of age—includes the period of skeletal maturation and was chosen to represent human exposures by ingestion of $^{90}$Sr-contaminated foods during early life.

Two additional small groups of dogs (S dogs, Table 2) were given their dosages of $^{90}$Sr as a single intravenous injection at 540 days of age to serve in the comparison of the two modes of administration and to act as a common link between this project and another related study at the University of Utah.

## Current Status of the Study

A total of 862 beagles entered the lifetime study, with 335 in the $^{226}$Ra series, 471 in the $^{90}$Sr ingestion series, and 46 in the $^{90}$Sr injection series. Of those dogs, as of Sept. 30, 1979, 89 remained alive in the $^{226}$Ra series (Table 3), 133 remained alive in the $^{90}$Sr ingestion series, and 15 remained alive in the $^{90}$Sr injection series (Table 4).

TABLE 3

STATUS OF $^{226}$Ra TOXICITY STUDY IN BEAGLES AS OF SEPTEMBER 1979

| Treatment code | Original number of dogs | Number alive as of September 1979 | Primary bone cancer deaths | Pathologic fractures |
|---|---|---|---|---|
| R00 | 84 | 44 | 0 | 0 |
| R05 | 46 | 21 | 0 | 0 |
| R10 | 40 | 19 | 2 | 0 |
| R20 | 42 | 5 | 21 | 0 |
| R30 | 41 | 0 | 32 | 2 |
| R40 | 41 | 0 | 38 | 1 |
| R50 | 41 | 0 | 27 | 6 |

## Metabolism and Dosimetry

Basic to the measurement of retained radioactivity in the $^{90}$Sr- or $^{226}$Ra-exposed beagle and the subsequent calculation of radiation dose is a

TABLE 4

STATUS OF $^{90}$Sr TOXICITY STUDY IN BEAGLES AS OF SEPTEMBER 1979

| Treatment code | Original number of dogs | Number alive as of September 1979 | Primary bone cancer deaths | Myeloproliferative syndrome deaths | Squamous-cell carcinoma (gingiva) |
|---|---|---|---|---|---|
| | | $^{90}$Sr Ingestion Series | | | |
| D00 | 81 | 42 | 0 | 0 | 0 |
| D05 | 79 | 27 | 0 | 0 | 0 |
| D10 | 40 | 12 | 0 | 0 | 0 |
| D20 | 66 | 23 | 0 | 0 | 0 |
| D30 | 65 | 20 | 1 | 2 | 0 |
| D40 | 61 | 9 | 4 | 7 | 5 |
| D50 | 60 | 0 | 15 | 24 | 8 |
| D60 | 19 | 0 | 10 | 7 | 1 |
| | | $^{90}$Sr Injection Series | | | |
| S20 | 20 | 7 | 1 | 1 | 0 |
| S40 | 26 | 8 | 2 | 1 | 0 |

program of lifetime whole-body monitoring. Skeletal burdens of $^{90}$Sr and its decay product $^{90}$Y, assumed to be in equilibrium, were determined in living dogs by periodic whole-body bremsstrahlung radiation counting with a 10- by 20-cm NaI(Tl) detector (later two detectors were used) in conjunction with a pulse-height analyzer.[5] A terminal count was made at death or just prior to death for dogs that were moribund. The bremsstrahlung spectrum was summed and compared with a frozen standard with a known $^{90}$Sr skeletal burden.[6]

Skeletal burdens of $^{226}$Ra were determined from whole-body gamma spectroscopy of the $^{214}$Bi daughter of $^{222}$Rn by application of the radon/radium ratio function described by Parks, Pool, and Williams.[7] This function is different for the R50-, R30-, and R20-level dogs, which implies a radiation dose-related effect on the radon/radium ratio. The functional values determined by Parks, Pool, and Williams for R30 also were used for the R40 level, and the values determined for R20 were used for the lower levels as well.

Whole-body counts of radioactivity (in microcuries) are combined with the skeletal weight and time in an expression of radionuclide concentration in microcurie-days per gram of bone. That value for concentration, when combined with appropriate radiobiologic constants, allows the calculation of skeletal dose rate (in rads/day) or the cumulative dose (in rads). Radiation doses for animals in this study are shown in Table 5. The overlapping of adjacent dose groups, which differed by factors of 3 in dosage, is due to the variations in metabolism and life span among the dogs.

TABLE 5

RANGES OF TOTAL DOSE AND AVERAGE DOSE RATE TO BEAGLE SKELETONS FROM INJECTED $^{226}Ra$ (R) OR INGESTED (D) OR INJECTED (S) $^{90}Sr$

| Treatment code | Total dose, rads | Average dose rate, rads/day |
|---|---|---|
| | $^{226}Ra$ Injection Series | |
| R00 | Background | Background |
| R05 | 10 to 130 | 0.01 to 0.04 |
| R10 | 140 to 500 | 0.04 to 0.2 |
| R20 | 770 to 2,100 | 0.2 to 0.7 |
| R30 | 1,800 to 4,300 | 0.8 to 2 |
| R40 | 5,000 to 10,000 | 3 to 9 |
| R50 | 11,000 to 25,000 | 10 to 30 |
| | $^{90}Sr$ Ingestion Series | |
| D00 | Background | Background |
| D05 | 10 to 70 | 0.003 to 0.02 |
| D10 | 80 to 180 | 0.02 to 0.05 |
| D20 | 150 to 1,300 | 0.1 to 0.3 |
| D30 | 1,100 to 4,200 | 0.3 to 1 |
| D40 | 1,400 to 11,000 | 1 to 3 |
| D50 | 2,500 to 19,000 | 3 to 10 |
| D60 | 6,700 to 19,000 | 10 to 25 |
| | $^{90}Sr$ Injection Series | |
| S20 | 170 to 1,300 | 0.1 to 0.4 |
| S40 | 1,700 to 12,000 | 0.9 to 2 |

Various aspects of $^{226}Ra$ metabolism, more particularly retention as a function of age[8] or intraskeletal distribution,[9] have been published, and much of the $^{90}Sr$ metabolism data, especially information related to the early part of the study, has been reviewed.[10,11]

## Effects

**Life Shortening.** The animals entered the study in several "waves," but none of the animals in the $^{90}Sr/^{226}Ra$ study is younger than 12 years. On the basis of data from all control animals, a median survival time of 14.7 years for unirradiated beagles is anticipated with a maximum life span of more than 17

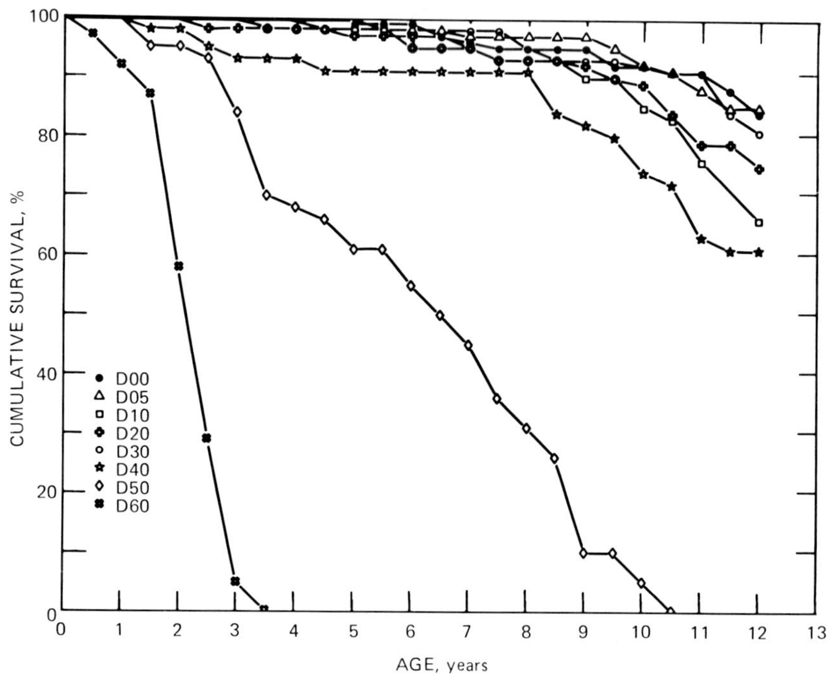


Fig. 1 Cumulative survival of $^{226}$Ra-injected dogs and controls through 12 years of age.

years. Most of our animals are younger than the projected median value, and a significant portion of the animals' longevity remains.

Beagles at risk and their cumulative survival rates indicate significant differences between dose levels. Cumulative survival rates of dogs exposed at the R05 and R10 levels appear indistinguishable from control animals through 12 years (Fig. 1). For other treatment groups, the Mantel–Haenszel[12] test for significance of differences of two cumulative survival rates shows significant differences ($p < 0.05$) from controls beginning from 8 years of age for R20 dogs, 6 years for R30's, 5 years for R40's, and 3.5 years for R50's.

Strontium-90-fed dogs exposed at D05, D10, D20, and D30 levels were not different from controls through 12 years of age (Fig. 2). Significant differences were reached at 10 years of age for D40 dogs, 3 years for D50's, and 1.5 years for D60's. Mortality of pups prior to weaning was the same in control and irradiated D groups.[13] Survival of S20 and S40 dogs was not different from controls through 12 years.

**Pathology of Radium-226-Induced Disease.** Radiation-induced mortality in beagles exposed to injected $^{226}$Ra has been largely due to osteosarcomas

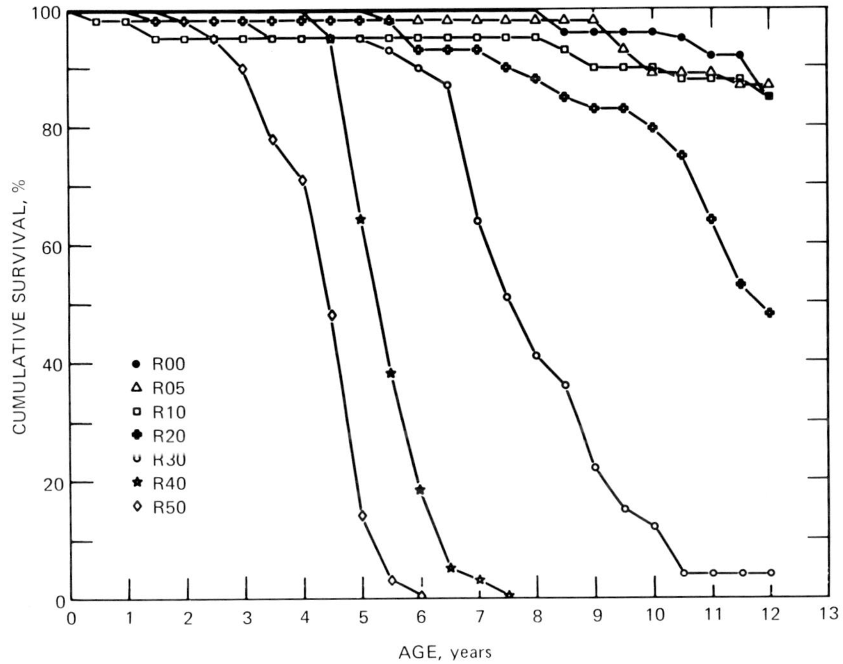


Fig. 2 Cumulative survival of $^{90}Sr$-fed dogs and controls through 12 years of age.

(Table 3).[14,15] A few animals receiving high dose levels of $^{226}Ra$ were euthanized because of radiation osteodystrophy or complications related to pathologic fractures. Three beagles (R50), all in one litter, manifested radiation nephritis, and one of those animals died as a direct result of that disease. Tapetal degeneration and several malignant melanomas of the eye have been found in dogs receiving radiation at all dose levels, but none have been found in controls.[16] Most malignant melanomas were treated by enucleation of the affected eye, but one R05 dog died of metastases. The radium dial painters did not show $^{226}Ra$-induced tumors of the eye, but humans do not have a structure equivalent to the dog tapetum lucidum, which concentrates $^{226}Ra$ by a factor of 10 over that in the beagle skeleton.[17]

The responses of dog and man to $^{226}Ra$ appear similar for both nonneoplastic and neoplastic bone damage.[18] Radium dial painters also had late occurring tumors adjacent to bones of the head, i.e., squamous-cell carcinoma of the mastoid and sinuses.[4] Our $^{226}Ra$-injected beagles have had a few radiation-induced tumors of the gingiva. In both species the time of appearance of bone tumors induced by $^{226}Ra$ clearly was postponed with decreasing dose and dose rates.

**Pathology of Strontium-90-Induced Disease.** Several different types of neoplastic disorders have caused the death of beagles that ingested $^{90}$Sr.[14,15] Some types of those tumors affected only dogs exposed to that radionuclide; others showed a greater incidence in dogs receiving doses above the D30 level.

The malignant disorders that have exclusively affected irradiated dogs include myeloproliferative disease and primary bone tumors (Table 4). We use the term "myeloproliferative disease" [or "myeloproliferative syndrome" (MPS)] as defined by Dameshek[19] to include both neoplastic hematopoietic disease and preleukemic conditions. Preleukemic conditions include refractory dysplastic anemias associated with bone-marrow erythroid hyperplasia and aplastic anemias associated with myelofibrosis and myeloid metaplasia in the spleen and liver. The spectrum ranges from those preleukemic disorders to overt fulminating leukemias of granulocytic and myelomonocytic types. The spectrum of this disease in $^{90}$Sr-fed dogs during the early part of this study was discussed by Dungworth et al.[20,21] Only cells of the granulocyte, megakaryocyte, or erythroid series were involved, although granulocyte disorders were by far the most common.

Most primary bone cancers were osteosarcomas, which to date have occurred only in dogs exposed to the D30 level or above. Other less-common primary bone tumors included hemangiosarcoma, fibrosarcoma, and chondrosarcoma. Squamous-cell carcinomas of the gingiva have been found only in dogs exposed to the D30 level or above. Dose and dose rates to the gingiva may not be represented well by the skeletal radiation dose since the teeth formed during the time of $^{90}$Sr feeding would show less subsequent loss of $^{90}$Sr than would the skeleton.[22]

Other types of tumors arising in tissues near to bone include fibrosarcomas arising adjacent to bones in the head, such as the nasal and maxillary bones. Some of those fibrosarcomas were detected clinically as soft-tissue tumors without bone involvement, whereas others had extensively eroded bone before they were first noted clinically. With both types, local invasion was extensive, but the tumors did not metastasize. That type of tumor is not uncommon in the general dog population (R. Schneider, Contra Costa–Alameda County Tumor Registry, School of Veterinary Medicine, University of California, Davis, Calif.), and one of those tumors has recently appeared in a control dog. Nasal carcinomas, not uncommon in the general dog population, also have had a greater incidence than expected in dogs receiving the intermediate dose levels (D30, D40). One D05 dog has had a nasal carcinoma. A number of other types of tumors arising near bone have occurred in dogs receiving radiation at the D30 and D40 levels, more than would be expected in a control population. The D40 level is particularly affected because the number of dogs at risk for tumors arising near bone has been significantly reduced by deaths due to myeloproliferative disease, osteosarcoma, and squamous-cell carcinoma of the gingiva.

Although tumors arising near bone, taken as a group, are not unique to dogs that ingested $^{90}$Sr, these tumors appear to be an important late-occurring oncologic hazard for dogs that ingested intermediate levels of $^{90}$Sr. With the exception of one squamous-cell carcinoma of the frontal sinus in the oldest survivor of dogs exposed to the D50 level, tumors arising adjacent to bone (other than squamous-cell carcinomas of the gingiva) did not occur in dogs exposed to higher dose levels (D50, D60), probably because most of those dogs died from other causes before those tumors could be manifested.

Most cases of myeloproliferative disease occurred sooner after $^{90}$Sr administration than did the other radiation-induced tumors. In animals fed or injected with smaller quantities of $^{90}$Sr (D30 or S20), the few cases of myeloproliferative disease have been in older individuals.

Dogs that received lower levels of $^{90}$Sr exposure have patterns of tumors similar to those of controls. The comparison of incidence of tumors in either irradiated or control animals in this beagle colony with historical data in a canine tumor registry is somewhat difficult because of the lengthened life spans of our colony beagles compared with those in a tumor registry derived from pet animals. In fact, the projected median age for control animals and those exposed to low dose levels in our colony is 2.2 years longer than that of beagles used in a previous life-span study in Davis.[23] Because of the lack of reliable tumor-incidence data for old pet dogs, the only valid comparison of tumor incidence between irradiated and nonirradiated beagles must be made within the study and with other beagle colonies.

Because the irradiation of precursor cells of the immune system of $^{90}$Sr-fed animals during development might reduce the immunosurveillance against tumors, we have examined the cumulative incidence rates of both malignant lymphomas and mammary carcinomas, two very common tumors, in our beagles. Age-specific tumor-incidence rates for those two types of neoplasms are similar for controls and $^{90}$Sr-irradiated beagles, which indicates no indirect effect of irradiation on the incidence of tumors that are not bone related.

## THE IDEAL EXPERIMENTAL SUBJECT: MAN OR LABORATORY ANIMAL?

Man may be the proper experimental subject for understanding problems related to human exposure,[4] but, for most radiation toxicology studies, human use is neither possible, feasible, nor desirable.

Direct information on radiation effects in humans has been derived generally from epidemiologic studies of human populations that have been exposed to radiation in wartime, by medical treatment, or as a result of their occupations, although the data from those "experimental" subjects in these cases are usually less than ideal because the dosimetry has not been without error. Nonetheless, epidemiological studies tend to point out radiation-related tumors of importance in a particular population.

At the other end of the experimental spectrum, small laboratory animal studies can give valuable data on effects of radiation, but those animals have relatively short life spans of 2 to 3 years. Rats and mice are used advantageously in large numbers for genetic work, which might require many generations of progeny. Usually those rodents are inbred, and strains vary widely in their response to a given radiation modality. Studies of cellular biology in small animals are advantageous for mechanistic studies but not for long-term effects. For studies involving longer exposures at reasonably low levels, with effects that are manifested 5 to 15 years after exposure, rodent species are not suitable.

For $^{90}$Sr, too few humans have been exposed to estimate the risk for neoplasia. A number of animal species, however, have been exposed to $^{90}$Sr. All species that were exposed to adequately high doses from $^{90}$Sr developed osteosarcomas; those species include, besides the dog,[24-26] the pig,[27] rabbit,[28] mouse,[29] and rat.[30] Mice developed lymphatic leukemia but not granulocytic leukemia.[29] An increase of the incidence of malignant lymphoma has been reported in swine;[26] however, a clear dose–response relationship is not evident. Myeloproliferative disease has been found in swine,[27] dogs,[20,21] and rats.[30] Tumors arising adjacent to bone have been noted in dogs, mice, rats, and rabbits but not in swine. These tumors, although often confined to the head, apparently have different sites of origin in the different species. In mice squamous-cell carcinomas originated in the hard palate and a sebaceous gland of the external auditory meatus.[29] Squamous-cell carcinomas reportedly developed in the maxillary sinus in rats injected with $^{90}$Sr, but the dose was not given.[30] Rabbits developed squamous-cell carcinomas of the lining of the internal and external ear.[28] Radiation exposure at a younger age increased the number of those tumors. In contrast, in our study several sarcomas and carcinomas have developed in dogs at various sites of origin. Continuous exposure of these dogs as young animals to $^{90}$Sr in the food may account for the greater number of those tumors. The long life span of dogs in this study also may help to explain the obvious increase in late-occurring tumors arising adjacent to bone.

For $^{226}$Ra, human exposures have been sufficient to determine the risk for neoplasia,[4] although many estimates of dosimetry and other data suffer from the likely inaccuracies of a retrospective study. Animals exposed to $^{226}$Ra include mice[31] and dogs in our study[18,32,33] and in others.[34,35] The value of animal experimentation, even when data on people are available, is that evaluation of the animal data can provide insight into explaining the less than perfect human data. We recently were able to use our data on osteosarcoma from $^{226}$Ra in dogs to interpret the human and murine data.[33]

In addition to the value of dogs as surrogates for people in toxicologic research, studies on dogs also provide valuable information on mechanisms of damage, the relationship of one effect to another, and the predictability of one effect from another. We have used, for example, canine data to "predict" osteosarcoma, a late effect, from neutropenia, an early effect,[36] and have related the two by considering the type or types of cells that may be affected by the various bone-seeking radionuclides.

The dog is also invaluable for the study of the pathogenesis of radiation-induced disease, particularly leukemogenesis. The extensive foundation of data on immunologic and hematologic values, in concert with new research techniques and a large body of information about neoplasia in dogs, both spontaneous and radiation-induced, make the canine model ideal for the study of leukemogenesis (see Chapter 2).

Beagles continue to be used in large numbers at several federally funded radiobiologic research laboratories as follows:

- The University of California at Davis (Davis, Calif.), where animals exposed to $^{90}Sr$ and $^{226}Ra$ continue to be followed in lifetime studies and where animals are exposed externally to gamma radiation from $^{60}Co$ are used in leukemogenesis studies.
- The University of Utah (Salt Lake City, Utah), where animals injected with single doses of various bone-seeking radionuclides to provide comparative radiobiologic data are followed in lifetime studies.
- The Lovelace Biomedical and Environmental Research Institute (Albuquerque, N. Mex.), where beagles are exposed via inhalation to fission products and transuranic elements, some of which translocate to liver and bone.
- Battelle Pacific Northwest Laboratories (Richland, Wash.), where the effects of inhaled transuranic elements are under study.
- Argonne National Laboratory (Argonne, Ill.), where beagles are receiving continuous irradiation externally from $^{60}Co$.
- Colorado State University (Fort Collins, Colo.), where beagles have been irradiated in utero or in early life to examine the age-related effects of single exposures to $^{60}Co$ gamma radiation.

## CONCLUSIONS

The beagle dog has demonstrated its value in a large study lasting two decades on the effects of long-term low-level irradiation. By careful evaluation of the effects in the beagle dog of irradiation in terms of dose and time, the results of these research efforts can be "scaled" across species barriers to people to predict the appropriate response in a human population exposed to low levels of chronic irradiation. Long-term studies such as ours on the beagle dog provide an immeasurably important intermediary between the shorter term studies of the cellular biologist, the short-term toxicologic studies on rodents, and the epidemiologic evaluations by the biostatistician.

## ACKNOWLEDGMENTS

The comments of Dr. C. E. Chrisp regarding pathology are gratefully acknowledged, as are the efforts of the staff of the Laboratory for Energy-Related Health Research, both past and present. This work was supported by the U. S. Department of Energy.

## REFERENCES

1. A. C. Andersen and L. C. Good (Eds.), *The Beagle as an Experimental Dog,* The Iowa State University Press, Ames, Iowa, 1970.
2. L. K. Bustad, K. A. Stitzel, E. K. Haro, and M. Goldman, The Choice of the Beagle for Radiobiologic Studies, in *Radiobiology of Plutonium,* B. J. Stover and W. S. S. Jee (Eds.), pp. 203-211, J. W. Press, Salt Lake City, Utah, 1972.
3. International Commission on Radiological Protection, *Report of Committee II on Permissible Dose for Internal Radiation,* ICRP Publication 2, Pergamon Press, Oxford, 1959.
4. T. C. Evans, Radium in Man, *Health Phys.,* 27: 497-510 (1974).
5. M. Goldman, L. G. Young, and T. J. Powell, Skeletal Dosimetry in Growing Beagles Continually Fed $^{90}$Sr, in *The Effects of Sr-90 and Ra-226 on the Beagle,* USAEC Report UCD-472-110, pp. 114-120, Radiobiology Laboratory, University of California, Davis, 1964, NTIS.
6. M. Goldman, R. J. Della Rosa, and M. H. Momeni, Radiation Dose to Beagles from Continuous $^{90}$Sr Exposure, in *Biomedical Implications of Radiostrontium Exposure,* AEC Symposium Series, Davis, Calif., Feb. 22–24, 1971, M. Goldman and L. K. Bustad (Eds.), pp. 103-113, CONF-710201, NTIS, 1972.
7. N. J. Parks, R. R. Pool, and J. R. Williams, Variation of Radon Retention Parameters for Radium-Burdened Dog Skeletons as a Function of Exposure Age and Dosage Level, *Radiat. Res.,* 73: 274-287 (1978).
8. N. J. Parks, R. R. Pool, J. R. Williams, and H. G. Wolf, Age and Dosage Level Dependence of Radium Retention in Beagles, *Radiat. Res.,* 73: 617-632 (1978).
9. N. J. Parks, Lifespan Dynamics of Intra-Skeletal Radionuclide Distribution in Radium-Injected Beagles, *Health Phys.,* 38: 11-20 (1980).
10. R. J. Della Rosa, M. Goldman, H. G. Wolf, and L. S. Rosenblatt, Application of Canine Metabolic Data to Man, in *Biomedical Implications of Radiostrontium Exposure,* AEC Symposium Series, Davis, Calif., Feb. 22–24, 1971, M. Goldman and L. K. Bustad (Eds.), pp. 52-67, CONF-710201, NTIS, 1972.
11. R. J. Della Rosa, D. H. McKelvie, M. Goldman, and L. S. Rosenblatt, Metabolism and Toxicity of $^{90}$Sr in the Immature Beagle, in *Radiation Biology of the Fetal and Juvenile Mammal,* AEC Symposium Series, Richland, Wash., May 5–8, 1969, M. R. Sikov and D. D. Mahlum (Eds.), pp. 529-541, CONF-690501, NTIS, 1969.
12. N. Mantel and W. Haenszel, Statistical Aspects of the Analysis of Data from Retrospective Studies of Disease, *J. Natl. Cancer Inst.,* 22: 719-748 (1959).
13. L. S. Rosenblatt, W. Bielfelt, and R. J. Della Rosa, Effects of Radiostrontium on Puppy Mortality and Litter Size in Dogs, in *Biomedical Implications of Radiostrontium Exposure,* AEC Symposium Series, Davis, Calif., Feb. 22–24, 1971, M. Goldman and L. K. Bustad (Eds.), pp. 334-344, CONF-710201, NTIS, 1972.

14. C. E. Chrisp, W. L. Spangler, R. R. Pool, and S. A. Book, Cumulative Causes of Death in Radionuclide-Treated Beagles, in *Radiobiology Laboratory Annual Report*, ERDA Report UCD-472-125, pp. 130-141, University of California, Davis, NTIS, 1979.

15. S. A. Book et al., Comparative Toxicity of Strontium-90 and Radium-226: Experimental Design and Current Status, in *Radiobiology Laboratory Annual Report*, ERDA Report UCD-472-125, pp. 200-231, University of California, Davis, NTIS, 1979.

16. R. G. White, Tapetum Lucidum Changes in $^{226}$Ra Treated Beagles, in *Radiobiology Laboratory Annual Report*, ERDA Report UCD-472-121, pp. 74-75, University of California, Davis, NTIS, 1974.

17. G. L. Fisher, C. P. Nash, N. J. Parks, and N. A. Jee, Chemical Quantitation of the Tapetum Lucidum with Application to $^{226}$Ra Dosimetry of the Canine Eye, in *The Health Effects of Plutonium and Radium*, W. S. S. Jee (Ed.), pp. 373-387, J. W. Press, Salt Lake City, Utah, 1976.

18. R. R. Pool, J. R. Williams, M. Goldman, and L. Rosenblatt, Comparison of Bone-Tumor Sites in Beagles Continually Fed $^{90}$Sr or Injected with $^{226}$Ra as a Means of Scaling Risk to Humans, in *Radionuclide Carcinogenesis*, AEC Symposium Series, Richland, Wash., May 11–13, 1972, C. L. Sanders, R. H. Busch, J. E. Ballin, and D. D. Mahlum (Eds.), pp. 475-486, CONF-720505, NTIS, 1973.

19. W. Dameshek, Some Speculations on the Myeloproliferative Syndromes, *Blood*, 6: 372-375 (1951).

20. D. L. Dungworth, M. Goldman, J. W. Switzer, and D. H. McKelvie, Development of a Myeloproliferative Disorder in Beagles Continuously Exposed to $^{90}$Sr, *Blood*, 34: 610-632 (1969).

21. D. L. Dungworth, M. Goldman, D. H. McKelvie, and J. W. Switzer, Development of a Form of Myelogenous Leukemia in Beagles Exposed Continuously to $^{90}$Sr, in *Myeloproliferative Disorders of Animals and Man*, AEC Symposium Series, Richland, Wash., May 20–23, 1968, W. J. Clarke, E. B. Howard, and P. L. Hackett (Eds.), pp. 272-289, CONF-680529, NTIS, 1970.

22. R. J. Della Rosa, H. Wolf, and G. Peterson, Strontium-90 in Teeth and Bones of Beagles, in *The Effects of Continued $^{90}$Sr Ingestion During the Growth Period of the Beagle and Its Relation to $^{226}$Ra Toxicity*, USAEC Report UCD-472-110, pp. 60-74, University of California, Davis, NTIS, 1964.

23. A. C. Andersen and L. S. Rosenblatt, The Effect of Whole-Body X-Irradiation on the Median Lifespan of Female Dogs (Beagles), *Radiat. Res.*, 39: 177-200 (1969).

24. R. R. Pool, J. R. Williams, and M. Goldman, Strontium-90 Toxicity in Adult Beagles After Continuous Ingestion, in *Biomedical Implications of Radiostrontium Exposure*, AEC Symposium Series, Davis, Calif., Feb. 22–24, 1971, M. Goldman and L. K. Bustad (Eds.), pp. 277-284, CONF-710201, NTIS, 1972.

25. J. H. Dougherty, G. N. Taylor, and C. W. Mays, Strontium-90 Toxicity in Adult Beagles After Acute Exposure, in *Biomedical Implications of Radiostrontium Exposure*, AEC Symposium Series, Davis, Calif., Feb. 22–24, 1971, M. Goldman and L. K. Bustad (Eds.), pp. 259-276, CONF-710201, NTIS, 1972.

26. M. P. Finkel, B. O. Biskis, I. Greco, and R. W. Camden, Strontium-90 Toxicity in Dogs: Status of Argonne Study on Influence of Age and Dosage Pattern, in *Biomedical Implications of Radiostrontium Exposure*, AEC Symposium Series, Davis, Calif., Feb. 22–24, 1971, M. Goldman and L. K. Bustad (Eds.), pp. 285-312, CONF-710201, NTIS, 1972.

27. W. J. Clarke et al., Strontium-90 Effects in Swine: A Summary to Date, in *Biomedical Implications of Radiostrontium Exposure*, AEC Symposium Series, Davis, Calif., Feb. 22–24, 1971, M. Goldman and L. K. Bustad (Eds.), pp. 242-258, CONF-710201, NTIS, 1972.

28. J. Vaughn and M. Williamson, $^{90}Sr$ in the Rabbit: The Relative Risks of Osteosarcoma and Squamous Cell Carcinoma, in *Delayed Effects of Bone-Seeking Radionuclides*, C. W. Mays, W. S. S. Jee, R. D. Lloyd, B. J. Stover, J. H. Dougherty, and G. N. Taylor (Eds.), pp. 337-355, University of Utah Press, Salt Lake City, Utah, 1969.

29. A. Nilsson, Strontium-90-Induced Malignancies in Mice, in *Biomedical Implications of Radiostrontium Exposure,* AEC Symposium Series, Davis, Calif., Feb. 22–24, 1971, M. Goldman and L. K. Bustad (Eds.), pp. 207-241, CONF-710201, NTIS, 1972.

30. Y. I. Moskalev, V. N. Streltsova, and L. A. Buldakov, Late Effects of Radionuclide Damage, in *Delayed Effects of Bone-Seeking Radionuclides,* C. W. Mays, W. S. S. Jee, R. D. Lloyd, B. J. Stover, J. H. Dougherty, and G. N. Taylor (Eds.), pp. 489-509, University of Utah Press, Salt Lake City, Utah, 1969.

31. M. P. Finkel, B. O. Biskis, and P. B. Jinkins, Toxicity of Radium-226 in Mice, in *Radiation-Induced Cancer,* Symposium Series, Athens, 1969, pp. 369-391, STI/PUB/228, International Atomic Energy Agency, Vienna, 1969.

32. M. Goldman, D. L. Dungworth, M. S. Bulgin, L. S. Rosenblatt, W. P. C. Richards, and L. K. Bustad, Radiation-Induced Neoplasms in Beagles After Administrations of $^{90}Sr$ and $^{226}Ra$, in *Radiation-Induced Cancer,* Symposium Series, Athens, 1969, pp. 345-360, STI/PUB/228, International Atomic Energy Agency, Vienna, 1969.

33. O. G. Raabe, S. A. Book, and N. J. Parks, Cancer in Bone from Radium: Canine Dose-Response Explains Mouse and Human Data, *Science* , 208: 61-64 (1980).

34. C. W. Mays et al., Radiation-Induced Bone Cancer in Beagles, in *Delayed Effects of Bone-Seeking Radionuclides,* C. W. Mays, W. S. S. Jee, R. D. Lloyd, B. J. Stover, J. H. Dougherty, and G. N. Taylor (Eds.), pp. 387-408, University of Utah Press, Salt Lake City, Utah, 1969.

35. T. F. Dougherty and C. W. Mays, Bone Cancer Induced by Internally-Deposited Emitters in Beagles, in *Radiation-Induced Cancer,* Symposium Series, Athens, 1969, pp. 361-367, STI/PUB/228, International Atomic Energy Agency, Vienna, 1969.

36. M. Goldman, F. D. Wilson L. S. Rosenblatt, and S. A. Book, Early Radiation Effects on Bone and Blood as 'Predictors' of Osteosarcoma, a Late Effect, in *Biological and Environmental Effects of Low-Level Radiation,* Symposium Series, Chicago, 1975, pp. 335-342, STI/PUB/409(Vol. 1), International Atomic Energy Agency, Vienna, 1976.

# The Malignant Canine Mammary Tumor as a Model for the Study of Human Breast Cancer

# 17

CLARENCE E. CHRISP
*and*
WILLIAM L. SPANGLER
*Laboratory for Energy-Related Health Research, School of Veterinary Medicine, University of California, Davis*

Malignant mammary tumors are the most common cancers in both canine[1] and human[2] females. An estimated 1 in 15 women will develop a malignancy of the mammary gland during her lifetime.[2] Despite that high incidence, the prognosis has not improved for several decades;[2] approximately 50% of women contracting this malignancy will die of the disease.[3] So that treatment of human breast cancer can be improved, an animal model of this disease is needed to supplement clinical trials in human patients with innovative methods of therapy, including immunotherapy. Others have suggested mammary cancer in the dog as a suitable model for the study of mammary cancer in women.[1,4,5] In this chapter we address the malignant canine mammary tumor as a model by comparing the epidemiological, morphological, behavioral, hormonal, and immunological aspects of human and canine mammary neoplasia.

Many of our ideas concerning mammary cancer in the dog have developed from our long clinical and pathologic study of the natural history of this disease in a colony of beagles during life-span studies of the effects of bone-seeking radionuclides.[6] That study has offered a unique opportunity to observe the behavior of canine mammary tumors under controlled conditions not available to most pathologists. In that study,[6] 800 beagles of both sexes were exposed to the bone-seeking radionuclides, $^{226}$Ra and $^{90}$Sr. A clinical protocol was established for quarterly examinations of the reproductive system, including the mammary glands. Any nodules approximately 1 cm or greater in diameter, as determined by palpation, are removed, and tissue sections are examined by a veterinary pathologist. If the pathologist believes the tumor is malignant, the mammary gland containing the tumor plus the glands in the path of lymphatic drainage of the tumor are removed. The surgical protocol is illustrated in Fig. 1.

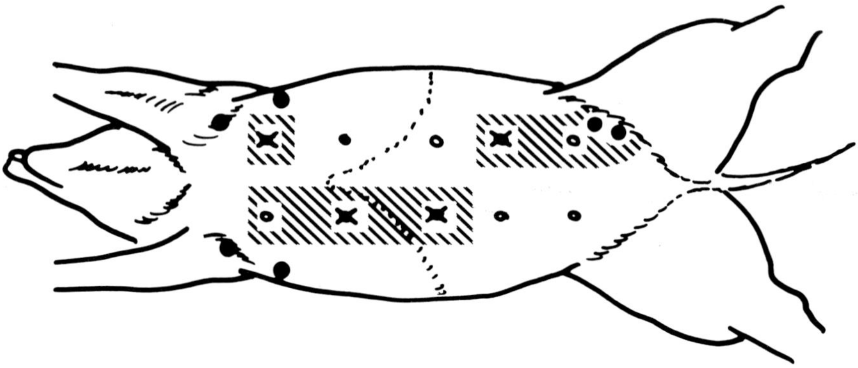

Fig. 1 Diagram for illustration of mammary tumor surgical protocol. ○, nipples. X, primary tumor sites. ●, lymph nodes. Shaded areas represent mastectomy sites.

Further surgery is not done if the tumors are designated as benign by a veterinary pathologist. Of the 390 female dogs originally at risk at 4 years of age or older (age of the first tumor), 116 bitches currently survive. We have removed 652 tumors from 179 animals. Seventy percent of the tumors have been designated benign, and 30% have been termed malignant. Of the 179 bitches, 112 had malignant tumors; 40 of those 112 have died or have been euthanized with terminal clinical signs of metastatic mammary tumors or local recurrences of inoperable malignancies. Nine more have probable lung metastases, as indicated by quarterly radiographs after mastectomy. Mammary tumor metastases have been found at necropsy in 12 additional animals that died of other causes. Thus at least 57% of the dogs with tumors diagnosed as malignant have had observable metastases. Since many of those females are still alive, several more bitches will develop metastases. Many other studies report a higher incidence of malignant tumors[7-10] and a lower incidence of metastases. The removal of primary tumors at an earlier age in our study may account for some of that discrepancy. Only one dog has died of a tumor recorded as benign by a pathologist. In retrospect, examination of this primary tumor biopsy indicated clear malignancy as evidenced by local invasion. Cumulative incidence rates of benign (Fig. 2) and malignant (Fig. 3) mammary tumors are similar in radionuclide-treated and in untreated dogs.

## COMPARATIVE EPIDEMIOLOGY

Many published epidemiology studies on factors affecting mammary tumor incidence in the dog are confounded by the fact that they are based on tumors from dogs referred to veterinary school teaching hospitals. Those studies are not

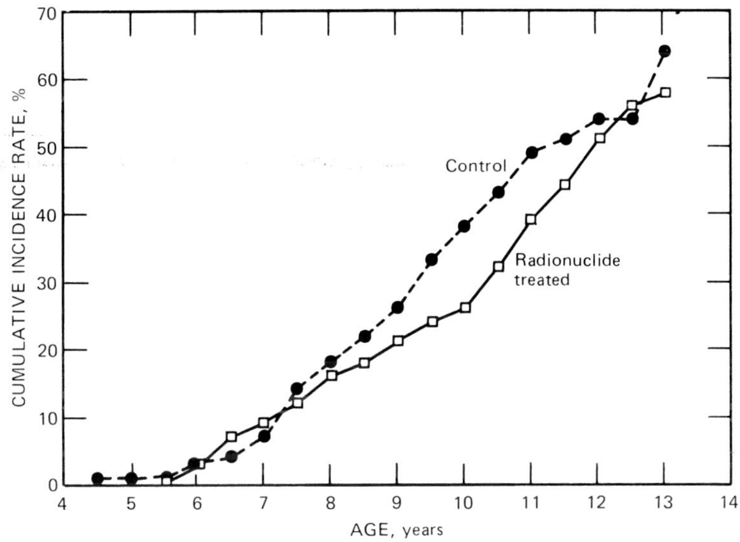


Fig. 2 Cumulative incidence rates of benign mammary tumors in control and radionuclide-treated beagles.

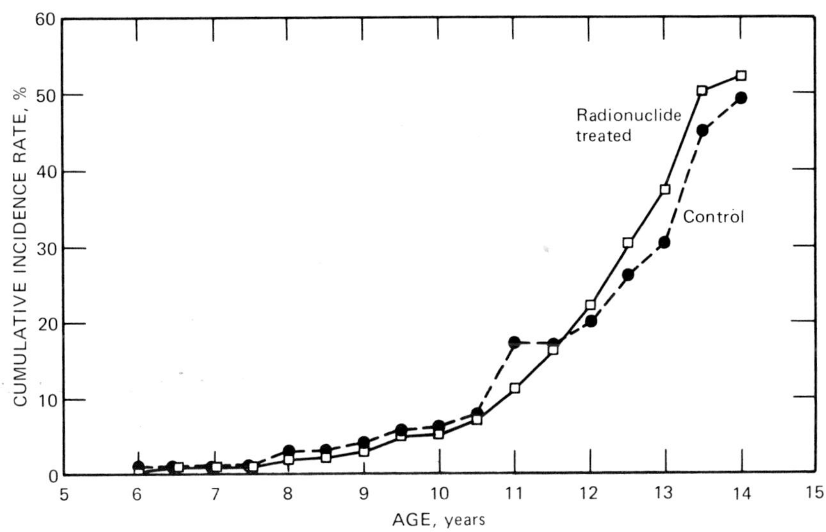


Fig. 3 Cumulative incidence rates of malignant mammary tumors in control and radionuclide-treated beagles.

supported by a defined population base of known characteristics such as age, size, breed, sex, and reproductive status. Exceptions are the population-based studies of Dorn et al.[11] and Schneider, Dorn, and Taylor,[1] and the epidemiological data derived from those two studies should be regarded as the most reliable. Taylor et al.[12] have published some information on mammary tumors in a colony of laboratory beagles. Our beagle study is even more tightly controlled statistically, but the population is not large, only one strain of beagles (with some possible linebreeding) is involved, environmental conditions may be more restricted than those encountered by pet dogs from a variety of environments, and all the bitches in our study are nulliparous.

The incidence of mammary tumors in male dogs is quite low.[13] We have seen no mammary tumors in approximately 400 male beagles at risk. We have the impression that benign mammary tumors are much more common in nulliparous females in our radionuclide study than they are in breeding colony females of the same age that have whelped multiple litters. In humans, breast cancer occurs more commonly in single women, infertile married women, and late-married women.[14] The effect of the time of first pregnancy has not been sufficiently studied in the dog. There is contradictory information on the effect of pseudopregnancy on the subsequent development of mammary tumors in the bitch.[7,15]

The effect of ovariectomy in the dog is most striking; spayed bitches exhibit approximately 12% of the mammary cancer risk found in intact animals.[1] Neutering after two estrous cycles but before 2.5 years of age significantly reduced mammary cancer risk. Animals neutered after 2.5 years of age did not benefit. Animals spayed by 2.5 years would be expected to have experienced a maximum of four estrous cycles. Schneider, Dorn, and Taylor[1] pointed out that data on the sparing effect of castration in humans indicate that the effect would be similar to that in dogs if ovariectomy were done early enough.

Schneider[1] noted that incidence of mammary cancer in dogs is similar to that in humans when dog ages are converted to human equivalents according to the method of Lebeau.[16] That formula assumes that a 1-year-old dog is equivalent to a 1-year-old human; that a dog of 2 is like a human of 24 years; and that, beyond the 2 and 24 years, each year of a dog's life is equal to 4 years of the human life.

There is a familial increase in mammary cancer in humans. The risk for a daughter to acquire mammary cancer becomes as great as 1 in 5 if her mother has had mammary cancer.[2] In dogs, information about the influence of heredity is contradictory. Significant increases in mammary tumor incidences were not seen in two studies, but the populations were small.[1,17] Conversely, another study has shown a higher incidence in purebred dogs.[10]

In summary, the influence of epidemiological factors, such as sex, incidence, and age of occurrence, apparently is similar for mammary neoplasia in man and the dog. The effect of early castration may also be similar. Failure to show any effect of parity in dogs[1] may be attributable to the rather sparse data available for dogs as compared with that derived from much larger human populations.

## COMPARATIVE MORPHOLOGY

There is probably no problem in veterinary oncology or pathology which has resulted in more controversy and confusion than the histopathology of canine mammary tumors. We hesitate to say that we see the "light at the end of the tunnel," but we believe our experience with these tumors in beagles does qualify us to comment on the subject. We believe most of the confusion has resulted from attempts to classify canine mammary tumors on a histogenetic basis and lack of attempts to correlate morphologic diagnosis and clinical prognosis. A bewildering array of taxonomic systems for naming these tumors has resulted.[9,18–25]

The basis for the varied opinions regarding the histogenesis of mammary tumors has been clouded by time. An exhaustive review by Hamilton[26] addresses that question in a historical context. Apparently some misunderstanding of the histogenetic origins of the canine "mixed tumors" has originated from attempts to compare them with the benign mammary fibroadenoma in humans. Although some pathologists still believe that the origin of these two tumors is the same,[24,25] that opinion is outdated and has been refuted by many excellent recent studies.[27–34] The human fibroadenoma is apparently composed of both epithelial cells and connective-tissue stroma, whereas the canine counterpart has been shown to be composed primarily of proliferative epithelial and myoepithelial cells. Both Cotchin[18] and Moulton et al.[9] believed that large amounts of myoepithelium were present in these so-called "mixed tumors." Their opinion has been supported by using histochemical and electron microscopy studies in the United States,[28] Germany,[29–33] and England.[34] In many canine mixed tumors, areas of cartilage or bone or both are juxtaposed with areas of proliferative myoepithelium. Histochemical and electron microscopic studies of cells with characteristics intermediate between myoepithelium and mature cartilage at the juncture of these two cell types have demonstrated that these cells do not have the characteristics of fibrocytes. Clearly the myoepithelial cell, which is thought to be derived embryologically from ectoderm,[27,35] is undergoing metaplasia to a cell that is thought to be derived from mesoderm. Although that process seemingly contradicts a basic biologic principle, the genetic information for any one somatic cell is contained in all somatic cells. This metaplastic phenomenon is unusual, but not impossible, for a tumor cell that may be undifferentiated. In contrast, in one recent report the investigator[24] not only denies the existence of abundant myoepithelium within the canine mixed mammary tumor but also that it is neoplastic, even when found within metastatic foci in the lymph nodes or lungs. Neither the evidence nor our experience with beagle mammary tumors supports that view.

Another histogenetic misunderstanding is one that has been perpetuated by both human and veterinary pathologists, who have classified mammary carcinomas as either ductal or lobular, depending on their assumed histogenetic origin. Recently, Wellings and Jensen[36] carefully correlated subgross anatomy and histopathology in human breasts and showed that so-called "ductal carcinomas" apparently arise in ductules and lobules. The distension of lobular components results in duct-like morphology in two-dimensional histology slides. Because Wellings and Jensen's study shows that the site of origin (ductal or glandular) of human mammary epithelial tumors cannot be readily determined, we question the use of the taxonomic terms ductal and lobular, which have been used in several classifications applied to dogs.[9,18,19,24,25] A classification scheme for canine mammary tumors developed by the World Health Organization (WHO)[20–23] has eliminated those terms. A series of meetings held for the purpose of developing a common classification for tumors in the beagle were held in Pacific Grove, Calif.; Seattle, Wash.; Estes Park, Colo.; and Albuquerque, N. Mex. These were attended by veterinary pathologists from the following laboratories that are funded by the U. S. Department of Energy: Radiobiology Laboratory, University of California at Davis, Calif.; Battelle–Northwest at Richland, Wash.; University of Utah at Salt Lake City, Utah; Argonne National Laboratory, Argonne, Ill.; and the Inhalation Toxicology Research Institute, Albuquerque, N. Mex. In addition, representatives from the Collaborative Radiobiological Health Laboratory at Fort Collins, Colo. (funded by the Food and Drug Administration), and the Armed Forces Institute of Pathology at Washington, D. C., attended these workshops. These pathologists came to the conclusion that the WHO classification scheme, with slight modifications, was the most useful system for the classification of mammary tumors. This classification scheme is as follows:

**I. Benign Mammary Tumors**

- A. Duct papilloma
  - 1. Simple
  - 2. Complex
- B. Adenoma
  - 1. Simple
  - 2. Complex
- C. Benign mixed tumor

**II. Carcinomas**

- A. Adenocarcinoma (tubulo/papillary cystic)
  - 1. Simple
  - 2. Complex
- B. Solid carcinoma
  - 1. Simple
  - 2. Complex

C. Anaplastic carcinoma

D. Squamous carcinoma

E. Spindle (can be classified as solid carcinoma)

F. Mucinous carcinoma

III. Carcinomas Associated with the Benign Mixed Tumor

IV. Carcinosarcoma

V. Sarcomas

A. Osteosarcoma (neoplasms containing malignant cartilage and bone may be designated osteochondrosarcomas)

B. Fibrosarcoma

C. Other types of sarcoma

These modifications were based largely on the collective experience of the participants and that of veterinary pathologists at the Armed Forces Institute of Pathology (H. D. Casey, Armed Forces Institute of Pathology, personal communication). In practice we have found the classification scheme to be quite useful for describing and naming both benign and malignant canine mammary tumors.

In the modified WHO classification system, histogenesis is used only to determine types of cells in the tumor with no reference to the anatomical site of origin. Tumors composed of either epithelium or myoepithelium are designated simple tumors. If the tumor contains both myoepithelial cells and epithelial cells, the tumor is called complex. If cartilage or bone is being formed in a tumor containing epithelial or both epithelial and myoepithelial cells, the tumor is termed a benign mixed tumor. A malignant tumor containing both epithelium and myoepithelium, where only one or both cell types are malignant, is labeled a complex adenocarcinoma. If epithelium and either cartilage or bone are malignant, the tumor is designated a carcinosarcoma. Sarcomas are described as osteosarcoma, chondrosarcoma, or fibrosarcoma, depending on their tissue composition. Although adjectives, such as tubular, solid, papillary, cystic, spindle cell, anaplastic, mucinous, and squamous-cell, are used to describe carcinomas, the terms "infiltrating" and "noninfiltrating" are not used.

The division between benign complex adenomas and benign mixed mammary tumors may seem artificial in light of the research on myoepithelial metaplasia to cartilage, but such a distinction may be useful. For example, we see fewer mixed benign tumors containing cartilage and bone in laboratory beagles than in pet dogs. That difference may result from the earlier removal of our beagle tumors because of our quarterly palpation schedule. Therefore benign mixed tumors may arise from complex adenomas. Since carcinosarcomas have been less common in our study than in studies using pet animals, they may arise from benign mixed tumors.

Another morphological problem causing confusion is the general lack of uniform criteria for the distinction between benign and malignant canine mammary tumors. Often, benign mammary tumors, particularly those containing both epithelium and myoepithelium, may show traditional indicators of malignancy; the epithelial cells will be pleomorphic, will be piled up in several layers, will show loss of polarity, and even will appear to have lost contact with the basement membranes, as determined by examination of hematoxylin and eosin stained sections. Those criteria, however, may not be reliable indicators of malignancy in these benign tumors. As an example, during a recent examination of wedge biopsies from complex adenomas, without the presence of tumor borders, our application of those criteria suggested that the epithelial portions of some tumors might be malignant. When the whole mass of the tumor was examined, however, we could more easily distinguish benign from malignant tumors. Carcinoma cells may be pleomorphic but generally are much more anaplastic and basophilic than those epithelial cells in the benign tumors. A meaningful histologic grading system for canine mammary tumors that correlates morphology with behavior needs to be developed.

In our study, in dogs dying with metastatic mammary carcinoma, local infiltration has been the best indicator of malignancy (Table 1). Tumors from 95% of those dogs showed local invasion; tumors from 70% showed unequivocal lymphatic permeation and/or vessel invasion with approximately 10% showing vessel invasion. Some have used aspiration of mammary tumors and subsequent cytology to diagnose malignant mammary tumors.[24,37] Interpretation of those studies is difficult because of the lack of reports on the subsequent biologic behavior of tumors diagnosed by that method. The rather bizarre cytologic changes and wide range of epithelial-cell variations that can occur in complex adenomas connotes that cytology will be a less useful tool in the diagnosis of canine mammary malignancies than it has been for certain other malignant tumors in the dog. Studies should be done to determine if cytology of mammary tumors can be correlated with their biologic behavior.

Carcinosarcomas and sarcomas compose a small percentage of malignant tumors in beagles (approximately 1 to 2% of malignant tumors in our study), but the most common type of malignant tumor causing death is the tubular or solid carcinoma (Table 2). Often by the time of death from metastases, a tumor diagnosed as tubular at biopsy is primarily a solid tumor at necropsy. The simple tubular adenocarcinoma, the most common lethal form of mammary adenocarcinoma in the dog, is morphologically similar to the most common malignant mammary tumor in human females.[38] The less commonly lethal complex mammary carcinomas in the dog occur very rarely in the human female.[38] Most benign tumors in human females are fibroadenomas, whereas complex adenomas or benign mixed tumors predominate in dogs. In our beagle study benign tumors had an earlier onset than malignant tumors (Fig. 4). However, most carcinomas apparently do not originate in benign tumors, which supports other evidence that hyperplastic alveolar nodules may be preneoplastic[38,39] in the dog. The

TABLE 1

HISTOPATHOLOGIC CHARACTERISTICS OF LETHAL MALIGNANT MAMMARY TUMORS IN FEMALE BEAGLES

| Biopsy diagnosis | Number of tumors locally invasive | Number of tumors with permeation of lymphatics and/or vessels |
|---|---|---|
| Simple tubular adenocarcinoma | 23/23 | 21/23 |
| Simple papillary adenocarcinoma | 2/3 | 1/3 |
| Simple solid carcinoma | 3/3 | 2/3 |
| Complex tubular adenocarcinoma | 7/8 | 2/8 |
| Spindle cell carcinoma | 1/1 | 1/1 |
| Anaplastic carcinoma | 1/1 | 1/1 |
| Carcinosarcoma | 1/1 | 0/1 |
| Total | 38/40 = 95% | 28/40 = 70% |

TABLE 2

HISTOPATHOLOGY AND SURVIVAL TIME OF BEAGLES WITH LETHAL MALIGNANT MAMMARY TUMORS

| Histologic type | Number of dogs | Median age at biopsy, years | Median age at death, years | Median survival time, years |
|---|---|---|---|---|
| Simple tubular adenocarcinoma* | 23 | 12.6<br>7.8–14.9† | 13.4<br>8.6–14.9† | 0.4<br>0.0–2.4† |
| Simple papillary adenocarcinoma | 3 | 10.9<br>9.5–11.9† | 13.0<br>12.0–13.7† | 2.2<br>1.9–2.4† |
| Simple solid carcinoma | 3 | 12.3<br>12.0–13.1† | 12.5<br>12.3–13.7† | 0.3<br>0.2–0.6† |
| Complex tubular adenocarcinoma | 8 | 12.5<br>9.1–13.4† | 13.4<br>9.5–15.0† | 1.5<br>0.4–1.9† |
| Spindle cell carcinoma | 1 | 11.1 | 12.1 | 1.0 |
| Anaplastic carcinoma | 1 | 11.2 | 11.3 | 0.1 |
| Carcinosarcoma | 1 | 12.6 | 12.7 | 0.1 |
| Total | 40 | 12.4<br>7.8–14.9† | 13.1<br>8.6–15.0† | 0.5<br>0.0–2.4† |

*Seven animals died before the primary tumor was large enough for biopsy (<1.0 cm in diameter).

†Range in age, years.

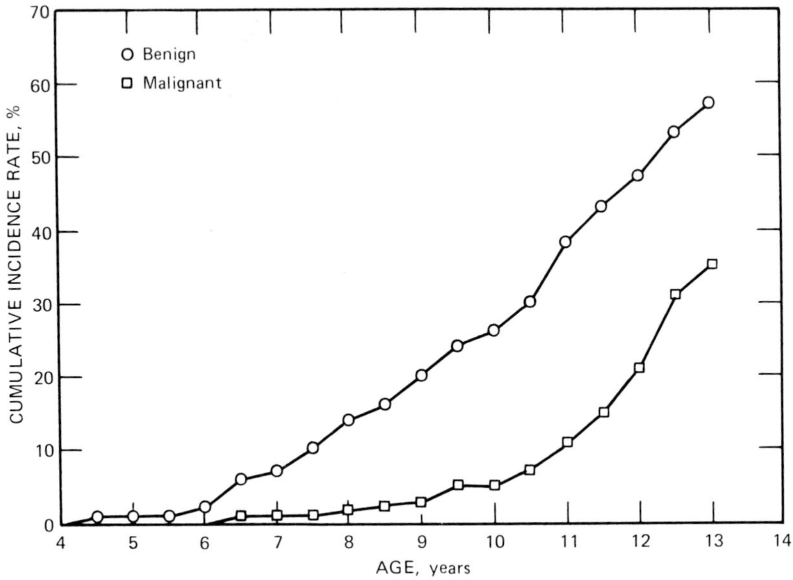


Fig. 4 Cumulative incidence rates of benign and malignant mammary tumors in beagles.

presence of atypical nodules has been correlated with early carcinoma in the human breast.[40]

## COMPARATIVE BEHAVIOR

Tumor behavior is defined as those events which take place between the time of onset and the time when the ultimate outcome has been determined. That outcome may be signified by survival time until death or by a designated arbitrary survival time after diagnosis.

Several studies involving the general dog population have attempted to correlate the biological behavior of malignant canine mammary tumors with morphology[8,9,18,19,24,41,42] by measuring the duration time of the tumor either from time of detection until death[9,19] or from time of tumor removal until death. All those studies have the disadvantage that the time of onset was either not determined or not accurately determined. In our opinion objective estimates of tumor duration cannot be obtained from most owners of pet dogs brought to veterinary clinics for treatment. Therefore only the survival time after surgery can be used as an accurate index for prognosis in most of these studies.

Our experience with beagles indicates that metastatic mammary tumors may present as late as 3 years after mastectomy. A few animals will survive for many months with radiographic evidence of lung metastases. Multiple tumors may complicate studies on the behavior of mammary tumors in dogs. Most of the multiple tumors are benign, but in our studies an average of 1.8 malignant tumors per dog occurred in 107 animals. Often the tumors that result in metastases can be distinguished by comparison of primary tumor morphology with that of the metastases. In other cases the source of metastases among two or more primary tumors cannot be determined. For those few cases, survival time cannot be used as an accurate index of behavior.

Misdorp[5] has compared the behavior of human and canine malignant mammary tumors (Table 3). Delaying mastectomy did not significantly affect

TABLE 3

EFFECT OF VARIOUS INFLUENCES ON SURVIVAL OF DOGS AND WOMEN WITH MAMMARY CARCINOMA*

| Influence | Dog | Woman |
|---|---|---|
| Delay in operating | — | — |
| Localization | — | — |
| Clinical stage of the primary tumor | + | + |
| Size, volume | + | + |
| Histologic type | + | + |
| Histologic grade | ± | + |
| Type of growth | + | + |
| Type of treatment | — | ? |
| Regional lymph node involvement | — | + |

*From W. Misdorp, Histologic Classification and Further Characterization of Tumors in Domestic Animals, *Adv. Vet. Sci. Comp. Med.,* 220 (1976).

survival time, which is in agreement with an earlier study in humans.[43] Localization of the canine malignant tumors (presence in anterior or posterior mammary glands) was not associated with prognosis.[5] Likewise, there was no association of tumor localization within the gland with prognosis in humans.[44]

Methods for clinical staging of mammary malignancies have been developed for humans.[44] The definition of stage is based on characteristics of the primary tumor (T) and regional lymph nodes (N) and the existence of distant metastases (M). In man a more-advanced clinical stage in this TMN system is correlated with a decreased survival time. As yet, prognosis for dogs with mammary tumors has not been correlated with a complete clinical staging system. However, Misdorp and Hart[41,42] have shown a significant association of prognosis with a more advanced stage as defined by characteristics of the primary tumor. Those features included size, degree of involvement of the skin, and degree of involvement of underlying tissues. In our study, which features early detection, few animals have been in an advanced clinical stage at the time of mastectomy. Exceptions would be those few animals which died of tumor metastases from a primary tumor that had not grown to 1 cm in diameter by the last quarterly examination. In every case the primary tumors were simple tubular or solid carcinomas (Table 2). Tumor size as measured by maximal diameter and volume (when corrected for clinical stage, of which size is a part) was associated with prognosis; dogs with larger tumors had a shorter survival time.[41] That association has also been noted in the human female.[44]

Moulton et al.,[9] in a study where primary tumors were left untreated, stated that carcinomas were associated with a shorter average survival time (2 to 4 years) than were other mammary neoplasms (3 to 4 years). Fowler, Wilson, and Koestner[19] attempted to correlate morphology of mammary neoplasms with behavior in a retrospective study, but a histogenetic classification made the results difficult to interpret. In the study by Fowler, Wilson, and Koestner, both lobular and ductal infiltrating carcinomas had the same prognosis.

The most convincing studies of behavior of canine malignant mammary tumors have been done by Bostock,[8] Misdorp,[5] and Misdorp and Hart.[41,42] Misdorp and Hart's studies show that canine complex adenocarcinomas have a better prognosis than simple carcinomas. These findings agree with our observations (Table 2) but disagree with those of Bostock.[8] Papillary adenocarcinomas, which are generally more differentiated, show a better prognosis than simple tubular or solid carcinomas in our study (Table 2) and the study of Bostock.[8] Bostock[8] believed that simple solid carcinomas had a worse prognosis than simple tubular carcinomas, but we have found no merit in attempting to separate these two types (Table 2).

In humans, special types of carcinoma, such as comedo, medullary, and colloid carcinoma, have a better prognosis.[5] Future work with dogs should include studies of subtypes of simple tubular or solid mammary carcinomas for a possible correlation with prognosis. Also, the simple carcinoma arising from myoepithelium in the dog should be examined for possible increased survival time. Electron microscopy could play a role in the further characterization of simple carcinoma subtypes.

Misdorp[5] and Misdorp and Hart[41,42] attempted to correlate prognosis with histologic grade of malignancy in canine mammary cancers by applying human criteria, including degree of differentiation, tubule formation, hyperchromatism and mitoses, anaplasia, and irregularity of size, shape, and staining of nuclei. A higher histological grade of malignancy was associated with a worse prognosis, but none of the individual components of that grading system was associated with a significantly worse prognosis. A grading system has been useful in predicting prognosis in humans.[2]

The mode of growth of malignant canine mammary tumors was found to be of prognostic value by Misdorp[5] and Misdorp and Hart.[41,42] Survival time was much shorter in dogs with extensive local infiltration than in those with expansive noninvasive growth. In our study (Table 1), tumors with noninvasive growth have accounted for only 5% of tumors that caused death. Mode of growth is also an important characteristic for determining the prognosis for mammary carcinomas in women.[5]

Misdorp[5] noted that there was no difference in survival time whether malignant tumors were surgically treated by block dissection or mastectomy. However, dogs with noninfiltrative clinical stage I tumors (less than 5 cm in diameter and restricted to mammary tissue) responded significantly better with mastectomy, whereas infiltrative stage I tumors responded better with block dissection. Interestingly, the value of radical mastectomy is an area of controversy among human surgeons; many are recommending fewer radical mastectomies for women.[45]

Metastases to the regional lymph node in women are unanimously regarded as one of the most unfavorable prognostic signs;[43] for dogs, tumor cells in the superficial inguinal lymph nodes did not adversely affect survival.[5] In our beagle study, however, a preliminary analysis indicates that tumor cells in the regional nodes do correlate with decreased survival. The earlier detection of tumors in our study may account for that difference. For example, Misdorp and Hart[41,42] found that approximately 50% of the dogs treated by block dissection showed involvement of regional nodes, whereas in our study only 10 to 20% of the dogs have had positive nodes. The clinical stage in our beagles may more nearly approximate that in women.

## COMPARATIVE ENDOCRINOLOGY

Mammary carcinomas in women may be either hormone dependent or independent according to their response to endocrine therapy.[46] Steroid hormones achieve their action in responsive organs, such as the mammary gland, by initially binding to protein molecules in the cytosol fraction of the cell. Those hormone-receptor complexes are translocated to the nucleus where they interact with chromatin, leading to qualitative and quantitative changes in translatable messenger ribonucleic acid. Protein hormone receptors for estrogen, progesterones, androgens, and glucocorticoids have been detected in human mammary carcinomas.[46]

Recently, several workers have found hormone protein receptors in canine mammary tumors.[47–49] Hamilton and Foreshaw[49] have shown that 39% of 129 canine mammary tumors contained estrogen receptors (ER). Twenty-five percent of the benign and forty-two percent of the malignant tumors contained this receptor protein. Estrogen receptors were not correlated with histologic type in either benign or malignant tumors. Conversely, Monson, Malbica, and Hubben[48] found a greater amount of ER in mammary adenocarcinomas than in mixed mammary tumors.

In some women breast tumors are positive for ER but do not respond to anti-estrogen therapy. Other hormone receptors, which could be responsible for that anomaly, have been found in breast tumors in women.[46] Recently, D'Arville and Pierrepoint[50] detected androgen and progesterone receptors in canine mammary tumors as well.

There is no clear-cut evidence in the literature that malignant canine mammary tumors have responded to anti-estrogen therapy, such as ovariectomy or adrenalectomy.[12] We have observed a few simple tubular adenocarcinomas where metastases are confined to the mammary glands without systemic involvement. These tumors possibly require hormones trophic for the mammary gland for their support. As yet there are no studies that have attempted to correlate the presence of hormone receptors with response to anti-estrogen therapy of malignant canine mammary tumors. The detection of those hormone receptors enhances the possibility that some canine mammary carcinomas may respond to hormone therapy. Studies on possible correlations of hormone receptor activity with subsequent response to chemotherapy or immunotherapy may also be productive.

Beagles have been used as test animals for the possible induction of mammary tumors by human contraceptive steroids.[51,52] Owen and Briggs[51] and Casey, Giles, and Kwapien[52] have reviewed those studies and have discussed the advantages and disadvantages of the dog as a test animal. One additional criticism of those studies is that they are terminated when the dogs are only 7 years of age. As Illustrated in Fig. 4, our beagles had a low incidence of benign mammary tumors and malignant mammary tumors at that age, and the effect of those drugs might not appear until the animals are older. For example, whole-body gamma irradiation decreased the latent period for malignant tumors in female beagles but did not increase the absolute incidence in a life-span study.[53] However, even in irradiated females, that effect was not apparent in dogs younger than 10 years of age.

## COMPARATIVE IMMUNOLOGY

The malignant canine mammary tumor may be a suitable model for the study of immunity to cancer.[54] The literature dealing with the measurement of immunological responses to canine mammary tumors is scant and generally preliminary. Uvland[55] purported to demonstrate cellular immunity to cells from malignant canine mammary tumors by use of a technique that measured inhibition of migration of leukocytes incubated with tumor cells. Nonspecific inhibition was greater in the autologous system (tumor cells and leukocytes from the same dog) than in the allogeneic system (tumor cells and leukocytes from different dogs), which indicates a greater cellular response to the dog's own tumor cells. Histoincompatibility reactions were not thought to be a factor in the allogeneic system because supernates from mixed lymphocyte cultures of histoincompatible donors did not inhibit migration within the first 24 hr. In contrast, the migration inhibition effect in the canine tumor study was seen within 20 hr. Defects in this study included the lack of normal mammary tissue controls, differences in tumor cell numbers, and use of primary tumor cells from some animals vs. those cells passaged in tissue culture from other animals.

Neuraminidase-treated autologous canine mammary tumor cells reportedly have an immunotherapeutic effect.[56] In that experiment cells from one tumor mass were treated with neuraminidase and then injected subcutaneously. A second tumor mass was left intact and measured for regression or progression. Both benign and malignant mammary tumors of variable morphological types were included in the study, but histology of each of the tumors used for injection was not recorded for each dog. The number of dogs in each dose group was small and quite variable, and the results were not statistically evaluated.

In the general reexamination of the field of tumor immunology, previous suppositions about the role of immune surveillance in the response of the host to autologous tumors are now questioned.[57] Some in vitro immunologic studies have examined the relationship between tumor cells and lymphocytes from human breast cancer patients,[58] but much more work must be directed toward the understanding of the basic mechanisms of host response. The dog could be a good model for those in vitro studies with certain advantages over such studies often done in rodents. The dog's larger size is better suited for serial invasive and noninvasive studies. Also, the canine has metastatic malignant mammary tumors comparable to those in women, whereas rodent malignant mammary tumors, although locally invasive, almost never metastasize.[52] In studies in which dogs are used as a model for human mammary carcinoma, the homogeneous character of simple and solid carcinomas is most compatible with the harvest of tumor cells. In addition, those tumors are the most common metastatic tumors in both dogs and women. The use of complex carcinomas and carcinosarcomas with their multiplicity of cell types would be much more technically difficult, and the results would be harder to interpret.

In vivo evidence points to an immunologic response to human mammary carcinoma.[59] Prognostically favorable characteristics include diffuse lymphoid infiltration of the primary tumor and perivenous lymphoid cell infiltrations around the tumor. Sinus histiocytosis of the draining lymph nodes is another factor associated with a positive prognosis. Studies of that type are needed for malignant canine mammary tumors.

Malignant mammary tumors in the dog can be most useful for evaluation of new chemotherapeutic and/or immunotherapeutic protocols. Bostock and Gorman[60] used dogs to study the effect of nonspecific immunotherapy of mammary carcinomas in the bitch. In a prior separate study, Bostock[8] found that dogs with mammary carcinomas had a median postmastectomy survival time of 70 weeks, whereas dogs with benign mammary tumors had a median survival time of at least 114 weeks.[8] Dogs with invasive carcinomas had a significantly lower survival time. Since most dogs that died from mammary carcinoma did so within the first year after surgery, Bostock suggested that any therapy regimen used in dogs could be evaluated much sooner than could be expected from human clinical trials. Bostock then proceeded to study the effect of multiple intravenous injections of *Mycobacterium tuberculosis* var. bacillus Calmette-Guérin (BCG) on the postsurgical survival time of dogs with invasive mammary carcinoma.[60] Thirty-four bitches with invasive mammary carcinomas were given either no postsurgical treatment (11 dogs), placebo treatment (10 dogs), or intravenous BCG therapy (13 dogs). The median survival time of the combined control group was 24 weeks vs. at least 100 weeks for the BCG-treated group. Those survival times were significantly different ($P < 0.05$), as calculated by the chi-square. Although those results are encouraging, an abnormally long survival time by a few animals in the treated group would have biased the results. For example, we have noted a much longer median postsurgical survival time in bitches with complex mammary carcinomas than in dogs with simple mammary carcinomas (Table 2), but that difference was not noted by Bostock.[8] The results of that study of BCG immunotherapy, although preliminary, demonstrate the feasibility of clinical trials in dogs with mammary carcinomas as a model for possible breast cancer therapy in humans. Studies with simple carcinomas rather than complex carcinomas would be much more productive since the survival times are considerably shorter and the tumors are most similar in behavior and morphology to breast cancer in humans.

## SUMMARY

The most common benign mammary tumors in the dog, i.e., complex adenomas and benign mixed mammary tumors, do not have a common counterpart in humans. Malignant mammary tumors in women and dogs, especially carcinomas derived from epithelium, have been shown to have many similar characteristics, including epidemiologic, morphologic, behavioral, and endocrinologic similarities. Those similarities encourage the use of canine mammary carcinomas as models for the study of human breast cancer. Preliminary studies indicate that canine mammary cancer may be a good immunotherapeutic model. Chemotherapeutic studies should be done to evaluate the potential of the canine malignant mammary tumor model.

## ACKNOWLEDGMENT

This work was supported in part by the U. S. Department of Energy.

## REFERENCES

1. R. Schneider, Comparison of Age, Sex, and Incidence Rates in Human and Canine Breast Cancer, *Cancer,* 26: 419-425 (1970).

2. R. W. Scarf and H. Torlonni, International Histological Classification of Tumors No. 2, in *Histological Typing of Breast Tumors,* pp. 1-19, World Health Organization, Geneva, Switzerland, 1968.
3. N. Lane, H. Gioksel, R. A. Salerno, and C. D. Haagenson, Clinicopathologic Analysis of the Surgical Curability of Breast Cancers: A Minimum Ten-Year Study of a Personal Series, *Ann. Surg.,* 153: 483-498 (1961).
4. J. D. Strandberg and D. G. Goodman, Human Disease Breast Cancer; Animal Model: Canine Mammary Neoplasia, *Am. J. Pathol.,* 75: 225-228 (1974).
5. W. Misdorp, Histologic Classification and Further Characterization of Tumors in Domestic Animals, *Adv. Vet. Sci. Comp. Med.,* 20: 191-221 (1976).
6. D. L. Dungworth, M. Goldman, J. W. Switzer, and D. H. McKelvie, Development of a Myeloproliferative Disorder in Beagles Continuously Exposed to $^{90}$Sr, *Blood,* 34: 610-632 (1969).
7. I. J. Fidler and R. S. Brodey, The Biological Behavior of Canine Mammary Neoplasms, *J. Am. Vet. Med. Assoc.,* 151: 1311-1318 (1967).
8. D. E. Bostock, The Prognosis Following the Surgical Excision of Canine Mammary Neoplasms, *Eur. J. Cancer,* 11: 389-396 (1975).
9. J. E. Moulton, D. O. N. Taylor, C. R. Dorn, and A. C. Anderson, Canine Mammary Tumors, *Pathol. Vet.,* 7: 289-320 (1970).
10. W. A. Priester, Occurrence of Mammary Neoplasms in Bitches in Relation to Breed, Age, Tumor Type, and Geographical Region from Which Reported, *J. Small Anim. Pract.,* 20: 1-11 (1979).
11. C. R. Dorn, D. O. N. Taylor, R. Schneider, H. H. Hibbard, and M. R. Klauber, Survey of Animal Neoplasms in Alameda and Contra Costa Counties, California. II. Cancer Morbidity in Dogs and Cats from Alameda County, *J. Natl. Cancer Inst.,* 40: 307-318 (1968).
12. G. N. Taylor, L. Shabestari, J. Williams, C. W. Mays, W. Angus, and S. McFarland, Mammary Neoplasia in a Closed Beagle Colony, *Cancer Res.,* 36: 2740-2743 (1976).
13. A. G. Jabara, Two Cases of Mammary Neoplasms Arising in Male Dogs, *Aust. Vet. J.,* 45: 476-480 (1969).
14. J. F. Kuzma, Breast, in *Pathology,* Vol. 2, W. A. D. Anderson (Ed.), pp. 1578-1605, C. V. Mosby Co., St. Louis, 1971.
15. R. W. Else and D. Hannant, Some Epidemiological Aspects of Mammary Neoplasia in the Bitch, *Vet. Rec.,* 104: 296-304 (1979).
16. A. Lebeau, L'age du chien et celui de l'homme: essai de statistique sur la mortalite canine, *Bull. Acad. Vet.,* 26: 229-232 (1953).
17. C. R. Dorn and C. R. Schneider, Inbreeding and Canine Mammary Cancer: A Retrospective Study, *J. Natl. Cancer Inst.,* 47: 545-548 (1976).
18. E. Cotchin, Mammary Neoplasms of the Bitch, *J. Comp. Pathol.,* 68: 1-26 (1958).
19. E. H. Fowler, G. P. Wilson, and A. Koestner, Biologic Behavior of Canine Mammary Neoplasms Based on a Histogenic Classification, *Vet. Pathol.,* 11: 212-229 (1974).
20. J. F. Hampe and W. Misdorp, Chap. IX: Tumors and Dysplasias of the Mammary Gland, *Bull. World Health Organ.,* 50: 111-133 (1974).
21. W. Misdorp, E. Cotchin, J. F. Hampe, A. G. Jabara, and J. von Sandersleben, Canine Malignant Mammary Tumors. I. Sarcomas, *Vet. Pathol.,* 8: 99-117 (1971).
22. W. Misdorp, E. Cotchin, J. F. Hampe, A. G. Jabara, and J. von Sandersleben, Canine Malignant Mammary Tumors. II. Adenocarcinomas, Solid Carcinomas and Spindle Cell Carcinomas, *Vet. Pathol.,* 9: 447-470 (1972).
23. W. Misdorp, E. Cotchin, J. F. Hampe, A. G. Jabara, and J. von Sandersleben, Canine Malignant Mammary Tumors. III. Special Types of Carcinomas, Malignant Mixed Tumors, *Vet. Pathol.,* 10: 241-256 (1973).

24. A. W. Monlux, J. F. Roszel, D. W. MacVean, and T. W. Palmer, Classification of Epithelial Canine Mammary Tumors in a Defined Population, *Vet. Pathol.*, 121: 194-197 (1977).
25. R. M. Mulligan, Mammary Cancer in the Dog: A Study of 120 Cases, *Am. J. Vet. Res.*, 36: 1391-1396 (1975).
26. J. M. Hamilton, Comparative Aspects of Mammary Tumors, *Adv. Cancer Res.*, 19: 1-45 (1974).
27. L. T. Pulley, Ultrastructural and Histochemical Demonstration of Myoepithelium in the Normal Canine Mammary Gland, *Am. J. Vet. Res.*, 34: 1505-1512 (1973).
28. L. T. Pulley, Ultrastructural and Histochemical Demonstration of Myoepithelium in Mixed Tumors of the Canine Mammary Gland, *Am. J. Vet. Res.*, 34: 1513-1522 (1973).
29. D. V. Bomhard and J. V. Sandersleben, The Ultrastructure of Mixed Mammary Gland Tumors in Bitches. I. The Incidence of Myoepithelial Cells in Myxoid Areas, *Virchows Arch., A*, 359: 87-96 (1973).
30. D. V. Bomhard and J. V. Sandersleben, The Ultrastructure of Mixed Mammary Gland Tumors in Bitches. II. The Incidence of Myoepithelial Cells in Chondroid Areas, *Virchows Arch., A*, 362: 157-167 (1974).
31. D. V. Bomhard and R. Raddatz, Electron Microscopic Studies on Solid Mammary Carcinomas in Bitches, *Z. Krebsforsch.*, 83: 129-143 (1975).
32. D. V. Bomhard and J. V. Sandersleben, Uber die Feinstruktur von Mammamischtumoren der Lundin. III. Die Anfangsstadien der Myoepithelialen Proliferation, *Virchows Arch., A*, 367: 219-229 (1975).
33. D. V. Bomhard and J. V. Sandersleben, The Ultrastructure of Mixed Mammary Gland Tumors in Bitches. IV. The Incidence of Myoepithelial Cells in Formations of Spindle Cells, *Virchows Arch., A*, 371: 219-226 (1976).
34. S. Tateyama and E. Cotchin, Alkaline Phosphatase Reaction of Canine Mammary Mixed Tumors: A Light and Electron Microscopic Study, *Res. Vet. Sci.*, 23: 356-364 (1977).
35. H. Hamperl, The Myothelia (Myoepithelial Cells). Normal State; Regressive Changes; Hyperplasia; Tumors, *Curr. Top. Pathol.*, 53: 161-220 (1970).
36. S. R. Wellings and H. M. Jensen, On the Origin and Progression of Ductal Carcinoma in the Human Breast, *J. Natl. Cancer Inst.*, 50: 1111-1118 (1973).
37. J. F. Roszel, Exfoliative Cytology in Diagnosis of Malignant Canine Neoplasms, *Vet. Scope*, 12: 14-18 (1967).
38. R. D. Cardiff, S. R. Wellings, and L. J. Faulkin, Biology of Breast Preneoplasia, *Cancer, Suppl.*, 39: 2734-2746 (1977).
39. A. M. Cameron and L. J. Faulkin, Jr., Hyperplastic and Inflammatory Nodules in the Canine Mammary Gland, *J. Natl. Cancer Inst.*, 47: 1277-1287 (1971).
40. S. R. Wellings, H. M. Jensen, and R. G. Marcum, An Atlas of Subgross Pathology of the Human Breast with Special Reference to Precancerous Lesions, *J. Natl. Cancer Inst.*, 55: 231-273 (1975).
41. W. Misdorp and A. A. M. Hart, Prognostic Factors in Canine Mammary Cancer, *J. Natl. Cancer Inst.*, 56: 779-786 (1976).
42. W. Misdorp and A. A. M. Hart, Canine Mammary Cancer. I. Prognosis, *J. Small Anim. Pract.*, 20: 385-394 (1979).
43. G. F. Robbins and I. Bross, The Significance of Delay in Relation to Prognosis of Patients with Primary Operable Breast Cancer, *Cancer*, 10: 338-344 (1957).
44. S. J. Cutler and M. H. Myers, Clinical Classification of Extent of Disease in Cancer of the Breast, *J. Natl. Cancer Inst.*, 39: 193-203 (1967).
45. R. S. Handley, Techniques of Surgical Treatment, in *The Treatment of Breast Cancer*, H. Atkins (Ed.), pp. 49-65, M.T.P. Publishing Co., Lancaster, England, 1974.

46. J. C. Allegra et al., Distribution, Frequency and Quantitative Analysis of Estrogen, Progesterone, Androgen, and Glucocorticoid Receptors in Human Breast Cancer, *Cancer Res.,* 39: 1447-1454 (1979).
47. C. R. Evans and C. G. Pierrepoint, Tissue-Steroid Interactions in Canine Hormone-Dependent Tumors, *Vet. Rec.,* 13: 464-467 (1975).
48. B. S. Monson, J. O. Malbica, and K. Hubben, Determination of Estrogen Receptors in Canine Tumors, *Am. J. Vet. Res.,* 38: 1937-1939 (1977).
49. J. M. Hamilton and P. Foreshaw, Oestrogen Receptors in Canine Mammary Tumors, *Vet. Rec.,* 101: 258-260 (1979).
50. C. N. D'Arville and C. G. Pierrepoint, The Demonstration of Oestrogen, Androgen, and Progestogen Receptors in the Cytosol Fraction of Canine Mammary Tumors, *Eur. J. Cancer,* 15: 875-883 (1979).
51. L. W. Owen and M. H. Briggs, Contraceptive Steroid Toxicology in the Beagle Dog and Its Relevance to Human Carcinogenicity, *Curr. Med. Res. and Opin.,* 4: 309-329 (1976).
52. H. W. Casey, R. C. Giles, and R. P. Kwapien, Mammary Neoplasia in Animals: Pathologic Aspects and the Effects of Contraception Steroids, *Recent Results Cancer Res.,* 79: 129-160 (1979).
53. C. E. Chrisp, R. Phemister, A. C. Andersen, L. S. Rosenblatt, and M. Goldman, *Pathology in a Life Time Study of X-Irradiated Female Beagles,* ERDA Report UCD-472-122, pp. 2-12, University of California, Davis, NTIS, 1975.
54. J. Ankerst, W. I. B. Beveridge, J. F. Dore, and T. Godal, Immunity to Cancer: Naturally Occurring Tumors in Domestic Animals as Models for Research, *Bull. World Health Organ.,* 49: 205-213 (1973).
55. M. F. Uvland, Cellular Immunity to Canine Mammary Tumor Cells Demonstrated by the Leukocyte Migration Technique, *Acta Vet. Scand.,* 16: 95-114 (1975).
56. H. H. Sedlacek and F. R. Seiler, Spontaneous Mammary Tumors in Mongrel Dogs. A Relevant Model to Demonstrate Tumor Therapeutic Success by Application of Neuraminidase-Treated Tumor Cells, in *International Symposium on Biological Preparations in the Treatment of Cancer,* London, Apr. 13–15, 1977, Vol. 38, pp. 399-412, International Association of Biological Standardization, Geneva, 1977.
57. V. S. Byers and R. W. Baldwin, Tumor Immunology, in *Basic and Clinical Immunology,* H. H. Fudenberg, D. P. Stites, J. L. Caldwell, and J. U. Wells (Eds.), pp. 283-307, Lange Med. Pubs., Los Altos, CA, 1978.
58. G. B. Cannon, J. C. McCoy, J. H. Dean, C. Alford, R. B. Herberman, and I. Nyirjesy, Immunologic Studies on Breast Disease, in *Proceedings of the First Annual Post-Graduate Course/Symposium on Breast Disease in Gynecological and Medical Primary Care Practice,* I. Nyirjesy (Ed.), pp. 213-251, Gynecological Society for Study of Breast Disease, Gaithersburg, Md., 1977.
59. M. M. Black, T. H. C. Barclay, and B. F. Hankey, Prognosis in Breast Cancer Utilizing Histologic Characteristics of the Primary Tumor, *Cancer,* 36: 2048-2055 (1975).
60. D. E. Bostock and N. T. Gorman, Intravenous BCG Therapy of Mammary Carcinoma in Bitches After Surgical Excision of the Primary Tumor, *Eur. J. Cancer,* 44: 879-883 (1978).

18

# Canine Cutaneous Histiocytoma

GARY L. COCKERELL
*and*
ROBERT W. O'DONNELL
*Department of Pathology,*
*New York State College of Veterinary Medicine,*
*Cornell University, Ithaca, New York*

As early as 1940 veterinary clinicians and pathologists recognized a peculiar tumor in the skin of dogs which had distinct morphologic, epizootiologic, and biologic characteristics. In 1948, Mulligan,[1] on the basis of the morphologic resemblance of the cells to tissue histiocytes and his belief that the condition was neoplastic, was the first to apply the term "histiocytoma" to the lesion. This terminology has persisted to the present time, but now the condition is more fully designated as canine cutaneous histiocytoma (CCH). Since CCH has no analogy in any other species of animals, including man, this name also differentiates the canine cutaneous lesion from similarly named conditions in man which are not limited to the skin and which have different morphologic and biologic characteristics. Because of its unique features, the CCH has continued to interest those involved in the diagnosis and treatment of canine dermatologic disorders.

The present state of knowledge of the CCH is based primarily on morphologic comparisons of naturally occurring tumors performed during the last 10 years. Little information, however, has become available on the etiology and pathogenesis of the CCH and its relationship to morphologically similar but malignant neoplasms in the dog.

In this chapter the past literature and recent information derived from studies initiated in our laboratory within the past year have been used to discuss the biological characteristics, histogenesis, and etiopathogenesis of the CCH. The results of our findings and implications of unresolved questions suggest the CCH as a biomedical model to study the origin and function of mononuclear phagocytes, disturbances of their growth patterns, and mechanisms of the spontaneous regression of neoplastic disease.

# BIOLOGICAL CHARACTERISTICS

## Epizootiology

The CCH is a commonly encountered tumor, particularly in younger dogs. In one study based on records from an animal neoplasm registry,[2] the CCH represented 17.9% of all canine cutaneous neoplasms and had an annual incidence of 117 per 100,000 dogs. Fifty percent of the reported cases have occurred in dogs that were less than 2 years of age.[2,3] Male dogs have a higher (but not statistically significant) incidence.[2,4] The boxer and dachshund breeds seem to be preferentially affected, whereas poodles are at decreased risk.[2,3] Purebred dogs, in general, are more apt to develop CCH than are mongrels.[2] The tumor always occurs in the skin, and nearly half of all cases involve the head, neck, and face, particularly the pinna of the ear. The next most common sites are the legs (especially the front) and feet, followed by the trunk and tail.[2–4]

## Clinical Behavior

The CCH almost invariably occurs as a single mass, which appears suddenly; it is often brought to the attention of the owner when it becomes ulcerated and secondarily infected by the dog's licking or otherwise irritating the surface of the tumor. Otherwise, the effect on dogs is minimal. The tumor regresses spontaneously over several months in a variable percentage of untreated cases.[2–8] Spontaneous regression would no doubt be more common if the histiocytomas were left untreated, but they are usually excised because they are subject to surface trauma and are similar in gross appearance to more malignant tumors of the skin of dogs. Surgical excision is curative in nearly all cases, with rare recurrences at the surgical site. Metastasis does not occur.

After either surgical removal or spontaneous regression, dogs seldom develop additional histiocytomas; this suggests that the first growth confers immunity against additional growth of the same tumor. These facts, together with morphologic evidence of intratumoral lymphocytic infiltrates (see the section on Microscopic Morphology in this chapter), form the basis for the suggestion that spontaneous regression of CCH may be immunologically mediated.[4]

It is important to differentiate CCH, with benign biologic behavior and excellent prognosis, from other cutaneous disorders in the dog which may mimic CCH in some respects but behave more aggressively or require different clinical management. The differential diagnosis must include transmissible venereal tumor (see Chap. 19), mast cell tumor, cutaneous histiocytic lymphosarcoma (reticulum cell sarcoma), and granulomatous dermatitis. Clinical histories are often helpful in differentiating CCH, but the most useful criterion is histologic examination of excised tissue. Other techniques, described later in this chapter, provide additional diagnostic evidence but also require somewhat more sophisticated techniques.

## Gross Morphology

The CCH usually presents as a round, dome-shaped to buttonlike, firm nodule in the dermis. The surface is largely hairless, often ulcerated, and red, and it oozes serum. The nodules are usually 1 to 2 cm in diameter, but they may be as small as 0.5 cm or as large as 4 cm in diameter. The skin over the surface of the tumor is unmoveable. The cut surface is solid, uniformly pale gray to light yellow, and well-delineated from the adjacent normal dermis but is nonencapsulated. The tumor extends from the subcutis up to, and is in intimate association with, the remaining portions of the overlying epidermis. Adnexal structures are excluded from the tumor mass and are compressed together in the dermis at edges of the tumor (Fig. 1).

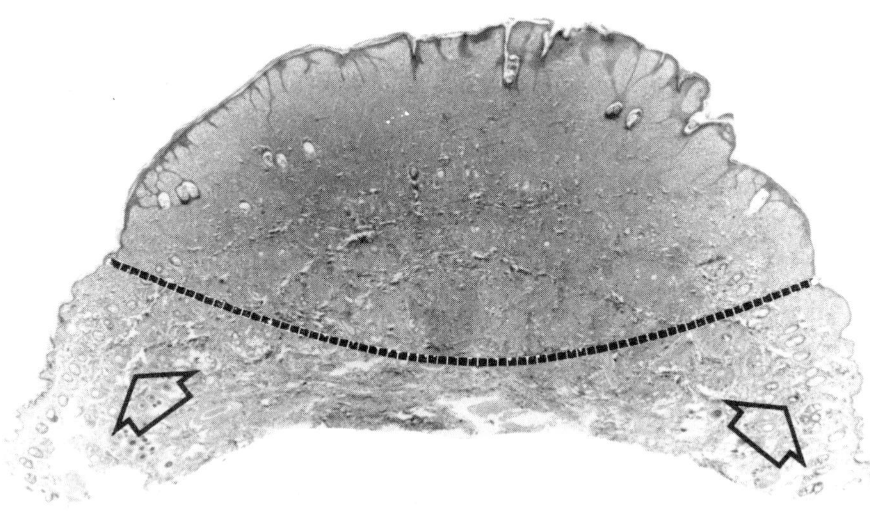

Fig. 1 Cross section of a CCH. The neoplasm extends from the subcutis (bottom) up to the epidermis, which is elevated and mildly thickened. The approximate limits of the tumor are shown by the broken line. Note the compressed hair follicles (arrows) at the peripheral margins of the tumor at its junction with normal dermis. (Hematoxylin and eosin stain; magnification, 7 x.)

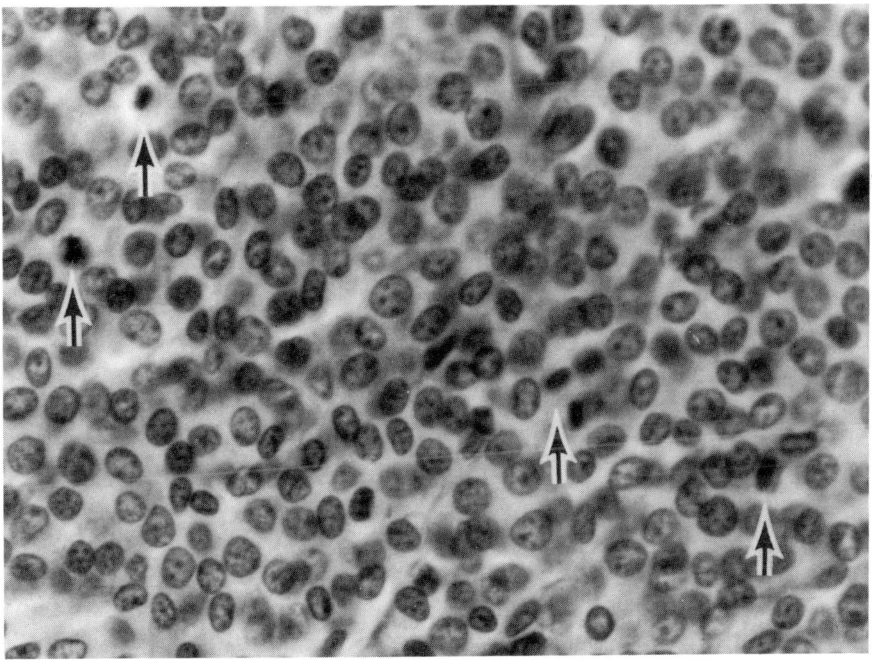

Fig. 2 Microscopic morphology of the CCH prior to infiltration by lymphocytes. The tumor cells are monomorphous, and many are in mitosis (arrows). (Hematoxylin and eosin stain; magnification, 770 x.)

## Microscopic Morphology

Despite its benign biologic behavior, the CCH looks histologically malignant. Medical pathologists who examine such lesions but are unaware of the canine entity will often render a poor prognosis. The neoplastic cell population may be quite pleomorphic, demonstrate a high mitotic index, and have variable amounts of inflammatory cell infiltrates, primarily lymphocytes.[1-3,7-10] We believe that these features are related to stages in the process of tumor progression to regression and, on that assumption, have classified CCH into four histological groups.[4] The most differential feature between individual tumors is the relative amount and pattern of intratumoral lymphocytic infiltrate. The tumor cell population in group 1 (progressive tumor, least amount of lymphocytic infiltrate) is characterized by a monomorphous, uniform sheet of cells with round to oval nuclei, frequent mitotic figures, and irregularly placed nucleoli (Fig. 2). The cytoplasm is abundant, finely granular, and eosinophilic, with indistinct borders between adjacent cells. In tumors with greater amounts of intratumoral lymphocytic infiltrate (groups 2, 3, and 4, successively), multiple

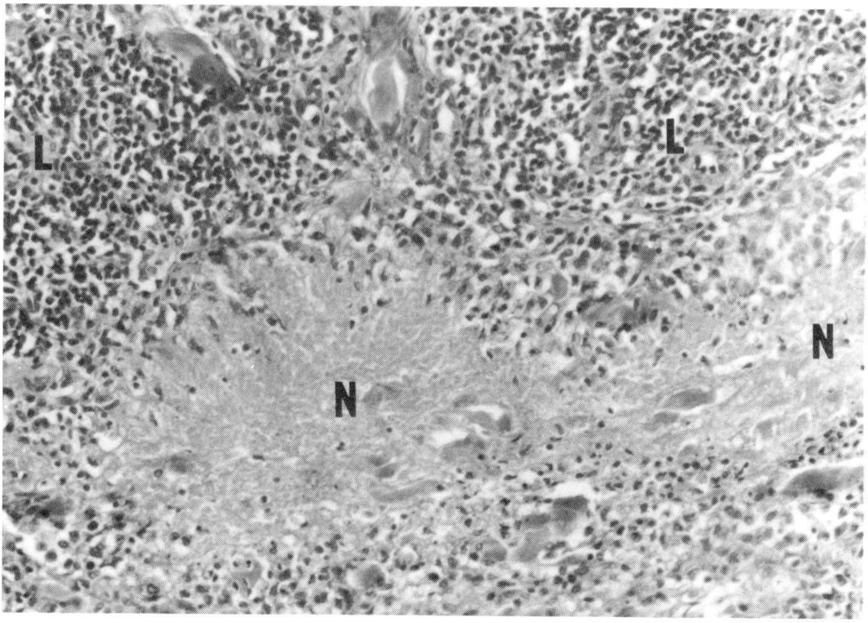


Fig. 3 Intratumoral necrotic foci (N) surrounded by aggregates of small, dark-staining lymphocytes (L) in a CCH. (Hematoxylin and eosin stain; magnification, 257 x.)

foci of necrosis appear within the tumor mass. These necrotic foci are always closely associated with and often rimmed by the lymphoid aggregates (Fig. 3). In group 4 tumors (regressive tumor) the lymphocytic population outnumbers the degenerating tumor cell population (Fig. 4), and large areas of the dermis are cleared of tumor cells.

Other histologic features of the CCH include ulceration of the overlying epithelium and necrosis of the subjacent tumor tissue. Opportunistic bacterial organisms and neutrophilic inflammatory cells are often seen in these areas but never gain access to deeper portions of the tumor. Tumor cells subjacent to areas of intact epithelium are frequently more individualized and may lie in rows perpendicular to the surface. This pattern probably results in part from intercellular edema as a consequence of mechanical trauma to the mass that protrudes above the normal surface.

## Ultrastructural Morphology

Several studies have detailed the ultrastructural characteristics of CCH, including comparative features of other canine cutaneous neoplastic disorders.[9-11] The neoplastic cells of CCH most often were large, rather

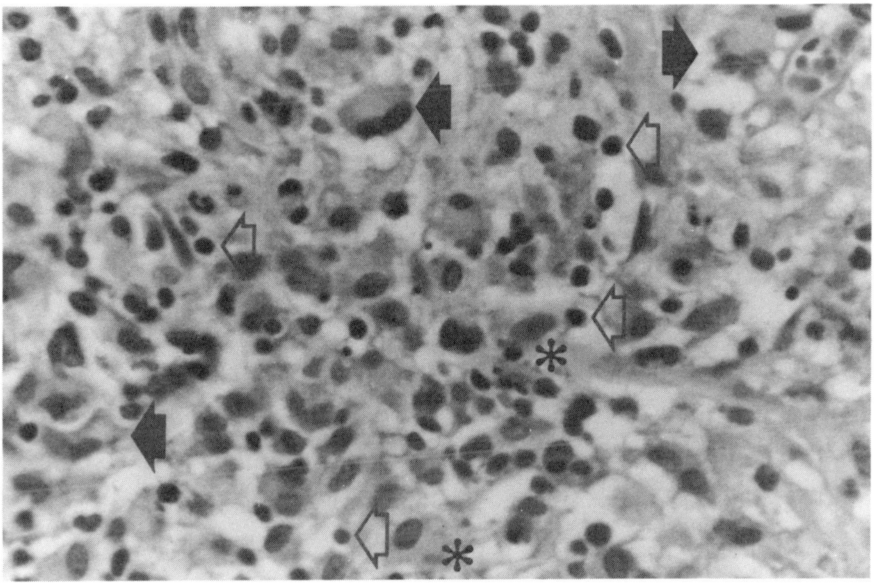

Fig. 4 Degenerative changes in neoplastic cells of a CCH associated with a lymphocytic infiltrate. Note the diffuse sprinkling of small, dark-staining lymphocytes (open arrows). Neoplastic cells have eccentric, irregular nuclei and vacuolated cytoplasm (solid arrows). There are scattered nuclear fragments (*). (Hematoxylin and eosin stain; magnification, 640 x.)

nondescript, irregularly shaped, closely apposed mononuclear cells measuring up to 10 μm in diameter (Fig. 5). Nuclei were frequently eccentrically located, often had irregular outlines of their limiting membranes, and were bordered by accumulations of perinuclear microfilaments. Cytoplasmic organelles were generally sparse and consisted of occasional mitochondria, several profiles of vesicular rough endoplasmic reticulum, free ribosomes, and a moderately developed Golgi apparatus. Primary lysosomal granules, pinocytotic vesicles, and lamellar deposits of cytoplasmic debris have also been observed. Cytoplasmic membranes between adjacent cells were closely apposed and usually smooth, but often they had irregular, interdigitating protrusions. Junctional complexes characteristic of epithelial cells were not observed.

Intratumoral lymphocytic infiltrates can alter the light microscopic appearance of the neoplastic cells. Therefore the extent of lymphoid invasion should be considered in the interpretation of ultrastructural findings.

## Histochemistry

Routine staining with hematoxylin and eosin is usually sufficient to establish the diagnosis of CCH. Special stains, however, can be used to differentiate

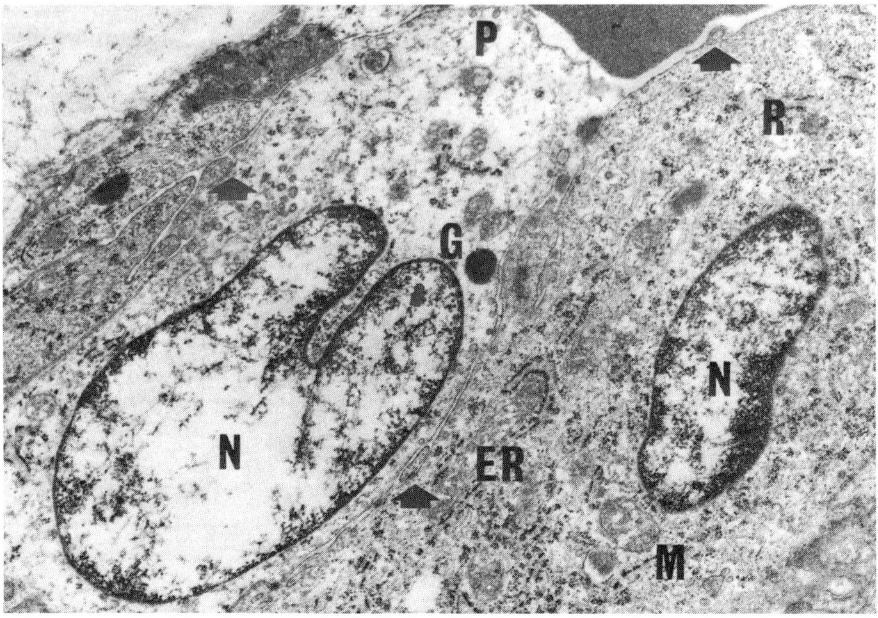


Fig. 5 Ultrastructural characteristics of two closely apposed neoplastic cells from a CCH. Nuclei (N) are oval or indented. Sparse cytoplasmic organelles include occasional mitochondria (M), profiles of rough endoplasmic reticulum (ER), free ribosomes (R), and lysosomal granules (G). Cytoplasmic membranes are generally smooth but with intermittent short, interdigitating protrusions (arrows). Occasional pinocytotic vesicles (P) are evident. (Uranyl acetate and lead citrate stain; magnification, 13,300 x.)

certain other tumors. For example, metachromatic granules that can be seen in mast cell tumors stained with toluidine blue are absent in the CCH.

Two studies have examined the enzyme histochemistry of CCH. In 1970, Kelly[10] reported the absence of acid phosphatase, a lysosomal enzyme, in tumor cells of the CCH. More recently, Glick, Holscher, and Campbell[11] found rather diffuse cytoplasmic staining for alpha-naphthyl acetate esterase (ANAE) on impression smears of CCH. This enzyme, although found in several types of cells, has been more specifically associated with phagocytic mononuclear cells.[12] We have confirmed the findings of Glick, Holscher, and Campbell[11] and further have identified ANAE-positive cells in frozen sections of CCH prepared according to the methods of Knowles and Holck.[13] Tumor cells have a dense, diffuse cytoplasmic reaction product which may be so intense that nuclear identification or detail is obscured (Fig. 6). Infiltrative lymphocytes have little, if any, ANAE activity. We do not know if dog T lymphocytes mark by a small, discrete focus of ANAE activity as has been reported for man and mouse.[13]

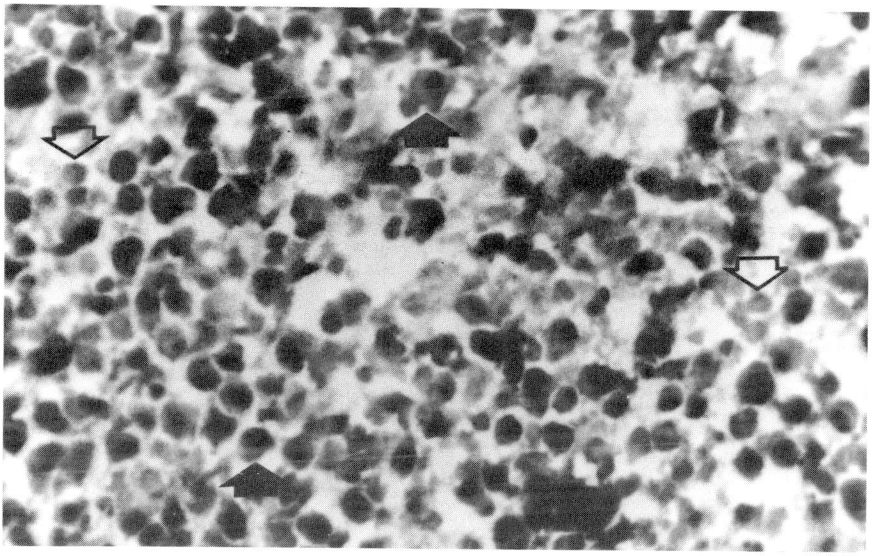

Fig. 6 Frozen section of a CCH stained for ANAE. The neoplastic histiocytic cells (solid arrows) have a large, dense cytoplasmic globule (reaction product) that partially obscures the less intensely counterstained (methyl green) nucleus. Occasional infiltrative lymphocytes (open arrows) react only with the counterstain. (Magnification, 640 x.)

## Cytogenetic Analysis

The karyotype of tumor cells from the CCH is similar to that from the normal dog.[7] Karyotype analysis has been used for diagnostic differentiation of CCH from the cytologically similar transmissible venereal tumor, which usually contains 59 chromosomes rather than the typical 78 chromosomes of normal dogs.[14]

## Transmission Studies

The most extensive attempt to experimentally transmit CCH was performed by Taylor, Dorn, and Luis.[2] Forty-one attempts to transmit viable cells, explants, or cell fragments of CCH to young dogs or fetuses by the subcutaneous, intradermal, intraocular, or intracranial routes were unsuccessful. Forty-two attempts to transmit CCH to the cheek pouch of cortisone-treated weanling hamsters were equally unsuccessful.

More recently we have attempted to transmit whole-cell preparations of CCH subcutaneously into nude mice. Preliminary results show no evidence of persistence of inoculated cells at the site of injection in mice killed and examined 3 weeks after inoculation.

## Microbiological Studies

Attempts to isolate or identify significant fungi, bacteria, or viruses from CCH have been unsuccessful.[2] Inoculation of ground CCH tissue onto cultures of dog kidney or onto hamster fetal lung or kidney cells failed to produce a cytopathic effect, and no noncytopathic agents were capable of interfering with subsequent infection of the same cells with infectious canine hepatitis virus.[2]

One ultrastructural study of tumor cells of the CCH revealed cytoplasmic inclusions suggestive of viral components.[11] These inclusions consisted of reticular aggregates of tubular arrays with crystalline regularity.

## Cell Culture

Attempts to propagate cells of CCH in vitro, either from individual cells or from tissue explants, have been only partially successful.[2] Attached, individualized cells with a "histiocytic" morphology were identified within 48 hr of initiation of cultures. By the third passage these attached cells were either replaced by or assumed a morphology similar to that of fibroblasts, and the new or altered cells persisted for up to 15 passages.

Other investigators have alluded to the problem of morphologic distinction between normal stromal fibroblasts and neoplastic mesenchymal precursor cells in culture.[15] We are currently attempting to correlate in vitro morphology of the cultured tumor cells with their ultrastructural appearance, presence of ANAE reactivity, and plasma membrane receptors for immunoglobulin or complement components. These features, which have been well described for lymphoid cells and mononuclear phagocytes,[16–19] should provide distinction of neoplastic precursors from normal cell elements.

In addition to these in vitro attempts to propagate CCH, more recently we have initiated studies to examine the growth and behavior of these tumor cells in several tumorigenicity assays, including the chick chorioallantoic membrane[20] and chick embryonic skin.[21] These assays, which compare the ability of tumor cells vs. normal cells to grow and invade the substrate, are similar in principle to the growth of inoculated tumor cells in nude mice but offer several advantages over the nude mouse system, including more rapid results, greater reliability, and less expense. Initial results indicate that CCH cells which have been passaged in tissue culture grow and establish themselves in both of the chick assay systems. However, as with tissue culture in vitro, the inoculated cells maintain their fibroblastic morphology and thus make identification of neoplastic histiocytes difficult. Our efforts, therefore, have been directed toward the identification and characterization of test cells prior to and after inoculation in culture. Similarly, we are comparing the behavior of CCH cells with normal canine fibroblasts and histiocytes.

## HISTOGENESIS

The cell of origin of the CCH has been disputed since the condition was first described, and the argument remains incompletely resolved. In the past the disorder has been designated as a round cell sarcoma,[22] reticulo or histiocytic sarcoma,[23] transmissible reticulum cell tumor,[5] transmissible lymphosarcoma,[24] and a facsimile or extragenital form of the transmissible venereal tumor of the dog.[1,5] Today the CCH falls into an incompletely defined group of canine tumors, which are rather loosely called "round cell tumors." In addition to those tumors listed in the differential diagnosis of CCH, this group includes tumors with a confusing array of cell types ranging from small, round, poorly differentiated mesenchymal cells to more typical histiocytic cells.[6]

The CCH is classified as a benign neoplasm without a malignant counterpart. However, the ultimate relationship of this tumor to more biologically aggressive, often multifocal, pleomorphic round cell neoplasms of the skin of the dog remains incompletely understood and awaits in-depth characterization beyond their light microscopic morphology.

Studies to date suggest that the origin of the CCH may be the mononuclear phagocyte system (MPS), which has been recently advanced by van Furth et al.[25] The MPS, which replaces the nomenclature and refines the definition of the reticuloendothelial system, constitutes a single cell lineage and is composed of promonocytes and their precursor cells in the bone marrow, monocytes, and fixed and free tissue macrophages. Cells of the CCH share many characteristics with those of the mononuclear phagocyte system, including cell size, nuclear to cytoplasmic ratio, nuclear and plasma membrane morphology, and cytoplasmic organellar content.

The functional criteria for cells of the MPS include phagocytosis in particular, pinocytosis, and the ability to adhere to glass. Cells of the CCH reportedly contain pinocytotic vesicles and grow as attached monolayers in tissue culture. However, avid phagocytosis is not a definite characteristic for CCH cells. Morphologic evidence of phagocytized material is rarely seen in cells of the CCH, but the identification of dense lysosomal-like granules and occasional lamellar cytoplasmic deposits in CCH cells suggests some residual or poorly developed phagocytic capacity. Furthermore, the cells are negative for the lysosomal enzyme acid phosphatase but strongly positive for ANAE, which has been tentatively localized within lysosomes.[26] The functional phagocytic capacity of freshly isolated CCH cells for exogenous material (e.g., latex beads and colloidal carbon) has not yet been tested. More important and reliable evidence for inclusion of CCH cells in the MPS would be the demonstration of plasma membrane receptors for the Fc portion of IgG, or C3b or C3d components, which indicates the capacity for immune phagocytosis.

## ETIOPATHOGENESIS

The etiology of CCH remains unknown, and lesions with similar morphologic and biologic behavior have not been produced experimentally. Cytoplasmic reticular aggregates observed in some CCH cells resemble structures that have been seen in virus-induced inflammatory and neoplastic diseases in man and animals.[11] Although further attempts to implicate an infectious etiology, either bacterial, fungal, or viral, have been uniformly unsuccessful, such past experiments do not preclude an infectious etiology. In what might be an analogous situation, a causative virus has not been isolated from lymphosarcoma in the dog; however, from lymphosarcoma in the cat and mouse, virus expression is readily manifested, and cell-free transmission is easily accomplished.[27,28] Furthermore, in the adult form of bovine lymphosarcoma, suspect causative C-type retroviruses are produced by malignant lymphocytes only after in vitro stimulation of the cells with such agents as phytohemagglutinin.[29] Similar studies need to be performed with CCH. Alternately, exposure of normal canine macrophages to known oncogenic viruses or activation with nonspecific stimulants, such as *Listeria monocytogenes*, followed by autochthonous dermal transplantation might provide useful clues to the etiology and pathogenesis of the CCH.

Some investigators have argued that the CCH is an inflammatory rather than a neoplastic condition.[7,30] This confusion stems partly from the variable appearance of the tumor cells which results from a progressive and destructive intratumoral lymphoid infiltrate. Prior to invasion by lymphocytes, the CCH appears as a well-delineated proliferation of a monomorphous population of cells with a high mitotic index. The pattern of growth at this stage is more compatible with a neoplastic process than with an inflammatory process. Further study of the behavior of CCH cells in organ culture tumorigenicity assays will provide additional and more definitive information for this distinction.

The sequential morphogenesis of naturally occurring CCH has never been proved, and the postulated mechanisms for its progression and regression remain speculative.[4] The close association of lymphoid infiltrates with intratumoral foci of necrosis and the spontaneous regression of the CCH in vivo suggest an immune-mediated anti-tumor host response. Abundant morphological and functional evidence of immunological reactivity against other experimentally induced and naturally occurring tumors exists[31-34] (cf. Chapter 5). The influence of the immune system in the regression of the CCH might be investigated through the identification of lymphoid subpopulations (B, T, or K cells) infiltrating the tumor, determination of immunoglobulin content of tumor tissue, and studies of mechanisms of anti-tumor immune response (direct cellular cytotoxicity, complement-mediated antibody cytotoxicity, or antibody-dependent cellular cytotoxicity).

## IMPORTANCE AS A BIOMEDICAL MODEL

The classification of phagocytic mononuclear cells and their histogenesis remains a recurring problem in pathobiology. Attempts to more completely define these cells have resulted in such designations as the macrophage system, reticuloendothelial system, or reticulohistiocytic system. The derivation of fixed macrophages (histiocytes) in connective tissue, including the skin, has been particularly obscure. The CCH provides the opportunity to obtain a relatively pure population of cells with characteristics of mononuclear phagocytes in a state of maturation arrest or dedifferentiation. Further characterization of the pathogenesis of the CCH would extend our knowledge of the histogenesis, morphological and functional potential, and manifestations that result from disturbances in the rate or pattern of growth of this cell line.

Spontaneous regression of neoplastic disease in humans has been debated for years and remains a difficult subject for study because of the paucity of documented cases, most of which are discovered retrospectively.[35-37] A variety of etiologic factors have been considered responsible for spontaneous tumor regression, but naturally acquired host resistance, particularly anti-tumor immune-mediated reactivity, is generally considered the most important. Much of our knowledge in tumor immunology has been generated from inbred rodent systems studied in vitro; yet human neoplasia remains a problem that occurs in vivo in a genetically heterogeneous species. The identification and study of potential models of spontaneous tumor regression in outbred animals are needed to further our understanding of the significance of this phenomenon. The CCH regresses in association with a progressive, tumor-destructive lymphoid infiltrate, occurs naturally in an outbred species of animals sharing the same environment with man, and is a potentially important model for the study of immune-mediated tumor regression.

## REFERENCES

1. R. M. Mulligan, Neoplastic Diseases of Dogs. II. Mast Cell Sarcoma, Lymphosarcoma, Histiocytoma, *Arch. Pathol.*, 46: 477-492 (1948).
2. D. O. N. Taylor, C. R. Dorn, and O. H. Luis, Morphologic and Biologic Characteristics of the Canine Cutaneous Histiocytoma, *Cancer Res.*, 29: 83-92 (1969).
3. E. B. Howard and S. W. Nielsen, Cutaneous Histiocytomas of Dogs, *Natl. Cancer Inst. Monogr.*, 32: 321-327 (1969).
4. G. L. Cockerell and D. O. Slauson, Patterns of Lymphoid Infiltrate in the Canine Cutaneous Histiocytoma, *J. Comp. Pathol.*, 89: 193-203(1979).
5. M. Orkin and R. M. Schwartzman, Comparative Study of Canine and Human Dermatology. II. Cutaneous Tumors—Introduction and Discussion of Transmissible Reticulum-Cell Tumor, *Am. Med. Assoc. Arch. Dermatol.*, 81: 347-358 (1960).
6. K. V. F. Jubb and P. C. Kennedy, The Skin and Appendages, in *Pathology of Domestic Animals*, 2nd ed., vol. 2, p. 649, Academic Press, Inc., New York, 1970.
7. H. A. Smith, T. C. Jones and R. D. Hunt, Neoplasia, in *Veterinary Pathology*, 4th ed., pp. 238-239, Lea & Febiger, Philadelphia, 1972.

8. A. A. Stannard and L. T. Pulley, Tumors of the Skin and Soft Tissues, in *Tumors in Domestic Animals,* 2nd ed., J. E. Moulton (Ed.), pp. 24-26, University of California Press, Berkeley, 1978.
9. W. Drommer and L.-Cl. Schulz, Vergleichende Licht- und Elektronenmikroskipische untersuschungen am ubertragbaren venerischen sarkom und histiozytom des hundes, *Pathol. Vet.,* 6: 273-286 (1969).
10. D. F. Kelley, Canine Cutaneous Histiocytoma. A Light and Electron Microscopic Study, *Pathol. Vet.,* 7: 12-27 (1970).
11. A. D. Glick, M. Holscher, and G. R. Campbell, Canine Cutaneous Histiocytoma: Ultrastructural and Cytochemical Observations, *Vet. Pathol.,* 13: 374-380 (1976).
12. L. T. Yam, C. Y. Li, and W. J. Crosby, Cytochemical Identification of Monocytes and Granulocytes, *Am. J. Clin. Pathol.,* 55: 283-290 (1971).
13. D. M. Knowles and S. Holck, Tissue Localization of T-Lymphocytes by the Histochemical Demonstration of Acid α-Naphthyl Acetate Esterase, *Lab. Invest.,* 39: 70-76 (1978).
14. M. Murray, H. James, and W. B. Martin, A Study of the Cytology and Karotype of the Canine Transmissible Venereal Tumor, *Res. Vet. Sci.,* 10: 565-568 (1969).
15. J. Sethi, Y. Hirshaut, S. I. Hajdu, and L. G. Clements, Growing Human Sarcomas in Culture, *Cancer,* 40: 744-755 (1977).
16. E. M. Shevach, E. S. Jaffee, and I. Green, Receptors for Complement and Immunoglobulin on Human and Animal Lymphoid Cells, *Transplant. Rev.,* 16: 3-28 (1973).
17. Th. J. L. M. Goud, C. Schotte, and R. van Furth, Identification and Characterization of the Monoblasts in Mononuclear Phagocyte Colonies Grown In Vitro, *J. Exp. Med.,* 142: 1180-1199 (1975).
18. E. R. Unanue, Secretory Function of Mononuclear Phagocytes, *Am. J. Pathol.,* 83: 396-417 (1976).
19. T. L. Whiteside and D. T. Rowlands, T-Cell and B-Cell Identification in the Diagnosis of Lymphoproliferative Disease, *Am. J. Pathol.,* 88: 754-790 (1977).
20. C. D. Scher, C. Haudenschild, and M. Klagsbrun, The Chick Chorioallantoic Membrane as a Model System for the Study of Tissue Invasion by Viral Transformed Cells, *Cell,* 8: 373-382 (1976).
21. P. D. Noguchi, J. B. Johnson, R. O'Donnell, and J. C. Petricciani, Chick Embryonic Skin as a Rapid Organ Culture Assay for Cellular Neoplasia, *Science,* 199: 980-983 (1978).
22. R. B. McClelland, Neoplasms in Dogs, *Cornell Vet.,* 30: 161-166 (1940).
23. H. E. Ottosen, Studies on Sarcomata of the Skin in Dogs, *Nord. Veterinaermed.,* 1: 7-30 (1949).
24. J. V. Lacroix and W. H. Riser, Transmissible Lymphosarcoma of the Dog, *N. Am. Vet.,* 28: 451-453 (1947).
25. R. van Furth, Z. A. Cohn, J. G. Hirsch, J. H. Humphrey, W. G. Spector, and H. L. Langevoort, The Mononuclear Phagocyte System: A New Classification of Macrophages, Monocytes, and Their Precursor Cells, *Bull. World Health Organ.,* 46: 845-852 (1972).
26. B. D. Lake, Histochemical Detection of the Enzyme Deficiency in Blood Films in Wolman's Disease, *J. Clin. Pathol.,* 24: 617-620 (1971).
27. M. Essex, Neoplasms of Cats Caused by Oncornaviruses, *Pathobiol. Ann.,* 5: 169-195 (1975).
28. G. J. Todaro, Evolution and Modes of Transmission of RNA Tumor Viruses, *Am. J. Pathol.,* 81: 590-606 (1975).
29. J. M. Miller, L. D. Miller, C. Olson, and K. G. Gillette, Virus Like Particles in Phytohemagglutinin-Stimulated Lymphocyte Cultures with Reference to Bovine Lymphosarcoma, *J. Natl. Cancer Inst.,* 43: 1297-1305 (1969).
30. E. Cotchin, Skin and Related Structures, in *Neoplasms of Domestic Animals,* pp. 4-6, Commonwealth Agricultural Bureaux, England, 1956.

31. J. C. Cerottini and K. T. Brunner, Cell-Mediated Cytotoxicity, Allograft Rejection and Tumor Immunity, *Adv. Immunol.,* 18: 67-137 (1974).
32. S. W. Russell and C. G. Cochrane, The Cellular Events Associated with Regression and Progression of Murine (Moloney) Sarcomas, *Int. J. Cancer,* 13: 54-63 (1974).
33. D. O. Slauson, B. I. Osburn, M. Shifrine, and D. L. Dungworth, Regression of Feline Sarcoma Virus-Induced Sarcomas in Dogs. I. Morphologic Investigations. II. Immunologic Investigations, *J. Natl. Cancer Inst.,* 54: 361-377 (1975).
34. H. L. Ioachim, The Stromal Reaction of Tumors: An Expression of Immune Surveillance, *J. Natl. Cancer Inst.,* 57: 465-475 (1976).
35. W. Boyd, *The Spontaneous Regression of Cancer,* Charles C Thomas, Inc., Springfield, 1966.
36. T. C. Everson and W. H. Cole, *Spontaneous Regression of Cancer,* W. B. Saunders Co., Philadelphia, 1966.
37. E. F. Lewison (Ed.), Conference on Spontaneous Regression of Cancer, *Natl. Cancer Inst. Monogr.,* 44: 1-148 (1976).

19

# The Transmissible Venereal Tumor

DAN COHEN
*Unit of Comparative Medicine, Faculty of Health Sciences, Ben-Gurion University of the Negev, Beer-Sheva, Israel*

The transmissible venereal tumor (TVT) of the dog is a unique, naturally occurring, neoplastic disease. The tumor is transmitted by coitus and can be induced experimentally in adult outbred dogs by the transplantation of living tumor cells. The naturally occurring neoplasm as well as the experimentally induced tumor often regress spontaneously after initial progressive growth. Both the transmissibility of the tumor and the high rate of spontaneous regressions have attracted the attention of numerous investigators since Novinsky,[1] in the year 1876, used the TVT in the first recorded successful experimental transplantation of a tumor.

Because this neoplasm is transplantable to allogeneic hosts and since the transmissibility of the tumor seemed to support the concept that cancer is "infectious," the experimentally induced disease was subject to several investigations around the turn of the century.[2–14] During this period the tumor was designated by such names as infective sarcoma and transplantable lymphosarcoma. Sticker,[11] a contemporary and co-worker of Paul Ehrlich, carried out a detailed study of the TVT and demonstrated that spontaneous regression of the tumor is followed by transplantation immunity. The tumor has since been called "Sticker's sarcoma" by some investigators.

A breakthrough in the understanding of the nature of the disease was achieved when Takayama and Makino,[15] in the year 1961, first reported their cytogenetic studies of the tumor. They showed that the TVT is characterized by consistent highly specific chromosome aberrations, which are suggestive of a cellular mechanism of transmission of the disease. The results of subsequent immunological investigations of the TVT also indicate that the tumor is a naturally occurring allograft.[16,17]

Current interest in the TVT stems from its clinical importance and from its usefulness as a model for tumor and transplantation immunology studies. The TVT is a common disease of the dog population of several countries, and about 10% of the tumor cases in some veterinary hospitals[18,19] are TVT cases. Although the tumor often regresses spontaneously, recurrence and metastatic spread are not uncommon. Further immunological investigations of this disease could lead to the development of effective therapeutic and preventive procedures.

In addition to the clinical importance of the disease, the TVT is a most attractive model for tumor immunology and transplantation immunology studies. Although experimentally induced tumors of laboratory rodents are the most commonly used models for investigations of the immune response toward neoplasia, spontaneous tumors of domestic animals, such as dogs and cats, recently have received increased attention.[20-22] Spontaneous neoplasia of domestic animals are valuable models for tumors in man because (1) both domestic animals and man are outbred species and are exposed to similar environmental influences, (2) tumors of domestic animals arise spontaneously and often have a natural history that is similar to their counterparts in man, and (3) the amount of tumor mass in relation to the size of the host reflects the quantitative relationships that exist in man (in contrast to tumor models of laboratory rodents). These quantitative relationships might be of special significance in studies involving the effects of circulating serum factors on the immune response toward neoplasia.

The transmissible venereal tumor is a naturally occurring allograft and therefore resembles choriocarcinoma in man.[23,24] Immunological investigations of metastatic TVT could contribute to the understanding of the immune response toward choriocarcinoma. The clinical course of the TVT is similar to the biological behavior of some widely used, spontaneously regressing, syngeneic murine tumor models.[25-28] Although the TVT is not a syngeneic or autochthonous neoplasm, detailed investigations of the immune response during progressive growth and regression could also yield valuable information about those immune mechanisms involved in the destruction of neoplastic cells in malignant tumors.

## GEOGRAPHIC DISTRIBUTION

The TVT occurs throughout the world, but its incidence seems to have fluctuated in some countries since it was first described. In Germany and Great Britain, for example, the disease was prevalent around the turn of the century,[10,11] but the incidence has been low in more recent years[29] (also D. Cohen, unpublished observations). The tumor seems to be prevalent in areas where stray dogs serve as a reservoir for the disease[30] and in those countries where the breeding activity of the dog population is not strictly controlled.

In recent years the occurrence of the tumor has been recorded in several European countries, including France,[31] Yugoslavia,[32] Italy,[33] Poland,[34] and the USSR.[19] TVT cases have also been recently recorded in Japan,[35] Iran,[36] Israel,[37] India,[38] and Australia.[39] The tumor is known to occur in African countries.[18,40] In Kenya, out of a series of 1535 dog tumors that were examined histologically, about 12% were TVT cases.[18] The TVT has also been recorded in recent years in the United States[17,41] and Mexico.[42]

## CLINICAL ASPECTS

The TVT is transmitted by coitus and usually affects the external genital organs of both sexes. Smith and Washbourn[5] observed the transmission of the tumor by coitus from one affected male dog to 11 out of 12 bitches. Few, if any, symptoms exist in the early stages of the disease. There might be a serous or sanguineous discharge, and the dog might lick the affected organs. In the male, examination reveals single or multiple small, gray–red firm nodules on the penis.[43] As the tumor increases in size, it appears as a gray–red firm mass, which frequently bleeds, ulcerates, and eventually undergoes necrosis. The neoplasm often has a cauliflower-like shape but also might be pedunculated, nodular, papillary, or multilobulated. The tumor mass can reach a diameter of 10 cm or more. In females, in the early stages of the disease, a serosanguineous discharge from the vulva might be the only indication of the existence of the lesion. As the disease progresses, the symptoms in the female are essentially similar to those in the male. The TVT can metastasize to the inguinal and iliac lymph nodes, spleen, liver, and brain.[30,41,44–47] In some cases this neoplasm is found in extragenital sites, appearing as nasal, oral, or subcutaneous tumors with or without involvement of the external genitalia.[18,48,49] The general clinical condition of dogs affected with the TVT is usually good, but widespread metastasis can result in emaciation and eventual death of the animal.[30,46] Spontaneous regression of the naturally occurring TVT may occur,[30,46] but no data concerning frequency of regression are available. The TVT can affect any breed of dog and appears most frequently in adult animals.

Because the TVT can regress spontaneously, it is difficult to evaluate the effect of therapeutic procedures on the course of the disease. The TVT reportedly can be successfully treated with vinblastin[34] and with a combination of cyclophosphamide and prednisone,[50] but the method of choice for therapy is surgical removal of the tumor[33,51] followed by X-irradiation of the tumor bed.[19,52] The recent introduction of bacillus Calmette-Guérin (BCG) immunotherapy is discussed in the section dealing with the immunological aspects of the tumor.

## HISTOLOGY AND ULTRASTRUCTURE

The TVT is an undifferentiated round cell neoplasm of reticuloendothelial origin. The tumor cells are spherical, polyhedral, or slightly oval, uniform in size and appearance, and have a diameter of 9 to 12 μm. The nuclei are large, round, and vesicular, usually with one eccentrically located nucleolus. Stained with hematoxylin and eosin, the nucleus is hyperchromatic, whereas the cytoplasm is pale pink or blue and granular. The cells are closely packed and arranged in diffuse masses, in clusters, or in cords along strands of fibrous stroma; the amount of stroma varies with the phase of tumor growth but is usually scanty. Silver stains reveal argyrophilic fibers, which surround groups of tumor cells. Many mitotic figures are found, sometimes six in one high-power field. Lymphocytes, plasma cells, macrophages, and polymorphonuclear cells can be seen throughout regressing tumors.[44,49,53,54] Despite several histochemical studies,[44,53-55] histogenesis of the TVT is uncertain. Bloom, Paff, and Noback[53] classified the tumor cells as mature end cells of reticuloendothelial origin, and Samso[56] defined the TVT as a reticulo-histiocytosarcoma.

Ultrastructure studies of the TVT were carried out by several investigators.[57-65] The cells are usually round with a high nucleoplasmic/cytoplasmic ratio, and the nucleus contains a large nucleolus. The cytoplasm contains many free ribosomes, often grouped in rosettes, as well as a Golgi apparatus. Few swollen mitochondria and granular endoplasmic reticulum are present. The membrane of tightly clustered cells forms numerous microvilli and extensive interdigitations with adjacent cells. The ultrastructural characteristics of the TVT also suggest that this neoplasm is of reticuloendothelial origin.[63,65,66] The ultrastructure of the tumor has also been studied at various stages of growth and regression.[60,63,64] Regressing tumors contain an increased number of degenerating neoplastic cells. Alterations in degenerating tumor cells consist of decreased amounts of endoplasmic reticulum and ribosomes, swelling and vacuolation of mitochondria, clumping of nuclear chromatin along the nuclear envelope, and a widened perinuclear space. Kennedy, Yang, and Allen[60] observed that the number of irregularly shaped round cells and fibroblast-like cells increased with increasing tumor mass and suggested that the round TVT cells differentiated into fibroblast-like cells.

## EXPERIMENTAL TRANSPLANTATION

One of the outstanding characteristics of the TVT is the facility with which it can be transplanted experimentally to allogeneic animals. In experimental studies subcutaneous transplantation of tumor fragments or cell suspensions is the method most often used,[43,67] but various other routes have been used successfully. DeMonbreun and Goodpasture[44] observed a general spread of the TVT in animals that had been injected intravenously with the tumor cells. Tumor cells placed on the scarified penis of a dog result in tumor growth, but the tumor does not grow when placed on the intact conjunctiva.[68] Karlson and Mann[43] transplanted the TVT through 40 generations, which consisted of a total of 564 dogs. Tumors developed in 68% of the animals, and the proportion of "takes" was not significantly altered during the course of the experiment. Most TVT nodules could be detected by palpation 3 weeks after transplantation, and the TVT regressed spontaneously within 180 days after its appearance in 87% of the dogs that developed the tumor. No recurrences were observed in animals in which the TVT had regressed spontaneously, but 5 out of 385 animals that developed the tumor had evidence of metastases.

Heterotransplantation of the TVT has been attempted in several animal species. The tumor can be transplanted successfully to the fox,[11,14] jackal,[56] and nude mice.[69] Furthermore, the TVT grows in the cheek pouch of cortisone-treated hamsters[56] and temporarily in X-irradiated mice.[68] Attempts to transplant the tumor to rats, normal mice, chickens, monkeys, guinea pigs, rabbits, and cats were unsuccessful.[9,44,56]

## ETIOLOGY

The fact that the TVT can be transmitted to outbred dogs, either by coitus or by inoculation of tumor cells, prompted many workers to search for a causative infectious agent. At the beginning of this century, several investigators[10,12,14] tried, using histological techniques, to determine whether a TVT induced by the inoculation of tumor cells originates from the host cells or from the transplanted tissue, but these studies gave conflicting results. More recently, some electron-microscope studies of the TVT revealed virus-like particles in the tumor cells,[60,61,70–74] whereas in other studies such particles could not be detected.[57–59,64]

Several researchers attempted to extract an oncogenic agent from TVT cells. Freezing and thawing the neoplasm or lyophilization destroyed the capacity of TVT cell suspensions to transmit the tumor.[44,68] Attempts to transmit the TVT by using cell-free extracts gave conflicting results. Although some investigators[75–77] claim success in transmitting the disease by the inoculation of cell-free material, others failed to do so.[56,63]

Although the agent(s) that induced the TVT in the original host(s) is unknown, cytogenetic and immunological studies strongly suggest that this neoplasm is transmitted by the transplantation of cells and is, in fact, a naturally occurring allograft. Highly significant and consistent karyotypic differences exist between normal dog cells and the cells of the tumor. The karyotype of normal dog cells consists of 78 chromosomes. All the autosomes are acrocentric, whereas the X and probably the Y chromosomes are metacentric. In contrast, Takayama and Makino[15] and Makino[78] demonstrated that TVT cells are characterized by a stemline chromosome number of 59. Of these 59 chromosomes, 16 or 17 are metacentric and 43 or 42 are acrocentric. These findings have been, with minor variations, confirmed by cytogenetic studies of the TVT in the United States,[79] France,[80] Uganda,[81] USSR,[82] Jamaica,[83] and Nigeria.[40] Although TVT cells contain fewer chromosomes than do normal dog cells, the total amount of deoxyribonucleic acid (DNA) and the number of chromosomal arms in the tumor cells are close to those of the normal cells.[80,81] A recent comparative banding pattern analysis in normal and TVT cells revealed extensive structural rearrangements in the tumor chromosomes.[35] After prolonged propagation of TVT cells in tissue culture, the karyotype was that of normal dog cells.[56] In another study[84] the karyotype of fresh TVT material and of TVT grown in tissue culture was investigated. Although fresh TVT cells as well as tumor cells grown 24 hr in tissue culture showed a karyotype typical of the TVT, most of the cells obtained from the TVT after 72 or 192 hr in tissue culture showed the chromosome pattern that is typical of normal dog cells. It seems that normal dog cells overgrew the TVT cells in tissue culture after 72 hr and that TVT cells stopped dividing or died. In contrast, Adams, Carter, and Sapp[85] grew the TVT in tissue culture for over 18 months and found that the tumor cells had the karyotype that is typical of the TVT.

The cytogenetic findings of profound and consistent chromosome aberrations of the TVT suggest that this neoplasm is not transmitted by an oncogenic agent[79] but support the concept that the TVT is transmitted by transplantation of cells. Further evidence suggestive of the cellular mode of transmission of the TVT was obtained by transplanting tumor cells labeled with $^3$H–thymidine.[86] The new tumors developed by multiplication of the transplanted cells. Immunological investigations provide additional evidence which indicates that the TVT is an allograft.

## IMMUNOLOGICAL ASPECTS

### The Immune Response In Vivo

Spontaneous regressions of autochthonous neoplasia in man are uncommon.[87] Although most experimentally induced animal tumors are also malignant in the isogeneic hosts, spontaneous regressions can be observed in some potentially malignant tumors of laboratory animals. Under certain conditions, Moloney virus-induced sarcomas in mice,[25,28] Gross virus-induced lymphomas in rats,[88,89] and an adenocarcinoma of mice[27] regress spontaneously in the isogeneic host following the response of the immune system against the tumor cells.

After experimental transplantation, the TVT grows rapidly for the first few weeks and may double its volume every 4 days.[90] Subsequent tumor growth is slower, and in most cases the neoplasm regresses spontaneously within several months.

Sticker[11] reported that spontaneous regression occurred in 16% of dogs affected with an experimentally transplanted TVT. In some cases the TVT developed into a malignant condition, whereas in others, after the initial growth, the tumor regressed spontaneously. Furthermore, in some dogs he observed alternate periods of progressive growth and regression of the tumor. The frequency of spontaneous regression of TVT varies under different experimental conditions,[44,68,91] but it can be as high as 87% (Ref. 43). In the naturally acquired disease, spontaneous regression may occur,[30,46] but no data about its frequency are available. Following complete spontaneous regression of the TVT, the tumor does not recur,[43] and the animals are immune to further challenge with TVT transplants[11,44,91] except when challenged with large inocula of tumor cells.[44]

The immunogenicity of the TVT in the dog can also be demonstrated in passive transfer experiments of postregression sera. Such sera can inhibit the growth of the experimentally transplanted tumor and prevent tumor development if administered simultaneously with tumor transplantation.[91] Further evidence demonstrating the role played by the immune response in inhibiting the growth and spread of this potentially malignant neoplasm was obtained in a study of the biological behavior of the TVT in immunosuppressed dogs.[92] In most of the animals that were subjected to whole-body X-irradiation prior to transplantation, the tumor grew progressively and metastasized. In contrast, in the normal dogs the tumor showed a more benign and irregular biological behavior, which ranged from growth followed by regression in some dogs to metastases in others. In transplantation experiments of the TVT in newborn dogs,[93] the tumor grew progressively and metastasized regularly in puppies born to normal dams but was more benign in adult dogs and in puppies born to dams immunized with the TVT.

Nonspecific stimulation of the immune response with BCG can alter the course of the experimentally induced disease.[94] Pairs of littermates, identical for the major histocompatibility complex, were transplanted with the TVT. One member of each pair was inoculated with BCG into one of two tumor nodules. The tumor nodules injected with BCG and the noninjected tumors in the same hosts regressed more quickly than did those in the control animals.

The spontaneous regressions of the TVT and subsequent transplantation immunity, the inhibition of tumor growth by passive transfer of immune serum or intralesional administration of BCG, and the enhanced tumor growth in immunosuppressed animals and in newborn dogs clearly demonstrate that the course of this disease is influenced by the immune response toward the tumor.

## Detection of Humoral Antibodies In Vitro

Humoral antibody to TVT homogenates has been demonstrated with a tanned red cell hemagglutination technique,[95] the Ouchterloney technique,[91] and the conglutinating-complement adsorption test.[96] Cohen,[67] who used the indirect membrane immunofluorescence test, first demonstrated humoral antibodies against membrane-associated antigens of TVT cells. These predominantly IgG antibodies can be detected in almost all tumor-bearing animals at 40 days or more after transplantation regardless of the stage of the disease or whether the tumor regressed or metastasized. Moreover, sera of clinical cases from Ireland reacted with TVT cells that originated from Malaya. Gomez-Estrada et al.[42] confirmed these findings and showed that the anti-TVT antibodies are not cytotoxic in the presence of dog complement.

Bennett, Debelak-Fehir, and Epstein,[97] who investigated the effect of anti-TVT sera on tumor cell colony formation in semisolid agar, found that the formation of tumor cell colonies is inhibited in the presence of sera from dogs in which the tumor was regressing (regressors). In contrast, colony formation was not inhibited in the presence of sera from dogs in which the tumor was growing progressively (progressors). The inhibitory effect of the regressor sera is blocked when the TVT cells are preincubated with progressor sera or sera from metastatic cases. The mechanisms involved in the inhibitory and blocking effect of the sera have not yet been characterized.

## Immunological Studies of the Mechanism of Transmission

Complementing the cytogenetic findings, the results of immunological studies also suggest that the TVT is an allograft and is not transmitted by an oncogenic agent. If the TVT were transmitted by a virus, the histocompatibility antigens of the host should be expressed on the surface of the tumor cells.[98–100] However, Cohen[16] demonstrated that the isoantigens of the host cannot be detected on the cell surface of the TVT with isoimmune sera directed against host tissue antigens. Further evidence indicating that the TVT is an allograft was obtained by demonstrating that anti-TVT sera react with normal lymphocytes of some dogs[16] and display antibody activity against DL-A antigens.[17]

Epstein and Bennett[17] investigated the histocompatibility antigens of the TVT and showed, by means of DL-A typing sera, that two different TVT lines, originating from Malaya and the United States, had a similar complement of antigenic specificities. These specificities remained constant throughout 40 transplant generations. Moreover, Bennett, Taylor, and Epstein[101] showed that the segregation of TVT growth patterns in canine families correlates with DL-A segregation.

It would be difficult to reconcile the results of these immunological studies with viral transmission of the disease.

## In Vitro Cellular Immune Response

The cellular immune response toward the TVT has been investigated in vitro with the mixed lymphocyte tumor culture (MLTC) and a $^{51}Cr$ release cytotoxicity assay.

Hess et al.[102] demonstrated that DNA synthesis of peripheral blood lymphocytes (PBL) from normal and regressor dogs is equally stimulated by TVT cells, whereas the reaction is partially suppressed in those same animals during the progressive phase of tumor growth. The stimulation of lymphocytes from progressor dogs is further reduced in the presence of autologous progressor serum. The MLTC response during the progressive phase of tumor growth correlates with the clinical course of the TVT. High blastogenic responses were detected in dogs in which the tumor eventually regressed, whereas no lymphocyte reactivity could be detected in dogs in which the tumors eventually metastasized.

Using a $^{51}Cr$ release cytotoxic assay, Cohen (D. Cohen, unpublished data) found that PBL of dogs in which the TVT had regressed are cytotoxic to the tumor cells, whereas the PBL of normal dogs and of TVT bearing dogs during progressive tumor growth are not cytotoxic. Furthermore, postregression sera mediate antibody dependent cellular cytotoxicity (ADCC) with normal dog lymphocytes or cytotoxic PBL as effector cells. With cytotoxic lymphocytes, the ADCC effect can be observed in addition to the direct cytotoxic effect.

These investigations show that the cellular immune responses toward the TVT correlate with the clinical course of the disease. Detailed kinetic studies of in vitro cell mediated immunity toward the tumor, using separated subpopulations of immune cells, could contribute to the understanding of the effector mechanisms involved in allograft rejection and tumor immunity.

## In Situ Expressions of the Immune Response

Although systemic immune responses toward allografts and neoplasia have been studied in great detail in recent years, little information is available about the presence and activity of components of the immune system inside allografts and neoplasia. Results of recent studies suggest that different effector mechanisms may operate in the central lymphatic system and inside tumors and allografts.[103] Investigations of in situ immunity could contribute to the understanding of the mechanisms involved in allograft and tumor immunity.

The TVT is a useful tumor system for such studies. In 1908, Wade[14] observed "a flooding of the tumor tissue with lymphocytes, polyblasts and plasma cells" during regression of the TVT and concluded that the disappearance of the tumor results from "gradual cytolysis associated with the presence of and probably caused by lymphocytes and polyblasts." The presence of lymphocytes, plasma cells, and macrophages in the TVT has been confirmed by several investigators who carried out histological and ultrastructural studies of this neoplasm.[53,60,64] Yang, Roberts, and Jones[104] studied the infiltration of the TVT with lymphoreticular cells (lymphocytes, plasma cells, macrophages, polymorphs, and mast cells) at different stages of tumor growth. The percentage of lymphoreticular cells in tumor suspensions was inversely correlated with the logs of tumor mass because the volume of the necrotic center of the tumor increased more rapidly than did the volume of those parts of the tumor which are perfused with peripheral blood.

The immunological activity of the infiltrating cells has not yet been characterized, but some information regarding their nature is available. Yang and Jones[93] reported that about 23% of the cells in single-cell suspensions of the TVT were small lymphocytes as determined by morphological criteria. In addition, a proportion of the cells (3 to 35%) of TVT single-cell suspensions were shown to form rosettes with human erythrocytes.[66] The rosette-forming cells consisted, as determined by morphological criteria, of both lymphocytes and apparent TVT cells. The rosette-forming lymphocytes are probably T cells since a proportion of dog T lymphocytes are known to form rosettes with human erythrocytes.[105,106] Further investigations are needed to determine whether the tumor-like rosette-forming cells are indeed TVT cells. Cohen, Gurner, and Coombs[107] found that up to 30% of the cells in TVT cell suspensions form rosettes with antibody-coated sheep erythrocytes and concluded that the cells were probably TVT cells. However, our recent morphological and histochemical studies (D. Cohen, R. Bargum, and C. Greenzweig, unpublished data) indicate that these rosette-forming cells are macrophages and lymphocytes.

An additional expression of the in situ immune response to TVT is the in vivo coating of the tumor cells with antibody. Using indirect membrane immunofluorescence, Cohen[67] found that the TVT cells are coated in vivo with antibody at 40 days or more after transplantation, which is simultaneous with the appearance of antibodies in the serum. The antibodies that coat the tumor cells can be eluted at low pH and are IgG.[17] In vivo coating of tumor cells with antibodies occurs in primary and transplanted murine neoplasia as well as in tumors in man.[108] These antibodies might play a role in the destruction of tumor cells through ADCC[109] or could protect neoplastic cells from cytotoxic lymphocytes by blocking the antigenic sites at the target cell level.[110]

The studies on local immunity toward the TVT demonstrate that components of the immune system infiltrate into the tumor. The isolation and characterization of these lymphoid cells and antibodies as well as studies of their immunological activities should enable a better understanding of the mechanisms that bring about the spontaneous regressions of this tumor.

## CONCLUDING REMARKS

The most striking characteristics of the TVT are the facility with which this neoplasm can be transmitted to adult outbred dogs and the variable clinical course of the disease, which ranges from spontaneous regression to metastatic spread. The cytogenetic and the immunological data leave little doubt that the TVT is transmitted by the transplantation of cells. In nature the transfer of the tumor might be facilitated by injuries of the vaginal and penile mucosa, which often follow sexual intercourse in dogs. The successful transplantation of the tumor to allogeneic hosts would appear to depend on critical modifications of the histocompatibility antigens on the surface of the neoplastic cells. The tumor also may induce transient specific immunosuppression. The transplantation of the tumor is followed, in most cases, by initial progressive tumor growth without apparent inhibition by the immune response. Several weeks after transplantation, however, the immunological reaction of the host plays a major role in determining the course of the disease. The intensity of the immune response toward the tumor and consequently the course of the disease probably reflect the degree of disparity between the histocompatibility antigens of the tumor and those of the host.

In vitro studies show that the lymphocyte response in MLTC, cellular cytotoxicity, antibody activity in ADCC, and colony inhibition correlate with the clinical course of the disease and reflect in vivo events. Detailed kinetic studies of cellular and humoral effector mechanisms in dogs with benign and malignant TVT might clarify the role played by each of those mechanisms in the in vivo situation. Such studies could lead to a better understanding of the immune response toward the TVT and toward allografts and neoplasia in general.

## ACKNOWLEDGMENTS

Part of my experimental work was financed by a grant from the Israel Cancer Association and a grant in memory of Ada and Ernest Klein.

I am grateful to M. Tsjechover for assistance in the preparation of the manuscript.

## REFERENCES

1. M. A. Novinsky, Zur Frage über die Impfung der Krebsigen Geschwülste, *Zentralbl. med. Wiss.*, 14: 790-791 (1876).
2. Wehr, Demonstration der durch Impfung von Hund auf Hund erzeugten Carcinomknötchen, *Zentralbl. Chir.*, 15: 8-9 (1888).

3. Wehr, Weitere Mittheilungen über die positiven Ergebnisse der Carcinom-Ueberimpfungen von Hund auf Hund, *Arch. Klin. Chir.*, 39: 226-228 (1889).
4. Geissler, Gelungene Carcinom-Uebertragung beim Hunde, *Verh. Dtsch. Ges. Chir.*, 24: 87-92 (1895).
5. G. B. Smith and J. W. Washbourn, Infective Venereal Tumours in Dogs, *J. Pathol. Bacteriol.*, 5: 99-100 (1898).
6. G. B. Smith and J. W. Washbourn, Infective Sarcomata in Dogs, *Br. Med. J.*, 2: 1346-1347 (1898).
7. G. B. Smith and J. W. Washbourn, Infective Sarcoma in Dogs, *Br. Med. J.*, 2: 1807-1810 (1899).
8. A. Sticker, Transplantables Lymphosarkom des Hundes. Ein Beitrag zur Lehre der Krebsübertragbarkeit, *Z. Krebsforsch.*, 1: 413-444 (1904).
9. A. Sticker, Transplantables Lymphosarkom des Hundes, *Berl. tierärztl. Wschr.*, 21: 353-358 (1905).
10. E. F. Bashford, J. A. Murray, and W. Cramer, Comparison Between the Transmission of an Infective Granuloma of the Dog and Carcinoma of the Mouse, *Sci. Rep. Invest. Imp. Cancer Res. Fund.*, 2: 33-37 (1905).
11. A. Sticker, Transplantables Rundzellensarkom des Hundes, Ein Beitrag zur Lehre der Krebsübertragbarkeit, *Z. Krebsforsch.*, 4: 227-314 (1906).
12. S. P. Beebe, The Growth of Lymphosarcoma in Dogs, Summary of Recent Observations, *J. Am. Med. Assoc.*, 49: 1492-1493 (1907).
13. G. W. Crile and S. P. Beebe, Transfusion of Blood in the Transplantable Lymphosarcoma of Dogs, *J. Med. Res.*, 18: 385-406 (1908).
14. H. Wade, An Experimental Investigation of Infective Sarcoma of the Dog with a Consideration of Its Relationship to Cancer, *J. Pathol. Bacteriol.*, 12: 384-425 (1908).
15. S. Takayama and S. Makino, Cytological Studies of Tumors. XXXV. A Study of Chromosomes in Venereal Tumors of the Dog, *Z. Krebsforsch.*, 64: 253-261 (1961).
16. D. Cohen, The Mechanism of Transmission of the Transmissible Venereal Tumour of the Dog, *Transplantation*, 17: 8-11 (1974).
17. R. B. Epstein and B. T. Bennett, Histocompatibility Typing and Course of Canine Venereal Tumors Transplanted into Unmodified Random Dogs, *Cancer Res.*, 34: 788-793 (1974).
18. C. G. Ndiritu, S. W. Mbogwa, and P. D. Sayer, Extragenitally Located Transmissible Venereal Tumor in Dogs, *Mod. Vet. Pract.*, 58: 945-946 (1977).
19. N. E. Osipov and V. A. Golubeva, Diagnostika i lechenie transmissivnoi sarkomy, *Veterinariya*, 7: 97-98 (1976).
20. M. Essex, Horizontally and Vertically Transmitted Oncornaviruses of Cats, *Adv. Cancer Res.*, 21: 175-248 (1975).
21. D. Cohen, The Transmissible Venereal Tumor of the Dog, a Naturally Occurring Allograft? *Isr. J. Med. Sci.*, 14: 14-19 (1978).
22. I. J. Fidler and R. W. McWilliams, Immune Reactivity In Vitro to Spontaneous Canine Oral Tumors, *J. Med.*, 6: 323-336 (1975).
23. K. D. Bagshawe, Risk and Prognostic Factors in Trophoblastic Neoplasia, *Cancer*, 38: 1373-1385 (1976).
24. J. I. Brewer, E. E. Torok, B. D. Kahan, C. R. Stanhope, and B. Halpen, Gestational Trophoblastic Disease: Origin of Choriocarcinoma, Invasive Mole and Choriocarcinoma Associated with Hydatidiform Mole, and Some Immunologic Aspects, *Adv. Cancer Res.*, 27: 89-147 (1978).
25. E. W. Lamon, H. M. Skurzak, and E. Klein, The Lymphocyte Response to a Primary Viral Neoplasm (MSV) Through Its Entire Course in BALB/c Mice, *Int. J. Cancer*, 10: 581-588 (1972).

26. J. C. Leclerc, E. Gomard, F. Plata, and J. P. Levy, Cell-Mediated Immune Reaction Against Tumors Induced by Oncorna Viruses. II. Nature of Effector Cells in Tumor-Cell Cytolysis, *Int. J. Cancer,* 11: 426-432 (1973).
27. J. S. Haskill, Y. Yamamura, and L. Radov, Host Responses Within Solid Tumors: Non-Thymus-Derived Specific Cytotoxic Cells Within a Murine Mammary Adenocarcinoma, *Int. J. Cancer,* 16: 798-809 (1975).
28. M. Harada, G. Pearson, L. Redmon, E. Winters, and S. Kasuga, Antibody Production and Interaction with Lymphoid Cells in Relation to Tumor Immunity in the Moloney Sarcoma Virus System, *J. Immunol.,* 114: 1318-1322 (1975).
29. U. Heitgress, Zum Auftreten und zur Histologie des Sogenannten Sticker-Tumors (Veneral Tumor) beim Hund in Deutschland, *Jahresverz. Deutschen Hochschul. Schr.,* 82: 7653 (1966).
30. J. H. Rust, Transmissible Lymphosarcoma in the Dog, *J. Am. Vet. Med. Assoc.,* 114: 10-14 (1949).
31. X. Woimant and S. Chaffaux, Neoplasie vénérienne du chien et de la chienne (sarcome de Sticker), *Rec. Med. Vet.,* 153: 331-338 (1977).
32. K. Cermak, J. Marolt, M. Herceg, and Z. Maticic, Prenosivi venericni tumor pasa, *Prax. Vet.,* 24: 279-287 (1976).
33. P. Muzetto, V. Petruzzi, and S. Lepori, Il sarcoma di Sticker, Nota II, (Trattamento chirurgico nella cagna), *Clin. Vet.,* 98: 124-128 (1975).
34. A. Wasecki and O. Mazur, Zastosowanie preparatu Vinblastin w leczeniu guzow Stickera, *Med. Weter.,* 33: 142-143 (1977).
35. M. Oshimura, M. Sasaki, and S. Makino, Chromosomal Banding Patterns in Primary and Transplanted Venereal Tumors of the Dog, *J. Natl. Cancer Inst.,* 51: 1197-1203 (1973).
36. B. Ivoghli, Canine Transmissible Venereal Tumor in Iran, *Vet. Pathol.,* 14: 289-290 (1977).
37. D. Cohen, R. Yagil, and T. Waner, Two Cases of the Transmissible Venereal Tumor of the Dog in Israel, *Refuah Vet.,* 35: 12-13 (1978).
38. G. C. Mohanty and B. S. Rajya, Growth and Morphological Characteristics of Canine Venereal Tumor Cells In Vitro, *Vet. Pathol.,* 14: 420-425 (1977).
39. K. B. Locke, F. J. Yeh, and P. T. Hooper, Transmissible Venereal Tumour in Dogs in Australia, *Aust. Vet. J.,* 51: 449 (1975).
40. L. Idowu, Chromosomes of the Transmissible Venereal Tumour of Dogs in Ibadan, Nigeria, *Res. Vet. Sci.,* 22: 271-273 (1977).
41. E. W. Adams and L. J. Slaughter, A Canine Venereal Tumor with Metastasis to the Brain, *Pathol. Vet.,* 7: 498-502 (1970).
42. H. Gomez-Estrada et al., Anticuerpos en el sarcoma venereo del perro, *Arch. Invest. Med.,* 7: 1-8 (1976).
43. A. G. Karlson and F. C. Mann, The Transmissible Venereal Tumor of Dogs: Observations on Forty Generations of Experimental Transfers, *Ann. N.Y. Acad. Sci.,* 54: 1197-1213 (1952).
44. W. A. DeMonbreun and E. W. Goodpasture, An Experimental Investigation Concerning the Nature of Contagious Lymphosarcoma of Dogs, *Am. J. Cancer,* 21: 295-321 (1934).
45. J. E. Prier and J. H. Johnson, Malignancy in a Canine Transmissible Venereal Tumour, *J. Am. Vet. Med. Assoc.,* 145: 1092-1094 (1964).
46. D. A. Higgins, Observations on the Canine Transmissible Venereal Tumour as Seen in the Bahamas, *Vet. Rec.,* 79: 67-71 (1966).
47. O. O. Oduye, B. O. Ikede, G. O. Esuruoso, and J. U. Akpokodje, Metastatic Transmissible Venereal Tumour in Dogs, *J. Small Anim. Pract.,* 14: 625-637 (1973).

48. E. Arru, Localizzazioni multiple extragenitali del tumore venereo del cane (c.d. Sarcoma di Sticker), *Clin. Vet.*, 95: 403-409 (1972).
49. E. C. Weir, M. J. Pond, J. R. Duncan, and D. J. Polzin, Extragenital Occurrence of Transmissible Venereal Tumor in the Dog: Literature Review and Case Reports, *J. Am. Anim Hosp. Assoc.*, 14: 532-536 (1978).
50. A. P. Hernandez Jauregui, Efectos de la combinacion de ciclofosfamida y prednisona en el linfoma transmisible venereo canino, Estudio clinico e histelogico, *Vet. (Mex.)*, 5: 3-9 (1974).
51. P. Muzetto, V. Petruzzi, and S. Lepori, Il sarcoma di Sticker, Nota I, Trattamento chirurgico nel cane maschio, *Clin. Vet.*, 97: 357-362 (1974).
52. Nanta, Marques, Laserre, Bazex, Bru, Puget, Les tumeurs vénériennes du chien étude therapeutique, *Rev. Med. Vet. (Toulouse)*, 102: 321-337 (1951).
53. F. Bloom, G. H. Paff, and C. R. Noback, The Transmissible Venereal Tumor of the Dog—Studies Indicating that the Tumor Cells Are Mature End Cells of Reticulo-Endothelial Origin, *Am. J. Pathol.*, 27: 119-139 (1951).
54. C. Jackson, The Cytology of the Contagious (Venereal) Tumour of the Dog, *Onderstepoort J. Vet. Sci. Anim. Ind.*, 20: 97-118 (1944).
55. Ch. Lombard and J. J. Yulzari, Sur la cytochimie du sarcome de Sticker, *Rev. Med. Vet.*, 118: 85-93 (1967).
56. A. Samso, Recherches expérimentales sur le sarcome de Sticker, *Bibl. France*, 155: 3650(Suppl. D) (1966).
57. J. P. Carteaud, F. Lagneau, and M. Fontaine, Contribution à l'étude cytologique de la tumeur vénérienne du chien (sarcome de Sticker), *Bull. Soc. Fr. Dermatol. Syphiligr.*, 70: 793-799 (1963).
58. W. Drommer and L-Cl. Schulz, Vergleichende licht und elektronenmikroskopische Untersuchungen am übertragbaren venerischen Sarkom und Histiozytom des Hundes, *Pathol. Vet.*, 6: 273-286 (1969).
59. M. Murray, Z. H. James, and W. B. Martin, A Study of the Cytology and Karyotype of the Canine Transmissible Venereal Tumour, *Res. Vet. Sci.*, 10: 565-568 (1969).
60. J. R. Kennedy, T. J. Yang, and P. L. Allen, Canine Transmissible Venereal Sarcoma: Electron Microscopic Changes with Time After Transplantation, *Br. J. Cancer*, 36: 375-385 (1977).
61. M. Battistacci and F. Moriconi, Richerche ultrastrutturali sul sarcoma di Sticker, *Nuova Vet.*, 50: 226-236 (1974).
62. P. Hernandez Jauregui, A. Gonzalez-Angulo, and G. De La Vega, Ultrastructural and Histochemical Pattern of Regressing Canine Venereal Lymphoma After Cyclophosphamide Treatment, *J. Natl. Cancer Inst.*, 51: 1187-1196 (1973).
63. P. Cabanie, G. van Haverbeke, and J. P. Magnol, Etude ultrastructurale du sarcome de Sticker du chien a differents stades de son evolution, *Rev. Med. Vet. (Toulouse)*, 124: 1239-1253 (1973).
64. J. M. Cockrill and J. N. Beasly, Ultrastructural Characteristics of Canine Transmissible Venereal Tumor at Various Stages of Growth and Regression, *Am. J. Vet. Res.*, 36: 677-681 (1975).
65. C. Lombard and P. Cabanie, Considerations sur la nature et recherches sur l'ultrastructure du sarcoma de Sticker du chien, *Bull. Cancer*, 54: 357-365 (1967).
66. T. J. Yang, J. R. Kennedy, and R. B. Andrews, Rosette Formation of Human Erythrocytes on Canine Transmissible Venereal Sarcoma Cells, *Am. J. Pathol.*, 83: 359-366 (1976).
67. D. Cohen, Detection of Humoral Antibody to the Transmissible Venereal Tumour of the Dog, *Int. J. Cancer*, 10: 207-212 (1972).
68. E. L. Stubbs and J. Furth, Experimental Studies on Venereal Sarcoma of the Dog, *Am. J. Pathol.*, 10: 275-286 (1934).

69. S. M. J. Oughton and L. N. Owen, Transplantation of Dog Neoplasms into the Mouse Mutant Nude, *Res. Vet. Sci.*, 17: 414-416 (1974).
70. P. Ajello, Sulla presenza di inclusioni nelle cellule del tumore di Sticker, *Boll. Soc. Ital. Biol. Sper.*, 37: 247-248 (1961).
71. P. Ajello and A. Gimbo, Presenza di particelle virali nelle cellule del tumore di Sticker, *Atti. Soc. Ital. Sci. Vet.*, 19: 736-739 (1965).
72. P. Ajello and A. Gimbo, Il virus del tumore il Sticker—Alcuni suoi aspetti morfologici, *Biol. Lat.*, 18: 311-317 (1965).
73. C. Lombard, P. Cabanie, and J. Izard, Images évoquant l'aspect de virus dans les cellules du sarcome de Sticker, *J. Microsc. (Paris)*, 6: 81-84 (1967).
74. W. J. Sapp and E. W. Adams, C-Type Viral Particles in Canine Venereal Tumor Cell Cultures, *Am. J. Vet. Res.*, 31: 1321-1323 (1970).
75. P. Ajello, Transmissione del tumore di Sticker con materiale acellulare, *Nuova Vet.*, 36: 179-183 (1960).
76. P. Ajello and A. Gimbo, Contributo allo studio della etiologia del tumore di Sticker, *Arch. Vet. Ital.*, 16: 275-286 (1965).
77. G. Dozza and V. Torlone, Riproduzione sperimentale del tumore di Sticker con cellule trispinizzate e con frazioni cellulari, *Vet. Ital.*, 11: 647-658 (1960).
78. S. Makino, Some Epidemiologic Aspects of Venereal Tumors of Dogs as Revealed by Chromosome and DNA Studies, *Ann. N.Y. Acad. Sci.*, 108: 1106-1122 (1963).
79. W. T. Weber, P. C. Nowell, and W. C. D. Hare, Chromosome Studies of a Transplanted and a Primary Canine Venereal Sarcoma, *J. Natl. Cancer Inst.*, 35: 537-547 (1965).
80. G. Barski and Fr. Cornefert-Jensen, Cytogenetic Study of Sticker Venereal Sarcoma in European Dogs, *J. Natl. Cancer Inst.*, 37: 787-797 (1966).
81. D. H. Wright, S. Peel, E. H. Cooper, and D. T. Hughes, Transmissible Venereal Sarcoma of Dogs. A Histochemical and Chromosomal Analysis of Tumours in Uganda, *Rev. Eur. Etud. Clin. Biol.*, 15: 155-160 (1970).
82. E. S. Kakpakova, E. E. Pogosyanz, and V. I. Ponomarkov, Peculiarities of the Karyotype of the Transmissible Sarcoma Cells in the Dog, *Vopr. Onkol.*, 14: 43-50 (1968).
83. M. J. Thorburn, R. V. R. Gwynn, M. S. Ragbeer, and R. I. Lee, Pathological and Cytogenetic Observations on the Naturally Occurring Canine Venereal Tumour in Jamaica (Sticker's Tumour), *Br. J. Cancer*, 23: 720-727 (1968).
84. J. E. Prier, Chromosome Pattern of Canine Transmissible Sarcoma Cells in Culture, *Nature (London)*, 212: 724-726 (1966).
85. E. W. Adams, L. P. Carter, and W. J. Sapp, Growth and Maintenance of the Canine Venereal Tumor in Continuous Culture, *Cancer Res.*, 28: 753-757 (1968).
86. T. Kudo, T. Koike, K. Otomo, and T. Sakai, Transplantation of Canine Transmissible Venereal Sarcoma: Autoradiographic Study with $^{3}$H—Thymidine, *Jpn. J. Vet. Res.*, 22: 105-110 (1974).
87. T. C. Everson, Spontaneous Regression of Cancer, *Ann. N.Y. Acad. Sci.*, 114: 721-735 (1964).
88. M. O. De Landazuri, E. Kedar, and J. L. Fahey, Synergistic Cooperation Between Isoantiserum and Immune Lymphoid Cells: In Vitro Studies with a Syngeneic Rat Lymphoma, *J. Immunol.*, 112: 2102-2110 (1974).
89. M. Glaser, D. H. Larvin, and R. B. Herberman, In Vivo Protection Against Syngeneic Gross Virus-Induced Lymphoma in Rats: Comparison with In Vitro Studies of Cell-Mediated Immunity, *J. Immunol.*, 16: 1507-1511 (1976).
90. D. Cohen and G. G. Steel, Thymidine Labelling Studies in a Transmissible Venereal Tumor of the Dog, *Br. J. Cancer*, 26: 413-419 (1972).
91. R. D. Powers, Immunologic Properties of Canine Transmissible Venereal Sarcoma, *Am. J. Vet. Res.*, 29: 1637-1645 (1968).

92. D. Cohen, The Biological Behaviour of the Transmissible Venereal Tumour in Immunosuppressed Dogs, *Eur. J. Cancer,* 9: 253-258 (1973).

93. T. J. Yang and J. B. Jones, Canine Transmissible Sarcoma: Transplantation Studies in Neonatal and Adult Dogs, *J. Natl. Cancer Inst.,* 51: 1915-1918 (1973).

94. A. D. Hess, R. Catchatourian, A. R. Zander, and R. B. Epstein, Intralesional Bacillus Calmette-Guerin Immunotherapy of Canine Venereal Tumors, *Cancer Res.,* 37: 3990-3994 (1977).

95. G. Dozza and V. Torlone, Anticorpi emoagglutinanti in cani portatori del cosidetto tumore di Sticker, *Atti. Soc. Ital. Sci. Vet.,* 14: 531-536 (1960).

96. J. M. McKenna and J. E. Prier, Some Immunologic Aspects of Canine Neoplasms, *Cancer Res.,* 26: 137-142 (1966).

97. B. T. Bennett, K. M. Debelak-Fehir, and R. B. Epstein, Tumor-Blocking and Inhibitory Serum Factors in the Clinical Course of Canine Venereal Tumor, *Cancer Res.,* 35: 2942-2947 (1975).

98. G. Klein and E. Klein, Histocompatibility Changes in Tumors, *J. Cell. Comp. Physiol.,* 52(Suppl.): 125-168 (1958).

99. E. Klein and E. Möller, Relationship Between Host Range and Isoantigenic Properties in Different Sublines of the Same Sarcoma, *J. Natl. Cancer Inst.,* 31: 347-364 (1963).

100. F. M. Kourilsky, J. Dausset, N. Feingold, J. M. Dupuy, and J. Bernard, Leukocyte Groups and Acute Leukemia, *J. Natl. Cancer Inst.,* 41: 81-87 (1968).

101. B. T. Bennett, Y. Taylor, and R. Epstein, Segregation of the Clinical Course of Transmissible Venereal Tumor with DL-A Haplotypes in Canine Families, *Transplant. Proc.,* 7: 503-505 (1975).

102. A. Hess, B. Cunningham, B. T. Bennett, and R. B. Epstein, In Vitro Correlates of the In Vivo Course of the Canine Transmissible Venereal Tumor Studied by Mixed Lymphocyte-Tumor Cultures, *Transplant. Proc.,* 7: 507-511 (1975).

103. J. S. Haskill, P. Hayry, and L. A. Radov, Systemic and Local Immunity in Allograft and Cancer Rejection, *Contemp. Top. Immunobiol.,* 8: 107-170 (1978).

104. T. J. Yang, R. S. Roberts, and J. B. Jones, Quantitative Study of Lymphoreticular Infiltration into Canine Transmissible Venereal Sarcoma, *Virchows Arch., B,* 20: 197-204 (1976).

105. A. R. Zander, N. Boopalam, and R. B. Epstein, Surface Markers on Canine Lymphocytes, *Transplant. Proc.,* 7: 369-373 (1975).

106. C. A. Bowles, G. S. White, and D. Lucas, Rosette Formation by Canine Peripheral Blood Lymphocytes, *J. Immunol.,* 114: 399-402 (1975).

107. D. Cohen, B. W. Gurner, and R. R. A. Coombs, A Phenomenon Resembling Opsonic Adherence Shown by Disaggregated Cells of the Transmissible Venereal Tumour of the Dog, *Br. J. Exp. Pathol.,* 52: 447-451 (1971).

108. I. P. Witz, Tumor-Bound Immunoglobulins: In Situ Expressions of Humoral Immunity, *Adv. Cancer Res.,* 25: 95-148 (1977).

109. P. Perlmann, H. Perlmann, and H. Wigzell, Lymphocyte Mediated Cytotoxicity In Vitro. Induction and Inhibition by Humoral Antibody and Nature of Effector Cells, *Transplant. Rev.,* 13: 91-114 (1972).

110. K. E. Hellström, I. Hellström, and J. Nepom, Specific Blocking Factors—Are They Important? *Biochem. Biophys. Acta,* 473: 121-148 (1977).

# 20 Metastatic Models

I. R. HART
*and*
I. J. FIDLER
*Cancer Metastasis and Treatment Laboratory,*
*Frederick Cancer Research Center, Frederick, Maryland*

Cancer metastasis, the spread of a tumor from a primary site to a distant organ, is responsible for most failures in the treatment of neoplastic disease. Advances in surgical technique and general patient care have led to an increased success rate in the resection of primary neoplasms, and yet, in spite of this, many patients succumb to multiple secondary tumor growths.[1] The goal of clinicians must be the arrest, prevention, and treatment of such tumor spread. The formulation of successful therapeutic approaches to disseminated disease can be facilitated by a better understanding of the pathobiology of the metastatic sequence. In this context the role of animal tumor models in attempts to unravel the dynamic interplay between host and tumor properties that influence the outcome of the process has been of fundamental importance. However, the relevance of the many induced or transplantable animal tumor systems as all-encompassing models for the spontaneous disease in man has been questioned closely.[2,3] In the field of tumor biology and for the study of metastatic spread in particular, the choice of a suitable animal model is a necessary prerequisite for designing relevant experiments.[4] The failure of animal tumor systems to serve as predictive models for human cancer does not diminish their usefulness but might reflect the choice of an unsuitable model to answer a specific question. Our purpose in this chapter is not to advocate the replacement of experimentally induced tumor models in laboratory animals with naturally occurring malignancies in dogs; rather, we suggest that studies on canines may serve as a useful adjunct to those using rodent systems.

## THE PATHOGENESIS OF METASTASIS

Neoplasms can be subdivided into two major groups. Well-differentiated tumors, which closely resemble the tissue of origin, are encapsulated, grow slowly, and have few mitotic figures, are generally classified as benign. Malignant tumors are usually fast growing and consist of many aneuploid cells that are frequently pleomorphic with various degrees of anaplasia.[5] Such a classification is based on the supposition that morphologic criteria can be used to predict the biological behavior of tumors in vivo and thus serve as a prognostic tool. Although the cellular structure of a tumor may give a rough guide to its potential course, attempts at prognosis based on morphological criteria are, at best, arbitrary. Ultimately, the only major characteristics unique to malignant neoplasms are their invasive and metastatic capacities.

The development of cancer metastasis involves the following sequential steps:

1. The invasion of surrounding host stroma by cells from the primary tumor with the eventual penetration of blood and/or lymph vessels.
2. Detachment and release of tumor emboli into the circulation.
3. Arrest of emboli in the capillary bed of distant organs.
4. Extravasation or infiltration of the organ parenchyma.
5. Growth of vascularized host stroma into the new tumor.

The whole process can then be repeated following local invasion from the metastatic deposit.[6] Successful metastasis is dependent on the outcome of the interplay between host and tumor properties. We will attempt to give a broad overview of the phenomenon, pointing out areas where we feel the study of autochthonous tumors in dogs could be beneficial.

## TUMOR CELL INVASION

The invasion of normal host tissue by malignant cells composes the first step in the pathogenesis of metastasis (Figs. 1 to 3). Tumor cell invasion may be caused by rapid tumor multiplication, which would lead to infiltration by mechanical pressure; by reduced adhesiveness of tumor cells accompanied by an increased or excessive motility; or by the destruction of host tissues by lytic enzymes, either host or tumor derived, followed by expansion into those areas of destruction. Those mechanisms are not necessarily mutually exclusive, and their relative importance probably varies from tumor to tumor.

Injections of quick-setting fluids into jelly or isolated animal tissues have produced casts with shapes that closely resemble the invasive patterns of some tumors; rapidly proliferating tumor cells purportedly force finger-like processes of expanding cells along lines of least mechanical resistance.[7] The exceptions to

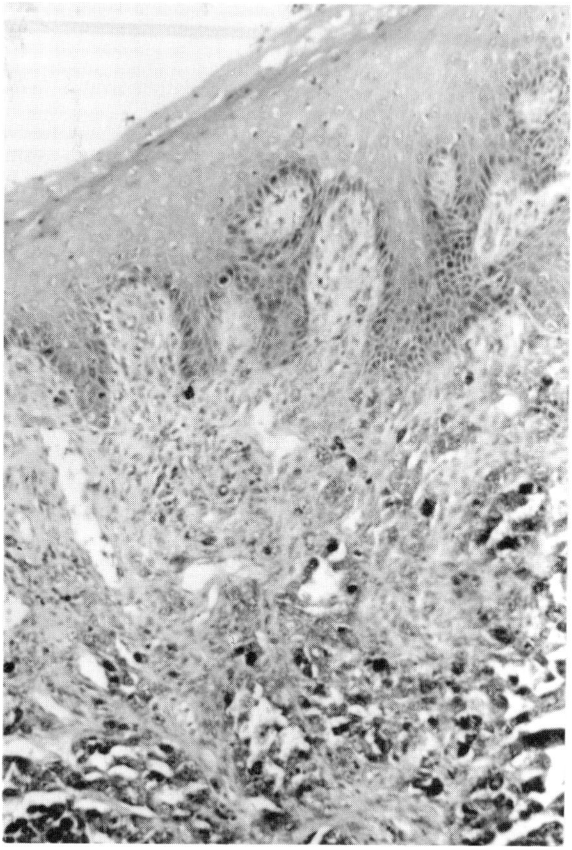

**Fig. 1 Primary oral melanoma in a 13-year-old male mixed breed. Note destruction of normal tissue architecture. (Magnification, 100 x.)**

that pattern, however, cast doubt that rapid tumor proliferation is a necessary prerequisite for tumor invasion. For example, slow-growing breast carcinomas are highly invasive, whereas rapidly growing breast fibroadenomas are noninvasive.[8] Tumor cell suspensions added to the surface of organ explants can infiltrate without a pressure factor,[9,10] whereas isolated tumor cells or clumps of tumor cells frequently have no continuity with the parent tumor.[11]

The invasive properties of malignant tumor cells have also been attributed to their increased motility, although little direct evidence supports such a contention. Rates of migration of tumor cells and homologous normal cells have seldom been compared.[12] Most observations on tumor cell motility have been performed on cells in monolayer tissue culture.[13-15] In recent studies of the

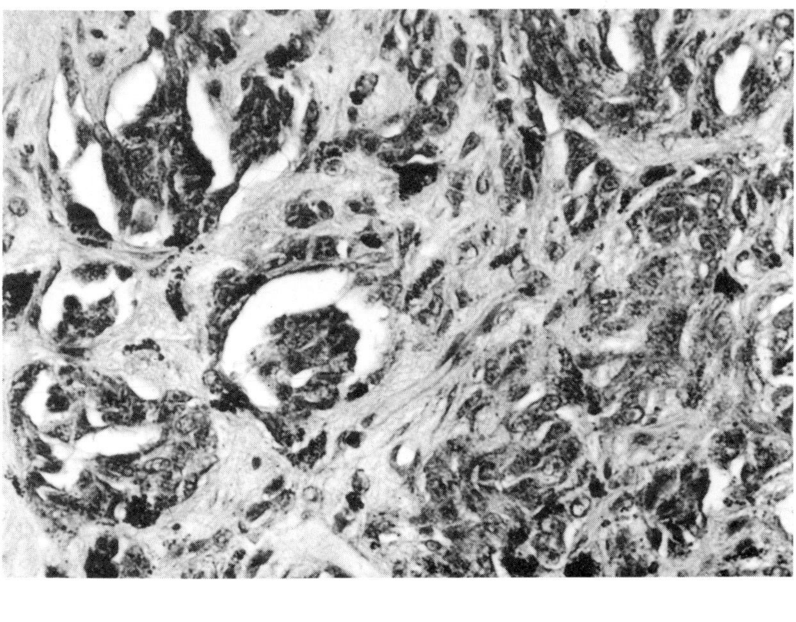

Fig. 2 Malignant oral melanoma. Note tumor cell invasion into small blood vessels. (Magnification, 400 x.)

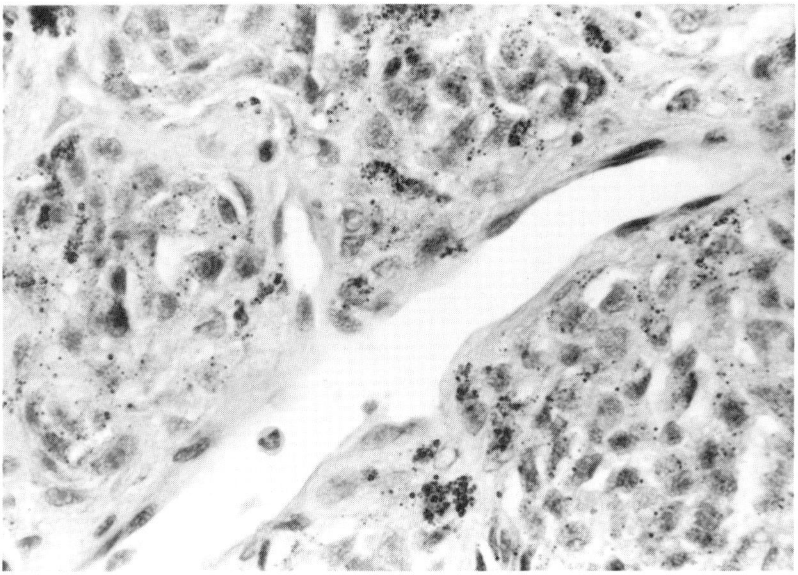

Fig. 3 Malignant oral melanoma. Note tumor cell invasion into small blood vessels. (Magnification, 400 x.)

rate of movement of radiolabeled tumor cells into homologous aggregates,[16,17] the increased in vivo tumorigenic potential of the hamster tumor cells was reflected by an increased cell motility.[18] The primary cultivation of spontaneous canine tumors[19-23] provides an opportunity for comparing migratory activities of such cells[15,24] with invasive behavior as expressed by the parent tumor. Also, migratory activities can be compared in cells from primary and secondary tumors derived from the same animal.

Many human and animal malignancies possess higher levels of proteolytic enzymes than do benign tumors or corresponding normal tissues.[25-28] Secretion of such enzymes could disrupt host stroma and facilitate invasion by tumor cells. Direct sampling of extracellular fluids from growing tumors has revealed marked increases in many enzymes.[29,30] The lysosomal proteinase, cathepsin B, is secreted at significantly higher levels by invasive breast carcinomas than by noninvasive breast fibroadenomas.[31] Again, the canine offers a chance to obtain material from both the primary and secondary tumor at the same time. Comparisons of enzyme levels in those two sites of tumor growth may reveal differences between primary and secondary growths, which could account for their different behavior.

Increased levels of proteolytic enzymes could cause the separation of tumor cells from one another[32,33] and facilitate their locomotion and subsequent penetration into normal host tissues. Susceptibility of host to tumor cell invasion varies with tissue type; cartilage, ligaments, arteries, and tendons are relatively resistant, whereas veins, lymphatics, soft tissues, and muscles are easily invaded. Tissues rich in ground substance or in elastic or dense collagen fibers seemingly provide a successful barrier to malignant cells[34] whether by the elaboration of anti-proteolytic factors[35] or by mechanical strength alone.

The chorioallantoic membrane (CAM) of the embryonated chicken egg has been used to study invasion.[9,36-38] Because of the absence of an immune reaction, such assays can be used to examine allogeneic tumor cells and their infiltrative capacity. Primary and secondary tumors derived from dogs and used in the CAM systems might provide some insight into the structural nature of invasion by spontaneous neoplasms.

## SPREAD BY DIRECT EXTENSION

Tumors that grow in or invade body cavities can release cells or fragments that migrate within the cavity, seed onto serosal or mucosal surfaces, and develop into new growths. In contrast, primary tumors of the central nervous system, although highly invasive, rarely develop metastases in organs outside the nervous system. The mode of spread appears to be by direct extension or via the cerebrospinal fluid.

## LYMPHATIC SPREAD

The division of tumor spread into lymphatic and hematogenous routes is somewhat arbitrary because tumor cells can readily pass from blood to lymphatic channels and back again,[39] blood can flow through lymphoid tissue,[40–42] and there are direct veno-lymphatic anastomoses.[43]

Clinical observations suggest that the lymphatic system provides the most common pathway for the initial spread of many carcinomas.[1] The pattern of lymph node involvement depends on the site of the primary tumor and on its lymphatic drainage. Lymphatic spread begins with the infiltration of a draining lymph node; metastasis to distant lymph nodes occurs later (Fig. 4). Tumor cells that bypass the chain of regional lymph nodes because of inflammatory changes, fibrosis following local X-irradiation, or obstruction by multiplying tumor cells may then be trapped to form distal nodal metastases.

Lymph nodes in the area of a primary neoplasm are often enlarged and clinically palpable. Histologically, the enlargement could be the result of either hyperplasia of lymph node follicles accompanied by proliferation of reticulum cells and sinus endothelium or active growth of tumor cells. Lymph node hyperplasia and sinus histiocytosis may indicate a host reactivity to autochthonous tumors. Berg et al.[44] suggested that the presence of any enlarged, nontumorous, apical (regional) lymph node in patients with breast cancer signified better prognosis than if the nodes were small or clinically absent. In those studies reactive nodes (better prognosis) consisted more of germinal center hyperplasia than sinus histiocytosis.

The lymph nodes of normal animals can be effective but temporary barriers to tumor spread. The filtration capacity of a lymph node can be altered by several mechanisms. Tumor proliferation and acute or chronic inflammatory reactions can lead to a decreased efficiency. Local X-irradiation, which is common in radiotherapy of neoplasms, can lead to fibrosis of the node and thus to cessation of filtration.

Undoubtedly, the regional lymph node (RLN) is immunologically involved in the host response to neoplasms. In 1965, Crile[45] challenged the classical concept of "en bloc" resection of primary tumor of the breast along with its regional lymphatics. He advocated simple mastectomy, which preserved the RLN, basing the choice on experiments that showed depressed immunity to tumor challenge after excision of the RLN. The retention of the RLN, if free of metastatic cells, apparently effected a high level of systemic tumor immunity, which was important in the prevention of growth of disseminated micrometastases. Indeed, the results of the early clinical data on survival of patients with breast cancer who underwent simple mastectomy seemed to confirm the hypothesis stating that the RLN should be preserved.[45,46]

Fisher and co-workers[47] have shown that the RLN is important for the initiation of immunity against a transplantable syngeneic tumor in mice. Also,

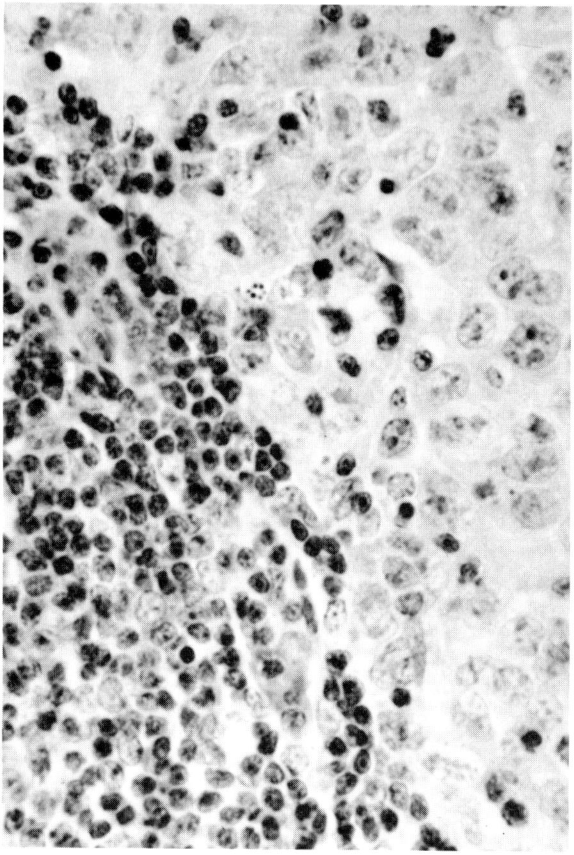

**Fig. 4 Melanoma metastasis in draining lymph node. (Magnification, 400 x.)**

lymphocytes from RLN of tumor-bearing mice were more cytotoxic in vitro and in vivo (neutralization assays) to the specific tumor than were lymphocytes from other anatomical sources.[48] However, such investigations that employ normal animals injected with transplantable tumors as models of immune reactivity to neoplasia produce an artificial situation where a normal lymph node or lymphatic system is suddenly challenged with a large number of tumor cells. Whether the results of such studies are relevant to clinical oncology is still questionable. The slow pathogenesis characteristic of clinical cases of neoplasia does not occur in normal animals challenged with a sudden and overwhelming tumor load. Those animals may die before developing an immune response. At best, experimental models such as those can only be used to investigate early events in the reaction of the RLN to transplantable tumors. Animals bearing naturally occurring tumors would be a more suitable model to investigate the role of the RLN in neoplasia.

Studies from our laboratories have attempted to overcome that problem by investigating the in vitro cell-mediated reactivity of lymphocytes to spontaneous neoplasms of dogs.[49] We compared the reactivities of lymphocytes obtained from RLN, distant lymph nodes, and peripheral blood. In dogs bearing spontaneous tumors of different histological types, the reactivities (cytotoxicity) of lymphocytes from the three sources were quite similar. Thus, in a model that accounts for the slow pathogenesis of cancer, the lymphocytes residing within the RLN were not unique in their in vitro reactivity against autochthonous tumor targets. Spontaneous tumors of dogs frequently parallel the biological behavior of human tumors; so these findings may be more relevant to human disease than findings from laboratory animal (rodent) models.

Whether the RLN's should be removed surgically must be determined by their degree of involvement in the metastatic process.[50,51] Crile[45] has suggested that, early in the course of cancer development, when the primary tumor is small and the RLN is not involved grossly, the removal of the RLN may increase the chance for metastasis. However, late in the course of the disease or when the tumor is already growing in the RLN, the RLN is unlikely to be of importance to host immunity and should be removed.

## HEMATOGENOUS SPREAD

Malignant tumor cells must invade blood vessels. The thin-wall veins (and lymphatic channels), in contrast to arteries with walls that contain elastic and collagen fibers, offer little resistance to penetration and thus provide the most common pathways for blood-borne tumor emboli. After infiltrating the vessels, tumor cells may be carried away passively in the bloodstream or they may remain localized and proliferate at the site of invasion.

Most malignant tumors have a well-established blood supply with multiple thin-wall vessels. Indeed, Folkman[52-54] has suggested that the vascular response of the host, or angiogenesis, is a prerequisite for progressive tumor growth. Liotta, Kleinerman, and Saidel[55] observed that, 5 days after the implantation of T-241 sarcoma in the thigh of C57BL/6 mice, all tumors contained small vessels at their periphery. No tumor cells were identified in the venous effluent before day 5; so tumor vascularization apparently must precede tumor cell dissemination.

The presence of tumor emboli in the blood does not constitute metastasis. The fate of circulating tumor cells is determined by their ability to survive the turbulence of the bloodstream, evade possible host defense mechanisms, attach firmly to the endothelium of distant small vessels, and gain entrance to the extravascular tissues. The overwhelming majority of circulating cancer cells (from solid tumors) die quickly in the blood,[56,57] and only very few survive to eventually form secondary tumors.[58] The hostile circulatory environment and trauma of transcapillary passage probably account for considerable tumor cell death.

Therapeutic interventions can result in the shedding of tumor cells into the bloodstream and lead to their dissemination, and tumor massage has resulted in increased experimental metastasis formation.[6] Although the prognostic value of determining the presence of circulating tumor cells in cancer patients is unclear,[56] the clinician should try to minimize tumor manipulations.

During circulatory transport tumor cells can undergo a variety of cellular interactions, including aggregation with other tumor cells (homotypic aggregation),[59–62] platelets,[63,64] and lymphocytes,[59] and also interactions with noncirculating host cells.[65] The larger the circulating embolus, the greater the rate of arrest and survival to form more secondary tumors.[55,66] Larger cell clumps or emboli are more effectively trapped in the capillaries than are single cells or small groups of cells.

The adhesion of tumor cells to endothelium may lead to damage in the vessel wall and subsequent accumulation of neutrophils. Tumor cells may then gain access to the extravascular tissues by following the pathway set by leukocytes that have traversed the vessel wall.[67] Platelets may aggregate at the site of tumor cell lodgement and release mediators that contribute to vascular spasm, increased permeability, and, perhaps, increased motility of tumor cells.[64]

## SITES OF METASTASES

Metastases are most frequently found in the lung and liver, possibly because most of the venous blood drains into those organs. All the systemic venous circulation returns through the right side of the heart and is filtered through the capillary bed of the lung. Similarly, the portal circulation passes through the liver. Metastases are also commonly found in the bones of many human cancer patients. Cancer cells in the venous system could reach the bones via the paravertebral plexus of veins, with its numerous interconnections from the inferior vena cava, pelvic veins, internal mammaries, and intercostal veins (Figs. 5 to 8). Infrequently, metastases are found in organs such as muscle, spleen, thyroid, and cartilage. Coman, DeLong, and McCutcheon[68] suggested that the number of metastases is proportional to the number of tumor emboli arrested within the capillary bed of an organ. In muscles, spleen, or thyroid, tumor cells are arrested in arterioles that present an effective barrier to extravascular invasion. The mechanisms responsible for the latency and organ selectivity of metastasis are not clear. In our quantitative analysis of the fate of circulating tumor cells,[58] metastatic growths of mouse melanoma were observed only in the lungs, but the lack of metastases elsewhere could not be attributed to the failure

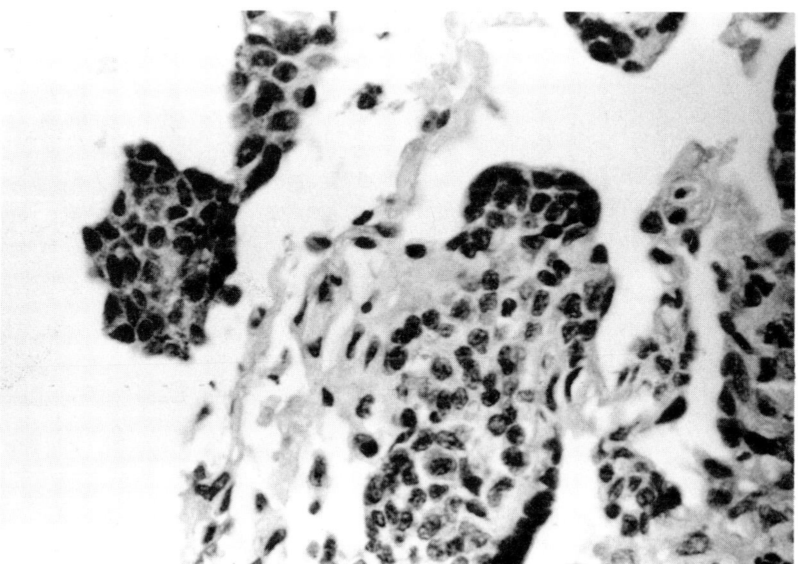

Fig. 5 Micrometastases of melanoma in lung parenchyma. (Magnification, 400 ×.)

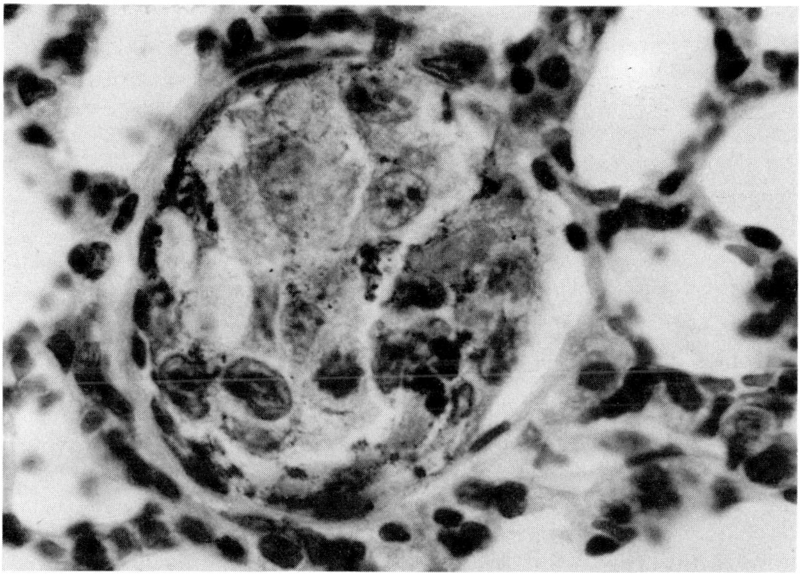

Fig. 6 Micrometastases of melanoma within a blood vessel. (Magnification, 630 ×.)

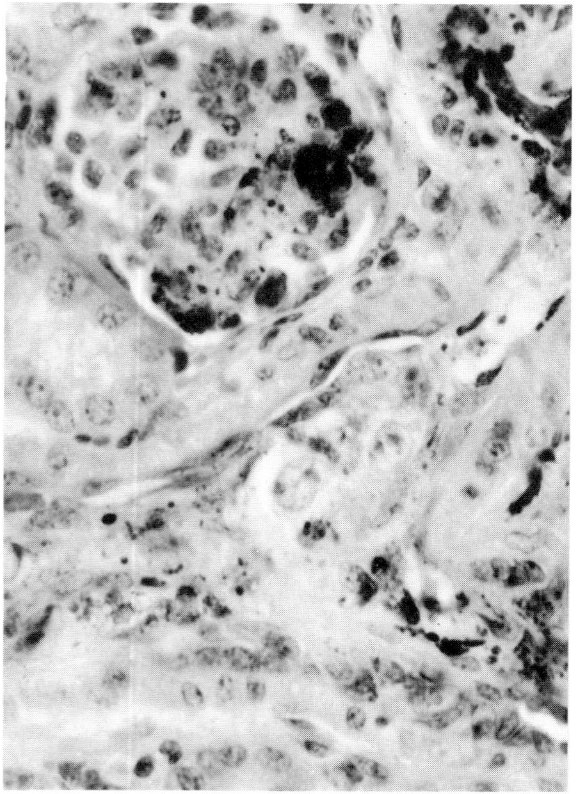

Fig. 7 Melanoma metastasis to kidney. Note melanoma cells in the glomeruli. (Magnification, 400 x.)

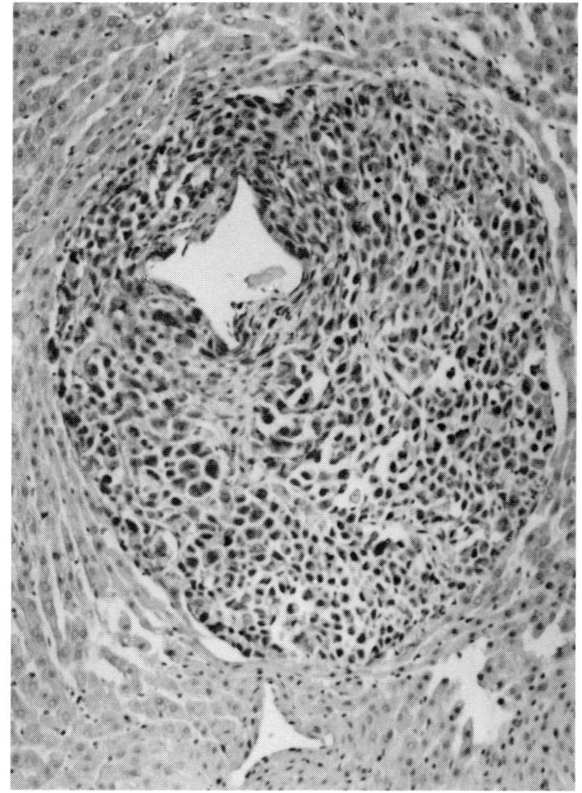

Fig. 8 Metastatic deposit of adenocarcinoma in hepatic parenchyma. (Magnification, 100 x.)

of arrest of tumor cells in those sites. Data from a variety of experimental tumor systems suggest that tumor cell properties, host "soil" factors, and mechanical distribution are all important factors in determining patterns of metastatic spread.[57]

## IMMUNE RESPONSE TO METASTASIS

Immune reactivity to neoplasia is one host factor that could influence metastatic spread. For many years investigators have suspected that some circulating tumor cells are destroyed by specific host immune mechanisms, but evidence for such a hypothesis is scant. In fact, some studies actually contradict the view that a patient's resistance to tumor is related to the activity of his reticuloendothelial system.

Many experimental and spontaneous tumors of animals possess specific antigens on their cell surfaces which differ from the host's normal cells. Such tumor-associated antigens have been demonstrated in various canine systems.[69,70] The capacity of the host to recognize these tumor-associated antigens as foreign and react against them constitutes the immune surveillance mechanism. One recent review[71] emphasized that "immune surveillance represents a different mechanism from the post factum immune response of the host against its own growing tumor"; so immune surveillance should be considered distinct from "tumor immunology." Any comparable surveillance mechanism against tumor dissemination will be modified by the coexistent response to the established primary malignancy. Although such a modified immune surveillance mechanism against embolic, metastatic cells may exist, the concept of the immune response and metastasis must also embrace the host's reactivity to solid secondary growths; the metastatic growths may represent a special population of tumor cells, which may also modify the immune response.

Evidence that the immune response is important in the regulation of metastatic events has been obtained from a number of studies. Kim,[72] using methylcholanthrene-induced mammary carcinomas in rats, found that the ability of those tumors to metastasize was inversely proportional to their immunogenicity. Experiments designed to test the effectiveness of the immune system in stopping the spread of tumors by monitoring the survival of tumor cells introduced into the circulation have been inconclusive. For example, treatment of animals with agents reported to depress immune responses, such as L-asparaginase[73] or anti-lymphocyte serum,[74] often resulted in greater numbers of metastatic foci, but that same outcome was also produced when host immune responses were stimulated or augmented by immunization[75] or transfer of a few immune cells.[59]

Contradictory results in studies of the relationship of lymphocyte/tumor cell interactions to the ultimate outcome of metastasis in experimental animals led to the proposition that two competing tumor cell/lymphocyte interactions might be involved.[76] The cell-mediated response in the early development of cancer or when tumors are weakly antigenic might directly stimulate tumor growth, whereas a strongly active cell-mediated immune response to a highly antigenic neoplasm could inhibit tumor growth.[77] Different tumor cell/immune cell interactions could characterize the pathogenesis of metastasis.

Our results for several systems, including spontaneous cancer of dogs, confirmed the hypothesis of Prehn[76] and Prehn and Lappé.[77] Additions of low ratios of sensitized lymph node or peripheral blood lymphocytes to autologous tumor cells often stimulated tumor growth in vitro.[78-80] Cultured tumor cells were destroyed by the addition of high numbers of lymphocytes,[79] but autologous serum in the culture often blocked such cytotoxicity or potentiated tumor growth.[78] Such serum blocking activity in dogs suffering from osteosarcoma decreases significantly after the surgical removal of the primary tumor.[81] Clearly, the interaction of host lymphocytes with invasive or circulating tumor cells determines the outcome of metastasis.

The existence of more than one subpopulation of thymus-dependent lymphocytes (T cells) has been proposed by a number of investigators[82-84] to explain the converse cell-mediated effects on tumor growth. One population of T cells might be tumoricidal, whereas another may have either no detrimental effect or perhaps even an enhancing effect on tumor cell growth. In vivo work by Umiel and Trainin[85] with the Lewis lung tumor (3LL) in mice demonstrated that the transfer of syngeneic lymphoid cells from tumor-bearing animals to intact mice receiving tumor transplants resulted in enhancement of tumor growth and metastatic spread. Later work showed that sensitized lymphoid cells could be separated into populations capable of inhibiting or enhancing tumor growth in vivo[86] and that these enhancing T cells were less mature than the inhibitory T cells.[87]

The search for such subpopulations of T lymphocytes in dogs bearing spontaneous tumors can proceed only when a more-detailed definition of normal circulating canine lymphocytes has been established. At present, the literature on that subject is somewhat confusing[88-92] (see also Chapter 6 in this volume). Nonetheless, as has been observed,[93] the role of cellular immunity in limiting metastatic spread needs investigation in natural animal models. The large size of the dog lends itself to such interferences as exchange transfusion and thoracic duct drainage. Meanwhile, the identification of serologically defined and lymphocyte-defined histocompatibility antigens in dogs[94,95] will help circumvent the problems associated with working with allogeneic systems.

## MACROPHAGES AND TUMOR SPREAD

Our in vitro studies demonstrated that, although lymphocytes from tumor-immune animals exhibit a dual response in their relationship to the cancer cells, macrophages are only cytotoxic, which suggests that macrophages are an important host defense against neoplasia.

Immune responses in which macrophages serve as effectors of tumor cell destruction have been demonstrated by many investigations.[96–99] The activation of macrophages against tumors may be either direct or mediated by lymphocytes or their products. The in vivo inhibitory effects of macrophages activated in vitro on established pulmonary metastases in mice were studied;[60] if untreated, the metastases would progress to kill the host. Mice treated with activated syngeneic macrophages had a dramatic decrease in pulmonary tumor nodules 2 to 3 weeks later.

Eccles and Alexander[100] studied the macrophage content of a group of chemically induced rat fibrosarcomata. The macrophage content of six tumors ranged from 8 to 54% and was inversely related to the incidence of metastasis, which suggests a role for tumor macrophages in the control of tumor dissemination.

We have found that peripheral blood monocytes isolated from normal dogs by continuous-flow filtration leukophoresis are cytotoxic to allogeneic tumor cells (unpublished observations). In another report, alveolar macrophages obtained from dogs were only cytotoxic to allogeneic tumor cells after the intravenous administration of BCG.[101] Future studies on canine mononuclear phagocytes should concentrate on whether animals with disseminated disease show defective mononuclear function, as has been reported in man,[102,103] and whether such depressions are dependent on such factors as tumor site, load, or type.

## SPONTANEOUS CANINE TUMORS AS MODELS FOR METASTASIS RESEARCH

Because of either their histological appearance, biological behavior, or rate of occurrence, the following canine models may be particularly suitable for the study of metastasis.

### Melanoma

Melanoma is a relatively common malignancy of older dogs. In a survey of over 1000 canine neoplasms, Cotchin[104] found that 8.8% were melanotic. In spite of the wide range of malignancy in these tumors and the high proportion of oral growths, a feature not shared by man, canine melanoma remains a potentially valuable model for metastasis research. The more-malignant tumors are rapidly dividing, highly anaplastic cancers with a poor prognosis. Metastasis is commonly spread by lymphatic and hematogenous routes, with frequent involvement of the lungs and other distant organs.[105]

Cell lines have been established in vitro and can be transplanted into immunologically depressed young animals.[106] This tumor has been successfully transplanted into nude mice,[107] and the use of that animal in the study of tumors derived from dogs may help circumvent the problems associated with implantation of neoplasms into immunocompetent hosts.

### Mammary Tumors

A complete review of the biology and histology of this complex group of tumors is presented in Chapter 17 of this volume. The tumor is common, accounting for 25 to 50% of all neoplasms in the bitch,[108–110] but the metastatic behavior can vary, depending on the morphological appearance.[111] Metastasis of carcinomas and sarcomas is common, and the simple type of carcinoma frequently shows penetration of lymphatics.[93]

### Osteosarcomas

Osteosarcoma is primarily a disease of large dogs[112,113] and most commonly occurs in the metaphysis of the long bones. The prognosis following surgical resection of the primary tumor is poor because in most patients metastasis to the lung has occurred.[114] Transplantation studies of canine osteosarcomas have shown that the tumor can be passaged into fetuses and immunosuppressed puppies[115] and nude mice.[107] Cell lines have been established, and cells from primary and secondary growths can be distinguished by karyotypic analysis.[116]

## SUMMARY

Experimental data dealing with mechanisms of tumor metastasis suffer major limitations. Many studies represent experiences with a one-tumor animal model or an in vitro system. Malignant tumor cells probably have some common properties, but there are many quantitative and qualitative differences in their biological behavior; the selection of an animal model for investigation presents a major difficulty. Experimental research should use a model as similar to clinical cancer as possible, but most investigations have been performed with transplantable or induced tumors in essentially normal rodents. Normal animals injected with tumor cells simply may not be good models for patients who develop cancer spontaneously. At present, the full potential for the use of the dog in investigations of spontaneous metastasis has been unrealized.

## ACKNOWLEDGMENT

This research was sponsored by the National Cancer Institute under contract No. N01-C0-75380 with Litton Bionetics, Inc.

## REFERENCES

1. E. V. Sugarbaker and A. S. Ketcham, Mechanism and Prevention of Cancer Dissemination, *Semin. Oncol.*, 4: 19-32 (1977).
2. G. L. Bartlett, J. W. Kreider, and D. M. Purnell, Immunotherapy of Cancer in Animals: Models or Muddles? *J. Natl. Cancer Inst.*, 56: 207-210 (1976).
3. P. Alexander, Back to the Drawing Board—The Need for More Realistic Model Systems for Immunotherapy, *Cancer*, 40: 467-470 (1977).
4. I. J. Fidler, General Considerations for Studies of Experimental Cancer Metastasis, in *Methods in Cancer Research*, Vol. XV, H. Busch (Ed.), pp. 399-439, Academic Press, Inc., New York, 1978.
5. R. T. Prehn, Neoplasia, in *Principles of Pathobiology*, M. F. Lavia and R. B. Hill (Eds.), pp. 191-232, Oxford University Press, New York, 1972.
6. E. V. Sugarbaker, A. M. Cohen, and A. S. Ketcham, Do Metastases Metastasize? *Ann. Surg.*, 174: 161-166 (1971).
7. G. Eaves, The Invasive Growth of Malignant Tumors as a Purely Mechanical Process, *J. Pathol.*, 109: 233-237 (1973).
8. L. M. Franks, Structure and Biological Malignancy of Tumors, in *Chemotherapy of Cancer Dissemination and Metastasis*, S. Garattini and G. Franchi (Eds.), pp. 71-78, Raven Press, New York, 1973.
9. D. M. Easty and G. C. Easty, Measurement of the Ability of Cells to Infiltrate Normal Tissue In Vitro, *Br. J. Cancer*, 29: 36-49 (1974).
10. P. D. Noguchi, J. B. Johnson, R. O'Donnell, and J. C. Petricciani, Chick Embryonic Skin as a Rapid Organ Culture Assay for Cellular Neoplasia, *Science*, 199: 980-983 (1978).
11. P. Strauli and L. Weiss, Cell Locomotion and Tumor Penetration, *Eur. J. Cancer*, 13: 1-12 (1977).
12. G. C. Easty and D. M. Easty, Mechanisms of Tumor Invasion, in *Scientific Foundations of Oncology*, T. Symington and R. L. Carter (Eds.), pp. 167-172, Heinemann, London, 1976.
13. M. Abercrombie and J. E. M. Heaysman, Invasive Behavior Between Sarcoma and Fibroblast Populations in Cell Culture, *J. Natl. Cancer Inst.*, 56: 561-569 (1976).
14. M. H. Gail and C. W. Boone, Density Inhibition of Mobility in 3T3 Fibroblasts and Their SV40 Transformants, *Exp. Cell Res.*, 64: 156-162 (1971).
15. J. Varani, W. Orr, and P. A. Ward, A Comparison of the Migration Patterns of Normal and Malignant Cells in Two Assay Systems, *Am. J. Pathol.*, 90: 159-171 (1978).
16. H. Gershman and J. Drumm, Mobility of Normal and Virus-Transformed Cells in Cellular Aggregates, *J. Cell Biol.*, 67: 419-435 (1975).
17. H. Gershman, J. Drumm, and J. J. Rosen, Dibutyryl Cyclic AMP Treatment of 3T3 and S40 Virus-Transformed 3T3 Cells in Aggregates. Effects on Mobility and Cell Contact Ultrastructure, *J. Cell Biol.*, 72: 424-440 (1977).
18. H. Gershman, W. Katzin, and R. T. Cook, Mobility of Cells from Solid Tumors, *Int. J. Cancer*, 21: 309-316 (1978).
19. E. I. Rosanoff, Continuous Cultivation of Malignant Cells from a Canine Testicular Tumor, *Proc. Soc. Exp. Biol.*, 100: 384-387 (1959).

20. G. R. Spencer and R. W. Leader, Effect of Cell-Free Tumor Materials on Irradiated and Newborn Animals, *Am. J. Vet. Res.*, 23: 587-591 (1962).
21. E. W. Adams, L. P. Carter, and W. J. Sapp, Growth and Maintenance of the Canine Venereal Tumor in Continuous Culture, *Cancer Res.*, 28: 753-757 (1968).
22. J. E. Prier, Chromosome Pattern of Canine Transmissible Sarcoma Cells in Culture, *Nature*, 212: 724-726 (1966).
23. G. C. Mohanty and B. S. Rajya, Growth and Morphological Characteristics of Canine Venereal Tumor Cells In Vitro, *Vet. Pathol.*, 14: 420-425 (1977).
24. M. Abercrombie and J. E. M. Heaysman, Social Behavior of Cells in Tissue Culture. II. "Monolayering" of Fibroblasts, *Exp. Cell Res.*, 6: 293-306 (1954).
25. M. H. Dresden, S. A. Heilman, and J. D. Schmidt, Collagenolytic Enzymes in Human Neoplasms, *Cancer Res.*, 32: 993-996 (1972).
26. K. Hashimoto, Y. Yamanishi, E. Maeyens, M. K. Dabbous, and T. Kanzaki, Collagenolytic Activities of Squamous Cell Carcinoma of the Skin, *Cancer Res.*, 33: 2790-2801 (1972).
27. B. Sylven, Biochemical and Enzymatic Factors Involved in Cellular Detachment, in *Chemotherapy of Cancer Dissemination and Metastasis*, S. Garattini and G. Franchi (Eds.), pp. 129-138, Raven Press, New York, 1973.
28. Y. Yamanishi, E. Maeyens, M. K. Dabbous, H. Ohyama, and K. Hashimoto, Collagenolytic Activity in Malignant Melanoma. Physiochemical Studies, *Cancer Res.*, 33: 2507-2512 (1973).
29. B. Sylven, Lysosomal Enzyme Activity in the Interstitial Fluid of Solid Mouse Tumour Transplants, *Eur. J. Cancer*, 4: 463-473 (1968).
30. B. Sylven and I. Bois, Protein Content and Enzymatic Assays of Interstitial Fluid from Some Normal Tissue and Transplanted Mouse Tumors, *Cancer Res.*, 20: 831-836 (1960).
31. A. R. Poole, K. J. Tiltman, A. D. Recklies, and T. A. M. Stoker, Differences in Secretion of the Proteinase Cathepsin B at the Edges of Human Breast Carcinomas and Fibroadenomas, *Nature*, 273: 545-547 (1978).
32. G. Poste, Sublethal Autolysis: Modification of the Cell Periphery by Lysosomal Enzymes, *Exp. Cell Res.*, 67: 11-16 (1971).
33. L. Weiss, Tumor Necrosis and Cell Detachment, *Int. J. Cancer*, 20: 87-92 (1977).
34. R. A. Willis, *Pathology of Tumors*, 3rd ed., Butterworths, Washington, D. C., 1960.
35. E. Eisenstein, K. E. Kuettner, C. Neapolitan, L. W. Soble, and N. Sorgente, The Resistance of Certain Tissues to Invasion. III. Cartilage Extracts Inhibit the Growth of Fibroblasts and Endothelial Cells in Culture, *Am. J. Pathol.*, 81: 337-347 (1975).
36. S. T. Palayoor and B. K. Batra, Behavioural Patterns of Tumor Cells on Chorioallantoic Membrane of Chick in Presence of Mouse Embryos, *Indian J. Exp. Biol.*, 15: 1-7 (1977).
37. C. D. Sher, C. Haudenschild, and M. Klagsbrum, The Chick Chorioallantoic Membrane as a Model System for the Study of Tissue Invasion by Viral Transformed Cells, *Cell*, 8: 373-382 (1976).
38. I. R. Hart and I. J. Fidler, An In Vitro Quantitative Assay for Tumor Cell Invasion, *Cancer Res.*, 38: 3218-3224 (1978).
39. B. Fisher and E. R. Fisher, The Interrelationships of Hematogenous and Lymphatic Tumor Cell Dissemination, *Surg. Gynecol. Obstet.*, 122: 791-798 (1966).
40. J. L. Gowans and E. J. Knight, The Route of Recirculation of Lymphocytes in the Rat, *Proc. R. Soc. (London), Ser. B*, 159: 275-282 (1964).
41. V. T. Marchesi and J. L. Gowans, The Migration of Lymphocytes Through the Endothelium of Venules in Lymph Nodes: An Electron Microscope Study, *Proc. R. Soc. (London), Ser. B*, 159: 283-290 (1964).

42. H. B. Stamper, Jr., and J. J. Woodruff, Lymphocyte Homing into Lymph Nodes: In Vitro Demonstration of Selective Effect Affinity of Recirculating Lymphocytes for High-Endothelial Venules, *J. Exp. Med.,* 144: 828-833 (1976).

43. J. A. del Regato, Pathways of Metastatic Spread of Malignant Tumors, *Semin. Oncol.,* 4: 33-38 (1977).

44. J. W. Berg, A. G. Huvos, L. M. Axtell, and G. F. Robbins, A New Sign of Favorable Prognosis in Mammary Cancer: Hyperplasia Reactive Lymph Nodes in the Apex of the Axilla, *Ann. Surg.,* 177: 8-16 (1973).

45. G. C. Crile, Rationale of Simple Mastectomy Without Radiation for Clinical Stage 1 Cancer of the Breast, *Surg. Gynecol. Obstet.,* 120: 975-982 (1965).

46. G. C. Crile, Possible Role of Uninvolved Regional Nodes in Preventing Metastasis from Breast Cancer, *Cancer,* 24: 1283-1285 (1969).

47. B. Fisher, N. Wolmark, J. Coyle, E. Saffer, and E. R. Fisher, Studies Concerning the Regional Lymph Node in Cancer. VIII. Effect of Two Asynchronous Tumor Foci on Lymph Node Cell Cytotoxicity, *Cancer,* 36: 521-527 (1975).

48. B. Fisher, E. A. Saffer, and E. R. Fisher, Studies Concerning the Regional Lymph Node in Cancer. IV. Tumor Inhibition by Regional Lymph Node Cells, *Cancer,* 33: 631-636 (1972).

49. I. J. Fidler, R. W. McWilliams, S. Bech-Nielsen, and M. B. Budmen, Role of the Regional Lymph Node in Neoplasia: Cellular-Mediated Reactivity In Vitro by Autologous Regional or Distant Lymph Nodes or Peripheral Blood Lymphocytes of Dogs with Spontaneous Neoplasms, *Immunol. Commun.,* 4: 325-335 (1975).

50. L. Weiss, Pathobiologic Overview of Metastasis, *Semin. Oncol.,* 4: 5-17 (1977).

51. C. J. H. Van de Velde, L. M. Van Putten, and A. Zwaveling, A New Metastasizing Mammary Carcinoma Model in Mice: Model Characteristics and Applications, *Eur. J. Cancer,* 13: 555-565 (1977).

52. J. Folkman, Tumor Angiogenesis, *Adv. Cancer Res.,* 19: 331-358 (1974).

53. J. Folkman, Tumor Angiogenesis. A Possible Control Point in Tumor Growth, *Ann. Intern. Med.,* 82: 96-100 (1975).

54. J. Folkman and K. Tyler, Tumor Angiogenesis: Its Possible Role in Metastasis and Invasion, in *Cancer Invasion and Metastasis: Biologic Mechanisms and Therapy,* S. B. Day, W. P. L. Myers, P. Stansly, S. Garattini, and M. G. Lewis (Eds.), pp. 95-103, Raven Press, New York, 1977.

55. L. A. Liotta, J. Kleinerman, and G. M. Saidel, Quantitative Relationships of Intravascular Tumor Cells, Tumor Vessels and Pulmonary Metastases Following Tumor Implantation, *Cancer Res.,* 34: 997-1004 (1974).

56. A. J. Salsbury, The Significance of the Circulating Cancer Cell, *Cancer Treatment Rev.,* 2: 55-72 (1975).

57. I. J. Fidler, Patterns of Tumor Cell Arrest and Development, in *Fundamental Aspects of Metastasis,* L. Weiss (Ed.), pp. 275-289, North-Holland, Amsterdam, 1976.

58. I. J. Fidler, Metastasis: Quantitative Analysis of Distribution and Fate of Tumor Emboli Labeled with $^{125}$I-5-iodo-2′-deoxyuridine, *J. Natl. Cancer Inst.,* 45: 773-782 (1970).

59. I. J. Fidler, Immune Stimulation-Inhibition of Experimental Cancer Metastasis, *Cancer Res.,* 34: 491-498 (1974).

60. I. J. Fidler, Inhibition of Pulmonary Metastasis by Intravenous Injection of Specifically Activated Macrophages, *Cancer Res.,* 34: 1074-1078 (1974).

61. G. L. Nicolson, J. L. Winkelhake, and A. C. Nussey, An Approach to Studying the Cellular Properties Associated with Metastasis. Some In Vitro Properties of Tumor Variants Selected In Vivo for Enhanced Metastasis, in *Fundamental Aspects of Metastasis,* L. Weiss (Ed.), pp. 291-303, North-Holland, Amsterdam, 1976.

62. J. L. Winkelhake and G. L. Nicolson, Determination of Adhesive Properties of Variant Metastatic Melanoma Cells to BALB/3T3 Cells and Their Virus-Transformed Derivatives by a Monolayer Attachment Assay, *J. Natl. Cancer Inst.,* 56: 285-291 (1976).
63. G. J. Gasic, T. B. Gasic, N. Galanti, T. Johnson, and S. Murphy, Platelet-Tumor Cell Interaction in Mice. The Role of Platelets in the Spread of Malignant Disease, *Int. J. Cancer,* 11: 704-718 (1973).
64. B. A. Warren, Environment of the Blood-Borne Tumor Embolus Adherent to Vessel Wall, *J. Med.,* 4: 150-177 (1973).
65. G. L. Nicolson and J. L. Winkelhake, Organ Specificity of Blood-Borne Tumour Metastasis Determined by Cell Adhesion, *Nature,* 255: 230-231 (1975).
66. I. J. Fidler, The Relationship of Embolic Homogeneity, Number, Size and Viability to the Incidence of Experimental Metastasis, *Eur. J. Cancer,* 9: 223-229 (1973).
67. S. Wood, Jr., Pathogenesis of Metastasis Formation Observed In Vivo in the Rabbit Ear Chamber, *Arch. Pathol.,* 66: 550-568 (1958).
68. D. R. Coman, R. P. DeLong, and M. McCutcheon, Studies on the Mechanism of Metastasis: The Distribution of Tumors in Various Organs in Relation to the Distribution of Arterial Emboli, *Cancer Res.,* 11: 648-654 (1951).
69. W. D. Winters and H. D. Snow, Immunodiffusion Studies of Canine Lymphosarcoma Antigens, *J. Natl. Cancer Inst.,* 53: 1027-1032 (1974).
70. D. Hannant, R. W. Else, and G. W. Crighton, Antigens Associated with Canine Spontaneous Mammary Carcinoma, *Vet. Rec.,* 103: 441-443 (1978).
71. O. Stutman, Immunodepression and Malignancy, *Adv. Cancer Res.,* 22: 261-422 (1976).
72. U. Kim, Metastasizing Mammary Carcinomas in Rats: Induction and Study of Their Immunogenicity, *Science,* 167: 72-76 (1970).
73. S. D. Deodhar, Enhancement of Metastasis by L-Asparaginase in a Mouse Tumour System, *Nature,* 231: 319-320 (1971).
74. E. R. Fisher, O. Soliman, and B. Fisher, Effect of Antilymphocyte Serum on Parameters of Growth of MCA-Induced Tumours, *Nature,* 221: 237-288 (1969).
75. R. Duff, E. Doller, and R. Rapp, Immunologic Manipulation of Metastasis due to Herpesvirus Transformed Cells, *Science,* 180: 79-81 (1973).
76. R. T. Prehn, The Immune Reaction as a Stimulator of Tumor Growth, *Science,* 176: 170-172 (1971)
77. R. T. Prehn and M. A. Lappé, An Immunostimulation Theory of Tumor Development, *Transplant. Rev.,* 7: 26-54 (1971).
78. I. J. Fidler, R. S. Brodey, and S. Bech-Nielsen, In Vitro Immune Stimulation-Inhibition to Spontaneous Canine Tumors of Various Histological Types, *J. Immunol.,* 3: 1051-1059 (1974).
79. I. J. Fidler and R. W. McWilliams, Immune Reactivity In Vitro to Spontaneous Canine Oral Tumors, *J. Med.,* 6: 323-336 (1975).
80. R. S. Brodey, I. J. Fidler, and S. Bech-Nielsen, Correlation of the In Vitro Immune Response in Dogs with Malignant Neoplasms to the Clinical Course of Their Disease, *Am. J. Vet. Res.,* 36: 75-80 (1975).
81. S. Bech-Nielsen, R. S. Brodey, I. J. Fidler, D. A. Abt, and J. S. Reif, The Effect of BCG on In Vitro Immune Reactivity and Clinical Course in Dogs Treated Surgically for Osteosarcoma, *Eur. J. Cancer,* 13: 33-41 (1977).
82. F. D. Burns, P. C. Marrack, J. W. Kappler, and C. A. Janeway, Functional Heterogeneity Among the T-Derived Lymphocytes of the Mouse. IV. Nature of Spontaneously Induced Suppressor Cells, *J. Immunol.,* 114: 1345-1347 (1975).
83. R. W. Dutton, Inhibitory and Stimulatory Effects of Concanavalin A on the Response of Mouse Spleen Cell Suspensions to Antigen. II. Evidence for Separate Stimulatory and Inhibitory Cells, *J. Exp. Med.,* 138: 1496-1505 (1973).

84. R. R. Rich and C. W. Pierce, Biological Expressions of Lymphocyte Activation. I. Effects of Phytomitogens on Antibody Synthesis In Vitro, *J. Exp. Med.*, 137: 205-223 (1973).
85. T. Umiel and N. Trainin, Immunological Enhancement of Tumor Growth by Syngeneic Thymus-Derived Lymphocytes, *Transplantation*, 18: 244-250 (1974).
86. M. Small and N. Trainin, Separation of Populations of Sensitized Lymphoid Cells into Fractions Inhibiting and Fractions Enhancing Syngeneic Tumor Growth In Vivo, *J. Immunol.*, 117: 292-297 (1976).
87. M. Small, Characteristics of the Immature Cells Involved in T Cell-Mediated Enhancement of Syngeneic Tumor Growth, *J. Immunol.*, 118: 1517-1523 (1977).
88. C. A. Bowles, G. S. White, and D. Lucas, Rosette Formation by Canine Peripheral Blood Lymphocytes, *J. Immunol.*, 114: 399-402 (1975).
89. A. R. Zander, N. Boopalam, and R. B. Epstein, Surface Markers on Canine Lymphocytes, *Transplant. Proc.*, 7: 369-373 (1975).
90. R. E. Esser, M. Cosgrove, G. Scott, and A. B. Cosimi, Rosette Formation by Canine Leukocytes with Human Erythrocytes, *Transplantation*, 24: 223-225 (1977).
91. S. Krakowka and D. J. Guyot, Rosette Formation Assays in Dogs: Lack of Specificity of E Rosettes for T-Lymphocytes, *Infect. Immun.*, 17: 73-77 (1977).
92. S. K. Dulta, M. N. Novilla, M. K. Bumgardner, and A. Ingling, Lymphocyte Responsiveness to Mitogens and Quantitation of T and B Lymphocytes in Canine Malignant Lymphoma, *Am. J. Vet. Res.*, 39: 455-458 (1978).
93. Memoranda. Immunity to Cancer. Naturally Occurring Tumours in Domestic Animals as Models for Research, *Bull. World Health Organ.*, 49: 205-213 (1973).
94. H. M. Vriesendorp, B. M. J. Smid-Mercx, T. P. Visser, R. E. W. Halliwell, and R. M. Schwartzman, Serological DL-A Typing of Normal and Atopic Dogs, *Transplant. Proc.*, 7: 375-377 (1975).
95. B. M. J. Smid-Mercx, B. Duyzer-Den Hartog, T. P. Visser, and H. M. Vriesendorp, Serological Studies of Canine Histocompatibility Antigens, *Transplant. Proc.*, 7: 361-364 (1975).
96. P. Alexander, R. Evans, and C. K. Grant, The Interplay of Lymphoid Cells and Macrophages in Tumor Immunity, *Ann. Inst. Pasteur*, 122: 645-658 (1972).
97. J. R. David, A Brief Review of Macrophage Activation by Lymphocyte Mediators, in *The Phagocytic Cell in Host Resistance*, J. A. Bellanti and D. H. Dayton (Eds.), pp. 143-153, Raven Press, New York, 1975.
98. W. F. Piessens, W. Hallowell Churchill, Jr., and J. R. David, Macrophages Activated In Vitro with Lymphocyte Mediators Kill Neoplastic but not Normal Cells, *J. Immunol.*, 114: 293-298 (1975).
99. R. Keller, Promotion of Tumor Growth In Vivo by Antimacrophage Agents, *J. Natl. Cancer Inst.*, 57: 1355-1361 (1976).
100. S. A. Eccles and P. A. Alexander, Macrophage Content of Tumours in Relation to Metastatic Spread, *Nature*, 250: 667-668 (1974).
101. N. T. Gorman, Alveolar Macrophage Cytotoxicity in the Dog, *Br. J. Cancer*, 36: 414 (Abstr.) (1977).
102. M. S. Hausman, S. Brosman, R. Snyderman, M. R. Mickey, and J. Fabrey, Defective Monocyte Function in Patients with Genitourinary Carcinoma, *J. Natl. Cancer Inst.*, 55: 1047-1052 (1975).
103. R. Snyderman and M. C. Pike, An Inhibitor of Macrophage Chemotaxis Produced by Neoplasms, *Science*, 192: 370-371 (1976).
104. E. Cotchin, Melanotic Tumours of Dogs, *J. Comp. Pathol.*, 65: 115-129 (1955).
105. D. E. Bostock, Prognosis After Surgical Excision of Canine Melanomas, *Vet. Pathol.*, 16: 32-40 (1979).

106. G. R. Betton and L. N. Owen, Allogeneic Grafts of Spontaneous Canine Melanomas and Their Cell Culture Strains in Neonatal Immunosuppressed Dogs, *Br. J. Cancer,* 34: 374-380 (1976).
107. S. M. J. Oughton and L. N. Owen, Transplantation of Dog Neoplasms into the Mouse Mutant Nude, *Res. Vet. Sci.,* 17: 414-416 (1974).
108. A. C. Anderson, Parameters of Mammary Gland Tumours in Aging Beagles, *J. Am. Vet. Med. Assoc.,* 147: 1653-1660 (1965).
109. I. J. Fidler, D. A. Abt, and R. S. Brodey, The Biological Behavior of Canine Mammary Neoplasms, *J. Am. Vet. Med. Assoc.,* 151: 1311-1318 (1967).
110. R. Schneider, C. R. Dorn, and D. O. N. Taylor, Factors Influencing Canine Mammary Cancer Development and Post Surgical Survival, *J. Natl. Cancer Inst.,* 43: 1249-1261 (1969).
111. A. W. Monlux, J. F. Roszel, D. W. MacVean, and R. W. Palmer, Classification of Epithelial Canine Mammary Tumors in a Defined Population, *Vet. Pathol.,* 14: 194-217 (1977).
112. L. N. Owen, *Bone Tumours in Man and Animals,* Butterworths, London, 1969.
113. L. N. Owen, Transplantation of Canine Osteosarcoma, in *Bone: Certain Aspects of Neoplasia,* C. H. G. Price and F. M. G. Ross (Eds.), pp. 327-337, Butterworths, London, 1973.
114. R. S. Brodey, Surgical Treatment of Canine Osteosarcoma, *J. Am. Vet. Med. Assoc.,* 147: 729-735 (1965).
115. D. E. Bostock and L. N. Owen, Transplantation and Tissue-Culture Studies of Canine Osteosarcomas, *Eur. J. Cancer,* 6: 499-502 (1970).
116. N. Taylor, M. Shifrine, H. G. Wolf, and A. Trommershausen-Smith, Canine Osteosarcoma Karyotypes from an Original Tumor, Its Metastasis and Tumor Cells in Tissue Culture, *Transplant. Proc.,* 7: 485-493 (1975).

# Contributors

**Betton, G. R.**, B.V.Sc., Ph.D.
Imperial Chemical Industries Ltd., Pharmaceuticals Division, Macclesfield, Cheshire, England

**Bijnen, A. Bart**, M.D.
Department of Surgery, University Hospital Dijkzigt and Laboratory of Experimental Surgery, Erasmus University, Rotterdam, The Netherlands

**Book, Steven A.**, Ph.D.
Laboratory for Energy-Related Health Research, University of California, Davis, California

**Chrisp, Clarence E.**, D.V.M.
Laboratory for Energy-Related Health Research, University of California, Davis, California

**Cockerell, Gary L.**, D.V.M., Ph.D.
Department of Pathology, New York State College of Veterinary Medicine, Cornell University, Ithaca, New York

**Cohen, Dan**, Dr. Med. Vet., Ph.D.
Unit of Comparative Medicine, Faculty of Health Sciences, Ben Gurion University of the Negev, Beer-Sheva, Israel

**Colgrove, Gary S.**, D.V.M., Ph.D.
Plum Island Animal Disease Laboratory, Greenport, Long Island, New York

**Fidler, Isaiah J.**, D.V.M., Ph.D.
Cancer Metastasis and Treatment Laboratory, Frederick Cancer Research Center, Frederick, Maryland

**Fritz, Thomas E.**, D.V.M.
Division of Biological and Medical Research, Argonne National Laboratory, Argonne, Illinois

**Hart, Ian R.**, B.V.Sc., Ph.D.
Cancer Metastasis and Treatment Laboratory, Frederick Cancer Research Center, Frederick, Maryland

**Holmberg, Charles A.**, D.V.M., Ph.D.
Department of Veterinary Pathology, College of Veterinary Medicine, Texas A&M University, College Station, Texas

**Jones, J. B.**, D.V.M.
Department of Medical Biology, The University of Tennessee Memorial Research Center and Hospital, The Center for the Health Sciences, Knoxville, Tennessee

**Lange, R. D.**, M.D.
Department of Medical Biology, The University of Tennessee Memorial Research Center and Hospital, The Center for the Health Sciences/ Knoxville Units, Knoxville, Tennessee

**Lewis, Robert M.**, D.V.M.
Department of Pathology, New York State College of Veterinary Medicine, Cornell University, Ithaca, New York

**MacKenzie, Malcolm R.**, M.D.
Department of Internal Medicine, School of Medicine, University of California, Davis, California

**Miller, Claramae H.**, Ph.D.
Department of Pathology and Internal Medicine, Section of Hematology and Oncology, School of Medicine, University of California, Davis, California

**Obertop, Huug**, M.D.
Department of Surgery, University Hospital Dijkzigt and Laboratory for Experimental Surgery, Erasmus University, Rotterdam, The Netherlands

**O'Donnell, Robert W.**, Ph.D.
Department of Pathology, New York State College of Veterinary Medicine, Cornell University, Ithaca, New York

**Pedersen, Niels C.**, D.V.M., Ph.D.
Department of Medicine, School of Veterinary Medicine, University of California, Davis, California

**Pool, Roy R.**, D.V.M., Ph.D.
Department of Pathology and Laboratory for Energy-Related Health Research, School of Veterinary Medicine, University of California, Davis, California

**Rose, Noel R.**, M.D., Ph.D.
Department of Immunology and Microbiology, Wayne State University, School of Medicine, Detroit, Michigan

**Shifrine, Moshe**, Ph.D.
Laboratory for Energy-Related Health Research, University of California, Davis, California

**Spangler, William L.**
Laboratory for Energy-Related Health Research, University of California, Davis, California

**Stormont, Clyde, Jr.**, Ph.D.
Department of Reproduction, School of Veterinary Medicine, University of California, Davis, California

**Suzuki, Yoshiko**, Dr. Ag.
Department of Reproduction, School of Veterinary Medicine, University of California, Davis, California

**Van Bekkum, D.**, M.D., Ph.D.
Radiobiological Institute GO-TNO and Laboratory of Experimental Surgery, Erasmus University, Rotterdam, The Netherlands

**Vriesendorp, H. M.**, M.D., Ph.D.
Radiobiological Institute GO-TNO and Laboratory of Experimental Surgery, Erasmus University, Rotterdam, The Netherlands

**Wilson, Floyd D.**, D.V.M.
Laboratory for Energy-Related Health Research, University of California, Davis, California

# Index

## DISCLAIMER

## ENERGY RESEARCH ABSTRACTS (ERA)

A semimonthly publication prepared by the Technical Information Center of the U.S. Department of Energy (DOE). Provides abstracts and indexing for all scientific and technical reports, journal articles, conference papers and proceedings, books, patents, theses, and monographs originated by DOE, its laboratories, energy centers, and contractors.

Subject matter includes DOE's research, development, demonstration, and technical programs resulting from its broad charter for energy systems, conservation, safety, environmental protection, physical research, and biology and medicine. *ERA* also covers other U. S. Government-sponsored energy information and the international literature on reactor technology, waste processing and storage, and fusion technology. Also covered is nonnuclear information obtained from foreign governments under agreements for cooperation.

*ERA* is available on an exchange basis to universities, research institutions, industrial firms, and publishers of scientific information. Federal, state, and municipal agencies may obtain *ERA* free of charge. Inquiries should be directed to the DOE Technical Information Center, P. O. Box 62, Oak Ridge, Tennessee 37830.

*ERA* is available to the public on a subscription basis from the Superintendent of Documents, U. S. Government Printing Office, Washington, D. C. 20402. The subscription rate for the 24 semimonthly issues including semiannual and annual cumulative indexes is $184.00 for domestic subscribers and $230.00 for foreign subscribers. The cost of the semiannual index is $18.00 (domestic) or $22.50 (foreign); the annual index is $33.00 (domestic) or $41.25 (foreign). A single issue costs $5.55 (domestic) or $6.95 (foreign).

## ENERGY ABSTRACTS FOR POLICY ANALYSIS (EAPA)

A monthly publication prepared by the Technical Information Center of the U.S. Department of Energy. Provides abstracts and indexing for nontechnological or quasi-technological literature relevant to energy-related analysis and evaluation. Designed primarily for those concerned with energy policy, *EAPA* draws pertinent material from congressional committee prints; reports from federal, state, and regional departments and agencies; publications from universities, private companies, and research groups; journal articles; news reports; books; and conference proceedings and papers.

*EAPA* offers a quick review of relevant literature in the following areas:

Conservation
Energy Analysis and Modeling
Supply, Demand, and Forecasting
Economics and Sociology
Policy, Legislation, and Regulation
Environment, Health, and Safety
Natural Resources
Fossil Fuels: Coal, Petroleum, Natural Gas, Oil Shales, and Tar Sands
Nuclear Energy
Hydrogen and Synthetic Fuels
Waste Heat Utilization
Transport and Storage
Unconventional Sources and Power Generation: Solar, Geothermal, Wind, Tidal, Waste Products
Electric Power: Generation, Transmission, and Distribution
Consumption and Utilization
Research, Development, Demonstration, and Commercialization

Subscriptions are available from the Superintendent of Documents, U. S. Government Printing Office, Washington, D. C. 20402. The annual rate for 12 issues and the annual index is $38.00 (foreign, $47.50). Bought separately, a single issue is $3.00 (foreign, $3.75) and the annual index is $4.00 (foreign, $5.00).